Principles of Digital Arts & Multimedia

Principles of Digital Arts & Multimedia

Editor
Joy Crelin

SALEM PRESS
A Division of EBSCO Information Services, Inc.
Ipswich, Massachusetts

GREY HOUSE PUBLISHING

Cover photo by scyther5/iStock.

Copyright © 2023, by Salem Press, A Division of EBSCO Information Services, Inc., and Grey House Publishing, Inc.

Principles of Digital Arts & Multimedia, published by Grey House Publishing, Inc., Amenia, NY, under exclusive license from EBSCO Information Services, Inc.

All rights reserved. No part of this work may be used or reproduced in any manner whatsoever or transmitted in any form or by any means, electronic or mechanical, including photocopy, recording, or any information storage and retrieval system, without written permission from the copyright owner. For information, contact Grey House Publishing/Salem Press, 4919 Route 22, PO Box 56, Amenia, NY 12501.

∞ The paper used in these volumes conforms to the American National Standard for Permanence of Paper for Printed Library Materials, Z39.48 1992 (R2009).

Publisher's Cataloging-In-Publication Data
(Prepared by Parlew Associates, LLC)

Names: Crelin, Joy, editor.
Title: Principles of digital arts & multimedia / Editor: Joy Crelin.
Description: Ipswich, MA : Salem Press, a division of EBSCO Information Services, Inc. ; Amenia, NY : Grey House Publishing, 2023. | Series: [Principles of science]. | Includes bibliographic references and index. | Includes b&w photos and illustrations.
Identifiers: ISBN 9781637004265 (hardback)
Subjects: LCSH: Art and technology. | Computer art. | Digital art. | Multimedia (Art). | BISAC: ART / Digital. | ART / Mixed Media. | ART / Reference.
Classification: LCC N7433.8 P75 2023 | DDC 776/03—dc23

First Printing
Printed in the United States of America

Contents

Publisher's Note

Digital Arts & Multimedia is the next volume in Salem's *Principles of Science* series, which includes *Microbiology*, *Energy*, *Marine Science*, *Geology*, *Information Technology*, and *Mathematics*.

This new resource delves into the inventive world of digital art creation and design, covering the history and evolution of related technology, explaining common terms and concepts, and addressing recent developments. The breadth of coverage makes this an indispensable resource for individuals interested in making digital art, music, building plans, films, and other technologically produced creations.

This work begins with a comprehensive Editor's Introduction to the topic of digital arts & multimedia written by Joy Crelin that discusses, among other things, how entertainment and media have been revolutionized by digital means of production, especially computer-generated imagery (CGI).

Following the Introduction, *Principles of Digital Arts & Multimedia* includes 120+ entries that follow a convenient alphabetical arrangement, making subjects easy to locate.

The wide variety of industries that employ digital media, from automotive manufacturing to architecture, are touched on, as well as what education and training is available. Augmented and virtual reality, nonfungible tokens (NFTs), and the increasing intersection of digital arts and multimedia and artificial intelligence round out the book. The ethical concerns associated with limiting the impact of AI tools on creative professionals, most recently brought to the fore by the SAG-AFTRA strike, are also covered.

Entries begin with a brief Abstract describing the key elements of the article, followed by a Background section and an Overview. All entries end with a helpful Further Reading section.

This work also includes the following helpful appendixes:

- Bibliography;
- Glossary;
- Organizations;
- Subject Index

Salem Press extends appreciation to all involved in the development and production of this work. Names and affiliations of contributors to this work follow the Editor's Introduction.

Principles of Digital Arts & Multimedia, as well as all Salem Press reference books, is available in print and as an e-book. Please visit www.salempress.com for more information.

Introduction

DEFINING DIGITAL ARTS AND MULTIMEDIA

According to the *Oxford English Dictionary*, the term *digital art* refers to "art created or modified using a computer or other digital medium," while the term *multimedia* describes a work or artist "that uses or combines a variety of artistic or communicative media." The relative broadness of those definitions calls attention to a crucial fact: the field of digital arts and multimedia is a far-reaching one that encompasses a host of artforms, technologies, and applications. Digital art and multimedia works may be created entirely through digital means, as in projects in which each step—for instance, from the initial sketch through the final rendering—has been completed on a computer. Other works may blend digital and analog tools and approaches; for example, an artist may create an initial sketch with pencil and paper before scanning the image and proceeding to modify it in a digital graphics program. Still other digital and multimedia works may bridge the digital and physical worlds, as in the case of immersive art installations that mix physical objects or artworks with digital audio, video, or other media. In some cases, a physical work of art may be accompanied by a complementary digital component; a painting of a musician, for instance, could be presented beside a video of the instrument being played or a quick response (QR) code leading to an interactive website that allows its visitors to experiment with synthesized sounds. Regardless of its specific characteristics, a work of digital art or a multimedia work is itself the result of decades of technological and artistic development spanning from the earliest experiments with computer art to the present day.

ENTERTAINMENT AND MEDIA

In the minds of many, the field of digital arts and multimedia is most closely associated with the entertainment industry, and for good reason. In the twenty-first century, digital technology has become the predominant means of creating media for entertainment, whether it be live action or animated, shortform or longform, and interactive or noninteractive. Some artists and other creative professionals still enjoy creating art and media through analog means such as film photography and hand-drawn animation, with minimal digital interventions. By and large, however, digital means of multimedia production have revolutionized the entertainment industry, allowing for the creation of inventive works of art while enabling artists to avoid some particularly laborious aspects of the creative process, such as the cutting and splicing of physical film that was once required when editing any filmed material.

A particularly significant development in the field's evolution was the rise of computer-generated imagery (CGI) and computer animation. Although most closely associated with the creation of fully animated films and television shows with distinct cartoon aesthetics, CGI technology has also contributed substantially to the twenty-first-century film industry's approach to special effects, costuming, and set design. Among other projects, digital technologies facilitated the creation of the effects-laden superhero films of the Marvel Cinematic Universe, the photorealistic animated lions of 2019's *The Lion King*, and the immersive digital world of Pandora in filmmaker James Cameron's *Avatar* series (2009-). Advances in animation technology also decreased the difficulty and cost of creating animated sequences, thus allowing relatively high-quality animation to appear in projects with lower budgets than films and television series, such as commercials and short web videos. Video games likewise benefited from advances in computer animation throughout the early twenty-first century as well as improvements to console hardware, which facilitated the development of games that were visually impressive, featured intriguing mechanics, and told engaging stories.

In addition to audiovisual media such as films and video games, two-dimensional (2D) art is likewise often created through the use of digital technology, with artists using graphic design software and other programs to create standalone art, illustrations for comic books and graphic novels, advertisements, and eye-catching graphics for use on social media. Desktop and electronic publishing reshaped the traditional publishing industry and presented opportunities for independent authors to share their work with readers. Digital technology likewise facilitated the recording, mixing, and editing of music, which could in turn be distributed through online means, including streaming-music services such as Spotify and online stores such as Bandcamp.

INDUSTRY APPLICATIONS

While most commonly associated by the public with entertainment and media, digital arts and multimedia tools and practices are in fact used in a variety of industries, as they are applicable to nearly any field in which a worker uses a computer to create a visual representation of an object or other visual content, regardless of whether that content will remain digital or transition into the physical form of a printed, manufactured, or constructed object. Digital graphic design tools are widely used by product designers in numerous fields. Consider, for example, an artist working as a designer for a musical instrument company. Tasked with creating a new guitar design for the company, the artist may create digital mockups of the proposed design, using graphic design software to draw elements such as decorative inlays around the instrument's sound hole and on the fretboard, and add colors and digital textures to indicate the different woods and other materials that will be used. The artist may also create diagrams detailing the guitar's structure and specifications, create printable marketing materials, or even design a new logo for the company.

The automotive industry is another industry to which digital art technologies have made a substantial contribution. In automotive design, computer-aided design (CAD) programs may be used to develop three-dimensional (3D) models of a new car and its individual components. This step of the design process enables stakeholders within the company to gain a greater understanding of what the car will look like, but such digital assets also play a significant role in the production of the physical car, as 3D models may be used in conjunction with computer-aided manufacturing (CAM) technology to produce the needed components. In nearly all industries, digital technology is also used to design product packaging, which must capture the attention of potential customers with its imagery and typography while also displaying mandatory information such as nutrition facts or safety warnings. Accompanying materials such as physical or digital user manuals and quick-start guides are also developed with the aid of digital publishing and graphics technologies.

Another major field to which digital art and design tools are deeply relevant is architecture. While some architects may prefer traditional, analog drafting tools, many make use of specialized design programs that enable them to create all stages of an architectural design, from initial sketches to realistic 3D renderings, using a computer. In some cases, architects and real estate development companies take a truly multimedia approach to their work, adding animated elements to the completed renderings of their proposed structures. Such animated renderings may serve as immersive tours or walkthroughs, showcase how elements such as water features would appear in motion, or illustrate how human beings would move through a space.

EDUCATION AND TRAINING

Digital art and multimedia content are also widely used in educational contexts. An educational program seeking to teach basic mathematics to kindergarteners, for instance, may incorporate a number of multimedia and artistic elements in addition to traditional text-based instruction. Such elements may include static or animated cartoons, live-action video segments, sound effects, music, and other additions that help to create an engaging learning experience for young children. The use of multimedia in education became increasingly important following the emergence of the COVID-19 pandemic in 2020, during which US students at all educational levels temporarily transitioned to a virtual learning environment. While many students had returned to in-person learning environments by 2023, the pandemic shed further light on the importance of digital media to education in the twenty-first century.

In addition to schools, digital multimedia content is used for training purposes in a wide range of fields. Virtual reality (VR) and augmented reality (AR) training programs, for instance, may be used to train pilots and other operators of specialized machinery as well as in the automotive industry, in health-care settings, and in the military. Multimedia training videos and interactive courses are also common.

DEVELOPING TECHNOLOGIES

In the third decade of the twenty-first century, a number of developing technologies represented the potential to usher in significant changes within the field of digital arts and multimedia. Improvements to both desktop and mobile computing technology enabled such devices to better run resource-intensive software such as that used for processes such as video and audio editing, animation, and digital design. Blockchain technology, best known for its use in the field of cryptocurrency, became tied to digital art with the development of nonfungible tokens (NFTs), which provided a new approach to the buying and selling of digital art.

One area of particularly significant development was that of augmented and virtual reality. AR combines the real and virtual worlds through the use of elements such as digital overlays, while VR, through the use of devices such as headsets, places the user within a fully virtual environment that may or may not resemble the real world. While numerous companies had explored AR and VR technology throughout the late twentieth and early twenty-first century, the 2020s saw increased investments in that space, some of them—such as the VR initiatives undertaken by technology company Meta—focused on creating an immersive metaverse in which users could socialize, play games, and even conduct business in a 3D animated environment.

Another area of interest during that period was the growing intersection between digital arts and multimedia and the developing fields of artificial intelligence (AI) and machine learning, as exemplified by AI tools such as the chatbot ChatGPT and art generators such as DALL•E 2, Stable Diffusion, and Midjourney. While technologies based on AI or machine learning, such as the content-aware fill tool included in Adobe Photoshop beginning in 2010, had long been used by digital artists as tools to simplify arduous image-editing processes, AI tools of the 2020s enabled users to generate text or art based on simple or complex prompts.

ETHICAL CONCERNS

Though developing technologies excited some professionals within the creative industries, others called the ethics of using such technologies into question. Efforts to develop and popularize VR technology, for instance, prompted some commentators to express concerns about the relationship between VR worlds and consumerism, the potential for discrimination in virtual environments, and the possibility that widespread adoption of VR could contribute to social isolation. NFTs were initially promoted as means for artists to profit from their work; however, they proved controversial in part because of the excessive energy consumption associated with the use of blockchain technologies. According to a 2021 report by *Wired* magazine, the 2019 sale of six pieces of NFT art created by the French artist Joanie Lemercier, for instance, used 8.7 megawatt-hours of energy, much to the dismay of the environmentally conscious artist.

Perhaps the most publicized ethical dilemma in the field of digital art by 2023, however, regarded the use of AI and machine learning technologies in the generation of creative works such as films, digital paintings, and screenplays. Such tools were not truly intelligent; rather, they generated content after being trained using large sets of data—such as text and images—that may have been accumulated and used without the creators' consent. As a result, AI art and writing tools generated content that was derived from existing works, thus introducing the possibility of plagiarism. In addition, critics of the use of AI and machine learning in creative contexts expressed concerns that corporations might attempt to use such tools to replace human workers—a concern that proved accurate when AI-generated articles appeared on several popular websites.

Limiting the impact of AI tools on creative professionals was a key priority for the Writers Guild of America (WGA), which in May of 2023 went on strike amid a labor dispute between the guild and the Alliance of Motion Picture and Television Producers (AMPTP). In addition to calling for improved pay and staffing in the list of proposals published to the guild's strike website, the guild sought to ensure that future agreements with the AMPTP "regulate use of artificial intelligence," for instance by prohibiting studios from using AI to "write or rewrite literary material" or "as source material." The Screen Actors Guild-American Federation of Television and Radio Artists (SAG-AFTRA) likewise went on strike in July of 2023 and joined the WGA in speaking out against AI technology, proposing the creation of "a comprehensive set of provisions to protect human-created work and require informed consent and fair compensation when a 'digital replica' is made of a performer, or when their voice, likeness, or performance will be substantially changed using AI."

THE FUTURE OF DIGITAL ARTS AND MULTIMEDIA

Like any field that is heavily reliant on computer technology, the field of digital arts and multimedia continually evolves alongside changes in technology, including the development of new programs and tools and the deprecation of older ones. In keeping with the field's propensity for such changes, professionals dedicated to creating digital art and multimedia content must work to remain up to date on their industries' tools and creation practices, although some holdouts may continue to use outdated software or other tools long after their peers cease to do so.

The means of distributing and publicizing digital and multimedia content likewise evolve considerably as new companies form and existing ones merge, go out of business, or make sweeping changes to their offerings. In the third decade of the twenty-first century, such changes were particularly common within the rapidly shifting realm of streaming video. Beginning in 2022, for instance, a merger between the services HBO Max and Discovery+ coincided with the removal of dozens of HBO Max series and films from the service, apparently carried out as a cost-cutting measure. In addition to confusing viewers, some of whom were surprised to find content missing from the newly rebranded Max, the service's business decision left many multimedia projects, including several in-progress television series, in a state of limbo that had serious professional ramifications for all who worked on them, from actors and directors to screenwriters, animators, composers, and other professionals.

Social media, another prominent venue for the distribution of digital art and multimedia, likewise proved itself to be a tumultuous industry during that period, which saw changes in the types and formats of multimedia content able to be shared on them. Independent creators, production companies, and advertisers seeking to post video content to the social networking service Instagram, for instance, were required to adjust their approaches with the introduction in 2020 and 2021 of Instagram Reels, short videos initially limited to fifteen seconds in length. That limit was increased substantially as Reels grew in popularity, thus placing pressure on creators to post correspondingly longer videos. Even the names of social networking services themselves changed; in July 2023, for example, Twitter owner Elon Musk announced his plan to rebrand the service as X. While the rebranding initiative began to be carried out in the days that followed, the final impact of Musk's decision remained unclear, not least because several other businesses—including Meta, owner of rival social network Facebook—owned potentially conflicting trademarks.

In addition to requiring digital and multimedia artists to display a certain level of adaptability, the rapid technological changes taking place in the twenty-first century likewise presented a substantial challenge in regard to the preservation of digital content. This was especially the case for content that was born digital—that is, originally created by digital means and in a digital format, as opposed to being created in an analog format and subsequently digitized. In some cases, changes in the prevailing programs, file extensions, and hardware in use make it difficult to open and view digital files created using older technologies; this can require archivists to transition the files into new formats for preservation, a process known as migration. According to the Smithsonian Institution Archives, one of many archives in the United States that work with born-digital materials, other strategies for preserving such files include emulation, which enables files to be accessed using software or hardware for which they were not intended, and bit-level preservation, which keeps the file itself intact but otherwise inaccessible. In addition to files preserved in archives, digital multimedia threatened by technological changes included the multitude of games, animated videos, and interactive websites created early in the twenty-first century for use with Adobe Flash Player, a product that ceased functioning in early 2021. Later developments such as the 2023 announcement that image-hosting website Imgur would delete a large number of images from its servers likewise demonstrated the potentially ephemeral nature of digital art and multimedia content.

Despite such challenges, digital arts and multimedia remains a fruitful field of study for those aspiring to work in, or informally explore, any of the myriad specialties within it. As of its September 2022 update, the US Bureau of Labor Statistics' *Occupational Outlook Handbook* projected that many occupations relevant to the field would experience growth between the years 2021 and 2031: employment within the category of arts and design was projected to grow by 2 percent during that period, with greater gains among graphic designers and special effects artists and animators, while employment within the media and communication category, which included video editors and broadcast technicians, was projected to increase by 6 percent. Employment within the category of entertainment and sports, which included professionals such as directors and musicians, was projected to grow by 13 percent. In light of such projections, the field of digital arts and multimedia appears poised to offer significant potential for professional growth and personal creative exploration for years to come.

—*Joy Crelin, MFA*

Further Reading

Barber, Gregory. "NFTs Are Hot. So Is Their Effect on the Earth's Climate." *Wired*, 6 Mar. 2021, www.wired.com/story/nfts-hot-effect-earth-climate. Accessed 3 Aug. 2023.

Brittain, Blake. "Meta, Microsoft, Hundreds More Own Trademarks to New Twitter Name." *Reuters*, 25 July 2023, www.reuters.com/technology/problem-with-x-meta-microsoft-hundreds-more-own-trademarks-new-twitter-name-2023-07-25/. Accessed 3 Aug. 2023.

Chan, Khee Hoon. "Tracing the Sprawling Roots of Flash Preservation." *Vice*, 18 Mar. 2021, www.vice.com/en/article/wx8y5y/tracing-the-sprawling-roots-of-flash-preservation. Accessed 3 Aug. 2023.

Mendez, Moises II. "Why Your Favorite Show Is Leaving HBO Max." *Time*, 14 Dec. 2022, time.com/6240853/hbo-max-shows-leaving. Accessed 3 Aug. 2023.

OED | Oxford English Dictionary, 2023, www.oed.com. Accessed 3 Aug. 2023.

"Preservation Strategies for Born-Digital Materials." *Smithsonian Institution Archives*, siarchives.si.edu/what-we-do/digital-curation/preservation-strategies-born-digital-materials. Accessed 3 Aug. 2023.

US Bureau of Labor Statistics. *Occupational Outlook Handbook*. 8 Sept. 2022, www.bls.gov/ooh/. Accessed 3 Aug. 2023.

"WGA Negotiations—Status as of May 1, 2023." *WGA on Strike*, 1 May 2023, www.wgacontract2023.org/the-campaign/wga-negotiations-status-as-of-5-1-2023. Accessed 3 Aug. 2023.

"Why We Strike." *SAG-AFTRA*, 13 July 2023, www.sagaftrastrike.org/why-we-strike. Accessed 3 Aug. 2023.

Contributors

Michael Adams
Horrmann Library, Wagner College

Michael Auerbach
Marblehead, MA

Craig Belanger, BA, MST
Journal of Advancing Technology

Daniel L. Berek, MA, MAT
Independent Scholar

Tyler Biscontini
Northeast Editing, Inc.

Mark Boccard
Independent Scholar

Sarah E. Boslaugh, PhD, MPH
Forest Park Community College

Leon James Bynum, MA
Columbia University

Cait Caffrey
Northeast Editing, Inc.

Josephine Campbell
Northeast Editing, Inc.

Patrick G. Cooper
Independent Scholar

Joy Crelin, MFA
Independent Scholar

Trevor Cunnington, MA, PhD
Independent Scholar

Kriselle de Leon
Independent Scholar

Richard De Veaux
Williams College

Joseph Dewey, BA, MA, PhD
Broward College

Sally Driscoll, PhD
Fairfield University

Mark Dziak
Northeast Editing, Inc.

Marlanda English, PhD
Capella University

Miroslav Farina
Independent Scholar

Andrew Farrell, MLIS
Independent Scholar

Ian Gill
Independent Scholar

Melissa A. Gill
David A Clarke School of Law

Kristen P. Goessling
Pennsylvania State University

Jim Greene, MFA
Jacksonville Public Library

Price Grisham
Independent Scholar

Allison Hahn
St. Norbert College

Stuart A. Hargreaves
Chinese University of Hong Kong

Angela Harmon
Independent Scholar

Kylie Grimshaw Hughes
University of South Australia

Micah L. Issitt
Kansas City, MO

Gayla Koerting
University of South Dakota

Jeanne L. Kuhler, BS, MS, PhD
Benedictine University

Dawn A. Laney
Atlanta, GA

Jack Lasky
Northeast Editing, Inc.

M. Lee, BA, MA
Independent Scholar

Laura L. Lundin, MA
Independent Scholar

J. N. Manuel
Independent Scholar

Michael Mazzei, JD
Ladder Capital

Roman Meinhold, MA, PhD
Assumption University

Julia M. Meyers, MA, PhD
Gibsonia, PA

Elizabeth Mohn
Northeast Editing, Inc.

Briana Nadeau
Independent Scholar

Jake D. Nicosia, MS
Stevens Institute of Technlogy

Laura Nicosia, PhD
Montclair State University

Mariam Orkodashvili, PhD
Independent Scholar

George R. Plitnik
Frostburg State University

Elena Popan, MA
Texas Tech University

John Pritchard
Richmond, VT

Luca Prono, PhD
Bologna, Italy

Wylene Rholetter
Auburn University

Elizabeth Rholetter Purdy, PhD
Georgia State University

Richard M. J. Renneboog, MSc
Independent Scholar

Mari Rich
NYU Tandon School of Engineering

Lindsay Rock Rohland
Penn Foster

Andrew Schenker
Hudson, NY

Teresa E. Schmidt
Independent Scholar

Lawrence Schwegler
Independent Scholar

Richard Sheposh
Northeast Editing, Inc.

Kerry Skemp
Enprecis, Inc.

Alexander Stingl
University of Warwick

Randa Tantawi, PhD
EBSCO Information Services

John M. Theilmann, PhD
Converse College

Bethany Thivierge, MPH
Technicality Resources

Janine Ungvarsky
MUSICare Project NEPA

Jennifer A. Vadeboncoeur
University of British Columbia

Maura Valentino, MSLIS
Central Washington University

Kathy Warnes, MA, PhD
University of Toledo

George M. Whitson III, BS, MS, PhD
University of Texas at Tyler

Max Winter
University of Iowa

Thomas A. Wikle
Oklahoma State University

Robin L. Wulffson, MD
American College of Obstetrics and Gynecology

Scott Zimmer, JD
Alliant International University

A

ACOUSTICS

ABSTRACT

Acoustics is the science dealing with the production, transmission, and effects of vibration in material media. If the medium is air and the vibration frequency is between 18 and 18,000 hertz (Hz), the vibration is termed "sound." Acoustics is also used in a broader context to describe sounds in solids and underwater and structure-borne sounds. Because mechanical vibrations, whether natural or human induced, have accompanied humans through the long course of human evolution, acoustics is among the most interdisciplinary sciences. For humans, hearing is a very important sense, and the ability to vocalize greatly facilitates communication and social interaction. Sound can have profound psychological effects; music may soothe or relax a troubled mind, and noise can induce anxiety and hypertension.

BACKGROUND

Acoustics arguably originated with human communication and music. The caves in which the prehistoric humans displayed their most elaborate paintings have resonances easily excited by the human voice, and stalactites emit musical tones when struck or rubbed with a stick. Paleolithic societies constructed flutes of bird bone, used animal horns to produce drones, and employed rattles and scrapers to provide rhythm.

In the sixth century BCE, Pythagoras was the first to correlate musical sounds and mathematics by relating consonant musical intervals to simple ratios of integers. In the fourth century BCE, Aristotle deduced that the medium that carries a sound must be compressed by the sounding body, and in the third century BCE, philosopher Chrysippus correctly depicted the propagation of sound waves with an expanding spherical pattern. In the first century BCE, the Roman architect and engineer Marcus Vitruvius Pollio explained the acoustical characteristics of Greek theaters, but when the Roman civilization declined in the fourth century CE, scientific inquiry in the West essentially ceased for the next millennium.

In the seventeenth century, modern experimental acoustics originated when the Italian mathematician Galileo explained resonance as well as musical consonance and dissonance, and theoretical acoustics got its start with Sir Isaac Newton's derivation of an expression for the velocity of sound. Although this yielded a value considerably lower than the experimental result, a more rigorous derivation by Pierre-Simon Laplace in 1816 obtained an equation yielding values in complete agreement with experimental results.

During the eighteenth century, many famous mathematicians studied vibration. In 1700, French mathematician Joseph Sauveur observed that strings vibrate in sections consisting of stationary nodes located between aggressively vibrating antinodes and that these vibrations have integer multiple frequencies, or harmonics, of the lowest frequency. He also noted that a vibrating string could simultaneously produce the sounds of several harmonics. In 1755, Daniel Bernoulli proved that this resultant vibration was the independent algebraic sum of the various harmonics. In 1750, Jean le Rond d'Alembert used calculus to obtain the wave equation for a vibrating string. By the end of the eighteenth century, the basic experimental results and theoretical underpinnings of acoustics were extant and in reasonable agreement, but it was not until the following century that theory and a concomitant advance of technology led to the evolution of the major divisions of acoustics.

Although mathematical theory is central to all acoustics, the two major divisions, physical and applied acoustics, evolved from the central theoretical core. In the late nineteenth century, Hermann von Helmholtz and Lord Rayleigh (John William Strutt), two polymaths, developed the theoretical aspects. Helmholtz's contributions to acoustics were primarily in explaining the physiological aspects of the ear. Rayleigh, a well-educated wealthy English baron, synthesized virtually all previous knowledge of acoustics and also formulated an appreciable corpus of experiment and theory.

Experiments by Georg Simon Ohm indicated that all musical tones arise from simple harmonic vibrations of definite frequency, with the constituent components determining the sound quality. This gave birth to the field of musical acoustics. Helmholtz's studies of instruments and Rayleigh's work contributed to the nascent area of musical acoustics. Helmholtz's knowledge of ear physiology shaped the field that was to become physiological acoustics.

Underwater acoustics commenced with theories developed by the nineteenth-century mathematician Siméon-Denis Poisson, but further development had to await the invention of underwater transducers in the next century.

Two important nineteenth-century inventions, the telephone (patented 1876) and the mechanical phonograph (invented 1877), commingled and evolved into twentieth- and twenty-first-century audio acoustics when united with electronics. Some products in which sound production and reception are combined are microphones, loudspeakers, radios, talking motion pictures, high-fidelity stereo systems, and public sound-reinforcement systems. Improved instrumentation for the study of speech and hearing has stimulated the areas of physiological and psychology acoustics, and ultrasonic devices are routinely used for medical diagnosis and therapy, as well as for burglar alarms and rodent repellants. Underwater transducers are employed to detect and measure moving objects in the water, while audio engineering technology has transformed music performance as well as sound reproduction. Virtually no area of human activity has remained unaffected by continually evolving technology based on acoustics.

OVERVIEW

Because acoustics is ubiquitous in human endeavors, it is perhaps the broadest and most interdisciplinary of sciences, and its most profound contributions have occurred when it is commingled with an independent field. The interdisciplinary nature of acoustics has often consigned it to a subsidiary role as a minor subdivision of mechanics, hydrodynamics, or electrical engineering. Certainly, the various technical aspects of acoustics could be parceled out to larger and better established divisions of science, but then acoustics would lose its unique strengths and its source of dynamic creativity. The main difference between acoustics and more self-sufficient branches of science is that acoustics depends on physical laws developed in and borrowed from other fields. Therefore, the primary task of acoustics is to take these divergent principles and integrate them into a coherent whole in order to understand, measure, and control vibration phenomena.

Acoustics may be subdivided into a number of main areas, the most important of which are ultrasonics, which examines high-frequency waves not audible to humans; psychological acoustics, which studies how sound is perceived in the brain; physiological acoustics, which looks at human and animal hearing mechanisms; speech acoustics, which focuses on the human vocal apparatus and oral communication; musical acoustics, which involves the physics of musical instruments; underwater sound, which examines the production and propagation of sound in liquids; and noise, which concentrates on the control and suppression of unwanted sound. Two other important areas of applied acoustics are architectural acoustics (including the acoustical design of concert halls and sound reinforcement systems) and audio engineering (recording and reproducing sound).

Ultrasonics. Dog whistles, which can be heard by dogs but not by humans, can generate ultrasonic frequencies of about 25 kilohertz (kHz). Two types of transducers, magnetostrictive and piezoelectric, are used to generate higher frequencies and greater power. Magnetostrictive devices convert magnetic energy into ultrasound by subjecting ferric material (iron or nickel) to a strong oscillating magnetic field. The field causes the material to alternately expand and contract, thus creating sound waves of the same frequency as that of the field. The resulting sound waves have frequencies between 20 Hz and 50 kHz and several thousand watts of power. Such transducers operate at the mechanical resonance frequency where the energy transfer is most efficient.

Piezoelectric transducers convert electric energy into ultrasound by applying an oscillating electric field to a piezoelectric crystal (such as quartz). These transducers, which work in liquids or air, can generate frequencies in the megahertz region with considerable power. In addition to natural crystals, ceramic piezoelectric materials, which can be fabricated into any desired shape, have been developed.

Physiological and Psychological Acoustics. Physiological acoustics studies auditory responses of the ear and its associated neural pathways, and psychological acoustics is the subjective perception of sounds through human auditory physiology. Mechanical, electrical, optical, radiological, or biochemical techniques are used to study neural responses to various aural stimuli. Because these techniques are typically invasive, experiments are performed on animals with auditory systems that are similar to the human system. In contrast, psychological acoustic studies are noninvasive and typically use human subjects.

A primary objective of psychological acoustics is to define the psychological correlates to the physical parameters of sound waves. Sound waves in air may be characterized by three physical parameters: frequency, intensity, and their spectrum. When a sound wave impinges on the ear, the pressure variations in the air are transformed by the middle ear to mechanical vibrations in the inner ear. The cochlea then decomposes the sound into its constituent frequencies and transforms these into neural action potentials, which travel to the brain where the sound is evidenced. Frequency is perceived as pitch, the intensity level as loudness, and the spectrum determines the timbre, or tone quality, of a note.

Another psychoacoustic effect is masking. When a person listens to a noisy version of recorded music, the noise virtually disappears if the music is being enjoyed. This ability of the brain to selectively listen has had important applications in digitally recorded music. When the sounds are digitally compressed, such as in MP3 (MPEG 1 audio layer-3) systems, the brain compensates for the loss of information; thus one experiences higher fidelity sound than the stored content would imply. Also, the brain creates information when the incoming signal is masked or nonexistent, producing a psychoacoustic phantom effect. This phantom effect is particularly prevalent when heightened perceptions are imperative, as when danger is lurking.

Psychoacoustic studies have determined that the frequency range of hearing is from 20 to about 20,000 Hz for young people, and the upper limit progressively decreases with age. The rate at which hearing acuity declines depends on several factors, not the least of which is lifetime exposure to loud sounds, which progressively deteriorate the hair cells of the cochlea. Moderate hearing loss can be compensated for by a hearing aid; severe loss requires a cochlear implant.

Speech Acoustics. Also known as acoustic phonetics, speech acoustics deals with speech production and recognition. The scientific study of speech began with Thomas Alva Edison's phonograph, which allowed a speech signal to be recorded and stored for later analysis. Replaying the same short speech segment several times using consecutive filters passing through a limited range of frequencies creates a spectrogram, which visualizes the spectral properties of vowels and consonants. During the first half of the twentieth century, Bell Telephone Laboratories invested considerable time and resources to the systematic understanding of all aspects of speech, including vocal tract resonances, voice quality, and prosodic features of speech. For the first time, electric circuit theory was applied to speech acoustics, and analogue electric circuits were used to investigate synthetic speech.

Musical Acoustics. A conjunction of music, craftsmanship, auditory science, and vibration physics, musical acoustics analyzes musical instruments to better understand how the instruments are crafted, the physical principles of their tone production, and why each instrument has a unique timbre. Musical instruments are studied by analyzing their tones and then creating computer models to synthesize these sounds. When the sounds can be recreated with minimal software complications, a synthesizer featuring realistic orchestral tones may be constructed. The second method of study is to assemble an instrument or modify an existing instrument to perform nondestructive (or on occasion destructive) testing so that the effects of various modifications may be gauged.

Underwater Sound. Also known as hydroacoustics, this field uses frequencies between 10 Hz and 1 megahertz (MHz). Although the origin of hydroacoustics can be traced back to Rayleigh, the deployment of submarines in World War I provided the impetus for the rapid development of underwater listening devices (hydrophones) and sonar (sound navigation ranging), the acoustic equivalent of radar. Pulses of sound are emitted and the echoes are processed to extract information about submerged objects. When the speed of underwater sound is known, the reflection time for a pulse determines the distance to an object. If the object is moving, its speed of approach or recession is deduced from the frequency shift of the reflection, or the Doppler effect. Returning pulses have a higher frequency when the object approaches and lower frequency when it moves away.

Noise. Physically, noise may be defined as an intermittent or random oscillation with multiple frequency components, but psychologically, noise is any unwanted sound. Noise can adversely affect human health and well-being by inducing stress, interfering with sleep, increasing heart rate, raising blood pressure, modifying hormone secretion, and even inducing depression. The environmental effects of noise are no less severe. The vibrations in irregular road surfaces caused by large rapid vehicles can cause adjacent buildings to vibrate to an extent that is intolerable to the buildings' inhabitants, even without structural damage. Machinery noise in industry is a serious problem because continuous exposure to loud sounds will induce hearing loss. In apartment buildings, noise transmitted through walls is always problematic; the goal is to obtain adequate sound insulation using lightweight construction materials.

Traffic noise, both external and internal, is ubiquitous in modern life. The first line of defense is to reduce noise at its source by improving engine enclosures, mufflers, and tires. The next method, used primarily when interstate highways are adjacent to residential areas, is to block the noise by the construction of concrete barriers or the planting of sound-absorbing vegetation. Internal automobile noise has been greatly abated by designing more aerodynamically efficient vehicles to reduce air turbulence, using better sound isolation materials, and improving vibration isolation.

Aircraft noise, particularly in the vicinity of airports, is a serious problem exacerbated by the fact that as modern airplanes have become more powerful, the noise they generate has risen concomitantly. The noise radiated by jet engines is reduced by two structural modifications. Acoustic linings are placed around the moving parts to absorb the high frequencies caused by jet whine and turbulence, but this modification is limited by size and weight constraints. The second modification is to reduce the number of rotor blades and stator vanes, but this is somewhat inhibited by the desired power output. Special noise problems occur when aircraft travel at supersonic speeds (faster than the speed of sound), as this propagates a large pressure wave toward the ground that is experienced as an explosion. The unexpected sonic boom startles people, breaks windows, and damages houses. Sonic booms have been known to destroy rock structures in national parks. Because of these concerns, commercial aircraft are prohibited from flying at supersonic speeds over land areas.

Construction equipment (such as earthmoving machines) creates high noise levels both internally and externally. When the cabs of these machines are not closed, the only feasible manner of protecting operators' hearing is by using earplugs. By carefully designing an enclosed cabin, structural vibration can be reduced and sound leaks made less significant, thus quieting the operator's environment. Although manufacturers are attempting to reduce the external noise, it is a daunting task because quieter components, such as the rubber tractor treads occasionally used to replace metal, are often not as durable.

APPLICATIONS AND PRODUCTS

Ultrasonics. High-intensity ultrasonic applications include ultrasonic cleaning, mixing, welding, drilling, and various chemical processes. Ultrasonic cleaners use waves in the 150 to 400 kHz range on items (such as jewelry, watches, lenses, and surgical instruments) placed in an appropriate solution. Ultrasonic cleaners have proven to be particularly effective in cleaning surgical devices because they loosen contaminants by aggressive agitation irrespective of an instrument's size or shape, and disassembly is not required. Ultrasonic waves are effective in cleaning most metals and alloys, as well as wood, plastic, rubber, and cloth. Ultrasonic waves are used to emulsify two nonmiscible liquids, such as oil and water, by forming the liquids into finely dispersed particles that then remain in homogeneous suspension. Many paints, cosmetics, and foods are emulsions formed by this process.

Although aluminum cannot be soldered by conventional means, two surfaces subjected to intense ultrasonic vibration will bond—without the application of heat—in a strong and precise weld. Ultrasonic drilling is effective where conventional drilling is problematic, for instance, drilling square holes in glass. The drill bit, a transducer having the required shape and size, is used with an abrasive slurry that chips away the material when the suspended powder oscillates. Some of the chemical applications of ultrasonics are in the atomization of liquids, in electroplating, and as a catalyst in chemical reactions.

Low-intensity ultrasonic waves are used for nondestructive probing to locate flaws in materials for which complete reliability is mandatory, such as those used in spacecraft components and nuclear reactor vessels. When an ultrasonic transducer emits a pulse of energy into the test object, flaws reflect the wave and are detected. Because objects subjected to stress emit ultrasonic waves, these signals may be used to interpret the condition of the material as it is increasingly stressed. Another application is ultrasonic emission testing, which records the ultrasound emitted by porous rock when natural gas is pumped into cavities formed by the rock to determine the maximum pressure these natural holding tanks can withstand.

Low-intensity ultrasonics is used for medical diagnostics in two different applications. First, ultrasonic waves penetrate body tissues but are reflected by moving internal organs, such as the heart. The frequency of waves reflected from a moving structure is Doppler-shifted, thus causing beats with the original wave, which can be heard. This procedure is particularly useful for performing fetal examinations on a pregnant woman; because sound waves are not electromagnetic, they will not harm the fetus. The second application is to create a sonogram image of the body's interior. A complete cross-sectional image may be produced by superimposing the images scanned by successive ultrasonic waves passing through different regions. This ultrasonography procedure, unlike an X-ray, displays all the tissues in the cross section and also avoids any danger posed by the radiation involved in X-ray imaging.

Physiological and Psychological Acoustics. Because the ear is a nonlinear system, it produces beat tones that are the sum and difference of two frequencies. For example, if two sinusoidal frequencies of 100 and 150 Hz simultaneously arrive at the ear, the brain will, in addition to these two tones, create tones of 250 and 50 Hz (sum and difference, respectively). Thus, although a small speaker cannot reproduce the fundamental frequencies of bass tones, the difference between the harmonics of that pitch will recreate the missing fundamental in the listener's brain.

Another psychoacoustic effect is masking. When a person listens to a noisy version of recorded music, the noise virtually disappears if the individual is enjoying the music. This ability of the brain to selectively listen has had important applications in digitally recorded music. When sounds are digitally compressed, as in MP3 systems, the brain compensates for the loss of information, thus creating a higher fidelity sound than that conveyed by the stored content alone.

As technology evolved, environmental noise increased concomitantly; lifetime exposure to loud sounds, commercial and recreational, has created an epidemic of hearing loss, most noticeable in the elderly because the effects are cumulative. Wearing a hearing aid, fitted adjacent to or inside the ear canal, is an effectual means of counteracting this epidemic. The device consists of one or several microphones, which create electric signals that are amplified and transduced into sound waves redirected back into the ear. More sophisticated hearing aids incorporate an integrated circuit to control volume, either manually or automatically, or to switch to volume contours designed for various listening environments, such conversations on the telephone or where excessive background noise is present.

Speech Acoustics. With the advent of the computer age, speech synthesis moved to digital processing, either by bandwidth compression of stored speech or by using a speech synthesizer. The synthesizer reads a text and then produces the appropriate phonemes on demand from their basic acoustic parameters, such as the vibration frequency of the vocal cords and the frequencies and amplitudes of the vowel formants. This method of generating speech is considerably more efficient in terms of data storage than archiving a dictionary of prerecorded phrases.

Another important, and probably the most difficult, area of speech acoustics is the machine recognition of spoken language. When machine speech recognition programs are sufficiently advanced, the computer is able to listen to a sentence in any reasonable dialect and produce a printed text of the utterance. Two basic recognition strategies exist, one dealing with words spoken in isolation and the other with continuous speech. In both cases, it is desirable to teach the computer to recognize the speech of different people through a training program. Because recognition of continuous speech is considerably more difficult than the identification of isolated words, very sophisticated pattern-matching models must be employed. Speech recognition technologies advanced greatly in the 2010s and 2020s, with examples included on many smartphones and computers along with virtual personal assistants, such as Apple's Siri. Such systems are able to process full sentences, enact commands, and even respond with synthesized speech in a realistic manner.

Musical Acoustics. The importance of musical acoustics to manufacturers of quality instruments is apparent. During the last decades of the twentieth century, fundamental research led, for example, to vastly improved French horns, organ pipes, orchestral strings, and the creation of an entirely new family of violins. Acoustics opens up the possibilities of advanced technologies such as software instruments and other innovative forms of music making and recording.

Underwater Sound. Applications for underwater acoustics include devices for underwater communication by acoustic means, remote control devices, underwater navigation and positioning systems, acoustic thermometers to measure ocean temperature, and echo sounders to locate schools of fish or other biota. Low-frequency devices can be used to explore the seabed for seismic research.

Although primitive measuring devices were developed in the 1920s, it was during the 1930s that sonar systems began incorporating piezoelectric transducers to increase their accuracy. These improved systems and their increasingly more sophisticated progeny became essential for the submarine warfare of World War II. After the war, theoretical advances in underwater acoustics coupled with computer technology have raised sonar systems to ever more sophisticated levels.

Noise. One system for abating unwanted sound is active noise control. The first successful application of active noise control was noise-canceling headphones, which reduce unwanted sound by using microphones placed in proximity to the ear to record the incoming noise. Electronic circuitry then generates a signal, exactly opposite to the incoming sound, which is reproduced in the earphones, thus canceling the noise by destructive interference. This system enables listeners to enjoy music without having to use excessive volume levels to mask outside noise and allows people to sleep in noisy vehicles such as airplanes. Because active noise suppression is more effective with low frequencies, most commercial systems rely on soundproofing the earphone to attenuate high frequencies. To effectively cancel high frequencies, the microphone and emitter would have to be situated adjacent to the user's eardrum, but this is not technically feasible. Active noise control has also been considered as a means of controlling low-frequency airport noise, but because of its complexity and expense, it was not yet commercially feasible by the early twenty-first century.

CAREERS AND COURSEWORK

Career opportunities occur in academia (teaching and research), industry, and national laboratories. Academic positions dedicated to acoustics are few, as are the numbers of qualified applicants. Most graduates of acoustics programs find employment in research-based industries in which acoustical aspects of products are important, and others work for government laboratories.

Although the subfields of acoustics are integrated into multiple disciplines, most aspects of acoustics can be learned by obtaining a broad background in a scientific or technological field, such as physics, engineering, meteorology, geology, or oceanography. Physics probably provides the best training for almost any area of acoustics. An electrical engineering major is useful for signal processing and synthetic speech research, and a mechanical engineering background is requisite for comprehending vibration. Training in biology is expedient for physiological acoustic research, and psychology coursework provides essential background for psychological acoustics. Architects often employ acoustical consultants to advise on the proper acoustical design of concert halls, auditoriums, or conference rooms. Acoustical consultants also assist with noise reduction problems and help design soundproofing structures for rooms. Although background in architecture is not a prerequisite for becoming this type of acoustical consultant, engineering or physics is.

Acoustics is rarely offered as a university major; therefore, specialized knowledge is best acquired at the graduate level. Many electrical engineering departments have at least one undergraduate course in acoustics, but most physics departments do not. Nevertheless, a firm foundation in classical mechanics (through physics programs) or a mechanical engineering vibration course will provide, along with numerous courses in mathematics, sufficient underpinning for successful graduate study in acoustics.

SOCIAL CONTEXT AND FUTURE PROSPECTS

Acoustics affects virtually every aspect of modern life; its contributions to societal needs are incalculable. Ultrasonic waves clean objects, are routinely employed to probe matter, and are used in medical diagnosis. Cochlear implants restore people's ability to hear, and active noise control helps provide quieter listening environments. New concert halls are routinely designed with excellent acoustical properties, and vastly improved or entirely new musical instruments have made their debut. Infrasound from earthquakes is used to study the composition of Earth's mantle, and sonar is essential to locate submarines and aquatic life. Sound waves are used to explore the effects of structural vibrations. Automatic speech recognition devices and hearing aid technology are constantly improving.

Many societal problems related to acoustics remain to be tackled. The technological advances that made modern life possible have also resulted in more people with hearing loss. Environmental noise is ubiquitous and increasing despite efforts to design quieter machinery and pains taken to contain unwanted sound or to isolate it from people. Also, although medical technology has been able to help many people with hearing and speech impairments, other individuals still lack appropriate treatments. For example, although voice generators exist, there is considerable room for improvement.

—George R. Plitnik

Further Reading

Bass, Henry E., and William J. Cavanaugh, editors. *ASA at Seventy-Five*. Acoustical Society of America, 2004.

Beyer, Robert T. *Sounds of Our Times: Two Hundred Years of Acoustics*. Springer-Verlag, 1999.

Everest, F. Alton, and Ken C. Pohlmann. *Master Handbook of Acoustics*. 7th ed, McGraw-Hill, 2021.

Filimowicz, Michael, editor. *Foundations in Sound Design for Embedded Media: A Multidisciplinary Approach*. Routledge, 2020.

Pierce, Allan D. *Acoustics: An Introduction to Its Physical Principles and Applications*. 3rd ed., Springer-International Publishing, 2019.

Pulkki, Ville, and Matti Karjalainen. *Communication Acoustics: An Introduction to Speech, Audio, and Psychoacoustics*. John Wiley & Sons, 2015.

Rossing, Thomas, and Neville Fletcher. *Principles of Vibration and Sound*. 2nd ed., Springer-Verlag, 2004.

Rumsey, Francis, and Tim McCormick. *Sound and Recording: An Introduction*. 8th ed., Routledge, 2021.

Strong, William J., and George R. Plitnik. *Music, Speech, Audio*. 4th ed., BYU Academic Publishing, 2013.

ADVERTISING

ABSTRACT

Advertising is the process of making potential consumers aware of a product, including goods and services, and describing the manifestation of the message. It may include print, broadcast, and social media advertising as well as nontraditional means of disseminating messages, such as novelty items and T-shirts. Advertising also includes such acts as product placement, contests to promote products, infomercials, and sponsorships. Although much advertising is related to commerce, social services, governments, and other groups also advertise to make people aware of events, services, and benefits they may wish to explore.

BACKGROUND

Early advertising was limited to verbal messages. Merchants shouted about what they had to sell, hoping to draw in passersby. This type of selling is still seen in some venues, such as sports arenas where hot dog and soft drink vendors shout out to spectators. Merchants hung signs and flags on their stalls, often with symbols indicating what was for sale, to alert customers. These methods were effective because in ancient times, many people were illiterate.

The first public notices were created in ancient Egypt about 2000 BCE. These were carved in steel and placed in public places. Much later, printers learned to create handbills, or small advertisements on paper or parchment that were hand distributed in public. Handbills became important to merchants in China during the Song dynasty (960-1276 CE). Chinese merchants used handbills with trademarks to differentiate their products from those sold by other merchants.

The invention of the movable-type printing press in 1440 created new opportunities. Inventor Johannes Gutenberg first used the press to mass-produce books, but as other printers acquired presses, they found new ways to use them. Some printers began producing handbills for merchants. The first handbill advertisement in England was produced in 1472 to sell prayer books.

Printed material was widely available by the eighteenth century. In addition to handbills, posters, pamphlets, and other materials were being distributed. The printing press also brought about the newspaper industry. In North America, the first newspaper advertisement appeared in the *Boston News-Letter* in 1704. One of the most famous newspaper publishers in history, Benjamin Franklin, launched the *Pennsylvania Gazette*, complete with numerous advertisements, in Philadelphia in 1729 and followed up thirteen years later with a new magazine. About three decades later, Franklin was at the

Photo via iStock/Shutthiphong Chandaeng. [Used under license.]

center of the American Revolution (1765-83), which was spurred at least in part by advertisements, including calls for men to enlist in the armed forces.

New printing processes led to the development of illustrated posters in the 1790s. To protect advertisements such as these from the weather, billposters began erecting boards on which to hang their bills in high-traffic areas. Large-scale posters originated in 1835 in New York. These ads of more than fifty square feet were printed by Jared Bell to promote circuses.

Advertising became a field in its own right during the late eighteenth and early nineteenth centuries. Most early advertising agencies did little more than sell advertising space with printers. During the mid-nineteenth century, however, agencies took on more creative roles. Copywriters began focusing on conveying a message to consumers. Many of these worked to promote businesses such as Wanamaker's department store and products such as Ivory soap. The rise in magazines, especially women's publications such as *Ladies' Home Journal*, increased the reach of advertising.

Sears, which made its name as a mail-order catalog business, created the first direct marketing campaign in 1892. Sears mailed eight thousand handwritten postcards to consumers. The effort was a success, generating two thousand orders.

Rapid developments in technology during the twentieth century led to broadcasting. Radio launched in 1920 and helped spread mass culture through music and programming. Many radio shows were sponsored by advertisers, a practice that continued with the development of television.

Society experienced a shift during the 1920s. The decade saw social upheaval, as women gained the vote and rejected many of the ideas of previous generations, and many people were looking for ways to enjoy themselves. Women became important consumers during the 1920s. Many went to work in white-collar office jobs, performing clerical and secretarial work. They had the means and interest in buying goods, in particular radios, home appliances, and ready-to-wear clothes.

During the 1920s, more people lived in urban settings than in rural communities for the first time in US history. Despite the poverty in many areas, a sizable population—mostly young urban dwellers—spent freely. Advertising capitalized on this culture. Companies increasingly promoted nonessential goods. National advertising campaigns and the growth of retail chains meant that consumers were aware of and able to buy the same products from coast to coast.

The heady days of advertising took a hit in 1929, however. The stock market crash left companies reeling. Advertising spending went from $3.5 billion to $1.5 billion in 1933. After World War II (1939-45), however, the purchasing power of Americans increased, and advertising helped usher in the rise of American consumerism. Having made its debut in 1941 with a brief commercial for the Bulova Watch Company, television advertising became a major force during the latter half of the twentieth century and encompassed not only television commercials but also program sponsorships.

Advertising became increasingly common in many public spaces. Ads were placed on train cars, and in 1962, a French company created the first bus shelter, which provided a place for commuters to sit but also generated revenue through advertising. The development of digital billboards allowed advertisers to change their messages almost instantly.

Late in the twentieth century and into the early twenty-first century, the internet and social media platforms, including applications (apps), became valuable advertising tools. Many newspapers also invested in developing advertising on their digital platforms.

OVERVIEW

Many businesses and advertising agencies develop advertising strategies, or blueprints for how to present products to potential customers. In developing a plan, the strategist must first understand the product and evaluate its purpose, strengths, and the ways in which it may meet consumers' needs better than other products on the market. Next, a market analysis identifies potential customers, studying demographics such as age, gender, and social status. Often, the next step is to identify which forms of media the desired customers are most likely to consume: do they watch certain television shows, read specific magazines, or use websites? The strategist must also consider whether the market is saturated with similar products.

Once the strategist understands the product, the consumers, and the market, the team or individual establishes goals for the advertising campaign, such as increases in overall sales or in sales to specific demographics, as well as a timeframe in which to meet the goals, which will help the business evaluate the success of the campaign. The company then decides where to advertise (selecting media, geography, television shows, etc.), develops ideas for how the ads should focus on the product's strengths, and establishes where the advertising budget will be spent.

Changes in society have shaped advertising strategies and tools. For example, many nineteenth-century ads were painted on the sides of barns and other buildings near roadways. These ads were for nearby businesses and touted their products. As the automobile grew to dominate transportation, outdoor advertising became increasingly important. An advertisement painted on the side of a building was semipermanent, but a printed large-scale advertisement could be changed as needed. The first billboard was leased in 1867 to take advantage of increased traffic. Although early billboards were of various shapes and sizes, in 1900 the Outdoor Advertising Association of America (OAAA) helped standardize billboards. This permitted companies to develop national billboard campaigns and produce ads that could be placed on billboards across the country. Coca-Cola, Kellogg's, and Palmolive began mass-producing billboards to market their products nationally.

ADVERTISING FLOPS

Some national advertising campaigns have failed very publicly. The Ford Motor Company began promoting its new automobile in early 1957. The campaign kept an air of mystery about the car, generating keen consumer interest with the statement "The Edsel is coming" but refraining from showing the vehicle. The campaign progressed to show quick glimpses of the car, from the silhouette to the hood ornament. The Edsel was finally revealed to the public on September 4, 1957, in showrooms. Crowds swept in to see the mysterious car, but few bought it. Ford had built up such high expectations with its advertising campaign that the car could never meet them. Although nothing was wrong with the Edsel, the public was disappointed by the reveal.

In 1985, Coca-Cola attempted to stave off competition from Pepsi with a new advertising campaign and a new cola. The company had conducted a number of taste trials while developing Diet Coke and discovered that people preferred a sweeter version of Coke. The new formula was announced with a splash of publicity in April of 1985, including a press conference, when the company announced it had replaced the old formula. The press concluded that New Coke tasted more like Pepsi. Consumers were divided—many liked the new formula, while others remained loudly loyal to traditional Coke. The company logged thousands of calls from angry customers. Although blind taste tests had proven that consumers preferred the new formula, the company had failed to consider the public's emotional connection to a century-old product. By July, the company announced it was bringing the old formula out of retirement and repackaging it as Coca-Cola Classic. The press conference drew national attention, with television news interrupting regular programming to report on the decision. The classic formula surged in popularity, and within months, Coke Classic was the top-selling sugar cola on the market. Although the advertising campaign had failed in its efforts to sell the new formula, in the end, Coca-Cola succeeded in claiming the major share of the soft drink market. New Coke was rebranded as Coke II in 1990 but faded away entirely in 2002.

ADVERTISING IN THE TWENTY-FIRST CENTURY

While advertising media such as print and television remained important means of marketing products and services during the twenty-first century, the rise of the internet forced many advertisers to reconsider their approaches to reaching the public. Internet advertising began in the 1990s and grew as an industry over the course of the decade. Companies spent $2 billion on internet advertising in 1999; by 2023, however, that number was projected to exceed $626 billion globally, according to a report by the market research firm Insider Intelligence.

Internet ad spending represented about 64 percent of all North American ad spending in 2021, according to the statistics platform *Statista*, while the second-largest market segment, television advertising, represented about 22 percent of spending in that market. The shift toward internet advertising created a number of new opportunities for advertisers, including the ability to tailor ads to increasingly specific demographics. Twenty-first-century advertising was often based on individuals' habits, including their internet browsing patterns, and tracking cookies enabled websites to serve visitors with advertisements for products they had seen previously while browsing. Social media likewise became essential to the advertising industry. In addition to enabling companies to purchase traditional online advertising space, social media services such as Instagram, YouTube, and TikTok led to the development of a population of online influencers and content creators willing to advertise products and services through sponsorships and brand partnerships. According to *Statista*, companies were projected to spend about $31 billion on influencer advertising worldwide over the course of 2023.

—Josephine Campbell

Further Reading

"Ad Age Advertising Century: Timeline." *Ad Age*, 29 Mar. 1999, adage.com/article/special-report-the-advertising-century/ad-age-advertising-century-timeline/143661. Accessed 13 July 2023.

"Advertising Expenditure in North America from 2017 to 2024, by Medium." *Statista*, Nov. 2022, www.statista.com/statistics/1347750/ad-spend-in-north-america-media. Accessed 13 July 2023.

"Commercial Advertisement." *University of Washington*, depts.washington.edu/chinaciv/graph/tcommain.htm. Accessed 13 July 2023.

Crain, Matthew. *Profit over Privacy: How Surveillance Advertising Conquered the Internet*. U of Minnesota P, 2021.

De Swaan Arons, Marc. "How Brands Were Born: A Brief History of Modern Marketing." *The Atlantic*, 3 Oct. 2011, www.theatlantic.com/business/archive/2011/10/how-brands-were-born-a-brief-history-of-modern-marketing/246012. Accessed 13 July 2023.

Fox, Zoe. "The Evolution of Advertising: From Stone Carving to the Old Spice Guy." *Mashable*, 26 Dec. 2011, mashable.com/2011/12/26/history-advertising/#6sdCJNghF5qK. Accessed 13 July 2023.

"Influencer Advertising—Worldwide." *Statista*, www.statista.com/outlook/amo/advertising/influencer-advertising/worldwide. Accessed 13 July 2023.

Klein, Christopher. "Why Coca-Cola's 'New Coke' Flopped." *History.com*, 23 May 2023, www.history.com/news/hungry-history/the-new-coke-flop-30-years-ago. Accessed 13 July 2023.

Lebow, Sara. "Worldwide Digital Ad Spend Will Top $600 Billion This Year." *Insider Intelligence*, 31 Jan. 2023, www.insiderintelligence.com/content/worldwide-digital-ad-spend-will-top-600-billion-this-year. Accessed 13 July 2023.

Mertes, Alyssa. "The History of TV Commercials: From Super Bowl Ads to Funny Ads We Still Love Watching." *Quality Logo*, 13 Apr. 2023,

www.qualitylogoproducts.com/promo-university/history-of-tv-ads.htm. Accessed 13 July 2023.
Vaux, Robert. "The Definition of an Advertising Strategy." *Chron*, smallbusiness.chron.com/definition-advertising-strategy-19332.html. Accessed 13 July 2023.

Algorithm

ABSTRACT

An algorithm is a set of steps to be followed in order to solve a particular type of problem. Having originated within the field of mathematics, algorithms make it easier for mathematicians to think of better ways to solve certain types of problems, because looking at the steps needed to reach a solution sometimes helps them to see where an algorithm can be made more efficient by eliminating redundant steps or using different methods of calculation. Algorithms are key to computer science and enable the automation of various processes, including the generation and manipulation of curves in digital graphics programs as well as the processing and compression of multimedia files. They are likewise essential to the creation and use of artificial intelligence and machine learning technologies.

BACKGROUND

The word "algorithm" originally came from the name of a Persian mathematician, Al-Khwarizmi, who lived in the ninth century CE and wrote a book about the ideas of an earlier mathematician from India, Brahmagupta. At first the word simply referred to the author's description of how to solve equations using Brahmagupta's number system, but as time passed it took on a more general meaning. First it was used to refer to the steps required to solve any mathematical problem, and later it broadened still further to include almost any kind of method for handling a particular situation. The concept has been analogized to a recipe for baking a cake; just as the recipe describes a method for accomplishing a goal (baking the cake) by listing each step that must be taken throughout the process, an algorithm is an explanation of how to solve a problem that describes each step necessary.

OVERVIEW

Algorithms are often used in mathematical instruction because they provide students with concrete steps to follow, even before the underlying operations are fully comprehended. There are algorithms for most mathematical operations, including subtraction, addition, multiplication, and division.

For example, a well-known algorithm for performing subtraction is known as the left to right algorithm. As its name suggests, this algorithm requires one to first line up the two numbers one wishes to find the difference between so that the units digits are in one column, the tens digits in another column, and so forth. Next, one begins in the leftmost column and subtracts the lower number from the upper, writing the result below. This step is then repeated for the next column to the right, until the values in the units column have been subtracted from one another. At this point the results from the subtraction of each column, when read left to right, constitute the answer to the problem.

By following these steps, it is possible for a subtraction problem to be solved even by someone still in the process of learning the basics of subtraction. This demonstrates the power of algorithms both for performing calculations and for use as a source of instructional support.

Algorithms are also important within the field of computer science. For example, without algorithms, a computer would have to be programmed with the exact answer to every set of numbers that an equation could accept in order to solve an equation—an impossible task. By programming the computer with the appropriate algorithm, the computer can follow the instructions needed to solve the problem, regardless of which values are used as inputs. Algorithms are crucial to a wide range of

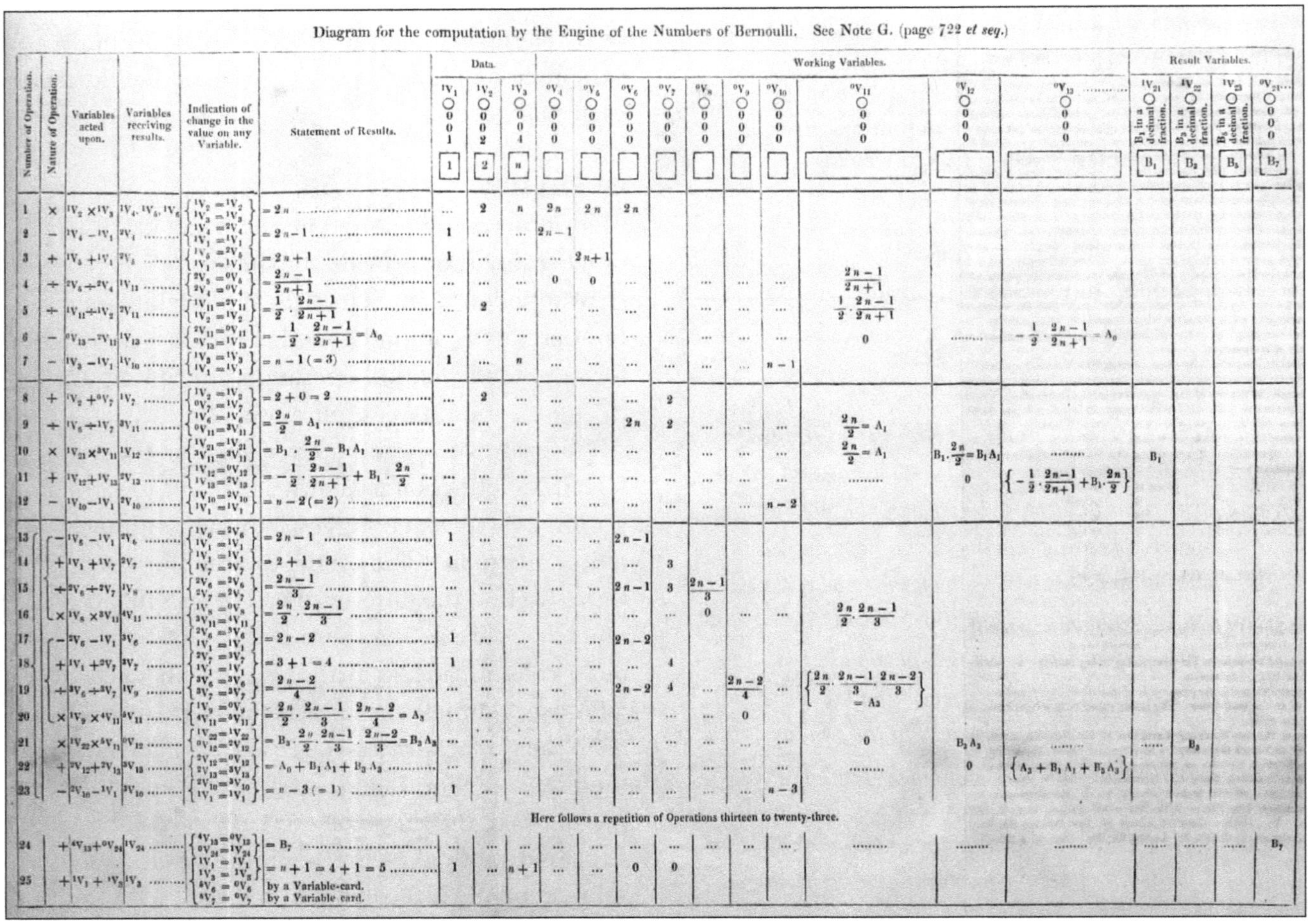

Ada Lovelace's diagram from "note G," the first published computer algorithm. Photo via Wikimedia Commons. [Public domain.]

computer technologies, facilitating the sorting and analysis of data, the use of automated decision-making processes, and other key functions. They are especially crucial to the fields of artificial intelligence and machine learning. In digital graphics and design programs, algorithms are particularly useful in automating the drawing of shapes that would otherwise require time-consuming calculations. For example, design programs use algorithms to generate curves and to enable designers to manipulate those curves in accordance with their needs. Algorithms are also widely used to process and compress multimedia files.

—*Scott Zimmer*

Further Reading

Cormen, Thomas H. *Algorithms Unlocked*. MIT Press, 2013.

Ferguson, R. Stuart. *Practical Algorithms for 3D Computer Graphics*. 2nd ed., AK Peters/CRC Press, 2013.

Fitter, Hetal N., Akash B. Pandey, Divyang D. Patel, and Jitendra M. Mistry. "A Review on Approaches for Handling Bezier Curves in CAD for Manufacturing." *Procedia Engineering*, vol. 97, 2014, pp. 1155-66.

MacCormick, John. *Nine Algorithms That Changed the Future: The Ingenious Ideas That Drive Today's Computers*. Princeton UP, 2012.

O'Leary, Timothy, Linda O'Leary, and Daniel O'Leary. *Computing Essentials 2023*. McGraw-Hill, 2022.

Parker, Matt. *Things to Make and Do in the Fourth Dimension: A Mathematician's Journey through Narcissistic Numbers, Optimal Dating Algorithms, at Least Two Kinds of Infinity, and More*. Farrar, 2014.

Rychagov, Michael N., Ekaterina V. Tolstaya, and Mikhail Y. Sirotenko, editors. *Smart Algorithms for Multimedia and Imaging*. Springer Cham, 2021.

Sarkar, Jayanta. *Computer Aided Design: A Conceptual Approach*. CRC Press, 2017.

Schapire, Robert E., and Yoav Freund. *Boosting: Foundations and Algorithms*. MIT Press, 2012.

Steiner, Christopher. *Automate This: How Algorithms Came to Rule Our World*. Penguin, 2012.

Valiant, Leslie. *Probably Approximately Correct: Nature's Algorithms for Learning and Prospering in a Complex World*. Basic, 2013.

Android OS

ABSTRACT

Introduced to consumers by the technology company Google, the Android operating system (OS) debuted on the mobile-device market with the release of the T-Mobile G1 (or HTC Dream) smartphone in 2008. The OS quickly became a major competitor in that market and as of 2023 was the dominant mobile OS both in the United States and worldwide.

BACKGROUND

Mobile computing is the fastest-growing segment of the tech market. As pricing has become more affordable, developing nations, particularly in Africa, are the largest growing market for smartphones. With smartphones, users shop, gather information, connect via social media such as Twitter and Facebook, and communicate—one of the uses more traditionally associated with phones.

By far the most popular operating system running on mobile phones is Android. Since its launch in 2008, Android has far and away overtaken the competition, which has included popular operating systems such as Apple's iPhone operating system (iOS). By mid-2023, more than 3 billion Android devices were active worldwide.

OVERVIEW

Android came about amid a transformative moment in mobile technology. Prior to 2007, slide-out keyboards mimicked the typing experience of desktop personal computers (PCs). In June of that year, Apple released its first iPhone, forever altering the landscape of mobile phones. Apple focused on multitouch gestures and touch-screen technology. Nearly concurrent with this, Google's Android released its first application program interface (API).

The original API of Google's new operating system (OS) first appeared in October 2008. The Android OS was first installed on the T-Mobile G1, also known as the HTC Dream. This prototype had a very small set of preinstalled apps, and as it had a slide-out QWERTY keyboard, there were no touch-screen capabilities. It did have native multitasking, which Apple's iOS did not yet have. Still, to compete with Apple, Google was forced to

Android logo. Image by Google Inc., via Wikimedia Commons.

replace physical keyboards and access buttons with virtual onscreen controls. The next iteration of Android shipped with the HTC Magic and was accompanied by a virtual keyboard and a more robust app marketplace. Among the other early features that have stood the test of time are the pull-down notification list, home-screen widgets, and strong integration with Google's Gmail service.

One later feature, the full-screen immersive mode, became quite popular as it reduces distractions. First released with Android 4.4, "KitKat," in 2013, it hides the navigation and status bars while certain apps are in use. It was retained for the release of Android 5.0, "Lollipop," in 2015, as well as for subsequent releases.

ANDROID CHANGES AND GROWS

Both of Google's operating systems—Android and its cloud-based desktop OS, ChromeOS—are based on the free open-source OS Linux, created by engineer Linus Torvalds and first released in 1991. Open-source software is created using publicly available source code. The open-source development of Android has allowed manufacturers to produce robust, affordable products that contribute to its widespread popularity in emerging and developing markets. This may be one reason why Android has captured many more new users than its closest rival, Apple's iOS. This strategy has kept costs down and has also helped build Android's app marketplace, the Google Play Store, which offers millions of apps, many free of charge. As of 2023, Android made up more than 70 percent of the global mobile operating system market.

This open-source development of Android has had one adverse effect: the phenomenon known as "forking," which occurs primarily in China. Forking is when a private company takes the OS and creates their own products apart from native Google services such as email. Google seeks to prevent this loss of control (and revenue) by not supporting these companies or including their apps in its marketplace.

Google's business model has always focused on rapid iteration. By contrast, rivals such as Microsoft and Apple have had a far slower, more deliberate pace due to hardware issues. One benefit of Google's faster approach is the ability to address issues and problems in a timely manner. A drawback is the phenomenon known as "cloud rot." As the cloud-based OS grows older, servers that were once devoted to earlier versions are repurposed. Since changes to the OS initially came every few months, apps that worked a month prior would suddenly lose functionality or become completely unusable. Later Android updates have been released on a timescale of six months or more.

Throughout many of Android's first years of existence, new versions of the OS were known by both version numbers and dessert-themed code names, such as Cupcake (version 1.5), Ice Cream Sandwich (version 4.0), and Oreo (versions 8.0 and 8.1). Following the release of Pie (version 9), however, Google moved away from that naming scheme with Android version 10, released in 2019. Android version 11, released as a public beta in June of 2020, was likewise referred to by its version number rather than by a themed nickname. Google continued that practice with Android version 12, released in 2021, as well as versions 12 L and 13, both released in 2022.

ANDROID'S FUTURE

One of the biggest concerns about Android's future is the issue of forking. Making the code available to developers at no cost has made Android a desirable and cost-effective alternative to higher-end makers such as Microsoft and Apple, but it has also made Google a target of competitors. Another consideration for Android's future is its link to ChromeOS, also a Google product. Google plans to keep the two separate. Further, Google executives have made it clear that Chromebooks (laptops that run Chrome) and Android devices have distinct purposes. Android's focus has been on touch-screen technology, multitouch gesturing, and screen resolution, making it a purely mobile OS for phones, tablets, wearable devices, and televisions. Meanwhile, Chrome has developed tools that are more useful in the PC and laptop environment, such as keyboard shortcuts. However, an effort to unify the appearance and functionality of Google's different platforms and devices called Material Design was introduced in 2014. Further, Google has ensured that Android apps can be executed on Chrome and has made the Google Play Store available on Chromebooks. Such implementations suggest a slow merging of the Android and Chrome user experiences.

—*Andrew Farrell*

Further Reading

Amadeo, Ron. "Report: Google Will Graciously Let Android OEMs Build Amazon Fire Devices." *Ars Technica*, 28 Oct. 2022, arstechnica.com/gadgets/2022/10/report-google-will-graciously-let-android-oems-build-amazon-fire-devices. Accessed 13 July 2023.

_____. "The (Updated) History of Android." *Ars Technica*, 31 Oct. 2016, arstechnica.com/gadgets/2016/10/building-android-a-40000-word-history-of-googles-mobile-os. Accessed 13 July 2023.

Curry, David. "Android Statistics (2023)." *Business of Apps*, 27 Feb. 2023, www.businessofapps.com/data/android-statistics. Accessed 13 July 2023.

Edwards, Jim. "Proof That Android Really Is for the Poor." *Business Insider*, 27 June 2014, www.businessinsider.in/Proof-That-Android-Really-Is-For-The-Poor/articleshow/37328668.cms. Accessed 13 July 2023.

"Mobile Operating System Market Share Worldwide—June 2023." *Statcounter*, June 2023, gs.statcounter.com/os-market-share/mobile/worldwide/#monthly-202206-202306. Accessed 13 July 2023.

Newman, Jared. "With Android Lollipop, Mobile Multitasking Takes a Great Leap Forward." *Fast Company*, 6 Nov. 2014, www.fastcompany.com/3038213/with-android-lollipop-mobile-multitasking-takes-a-great-leap-forward. Accessed 13 July 2023.

"Number of Available Applications in the Google Play Store from December 2009 to June 2023." *Statista*, June 2023, www.statista.com/statistics/266210/number-of-available-applications-in-the-google-play-store. Accessed 13 July 2023.

Raphael, J. R. "Android Versions: A Living History from 1.0 to 14." *Computerworld*, 26 June 2020, www.computerworld.com/article/3235946/android-versions-a-living-history-from-1-0-to-today.html. Accessed 13 July 2023.

Animation

ABSTRACT

Animation is a process in which still images appear to be in motion. A popular artform from the early twentieth century on, animation encompasses multiple distinct styles and techniques. Computer animation became the dominant form of animation in the early twenty-first century, although traditional and stop-motion animation remained in use.

BACKGROUND

Animation is a process in which still images appear to be in motion. When individual images flash quickly in sequence, the first image has not yet faded when the next one appears. This phenomenon is called persistence of vision. It is a trick of the eye. The brain begins to ignore the separate images as the speed of the changes increases. As a result, the individual images seem to merge together so that the person sees motion. This effect is named the illusion of motion. It happens at around eight to twelve frames per second (fps).

Early animation began in the 1830s. Devices such as the phenakistoscope and zoetrope were invented. Both were rotating tools, and their surfaces were lined with images. Theater animation debuted in 1876. Émile Reynaud painted pictures on film ribbons. These pictures were projected using mirrors. The first animator to use film was J. Stuart Blackton, who made *Humorous*

Phases of Funny Faces in 1906. For that project, each film frame was photographed individually.

An early animated character that became popular, Felix the Cat, first appeared in 1919. Two men claimed credit for this character—Otto Messmer and Pat Sullivan. Felix was the first animated character to appear in a movie. Felix inspired Walt Disney to create Oswald the Lucky Rabbit, Disney's first major character. Disney eventually lost his rights to Oswald, so he created a new character, Mickey Mouse. Micky Mouse has since become an animated icon. In 1928, Disney produced *Steamboat Willie*. It was the first animated film to use recorded music that played along with the action in the movie. *Snow White and the Seven Dwarfs* was his next big milestone. It became a widely distributed movie in 1937.

Max and Dave Fleischer also became famous in the 1930s. They produced Betty Boop and Popeye the Sailor cartoons. Another major studio at the time was Warner Bros. Its animation division produced cartoons featuring Bugs Bunny and friends. Television animation boomed in the 1950s and was led by Hanna-Barbera.

European animation differed from its US counterparts. It used line drawings, stop motion, puppets, and shadows. An example is the work of German filmmaker Lotte Reiniger. Reiniger created *The Adventures of Prince Achmed* (1926), which used animated silhouettes. Japanese animation, or anime, would also become well known throughout the latter half of the twentieth century.

There are many famous animation studios in the twenty-first century. They include Walt Disney Animation Studios, Disney subsidiary Pixar Animation Studios, DreamWorks Animation, Cartoon Network Studios, Nickelodeon Animation Studio, and Studio Ghibli.

OVERVIEW

Animation can be categorized based on media and techniques. Traditional media are still used. Digital media are a more recent development.

Computers are standard tools in animation. They are used for various steps of the process. Computer-generated imagery (CGI) contributes to two-dimensional (2D) and three-dimensional (3D) animation.

Mixed-media animation is drawn animation. It involves drawing different frames on cels or paper and then photographing each frame individually. A cel is a thin sheet that is transparent. Another form of animation involves placing images directly on film. This method is direct-to-film animation.

Stop-motion animation uses photography. Animated models of figures may be made of clay, wood, and other materials. The figures are photographed, moved slightly, and then photographed again to create the illusion of motion. A technique called pixilation uses real people or objects. One photo is taken to create every frame.

—Kriselle de Leon

Further Reading

Arnold, Gordon. *Animation and the American Imagination: A Brief History*. Praeger, 2017.

Brown, Blain. *The Basics of Filmmaking: Screenwriting, Producing, Directing, Cinematography, Audio, and Editing.* Routledge, 2020.

Furniss, Maureen. *A New History of Animation*. Thames & Hudson, 2016.

Heidmets, Rao. *The Animation Textbook*. CRC Press, 2022.

TED-Ed. "Animation Basics: The Optical Illusion of Motion." *TedEd*, ed.ted.com/lessons/animation-basics-the-optical-illusion-of-motion-ted-ed#watch. Accessed 13 July 2023.

Application

ABSTRACT

An application is a piece of software created to perform a task, such as word processing, web browsing, or chess playing. Each application is designed to run on a particular platform, which is a type of system software that is installed on desktop computers, laptops, or mobile devices such as tablet computers or smartphones.

BACKGROUND

Applications are software programs that perform particular tasks, such as word processing or web browsing. They are designed to run on one or more specific platforms. The term "platform" can refer to any basic computer infrastructure, including the hardware itself and the operating system (OS) that manages it. An OS is a type of system software that manages a device's hardware and other software resources. Application designers may create different versions of an application to run on different platforms. A cross-platform application is one that can be run on more than one platform.

In the context of mobile devices such as tablets and smartphones, the term "application" is typically shortened to app. Since the introduction of the iPhone in 2007, apps have taken center stage in the realm of consumer electronics. Previously, consumers tended to be attracted more to a device's hardware or OS features. A consumer might have liked a certain phone for its solid design or fast processor, or they might have preferred the graphical interface of Microsoft Windows or macOS to the command-line interface of Linux. These features have since become much less of a concern for the average consumer. Instead, consumers tend to be more interested in finding a device that supports the specific apps they wish to use.

OVERVIEW

Over the years, applications have become more and more specialized. Even basic utility programs that were once included with OSs are now available for purchase as separate applications. In some cases, these applications are a more advanced version of the utility software that comes with the OS. For example, an OS may come with free antivirus software, but a user may choose to purchase a different program that offers better protection.

Some software companies offer application suites of interoperating programs. Adobe Creative Cloud is a cloud-based graphic design suite that includes popular design and editing programs such as Photoshop and InDesign. Microsoft Office is an office suite consisting of a word processor (Word), a spreadsheet program (Excel), and other applications commonly used in office settings. These programs are expensive and can take up large amounts of storage space on a user's computer. Before broadband internet access became widely available, application suites were distributed on portable media such as floppy disks, CD-ROMs, or DVDs, because downloading them over a dial-up connection would have taken too long.

As high-speed internet access has become much more common, application developers have taken a different approach. Instead of investing in bulky application suites, users often have the option of using web applications. These applications run partly or entirely on remote servers, avoiding the need to install them on the computer's hard drive. In many cases, companies offering such applications make use of a subscription model, charging users a set annual or monthly fee to use the applications rather than a single lump sum.

TYPES OF APPLICATIONS

Many different types of software fall under the broad heading of applications. A large segment of the application market is focused on office and productivity software. This category includes email applications, word processors, spreadsheet software, presentation software, and database management systems. In an office environment, it is critical that users be able to create documents using these applications and share them with others. This often means that a business or organization will select a particular application suite and then require all employees to use it.

Other types of applications include games, audio-video editing and production software, and even software that helps programmers write new software.

Due to the complexity of software engineering, programmers have developed many applications to help them produce more polished, bug-free programs. Software developers may use multiple applications to code a single program. They might use a word processor or text editor to write the source code and explanatory comments, a debugging tool to check the code for errors, and a compiler to convert the code into machine language that a computer can execute. There is even a type of application that can emulate a virtual computer running inside another computer. These applications are often used by web-hosting companies. Instead of having to set up a new physical server for each customer that signs up, they can create another virtual server for the user to access.

SECURITY IMPLICATIONS

Applications must have certain privileges in order to use the resources of the computer they are running on. As a result, they can sometimes be a point of weakness for attackers to exploit. A clever attacker can take over a vulnerable application and then use its privileges to make the computer behave in ways it should not. For example, the attacker could send spam emails, host illegally shared files, or even launch additional attacks against other computers on the same network.

CAREERS IN APPLICATIONS

Applications are the focus of a variety of career options for those interested in working with software. Apart from the obvious role of computer programmer, there are several other paths one might take. One option is quality assurance. Quality assurance staff are responsible for testing software under development to make sure it performs as it should. Technical support is another option. Technical support specialists assist users with operating the software and fixing errors it might cause. Yet another path is technical writing. Technical writers create software user manuals and training materials. Finally, some applications are so complex that using them can be a career in itself.

—*Scott Zimmer*

Further Reading

Bell, Tom. *Programming: A Primer; Coding for Beginners.* Imperial College Press, 2016.

Calude, Cristian S., editor. *The Human Face of Computing.* Imperial College Press, 2016.

Campbell-Kelly, Martin, William F. Aspray, Jeffrey R. Yost, Honghong Tinn, and Gerardo Con Díaz. *Computer: A History of the Information Machine.* 4th ed., Routledge, 2023.

Goriunova, Olga, editor. *Fun and Software: Exploring Pleasure, Paradox, and Pain in Computing.* Bloomsbury, 2014.

Neapolitan, Richard E. *Foundations of Algorithms.* 5th ed., Jones, 2015.

O'Leary, Timothy, Linda O'Leary, and Daniel O'Leary. *Computing Essentials 2023.* McGraw-Hill, 2022.

ARCHITECTURE SOFTWARE

ABSTRACT

Architecture software is a category of specialized computer software intended for use by architects. Through the use of such software, architects may design, model, and modify plans for homes or other buildings digitally rather than on paper. As the type and scale of the structures being designed vary greatly, many types of architecture software are available to address the differing requirements of individual projects.

BACKGROUND

During the late twentieth and early twenty-first centuries, architecture became one of many fields to be dramatically transformed through the use of computer technology. For centuries, architects designed single structures such as homes as well as larger, multibuilding facilities by hand. They drafted plans

Photo via iStock/Warchi. [Used under license.]

on paper and created three-dimensional (3D) scale models out of materials such as cardboard, wood, and foam. These processes relied heavily on physical components. Therefore, it was relatively difficult to collaborate with fellow architects or other professionals who were not located nearby. Making changes to one's work, particularly to models that were already assembled, could be time-consuming and difficult. The advent of architecture software brought advances in areas such as drafting, modeling, and modifying models and floorplans as well as collaborating with other professionals.

Architecture software is designed to assist in one or more of the many tasks required for building information modeling (BIM). Depending on one's role and how one relates to the building process, certain tasks may be more important. Determining the proper architecture software first requires one to know what BIM tasks one hopes to take on.

Although typically designed for and marketed to architects, architecture software is similar in many ways to other types of software used in the broader field of art and design. Some architecture programs are essentially specialized versions of computer-aided design (CAD) software. For example, one particularly popular producer of CAD programs, Autodesk, offers an industry-specific architecture toolset for use with its flagship CAD software, AutoCAD. Architects began to adopt the use of CAD programs in the 1980s, and by the following decade, such software was widely used in the field. Some architects also make use of animation software such as

Autodesk's 3Ds Max and video game engines such as Unity to create animated renderings of their designs, thus further blurring the lines between the fields of architecture, art, and media.

OVERVIEW

Architecture software is created and sold by a number of different technology companies. Therefore, the available programs vary in terms of features, although there are certain common ones. The first and perhaps most essential feature is the ability to draft floor plans, a process that was historically done with paper and pencil. Architecture software also typically allows users to create 3D models of structures and easily insert elements such as doors and windows. Unlike physical models of proposed structures, digital models can be modified with just a few clicks of a mouse. Architecture software also sometimes includes features that support collaboration among architects and other professionals. These can include the ability to check files in and out, add notes to floor plans, and record measurements.

Specialized software devoted to BIM has much in common with standard architecture software. However, BIM software also allows the user to incorporate information such as expected time frames, costs, and the functions of a particular building or facility into the design. BIM software allows the various professionals involved in the design process to evaluate and modify the design prior to moving forward with the project.

RENDERING AND POSTPRODUCTION

In addition to creating floor plans and 3D models, architects are frequently tasked with creating images of their designs for clients. Transforming a 3D model into a two-dimensional (2D) image is known as "rendering." In postproduction, an architect may use graphic design software to add visual effects to the rendered image. For instance, if the architect has designed a waterfront apartment complex, he or she may use a graphic design program such as Adobe Photoshop to add photorealistic water textures to the image and increase its aesthetic appeal.

When creating floor plans and models, architects typically use vector images. Vector images are based on defined points and lines. This allows the images to remain true to scale no matter how much their overall size is modified. Graphic design software such as Photoshop, on the other hand, is a raster-based program. In a raster-based program, the images are based on individual pixels. As such, architects also sometimes use software capable of converting from vector to raster.

ADVANTAGES AND DISADVANTAGES

The use of specialized software in the field of architecture has a number of advantages and disadvantages. Chief among the advantages is a potential increase in productivity and time savings. Digital tools to draft floor plans and create models have largely replaced more time-consuming physical methods. The abilities to modify designs, test designs through digital modeling, and easily collaborate with fellow professionals on a project are cited as key benefits of architecture software.

However, the use of architecture software is not a substitute for architectural training. Some architects argue that the availability of such software can enable some users to produce substandard work. In addition, ongoing changes in the field require architects to maintain up-to-date knowledge of current technology. This can put architects accustomed to working with older systems at a disadvantage.

—Joy Crelin

Further Reading

Ambrose, Gavin, Paul Harris, and Sally Stone. *The Visual Dictionary of Architecture*. AVA, 2008.

Bacus, John. *Digital Sketching: Computer-Aided Conceptual Design*. Wiley, 2020.

Bergin, Michael S. "A Brief History of BIM." *ArchDaily*, 7 Dec. 2012, www.archdaily.com/302490/a-brief-history-of-bim. Accessed 13 July 2023.

Ching, Francis D. K. *Architectural Graphics*. 6th ed., Wiley, 2015.

Leach, Neil. *Architecture in the Age of Artificial Intelligence: An Introduction to AI for Architects*. Bloomsbury Visual Arts, 2021.

McMorrough, Julia. *The Architecture Reference & Specification Book*. Rev. ed., Rockport, 2018.

Wang, Tian. "Temporal Design: An Introduction to the 4 Most Popular Animation Software for Architects." *Architizer*, architizer.com/blog/practice/tools/animation-software-for-architects. Accessed 13 July 2023.

ART DIRECTOR

ABSTRACT

Art directors oversee the aesthetic direction of a project from its conception to completion. They work in a wide range of industries, from advertising to video games. Students aspiring to enter the profession should pursue studies in fields such as visual art, film studies, art history, English, theater, art administration, and advertising.

BACKGROUND

Art directors work in a variety of industries, including advertising, theater, film, video games, and publishing. While these fields involve different media, the essential task of an art director is the same: an art director oversees the aesthetic direction of a project from its conception to completion. The art director typically does not play an active role in the creation of the various elements of a project; rather, he or she works closely with artists and writers to reach a shared goal.

OVERVIEW

Depending on the industry and the size of the project, an art director can expect to work with varying numbers of artists and writers on a particular project. No matter what the industry, the art director has executive control of the work. All artistic decisions must be made with the art director's consent and approval before being made public. In a large advertising firm, an art director may report to an executive creative director. Art directors usually work during standard business hours. Long hours may be required to meet deadlines.

Working as an art director appeals to individuals with creative vision who are able to articulate and carry out that vision in an effective manner. Those drawn to this occupation have a firm grasp of the history of their media, are aware of cultural trends, and have creative minds. They can imagine the final product, whether it is an advertising campaign, a book, a magazine, or a film, and they can coordinate the various tasks and elements involved in the creative process. Art directors have strong people skills and work with a team to realize an idea. Those with an undergraduate degree in advertising, art history, or graphic design, or previous experience as a visual artist, actor, or filmmaker would be suited to the field. Successful art directors should be problem solvers, have strong communication skills, and must be well organized.

DUTIES AND RESPONSIBILITIES

An art director's daily duties vary by industry. Art directors commonly work for film, advertising, and publishing companies.

In large film productions, an art director meets with the prop master and costume and set designers to develop the overall "look" of a movie. He or she reports to the film's production director. The art director is often responsible for scheduling and hiring individuals working in construction, sound, and special effects and ensuring that the set construction and location are ready for filming. The art director also often manages a portion of a film's budget. Art directors for smaller productions may be required to take on more responsibilities and tasks.

In an advertising agency, an art director collaborates with one or more artists, such as graphic designers, illustrators and animators, and copywriters, to develop the overall concept for a project. The art director organizes face-to-face or virtual meetings to discuss the relationship between the textual and visual components of an advertisement and any related promotional material. Suggestions may be made from one department to another, with copywriters and artists exchanging ideas about visual and textual aspects of the advertising campaign. During these conversations, the art director acts as facilitator and executive decision maker. Once aesthetic decisions have been made, the art director may supervise the work itself.

An art director working in publishing performs a similar job function to that of an advertising art director. He or she works closely with writers, editors, and designers to establish an aesthetic approach for the layout of a book or magazine. The art director typically has the final say on matters such as the typeface of a book, the visual details of the book interior, and the jacket design that best fits the work.

WORK ENVIRONMENT

Physical Environment. Art directors working in advertising or publishing usually work in an office setting. A large creative department may have separate studio spaces for the various creative personnel who work there. Artists and directors working in theater or film work predominately in offices but also spend time on the set.

Human Environment. While art directors do not usually hold the top position in any industry, they have considerable control over their specific projects, acting in a guiding, executive role with the various artists, assistants, and writers they supervise.

Technological Environment. Art directors interact with clients and colleagues using standard telecommunications tools (email, phone, video conferencing) and in face-to-face meetings. Computers play a large role in their daily activities. Art directors should have familiarity with graphic and photo-imaging software, as they may adjust a project or need to demonstrate a compositional idea. Art directors in the film industry may be familiar with even more sophisticated graphic technologies depending on their field. Experience with web design and computer code can also be valuable as more companies expand their efforts in internet and social media marketing.

EDUCATION AND TRAINING

High School/Secondary. Students aspiring to become an art director should pursue a rigorous college preparatory program, with an emphasis on coursework in the arts, such as theater, media arts, computers, drafting, art history, visual art, and English. Students particularly interested in the financial and administrative aspects of art direction may also find advanced courses in arts management and economics helpful.

Interested students should research and apply to postsecondary schools that offer a relevant major. Some professional organizations provide career workshops for high school students, as well as scholarships for postsecondary studies in art direction.

College/Postsecondary. At the university level, students should consider a major in visual art, film studies, art history, English, theater, art administration, or advertising, depending on their industry of interest. An aspiring art director should major in art administration or pursue summer internships in an industry relevant to his or her interests.

Adult Job Seekers. Art direction is a highly competitive field; most employers hire those with experience in the industry. Entry-level positions that may lead to a career in art direction are often unpaid or low paying. Young adults may opt for internships to make connections in a particular industry. Artists may easily enter the field and advance to an art director position.

Professional Certification and Licensure. No certifications or licenses are needed to become an art director. In the film industry, however, some art directors do pursue specialized certificates, particularly if they are serving in the capacity of creative director or a similar role.

Additional Requirements. Owing to the competitive nature of the field, most art directors are extremely motivated, hardworking, efficient, organized, and creative individuals. Successful art directors are excellent communicators and comfortable working collaboratively. Working well under pressure can be a deciding factor for future success.

—*Mark Boccard*

Further Reading

Clarke, Andy. *Art Direction for the Web*. Smashing Media, 2019.

Costello, Vic. *Multimedia Foundations: Core Concepts for Digital Design*. 3rd ed., Focal Press, 2023.

Williams, Nathan. *The Eye: How the World's Most Influential Creative Directors Develop Their Vision*. Artisan, 2018.

Artificial Intelligence

ABSTRACT

Artificial intelligence is the design, implementation, and use of programs, machines, and systems that exhibit human intelligence, with its most important activities being knowledge representation, reasoning, and learning. Artificial intelligence encompasses a number of important subareas, including voice recognition, image identification, natural language processing, expert systems, neural networks, planning, robotics, and intelligent agents. Several important programming techniques have been enhanced by artificial intelligence researchers, including classical search, probabilistic search, and logic programming.

BACKGROUND

Although the concept of artificial intelligence probably has existed since antiquity, the term was first used by American scientist John McCarthy at a conference held at Dartmouth College in 1956. In 1955-56, the first artificial intelligence program, Logic Theorist, had been written in IPL, a programming language, and in 1958, McCarthy invented Lisp, a programming language that improved on IPL. *Syntactic Structures* (1957), a book about the structure of natural language by American linguist Noam Chomsky, made natural language processing into an area of study within artificial intelligence. In the next few years, numerous researchers began to study artificial intelligence, laying the foundation for many later applications, such as general problem solvers, intelligent machines, and expert systems.

In the 1960s, Edward Feigenbaum and other scientists at Stanford University built two early expert systems: DENDRAL, which classified chemicals, and MYCIN, which identified diseases. These early expert systems were cumbersome to modify because they had hard-coded rules. By 1970, the OPS expert system shell, with variable rule sets, had been released by Digital Equipment Corporation as the first commercial expert system shell. In addition to expert systems, neural networks became an important area of artificial intelligence in the 1970s and 1980s. Frank Rosenblatt introduced the Perceptron in 1957, but it was *Perceptrons: An Introduction to Computational Geometry* (1969), by Minsky and Seymour Papert, and the two-volume *Parallel Distributed Processing: Explorations in the Microstructure of Cognition* (1986), by Rumelhart, McClelland, and the PDP Research Group, that really defined the field of neural networks. Development of artificial intelligence has continued, with game theory, speech recognition, robotics, and autonomous agents being some of the best-known examples.

OVERVIEW

Artificial intelligence is a broad field of study, and definitions of the field vary by discipline. For computer scientists, artificial intelligence refers to the development of programs that exhibit intelligent behavior. The programs can engage in intelligent planning (timing traffic lights), translate natural languages (converting a Chinese website into English), act like an expert (selecting the best wine for dinner), or perform many other tasks. For engineers, artificial intelligence refers to building machines that perform actions often done by humans. The machines can be simple, like a computer vision system embedded in an automated teller machine (ATM); more complex, like a robotic rover sent to Mars; or very complex, like an automated factory that builds an exercise machine with little human intervention. For cognitive scientists, artificial intelligence refers to building models of human intelligence to better understand human behavior. In the early days of artificial intelligence, most models of human intelligence were symbolic and closely related to cognitive psychology and philosophy, the basic idea being that regions of the brain perform complex reasoning by processing symbols. Later, many models of human cognition were developed to mirror the operation of the brain as an electrochemical computer, starting with the simple Perceptron, an artificial neural network described by Marvin Minsky in 1969, graduating to the backpropagation algorithm described by David E. Rumelhart and James L. McClelland in 1986, and culminating in a large number of supervised and nonsupervised learning algorithms.

When defining artificial intelligence, it is important to remember that the programs, machines, and models developed by computer scientists, engineers, and cognitive scientists do not actually have human

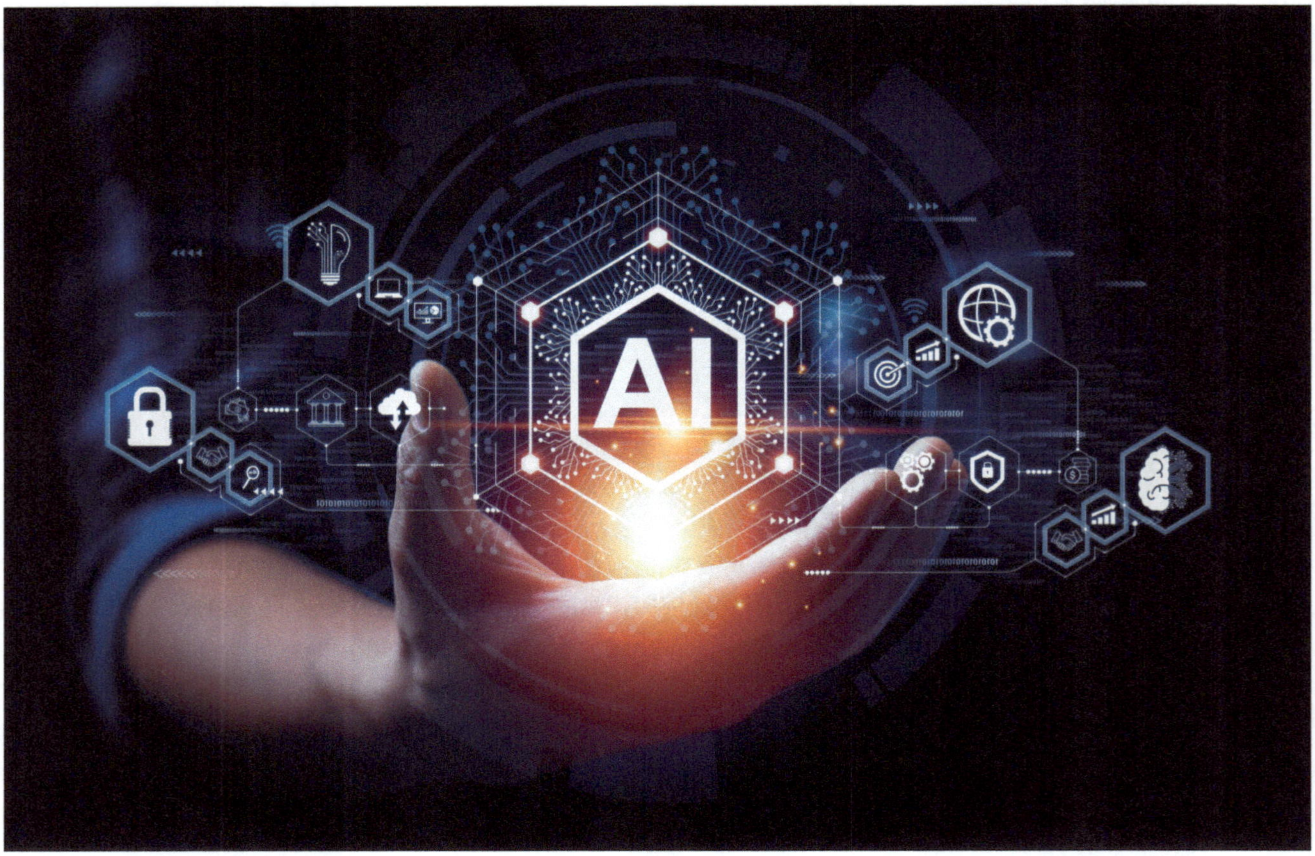

Photo via iStock/Shutthiphong Chandaeng. [Used under license.]

intelligence; they only exhibit intelligent behavior. This can be difficult to remember because artificially intelligent systems often contain large numbers of facts, such as weather information for New York City; complex reasoning patterns, such as the reasoning needed to prove a geometric theorem from axioms; complex knowledge, such as an understanding of all the rules required to build an automobile; and the ability to learn, such as a neural network learning to recognize cancer cells. Scientists continue to look for better models of the brain and human intelligence.

HOW IT WORKS

The first activity of artificial intelligence is to understand how multiple facts interconnect to form knowledge and to represent that knowledge in a machine-understandable form. The next task is to understand and document a reasoning process for arriving at a conclusion. The final component of artificial intelligence is to add, whenever possible, a learning process that enhances the knowledge of a system.

Knowledge Representation. Facts are simple pieces of information that can be seen as either true or false, although in fuzzy logic, there are levels of truth. When facts are organized, they become information, and when information is well understood, over time, it becomes knowledge. To use knowledge in artificial intelligence, especially when writing programs, it has to be represented in some concrete fashion. Initially, most of those developing artificial intelligence programs saw knowledge as represented symbolically, and their early knowledge representations were symbolic. Semantic nets, directed graphs of facts with added semantic content, were highly successful representations used in many of the early artificial intelligence programs. Later, the nodes of the semantic nets were expanded to contain more information, and the resulting knowledge representation was referred to as frames. Frame representation of knowledge was very similar to object-oriented data representation, including a theory of inheritance.

Another popular way to represent knowledge in artificial intelligence is as logical expressions. English mathematician George Boole represented knowledge as a Boolean expression in the 1800s. English mathematicians Bertrand Russell and Alfred Whitehead expanded this to quantified expressions in 1910, and French computer scientist Alain Colmerauer incorporated it into logic programming, with the programming language Prolog, in the 1970s. The knowledge of a rule-based expert system is embedded in the if-then rules of the system, and because each if-then rule has a Boolean representation, it can be seen as a form of relational knowledge representation.

Neural networks model the human neural system and use this model to represent knowledge. The brain is an electrochemical system that stores its knowledge in synapses. As electrochemical signals pass through a synapse, they modify it, resulting in the acquisition of knowledge. In the neural network model, synapses are represented by the weights of a weight matrix, and knowledge is added to the system by modifying the weights.

Reasoning. Reasoning is the process of determining new information from known information. Artificial intelligence systems add reasoning soon after they have developed a method of knowledge representation. If knowledge is represented in semantic nets, then most reasoning involves some type of tree search. One popular reasoning technique is to traverse a decision tree, in which the reasoning is represented by a path taken through the tree. Tree searches of general semantic nets can be very time-consuming and have led to many advancements in tree-search algorithms, such as placing bounds on the depth of search and backtracking.

Reasoning in logic programming usually follows an inference technique embodied in first-order predicate calculus. Some inference engines, such as that of Prolog, use a back-chaining technique to reason from a result, such as a geometry theorem, to its antecedents, the axioms, and also show how the reasoning process led to the conclusion. Other inference engines, such as that of the expert system shell CLIPS, use a forward-chaining inference engine to see what facts can be derived from a set of known facts.

Neural networks, such as backpropagation, have an especially simple reasoning algorithm. The knowledge of the neural network is represented as a matrix of synaptic connections, possibly quite sparse. The information to be evaluated by the neural network is represented as an input vector of the appropriate size, and the reasoning process is to multiply the connection matrix by the input vector to obtain the conclusion as an output vector.

Learning. Learning in an artificial intelligence system involves modifying or adding to its knowledge. For both semantic net and logic programming systems, learning is accomplished by adding or modifying the semantic nets or logic rules, respectively. Although much effort has gone into developing learning algorithms for these systems, all of them, to date, have used ad hoc methods and experienced limited success. Neural networks, on the other hand, have been very successful at developing learning algorithms. Backpropagation has a robust supervised learning algorithm in which the system learns from a set of training pairs, using gradient-descent optimization, and numerous unsupervised learning algorithms learn by studying the clustering of the input vectors.

APPLICATIONS AND PRODUCTS

There are many important applications of artificial intelligence, ranging from computer games to programs designed to prove theorems in mathematics. This section contains a sample of both theoretical and practical applications.

Expert Systems. One of the most successful areas of artificial intelligence is expert systems. Literally thousands of expert systems are being used to help both experts and novices make decisions. There are several categories of expert systems, but by far the most popular are the rule-based expert systems. Most rule-based expert systems are created with an expert system shell. The first successful rule-based expert system shell was the OPS 5 of Digital Equipment Corporation (DEC), and the most popular modern systems are CLIPS, developed by the National Aeronautics and Space Administration (NASA) in 1985, and its Java clone, Jess, developed at Sandia National Laboratories in 1995. All rule-based expert systems have a similar architecture, and the shells make it fairly easy to create an expert system as soon as a knowledge engineer gathers the knowledge from a domain expert. The most important component of a rule-based expert system is its knowledge base of rules. Each rule consists of an if-then statement with multiple antecedents, multiple consequences, and possibly a rule certainty factor. The antecedents of a rule are statements that can be true or false and depend on facts that are either introduced into the system by a user or derived as the result of a rule being fired. For example, a fact could be red-wine and a simple rule could be if (red-wine) then (it-tastes-good). The expert system also has an inference engine that can apply multiple rules in an orderly fashion so that the expert system can draw conclusions by applying its rules to a set of facts introduced by a user. Although it is not absolutely required, most rule-based expert systems have a user-friendly interface and an explanation facility to justify its reasoning.

Theorem Provers. Most theorems in mathematics can be expressed in first-order predicate calculus. For any particular area, such as synthetic geometry or group theory, all provable theorems can be derived from a set of axioms. Mathematicians have written programs to automatically prove theorems since the 1950s. These theorem provers either start with the axioms and apply an inference technique, or start with the theorem and work backward to see how it can be derived from axioms. Resolution, developed in Prolog, is a well-known automated technique that can be used to prove theorems, but there are many others. For Resolution, the user starts with the theorem, converts it to a normal form, and then mechanically builds reverse decision trees to prove the theorem. If a reverse decision tree whose leaf nodes are all axioms is found, then a proof of the theorem has been discovered.

Gödel's incompleteness theorem (proved by Austrian-born American mathematician Kurt Gödel) shows that it may not be possible to automatically prove an arbitrary theorem in systems as complex as the natural numbers. For simpler systems, such as group theory, automated theorem proving works if the user's computer can generate all reverse trees or a suitable subset of trees that can yield a proof in a reasonable amount of time. Efforts have been made to develop theorem provers for higher order logics than first-order predicate calculus, but these have not been very successful.

Computer scientists have spent considerable time trying to develop an automated technique for proving the correctness of programs, that is showing that any valid input to a program produces a valid output. This is generally done by producing a consistent model and mapping the program to the model. The first example of this was given by English mathematician Alan Turing in 1931, by using a simple model now called a Turing machine. A formal system that is rich enough to serve as a model for a typical programming language, such as C++, must support higher order logic to capture the arguments and parameters of subprograms. Lambda calculus, denotational semantics, von Neuman geometries, finite state machines, and other systems have been proposed to provide a model onto which all programs of a language can be mapped. Some of these do capture many programs, but devising a practical automated method of verifying the correctness of programs has proven difficult.

Intelligent Tutor Systems. Almost every field of study has many intelligent tutor systems available to assist students in learning. Sometimes the tutor system is integrated into a package. For example, in come versions of Microsoft Office developed during the late 1990s and early 2000s, an embedded intelligent helper provided popup help boxes to a user when it detected the need for assistance and full-length tutorials if it detected more help was needed. In addition to the intelligent tutors embedded in programs as part of context-sensitive help systems, there are a vast number of stand-alone tutoring systems in use.

The first stand-alone intelligent tutor was SCHOLAR, developed by J. R. Carbonell in 1970. It used semantic nets to represent knowledge about South American geography, provided a user interface to support asking questions, and was successful enough to demonstrate that it was possible for a computer program to tutor students. At about the same time, the University of Illinois developed its PLATO computer-aided instruction system, which provided a general language for developing intelligent tutors with touch-sensitive screens, one of the most famous of which was a biology tutorial on evolution. Of the thousands of later intelligent tutors, SHERLOCK, a training environment for electronic troubleshooting, and PUMP, a system designed to help learn algebra, are typical.

Electronic Games. Electronic games have been played since the invention of the cathode-ray tube for television. In the 1980s, games such as *Solitaire*, *Pac-Man*, and *Pong* for personal computers became almost as popular as the stand-alone game platforms. By the 2020s, multiuser internet games were enjoyed by young and old alike, and game playing on mobile devices had become an important application. In all of these electronic games, the user competes with one or more intelligent agents embedded in the game, and the creation of these intelligent agents uses considerable artificial intelligence. When creating an intelligent agent that will compete with a user or, as in *Solitaire*, just react to the user, a programmer has to embed the game knowledge into the program. For example, in chess, the programmer would need to capture all possible configurations of a chess board. The programmer also would need to add reasoning procedures to the game; for example, there would have to be procedures to move each individual chess piece on the board. Finally, and most important for game programming, the programmer would need to add one or more strategic decision modules to the program to provide the intelligent agent with a strategy for winning. In many cases, the strategy for winning a game would be driven by probability; for example, the next move might be a pawn, one space forward, because that yields the best probability of winning, but a heuristic strategy is also possible; for example, the next move is a rook because it may trick the opponent into a bad series of moves.

CAREERS AND COURSEWORK

A major in computer science is the most common way to prepare for a career in artificial intelligence. One needs substantial coursework in mathematics, philosophy, and psychology as a background for this degree. For many of the more interesting jobs in artificial intelligence, one needs a master's or doctoral degree. Most universities teach courses in artificial intelligence, neural networks, or expert systems, and many have courses in all three. Although artificial intelligence is usually taught in computer science, it is also taught in mathematics, philosophy, psychology, and electrical engineering. Taking a strong minor in any field is advisable for someone seeking a career in artificial intelligence because the discipline is often applied to another field.

Those seeking careers in artificial intelligence generally take a position as a systems analyst or programmer. They work for a wide range of companies, including those developing business, mathematics, medical, and voice recognition applications. Those obtaining an advanced degree often take jobs in industrial, government, or university laboratories developing new areas of artificial intelligence.

SOCIAL CONTEXT, ETHICS, AND FUTURE PROSPECTS

After artificial intelligence was defined by McCarthy in 1956, it has had a number of ups and downs as a discipline, but the future of artificial intelligence looks good. Almost every commercial program has a help system, and increasingly these help systems have a major artificial intelligence component. Health care is another area that has been poised to make major use of artificial intelligence to improve the quality and reliability of the care provided, as well as to reduce its cost by providing expert advice on best practices in health care. Smartphones and other digital devices employ artificial intelligence for an array of applications, syncing the activities and requirements of their users.

Ethical questions have been raised about trying to build a machine that exhibits human intelligence. Many of the early researchers in artificial intelligence were interested in cognitive psychology and built symbolic models of intelligence that were considered unethical by some. Later, many artificial intelligence researchers developed neural models of intelligence that were not always deemed ethical. The social and ethical issues of artificial intelligence are nicely represented by HAL, the Heuristically programmed Algorithmic computer, in Stanley Kubrick's 1968 film *2001: A Space Odyssey*, which first works well with humans, then acts violently toward them, and is in the end deactivated.

Another important ethical question posed by artificial intelligence is the appropriateness of developing programs to collect information about users of a program. Intelligent agents are often embedded in websites to collect information about those using the site, generally without the permission of those using the website, and many question whether this should be done.

In the mid-to-late 2010s, fully autonomous self-driving cars were developed and tested in the United States. In 2018, an Uber self-driving car hit and killed a pedestrian in Tempe, Arizona. There was a safety driver at the wheel of the car, which was in self-driving mode at the time of the accident. While the accident led Uber to suspend its driverless-car testing program for a time, by the next year testing had resumed, initially at a smaller scale. Even before the accident occurred, ethicists had raised questions regarding collision avoidance programming, moral and legal responsibility, among others. By mid-2020, companies such as Tesla had continued devoting resources to developing fully autonomous vehicles for eventual widespread use, but despite sustained technological advancements concerning artificial intelligence, projections about getting driverless cars on the road by that point had not been met. Commentators noted that though some related technology had been incorporated into cars in use, such as automatic braking, object sensitivity, and lane detection, a lack of training data meant that the proper technology had still not been perfected for a car to be able to drive on its own reliably. Still, Waymo, which had begun implementing a commercial self-driving ride-hailing service in the Phoenix area run through a smartphone application, had made efforts to further expand the service by 2020, including adding different vehicles to its fleet and testing the program in other areas. In 2021, companies such as Waymo and Apple continued to improve their self-driving cars. Also in 2021 was the deployment of delivery robots. The company Starship announced that it had made 2 million successful deliveries while Alibaba had made 1 million. During the same year, three companies in China began deploying robotaxis without safety drivers.

As more complex AI is created and imbued with general, humanlike intelligence (instead of concentrated intelligence in a single area, such as Deep Blue and chess), it will run into moral requirements as humans do. According to researchers Nick Bostrom and Eliezer Yudkowsky, if an AI is given "cognitive work" to do that has a social aspect, the AI inherits the social requirements of these interactions. The AI then needs to be imbued with a sense of morality to interact in these situations. If an AI has humanlike intelligence and agency, then Bostrom has also theorized that AI will need to also be considered both persons and moral entities. There is also the potential for the development of superhuman intelligence in AI, which would breed superhuman morality. The questions of intelligence and morality and who is given personhood are some of the most significant issues to be considered contextually as AI advance. Additionally, as AI technology such as facial recognition and data algorithms continued evolving and played even larger roles in society, some worried about a progressive erosion of privacy.

By 2021, following the declaration of the coronavirus 2019 (COVID-19) pandemic in early 2020, some health-care facilities had been experimenting more with incorporating artificial intelligence models and algorithms into their treatment and monitoring processes in an effort to cope with the surge of illness, particularly when little was still known about the novel coronavirus and the disease it caused. However, debates still existed around such use of artificial intelligence in clinical settings, and some argued that there were unsolved ethical issues and that the data used by algorithms would need consistent updating, meaning that they should be used cautiously and not fully relied upon.

In 2022, AI research group OpenAI released ChatGPT, an AI chatbot capable of generating complex responses to user prompts. In the months following its release, ChatGPT went viral for its potential real-world applications, including uses in business, research, and education. However, several flaws were soon identified with the program, including potential factual unreliability in its responses. In January 2023, *NBC News* reported that the New York City Department of Education had banned the use of ChatGPT in the classroom due to the program's potential impact on student learning. (Users discovered that ChatGPT can be used to write essays, solve complex problems, and generate computer code, among other uses, which caused fears among many that students could use the program to complete their coursework.) The following month, OpenAI announced ChatGPT Plus, a subscription to an enhanced version of the chatbot with expanded features. As ChatGPT continued to gain popularity in 2023, critics also raised concerns about the ethics of using ChatGPT and other AI-based tools to create written and visual works of art as well as the potential effects of such tools on the carers and livelihoods of creative professionals.

—George M. Whitson III

Further Reading

Audry, Sofian. *Art in the Age of Machine Learning*. MIT Press, 2021.

Basl, John. "The Ethics of Creating Artificial Consciousness." *American Philosophical Association Newsletters: Philosophy and Computers*, vol. 13., no. 1, 2013, pp. 25-30.

Berlatsky, Noah. *Artificial Intelligence*. Greenhaven Press, 2011.

Bostrom, Nick. "Ethical Issues in Advanced Artificial Intelligence." *NickBostrom.com*, 2003, nickbostrom.com/ethics/ai.html. Accessed 13 July 2023.

Bostrom, Nick, and Eliezer Yudkowsky. "The Ethics of Artificial Intelligence." *The Cambridge Handbook of Artificial Intelligence*, edited by Keith Frankish and William M. Ramsay, Cambridge UP, 2014, pp. 316-34.

Heikkilä, Melissa. "The Algorithm: AI-Generated Art Raises Tricky Questions about Ethics, Copyright, and Security." *MIT Technology Review*, 20 Sept. 2022, www.technologyreview.com/2022/09/20/1059792/the-algorithm-ai-generated-art-raises-tricky-questions-about-ethics-copyright-and-security. Accessed 13 July 2023.

Laalaoui, Yacine, and Nizar Bouguila, editors. *Artificial Intelligence Applications in Information and Communication Technologies*. Springer International Publishing, 2015.

Lee, Timothy B. "Why It's Time for Uber to Get Out of the Self-Driving Car Business." *Ars Technica*, 27 Mar. 2018, arstechnica.com/cars/2018/03/ubers-self-driving-car-project-is-struggling-the-company-should-sell-it. Accessed 13 July 2023.

Metz, Cade. "OpenAI to Offer New Version of ChatGPT for a $20 Monthly Fee." *New York Times*, 1 Feb. 2023, www.nytimes.com/2023/02/01/technology/openai-chatgpt-plus-subscription.html. Accessed 13 July 2023.

Miller, Arthur I. *The Artist in the Machine: The World of AI-Powered Creativity*. MIT Press, 2019.

Minsky, Marvin, and Seymour A. Papert. Perceptrons: An Introduction to Computational Geometry. Reissue ed., MIT Press, 2017.

Morrison, Jim. "How Doctors Are Using Artificial Intelligence to Battle Covid-19." *Smithsonian Magazine*, 5 Mar. 2021, www.smithsonianmag.com/science-nature/how-doctors-are-using-artificial-intelligence-battle-covid-19-180977124. Accessed 13 July 2023.

Nyholm, Sven, and Jilles Smids. "The Ethics of Accident-Algorithms for Self-Driving Cars: An Applied Trolley Problem?" *Ethical Theory & Moral Practice*, vol. 19, no. 5, 2016, pp. 1275-89.

Rosenblatt, Kalhan. "ChatGPT Banned from New York City Public Schools' Devices and Networks." *NBC News*, 5 Jan. 2023, www.nbcnews.com/tech/tech-news/new-york-city-public-schools-ban-chatgpt-devices-networks-rcna64446. Accessed 13 July 2023.

Rumelhart, David E., James L. McClelland, and the PDP Research Group. *Parallel Distributed Processing: Explorations in the Microstructure of Cognition*. 1986. MIT Press, 1989. 2 vols.

Russell, Stuart, and Peter Norvig. *Artificial Intelligence: A Modern Approach*. 4th ed. Pearson, 2020.

Templeton, Brad. "Self-Driving Cars 2021: Year in Review." *Forbes*, 3 Jan. 2022, www.forbes.com/sites/bradtempleton/2022/01/03/self-driving-cars-2021-year-in-review/?sh=4aa85563773b. Accessed 13 July 2023.

ASCII

ABSTRACT

The American Standard Code for Information Interchange (ASCII) is a character encoding system. It enables computers and other electronic communication devices to store, process, transmit, print, and display text and other graphic characters. Initially published in 1963, ASCII formed the basis for several other character encoding systems developed for use with personal computers (PCs) and the internet.

BACKGROUND

Written language, or text, is composed of a variety of graphic symbols called characters. In many languages, these characters include letters, numbers, punctuation marks, and blank spaces. Such characters are also called printable characters because they can be printed or otherwise displayed in a form that can be read by humans. Another type of character is a control character. Control characters affect the processing of other characters. For example, a control character might instruct a printer to print the next character on a new line. Character encoding is the process of converting characters into a format that can be used by an electronic device such as a computer or telegraph.

Originally designed for use with Samuel Morse's telegraph system, Morse code was one of the first character encoding schemes adopted for widespread use. Telegraphs transmit information by sending electronic pulses over telegraph wires. Morse code assigns each character to a unique combination of short and long pulses. For example, the letter *A* was assigned to the combination of one short followed by one long pulse, while the letter *T* was assigned to a single long pulse. Using Morse code, a telegraph operator can send messages by transmitting a sequence of pulses. The sequence, or string, of pulses represents the characters that comprise the message text.

Other character encoding systems were created to meet the needs of new types of electronic devices including teleprinters and computers. By the early 1960s, the use of character encoding systems had become widespread. However, no standard character encoding system existed to ensure that systems from different manufacturers could communicate with each other. In fact, by 1963, over sixty different encoding systems were in use. Nine different systems were used by International Business Machines (IBM) alone. To address this issue, the American Standards Association (ASA) X3.4 Committee developed a standardized character encoding scheme called the American Standard Code for Information Interchange, or ASCII.

OVERVIEW

The ASCII standard is based on English. It encodes 128 characters into integer values from 0 to 127. Thirty-three of the characters are control characters, and ninety-five are printable characters that include the upper- and lowercase letters from *A* to *Z*, the numbers zero to nine, punctuation marks, and a blank space. For example, the letter *A* is encoded as 65 and a comma as 44.

The encoded integers are then converted to bits, the smallest unit of data that can be stored by a computer system. A single bit can have a value of either zero or one. In order to store integers larger than one, additional bits must be used. The number of bits used to store a value is called the bit width. ASCII specifies a bit width of seven. For example, in ASCII, the integer value 65 is stored using seven bits, which can be represented as the bit string 1000001.

The ASCII seven-bit integer values for specific characters were not randomly assigned. Rather, the integer values of specific characters were selected to maximize the hamming distance between each value. Hamming distance is the number of bits set to different values when comparing two bit strings. For example, the bit strings 000000**1** (decimal value 1)

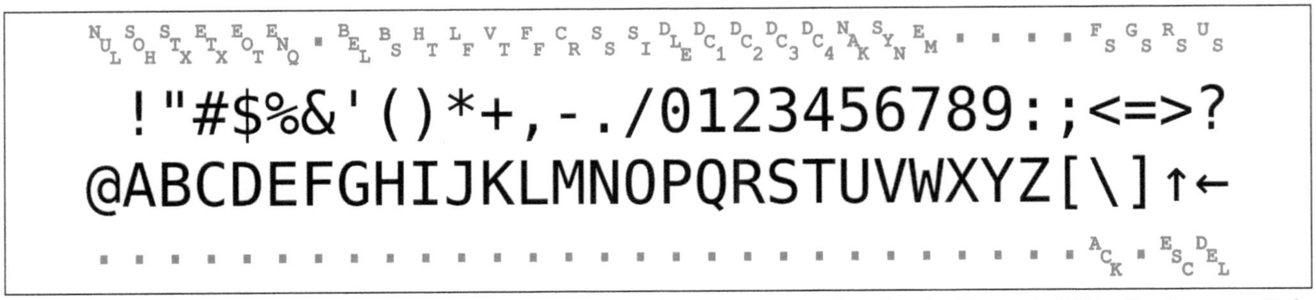

ASCII (1963). Control Pictures of equivalent controls are shown where they exist, or a grey dot otherwise. Image via Wikimedia Commons. [Public domain.]

and 00000**1**1 (decimal value 3) have a hamming distance of 1 as only the second to last bit differs between the two strings. The bit patterns 0000**11**1 (decimal value 7) and 0000**00**1 (decimal value 1) have a hamming distance of two as the bit in the third to last position also differs between the two strings. ASCII was designed to maximize hamming distance because larger hamming distances enable more efficient data processing as well as improved error detection and handling.

BEYOND ASCII

Following its introduction in 1963, ASCII continued to be refined. It was gradually adopted for use on a wide range of computer systems, including the first IBM PC. Other manufacturers soon followed IBM's lead. The ASCII standard was also widely adopted for use on the internet. However, as the need for more characters to support languages other than English grew, other standards were developed to meet this need. One such standard, Unicode, can encode more than 140,000 characters.

ASCII remains an important technology, however. Many systems still use ASCII. Character encoding systems such as Unicode incorporate ASCII to promote compatibility with existing systems. ASCII has likewise continued to be used in ASCII art, which consists of images—ranging from simple emoticons to detailed portraits—that are created through the strategic placement of ASCII characters.

—*Maura Valentino*

Further Reading

Anderson, Deborah. "Global Linguistic Diversity for the Internet." *Communications of the ACM*, vol. 48, no. 1, 2005, pp. 27-28.

McConnell, Robert, James Haynes, and Richard Warren. "Understanding ASCII Codes." *NADCOMM*, 14 May 2011, www.nadcomm.com/index3fbb.html?p=89. Accessed 13 July 2023.

Oppliger, Rolf. *Secure Messaging on the Internet*. Artech House, 2014.

Rognerud, Jon. *Ultimate Guide to Optimizing Your Website*. 3rd ed., Entrepreneur Media, 2014.

Silver, H. Ward. "Digital Code Basics." *Qst*, vol. 98, no. 8, 2014, pp. 58-59.

"Timeline of Computer History: 1963." *Computer History Museum*, www.computerhistory.org/timeline/1963. Accessed 13 July 2023.

Wagner, Karin. *From ASCII Art to Comic Sans: Typography and Popular Culture in the Digital Age*. MIT Press, 2023.

"What Is ASCII?" *Indiana University Knowledge Base*, 18 Jan. 2018, kb.iu.edu/d/afht. Accessed 13 July 2023.

Audio Engineering

ABSTRACT

Audio engineering is the capture, enhancement, and reproduction of sounds. It requires an aesthetic appreciation of music and sound quality, a scientific understanding of sound physics, and a technical familiarity with recording equipment and computer software. This applied science is essential to the music industry, film, television, and video game production, live television and radio broadcasting, and advertising. In addition, it contributes to educational services for the visually impaired and to forensic evidence analysis.

BACKGROUND

Audio engineering, also known as sound engineering and audio technology, is the recording, manipulation, and reproduction of sound, especially music. Sound capture and reproduction, and thus audio engineering, began with the invention of the phonograph by Thomas Alva Edison in 1877. Sound was recorded on cylinders; the first were wrapped in tin foil and later ones in wax. By 1910, cylinders were replaced with discs, which held longer recordings, were somewhat louder, and could be more economically mass-produced. The discs were spun on a turntable at standard speeds—initially 78 revolutions per minute (rpm). Larger discs were played at 33 rpm and smaller discs were played at 45 rpm. Discs were originally made of shellac and later made of vinyl. They were played with needles (styli) made of industrial diamond, which held a point.

Concurrently, RCA was creating microphones that improved recorded sound quality. In the 1940s, sound began being recorded on magnetic tape and could be reproduced in stereo and as mixed multiple tracks. Digital technology appeared in the 1980s and by the turn of the century, digital recordings were produced with computer technology. Using data compression, digital audio recordings can produce quality replication of the original music in a format that requires less data storage (computer memory); MP3 (MPEG Audio Layer-3) is one such format and is popular for portable consumer music systems.

OVERVIEW

Audio engineers run recording sessions, operate the equipment, and collaborate on the finished product. They are simultaneously technicians, scientists, and creative advisers. They work in recording studios, producing the following: instrumental and vocal music recordings; film and television soundtracks, syncing, and sound effects; music and voice-overs for radio and television commercials; and music and sound effects for video games.

A recording studio is a specialized environment designed to capture sounds accurately for enhancement and reproduction. The acoustic sounds produced by instruments and vocalists are picked up by strategically placed microphones and transmitted as analogue electrical signals to recording equipment, where they may be converted into digital data. Signals may be modified by the use of a mixing console, also called a mixing board or sound board, that changes the characteristics and balance of the input, which may be coming from multiple microphones or signals recorded in different sessions. The final product then undergoes mastering for commercial reproduction and compression for distribution in a digital format.

Audio engineers are not acoustic engineers: graduates of a formal university program in engineering who work with architects and interior designers to plan and install audio systems for large venues such as churches, school auditoriums, and concert halls. Audio engineers are engineers in the sense that they are needed to devise a creative solution to a complex sound challenge and oversee its implementation.

HOW IT WORKS

Sound. Sound is the waves of pressure a vibrating object emits through air or water. The three most meaningful characteristics of sound waves are wavelength, amplitude, and frequency. The wavelength is the distance between equivalent points on consecutive waves, such as peak to peak. A short wavelength means that more waves are produced per second, resulting in a higher sound. The amplitude is the strength of the wave; the greater the amplitude, the greater the volume (loudness). The frequency is the number of wavelengths that occur in one second; the greater the frequency, the higher the pitch because the sound source is vibrating quickly.

Hearing. Hearing is the ability to receive, sense, and decipher sounds. To hear, the ear must direct the sound waves inside, sense the sound vibrations, and translate the sensations into neurological impulses that the brain can recognize. The outer ear funnels sound into the ear canal. It also helps the brain determine the direction from which the sound is coming.

When the sound waves reach the ear canal, they vibrate against the eardrum. These vibrations are amplified by the eardrum's movement against three tiny bones (the malleus, incus, and stapes) located behind it. The stapes rests against the cochlea, and when it transmits the sound, it creates waves in the fluid of the cochlea.

The cochlea is a coiled, fluid-filled organ that contains 30,000 hairs of different lengths that resonate at different frequencies. Vibrations of these hairs trigger complex electrical patterns that are transmitted along the auditory nerve to the brain, where they are interpreted.

The frequency of sound waves is measured in hertz (Hz). Humans have a hearing range from 20 to 20,000 Hz. Another name for the frequency of a sound wave is the musical pitch. Pitches are often referred to as musical notes, such as middle C.

The relative loudness of a sound compared with the threshold of human hearing is measured in decibels (dB). Conversation is usually conducted at 40 to 60 dB, while a car passing at 32 feet (10 meters) may be 80 to 90 dB, a jet engine 328 ft (100 m) away may be 110 to 140 dB, and a rifle fired 1 meter away is 150 dB. Long-term (not necessarily continuous) exposure to sounds greater than 85 dB may cause hearing loss.

The human ear can discern between two musical instruments playing the same note at the same volume by the recognition of a sound characteristic called timbre. Often described by adjectives such as bright versus dark, smooth versus harsh, and regular versus random or erratic, timbre is often what distinguishes music from noise.

Sound Capture. Transducers are devices that change energy from one form into another. A microphone changes acoustical signals into electrical signals, while a speaker changes electrical signals into acoustical signals. The source of the incoming electrical signals may be immediate, such as a microphone or electrical musical instrument, or it may be a recording, such as a compact disc or an MP3 file.

Microphones come in many varieties, such as dynamic, ribbon, condenser, parabolic, and lavaliere. They also vary by their polar patterns, that is, their area of sensitivity to sounds coming in from different directions relative to the receiving membrane. They may be omnidirectional (sensitive to sounds coming from all directions), unidirectional (intended for directed sound reception), or cardioid (having a heart-shaped area of sensitivity). The choices of variety, polar pattern, and placement affect the quality and quantity of sound capture.

Signal Processing. The auditory electrical signal from a microphone is relatively weak, so it must be amplified before the sound can be deliberately modified through signal processing. Incoming sound may be modified in its analogue form or converted to digital data before alteration. Sound mixing is the process of blending sounds from multiple sources into a desired end product. It often starts with finding a balance between vocal and instrumental music or dissimilar instruments so that one does not overshadow the other. It involves the creation of stereo or surround sound from the placement of sound in the sound field to simulate directionality (left, center, or right). Equalizing adjusts the bass and treble frequency ranges. Effects such as reverberation may be added to create dimension. Signals may undergo gating and compression to remove unwanted noise and extraneous data selectively.

Sound Output. Auditory electrical signals may then be sent to speakers, where they are converted into acoustical signals to be heard by a live audience. Digital signals may be broadcast in real time over the internet as streaming audio. Otherwise, the processed signals may be stored for future reproduction and distribution. Analogue signals may be stored on magnetic tape. Digital signals may be stored on a compact disc or subjected to MP3 encoding for storage on a computer or personal music player.

SOCIAL CONTEXT AND FUTURE PROSPECTS

Audio engineering is a combination of technology, science, and art. Advancements in audio engineering will come in all three areas. On the technical front, classic (especially pre-1920) recordings will continue to be found, researched, and digitally restored. Improved transducer materials are being sought and new computer software applications for signal processing are being developed. Surround sound is being refined to accompany three-dimensional (3D) and high-definition television (HDTV) programs and films as well as video games.

Scientific research into psychoacoustics, the study of sound perception, is expanding. The eventual understanding of how music affects a person's brain will advance the field of music therapy, which seems to touch every facet of a person's being to restore and maintain health. Researchers are also exploring the connections between sound characteristics and the perceptions of timbre and spatial placement and between these perceived attributes and listening preference.

The artistic manipulation of sound is broadening the definition of music and musical instruments. Computer-mediated music has inspired the creation of mobile phone and laptop orchestras. Music enhancement by selectively masking undesired frequencies of instruments and highlighting others is introducing new sound combinations previously not experienced.

—Bethany Thivierge

Further Reading

Ballou, Glen, editor. *Handbook for Sound Engineers*. 5th ed., Focal Press, 2015.

Dittmar, Tim. *Audio Engineering 101: A Beginner's Guide to Music Production*. 2nd ed., Routledge, 2018.

Gaston-Bird, Leslie. *Women in Audio*. Focal Press, 2019.

Hampton, Dave. *The Business of Audio Engineering*. 2nd ed., Hal Leonard, 2013.

_____. *So, You're an Audio Engineer: Well, Here's the Other Stuff You Need to Know*. Outskirts, 2005.

Huber, David Miles, and Emiliano Caballero. *Modern Recording Techniques: A Practical Guide to Modern Music Production*. 10th ed., Focal Press, 2023.

Powell, John. *How Music Works: The Science and Psychology of Beautiful Sounds, from Beethoven to the Beatles and Beyond*. Little, Brown, 2010.

Winer, Ethan. *The Audio Expert: Everything You Need to Know about Audio*. Focal Press, 2013.

Audio Streaming

ABSTRACT

Music streaming services offer vast libraries of songs and albums and allow subscribers to listen to anything in the collection by way of an internet connection. Many services also allow a limited level of access at no charge, as a means of encouraging potential customers to try out the service before signing up as paid subscribers. Music streaming services are able to offer access to songs and albums through licensing agreements they make with the record labels that own the rights to those songs and albums. The music streaming services are permitted to share access to music with their subscribers, in exchange for monetary compensation paid to the music owners. Understanding how the music industry came to allow this sort of arrangement requires a review of how the industry has evolved since the late twentieth century.

BACKGROUND

Up until the late 1980s and early 1990s, the music industry was firmly under the control of record labels, most of which were large corporations or divisions of companies with other interests. Record labels would find and recruit musicians, record their music, and then market that music to consumers, who would purchase it on some type of physical medium, such as vinyl record albums, eight-track tapes, audiocassettes, or compact discs. Those who purchased music were licensed to listen to it themselves or to sell the physical album, tape, or disc to others, but not to make copies. Despite these limitations on consumers' rights, it was common for people to copy music onto blank media, either to make it more portable, or to share it with friends. While such actions technically infringed the record labels' copyright, it would have been virtually impossible to stop, so little was done by record companies to prevent or prosecute individuals from this type of small-scale sharing.

The music landscape began to change in the 1990s due to the emergence of the internet in a form that was usable even by people without advanced knowledge of computers and to advances made in digital compression technology. When compact discs became common in the marketplace, this caused a shift from analogue music storage (i.e., vinyl records and audiotapes) toward digital music storage (e.g., compact discs and audio files). With analogue storage media, recording media captured music as ranges of sound going up and down in wavelike patterns. Copying music to and from analogue media was of limited utility, because each time a copy was made, there was some physical wearing of the original physical medium, resulting in a deterioration in sound quality over the life of the master re-

Image via iStock/treety. [Used under license.]

cording. Commercial copies of the master were slightly degraded reproductions of the original master, the quality of which depended on the limitations of the storage medium, which also deteriorated with use, introducing noise. Copying a commercial copy, usually onto magnetic tape in the form of audiocassettes, further degraded sound quality and added background noise. Compact discs changed this, because they stored information digitally, as a series of ones and zeroes. This meant that each time a compact disc was copied, the copy was flawless—one could listen to the original compact disc and the copy and there would be no way to distinguish the two. This made unauthorized sharing of music a much greater potential source of trouble for record labels, because if people could obtain perfect copies of music from friends, there would be no need to buy music themselves.

These issues were compounded tenfold by the rise of the internet, a digital platform. With music available in digital form and the internet available as a means of transmitting digital information, record labels would have faced a nightmare scenario, but for one factor—file size. The average compact disc could store about seven hundred megabytes of information, and a typical album would take up at least half of that space. Since internet speeds at the time were a small fraction of what they came to be, sharing several hundred megabytes of information online was almost impossible and would take days or weeks even if it could be done. To the record companies' chagrin, however, this quickly changed. File compression software takes a digital file as input and transforms it according to compression algorithms so that it takes up less computer storage space. It does this by looking for patterns in the digital file and then coming up with what are essentially abbreviations for each pattern.

Thus, if the pattern 0101001101101 is found to occur frequently in a file, then the compression software might translate this string of thirteen characters as Am9, for example, thus reducing the amount of storage space needed from thirteen to three. The best-known compression software for audio files that emerged around this time is still in use: MP3. MP3 software can reduce file sizes by factors of ten or more. This, coupled with ever-increasing internet speeds, made online music sharing a reality that the music industry at last had to face.

Music piracy almost instantly became commonplace, once file compression and moderately fast internet access became the norm. A wide variety of programs and protocols for sharing files emerged, and the music industry responded to each one by using legal action, or the threat thereof, to discourage people from sharing. Many of the early file-sharing applications, such as Napster, Kazaa, and others, were also significant security risks for users, because they functioned by opening the part of the user's computer where music was stored, so that it could be browsed by other users of the file-sharing application. Even this limited, restricted form of access was sometimes enough for a determined hacker to take control of another person's computer, with devastating consequences.

Still, there were many users who felt that this could not happen to them because they were too sophisticated in their knowledge of computer security. It was the mindset of this group that was the target of the music industry's legal action. The industry worked with internet service providers (ISPs) to identify people who were using file-sharing software and then filed lawsuits against these individuals. Often, the music industry would have designated employees use the file-sharing networks, posing as ordinary users, to gather information about which users were sharing and what files were being shared.

Music industry lawsuits frequently alleged damages in the tens of thousands of dollars, even for sharing a single song, because they estimated the cost of producing and releasing the song, and then also estimated the lost revenue incurred because of all the people who downloaded the song instead of purchasing it legitimately. At first these lawsuits saw some success, and cases were widely reported in which a parent was suddenly liable for hundreds of thousands of dollars in damages because their child illegally downloaded music with the family computer. Gradually, however, courts became more skeptical of the record companies' claims about their financial losses, as well as the technologically questionable methods they used to connect an ISP customer's identity with a user account on a file-sharing platform.

What was needed was a commitment by record labels to modernize their distribution methods and channels, so that their customers would not continue to flock to file-sharing applications in order to obtain the music they wanted to hear, without the restrictions on usage that record companies had traditionally required through their use of digital rights management (DRM) software incorporated into the music they sold. People wanted to be able to buy individual songs instead of having to purchase an entire album only to listen to one song on the record. They also wanted to be able to buy a song once and listen to it on different computers and devices, without having to pay again and without having to go through complicated steps necessitated by DRM. However, this did not happen.

The music industry was too fearful that taking steps to adapt its offerings to customers' evolving expectations would open up the floodgates for music pirates and file sharers all over the world, leading to bankruptcy for record companies. This created a void in the music marketplace, an empty space that was unfilled by a product that consumers actually desired.

OVERVIEW

Into this empty space stepped technology giants such as Apple and Google. These companies attempted to provide consumers with a means of interacting with their music in the way they wanted to. Both companies began offering what amounted to online storage options for legitimately acquired music. Users would import their music collections into these services and would then be able to access the music from any device that Google or Apple (or another such company) recognized. These services solved some issues that had long plagued music lovers—online storage and management, access from multiple devices—but not others. Music remained relatively expensive despite the huge technological advances that have made its production and distribution less resource-intensive. People wanted to listen to music anywhere and anytime, but who did not possess their own digital collection and did not have the resources to acquire one. This was a relatively large percentage of the music buying public, and it represented a demographic that continued to have significant incentives to engage in file sharing and piracy in order to obtain their music.

In response to this persistent need, streaming music services emerged. These were launched by companies that arranged licensing and royalties with most major record labels and charged users subscription fees to have access to all of the music they wanted, all of the time. Streaming services collect the subscription fees from users and then make payments to record companies based on how many times each company's music was streamed by users. While some free accounts are available, they have limited functionality or require users to listen to advertisements between songs, as a way of paying for the service.

Streaming services do not have every piece of music ever recorded, but their collections tend to be so large that for the average user, the selection is fairly comprehensive. This allows users to avoid having to spend time and money collecting music from their favorite artists, and simply start listening. If at any time they wish to cancel their subscription, they can do so. Rates are modest, usually adding up to less than twenty dollars per month.

ISSUES

Music streaming does not solve every problem, although it does what it is intended to do quite well. One issue is fairly obvious: music streaming services are only as good as one's internet connection, so in rural areas where mobile phone signals are weak or nonexistent, or in places where buildings, terrain, or other features interfere with signals, it can be difficult or impossible to use them. Some services compensate for this by allowing a limited amount of content to be downloaded to a mobile device, but in general this is discouraged since permitting too much downloaded content might encourage piracy or violate the licensing agreements between the streaming service and the record labels. Another disadvantage is that for those who come to rely on streaming music subscription services, if they find it necessary to stop paying their subscription, they will instantly lose access to their music, regardless of how long they have subscribed to the service. There is no feature that allows users to build up equity in the service so that they acquire perpetual rights to some content; it is an all-or-nothing proposition.

On the more positive side, streaming music services have some social interaction features that provide their users with benefits not available elsewhere. Most have options to play random music and to allow users to either listen to each track or to skip over the ones they do not care for. Each time a user listens or skips, the streaming service remembers that preference and builds a profile for the user, which it can use in order to recommend new music that is likely to suit the user's taste. In this way, users can discover new music that they otherwise might not have ever encountered.

Some services also make it possible for users to share songs and playlists with one another. Playlists, collections of songs that are assembled by users according to their tastes or for a particular purpose, such as a birthday party or wedding, serve much the same purpose as old-fashioned "mix-tapes," from the audiocassette era, and like mix-tapes, playlists and music sharing encourage people within the service's user community to listen to still more streaming content. To a lesser extent, some streaming services allow users to share their musical tastes on social media, although this tends to be limited to sharing lists of song titles rather than recordings of entire songs, for copyright reasons.

—Scott Zimmer

Further Reading

Borja, K., and S. Dieringer. "Streaming or Stealing? The Complementary Features between Music Streaming and Music Piracy." *Journal of Retailing & Consumer Services,* vol. 32, 2016, pp. 86-95.

Eriksson, Maria, et al. *Spotify Teardown: Inside the Black Box of Streaming Music.* MIT Press, 2019.

Hagen, A. N. "The Metaphors We Stream By: Making Sense of Music Streaming." *First Monday*, vol. 21, no. 3, 2016, p. 8.

Hampton-Sosa, W. "The Impact of Creativity and Community Facilitation on Music Streaming Adoption and Digital Piracy." *Computers in Human Behavior,* vol. 69, 2017, pp. 444-53.

Kim, J., C. Nam, and M. H. Ryu. "What Do Consumers Prefer for Music Streaming Services?: A Comparative Study between Korea and US." *Telecommunications Policy*, vol. 41, no. 4, 2017, pp. 263-72.

Lee, W., C. Chen, J. Huang, and J. Liang. "A Smartphone-Based Activity-Aware System for Music Streaming Recommendation." *Knowledge-Based Systems*, vol. 131, 2017, pp. 70-82.

Passman, Donald S. *All You Need to Know about the Music Business.* 11th ed., Simon & Schuster, 2023.

Pongnumkul, S., and K. Motohashi. "A Bipartite Fitness Model for Online Music Streaming Services." *Physica A*, vol. 490, 2018, pp. 1125-37.

Sinclair, G., and J. Tinson. "Psychological Ownership and Music Streaming Consumption." *Journal of Business Research*, vol. 71, 2017, pp. 1-9.

Augmented Reality

ABSTRACT

Augmented reality (AR) refers to any technology that inserts digital interfaces or graphics into the real world. In addition to headsets and glasses, AR technology can also include cell phones and other devices. In time, AR technology could be used in contact lenses and other small wearable devices.

BACKGROUND

Augmented reality (AR) is related to, but separate from, virtual reality. Virtual reality attempts to create an entirely different reality that is separate from real life. AR, however, adds to the real world and does not create a unique world. Users of AR will recognize their surroundings and use the AR technology to enhance what they are experiencing. Both augmented and virtual realities have become better as technology has improved. A number of companies (including large tech companies such as Google and Apple) have made investments in AR in the hope that it will be a major part of the future of technology and will change the way people interact with technology.

In the past, AR was seen primarily as a technology to enhance entertainment (e.g., gaming, communicating, etc.); however, AR has the potential to revolutionize many aspects of life. For example, AR technology could provide medical students with a model of a human heart. It could also help people locate their cars in parking lots. AR technology has already been used in cell phones to help people locate nearby facilities (e.g., banks and restaurants) and to inform people about nearby locations, events, and the people they meet and interact with. AR is used by online retailers to help customers visualize home décor items such as furniture in their rooms. It has also been implemented in smartphone applications such as the photo sharing application Snapchat and the mobile game Pokémon Go.

The term "augmented reality" was developed in the 1990s, but the fundamental idea for AR was established in the early days of computing. Technology for AR developed in the early twenty-first century, but at that time AR was used mostly for gaming technology.

OVERVIEW

In the early 2010s, technology made it possible for AR headsets to shrink and for graphics used in AR to improve. Google Glass (2012) was one of the first AR devices geared toward the public that was not meant for gaming. Created by the large tech company Google, Google Glass was designed to give users a digital interface they could interact with in ways that were somewhat similar to the way people interacted with smartphones (e.g., taking pictures, looking up directions, etc.). Although Google Glass was not a success, Google and other companies developing similar products believed that eventually wearable technology would become a normal part of everyday life.

During this time, other companies were also interested in revolutionizing AR technology. The AR company Magic Leap, for example, worked to develop technology that would beam images directly into a wearer's retinas, fooling the brain so it could not tell the difference between light from the outside world and the light coming from the AR device. The company's first commercially released AR headset, called Magic Leap One, sought to allow the brain to process digital objects the same way it does physical objects and was made available to consumers in 2018. Magic Leap released its second commercial AR headset, Magic Leap 2, in 2022.

Microsoft's AR company, HoloLens, created technology that was similar to Magic Leap's, though the final products had many differences. HoloLens created a headset that blended the physical and digital worlds, allowing wearers to interact with holograms and use voice controls to complete tasks. It also included "spatial sound" so that visual images accompanied by sounds seemed to be closer or farther away, corresponding with the visuals. For example, a person could see an animal running toward them on the HoloLens glasses, and they would hear corresponding sounds that got louder as the animal got closer. The initial HoloLens headset debuted in 2016, and its successor, the HoloLens 2, appeared in 2019.

Other AR companies, such as Leap Motion (later known as Ultraleap), designed AR products to be used in conjunction with technology people already relied on. This company developed AR technology that worked with computers to change the type of display people used. Leap Motion's design allowed people to wear a headset to see the computer display in front of them. They could then use their hands to move the parts of the display seemingly through the air in front of them (though people not wearing the headset would not see the images from the display). Other companies also worked on making AR technology more accessible through mobile phones and other devices that people used frequently.

Having entered the field of virtual reality through its 2014 acquisition of Oculus VR and established itself as a major competitor in that space with the release of the Meta Quest VR headset line, the technology company Meta (formerly known as Facebook) likewise delved into AR technology during the third decade of the twenty-first century. In early 2023, the company announced that its forthcoming Meta Quest 3 headset would prominently feature AR, which Meta referred to as "mixed reality." The company also announced plans to release several generations of "smart glasses" with AR features.

THE FUTURE OF AR

Although multiple companies have developed wearable technology that is small enough, light enough, and powerful enough to provide users with the feeling of reality, the AR field still faces many obstacles. AR companies are developing new technologies to make AR performance better, but many experts agree that successfully releasing this technology to the public could take years.

Another hurdle for AR technology companies is affordability. The technology they sell has to be priced so that people will purchase it. Since companies are investing so much money in the development of high-tech AR technology, they might not be able to offer affordable AR devices for a number of years. Another problem that AR developers have to manage is the speed and agility of the visual display. Since any slowing of the image or delay in the process could ruin the experience for the AR user, companies have to make sure the technology is incredibly fast and reliable.

Cultural organizations such as museums have been early adopters of AR. The technology has been used to enhance visitors' experiences, for example by allowing a patron to see how an incomplete object such as a partial vase likely looked when it was whole. Virtual exhibitions and digital artworks are other popular uses of AR technology.

In the future, AR devices could be shrunk to smaller sizes, and people could experience AR technology through contact lenses or even bionic eyes. Yet, AR technology still has many challenges to overcome before advanced AR devices become mainstream. Technology experts agree that AR technology will likely play an important role in everyday life in the future.

—*Elizabeth Mohn*

Further Reading

Altavilla, Dave. "Apple Further Legitimizes Augmented Reality Tech with Acquisition of Metaio." *Forbes*, 30 May 2015, www.forbes.com/sites/davealtavilla/2015/05/30/apple-further-legitimizes-augmented-reality-tech-with-acquistion-of-metaio. Accessed 13 July 2023.

Cisternino, Doriana, et al. "Augmented Reality Applications to Support the Promotion of Cultural Heritage: The Case of the Basilica of Saint Catherine of Alexandria in Galatina." *Journal on Computing and Cultural Heritage*, vol. 14, no. 4, Dec. 2021, pp. 1-30.

Donally, Jaime. *The Immersive Classroom: Create Customized Learning Experiences with AR/VR*. International Society for Technology in Education, 2021.

Farber, Dan. "The Next Big Thing in Tech: Augmented Reality." *CNET*, 7 June 2013, www.cnet.com/news/the-next-big-thing-in-tech-augmented-reality. Accessed 13 July 2023.

Heath, Alex. "This Is Meta's AR/VR Hardware Roadmap for the Next Four Years." *Verge*, 28 Feb. 2023, www.theverge.com/2023/2/28/23619730/meta-vr-oculus-ar-glasses-smartwatch-plans. Accessed 13 July 2023.

Narula, Herman. *Virtual Society: The Metaverse and the New Frontiers of Human Experience*. Currency, 2022.

Vanhemert, Kyle. "Leap Motion's Augmented-Reality Computing Looks Stupid Cool." *Wired*, 7 July 2015, www.wired.com/2015/07/leap-motion-glimpse-at-the-augmented-reality-desktop-of-the-future. Accessed 13 July 2023.

AVATARS AND SIMULATION

ABSTRACT

Avatars and simulation are elements of virtual reality (VR), which attempts to create immersive worlds for computer users to enter. Simulation is the method by which the real world is imitated or approximated by the images and sounds of a computer. An avatar is the personal manifestation of a particular person. Avatars were around long before social media. As computers have become more powerful and their rendering capabilities more efficient, avatars have improved in detail and diversity. Simulation and VR are used for many applications, from entertainment to business.

BACKGROUND

Computer simulation and virtual reality (VR) have existed since the early 1960s. While simulation has been used in manufacturing since the 1980s, avatars and virtual worlds have yet to be widely embraced outside gaming and entertainment. VR uses computerized sounds, images, and even vibrations to model some or all of the sensory input that human beings constantly receive from their surroundings every day. Users can define the rules of how a VR world works in ways that are not possible in everyday life. In the real world, people cannot fly, drink fire, or punch through walls. In VR, however, all of these things are possible, because the rules are defined by human coders, and they can be changed or even deleted. This is why users' avatars can appear in these virtual worlds as almost anything one can imagine—a loaf of bread, a sports car, or a penguin, for example. Many users of virtual worlds are drawn to them because of this type of freedom.

Because a VR simulation does not occur in physical space, people can "meet" regardless of how far apart they are in the real world. Thus, in a company that uses a simulated world for conducting its meetings, staff from Hong Kong and New York can both occupy the same VR room via their avatars. Such virtual meeting spaces allow users to convey nonverbal cues as well as speech. This allows for a greater degree of authenticity than in telephone conferencing.

OVERVIEW

The animation of avatars in computer simulations often requires more computing power than a single workstation can provide. Studios that produce animated films use render farms to create the smooth and sophisticated effects audiences expect.

Before the rendering stage, a great deal of effort goes into designing how an animated character or avatar will look, how it will move, and how its textures will behave during that movement. For example, a fur-covered avatar that moves swiftly outdoors in the wind should have a furry or hairy texture, with fibers that appear to blow in the wind. All of this must be designed and coordinated by computer

Avatar in the Metaverse greeting a friend. Photo via iStock/luza studios. [Used under license.]

animators. Typically, one of the first steps is keyframing, in which animators decide what the starting and ending positions and appearance of the animated object will be. Then they design the movements between the beginning and end by assigning animation variables (avars) to different points on the object. This stage is called "in-betweening," or "tweening." Once avars are assigned, a computer algorithm can automatically change the avar values in coordination with one another. Alternatively, an animator can change "in-between" graphics by hand. When the program is run, the visual representation of the changing avars will appear as an animation.

In general, the more avars specified, the more detailed and realistic that animation will be in its movements. In an animated film, the main characters often have hundreds of avars associated with them. For instance, the 1995 film *Toy Story* used 712 avars for the cowboy Woody. This ensures that the characters' actions are lifelike, since the audience will focus attention on them most of the time. Coding standards for normal expressions and motions have been developed based on muscle movements. The MPEG-4 international standard includes 86 face parameters and 196 body parameters for animating human and humanoid movements. These parameters are encoded into an animation file and can affect the bit rate (data encoded per second) or size of the file.

Simulation has long been a useful method of training in various occupations. Pilots are trained in flight simulators, and driving simulators are used to prepare for licensing exams. Newer applications have included training teachers for the classroom and improving counseling in the military. VR holds the promise of making such vocational simulations much more realistic. As more computing power is added, simulated environments can include stimuli that better approximate the many distractions and detailed surroundings of the typical driving or flying situation, for instance. Though primarily used for purposes such as gaming in the early 2020s, three-dimensional (3D) VR headsets such as the Meta Quest may one day facilitate richer vocational simulations for military and emergency personnel, among others.

—*Scott Zimmer*

Further Reading

Chan, Melanie. *Virtual Reality: Representations in Contemporary Media*. Bloomsbury, 2014.

Donally, Jaime. *The Immersive Classroom: Create Customized Learning Experiences with AR/VR*. International Society for Technology in Education, 2021.

Gee, James Paul. *Unified Discourse Analysis: Language, Reality, Virtual Worlds, and Video Games*. Routledge, 2015.

Griffiths, Devin C. *Virtual Ascendance: Video Games and the Remaking of Reality*. Rowman, 2013.

Kizza, Joseph Migga. *Ethical and Social Issues in the Information* Age. 5th ed., Springer, 2013.

Narula, Herman. *Virtual Society: The Metaverse and the New Frontiers of Human Experience*. Currency, 2022.

Parisi, Tony. *Learning Virtual Reality: Developing Immersive Experiences and Applications for Desktop, Web, and Mobile*. O'Reilly, 2015.

B

Bandcamp

ABSTRACT

Bandcamp is a non-label music storage and streaming website that can be used by independent artists and established recording artists. The service enables musicians to sell digital albums and individual tracks, in addition to physical recordings and merchandise, directly to their fans and listeners. By 2023, users had paid $1.5 billion to musicians through Bandcamp.

BACKGROUND

The commercial music industry has followed a certain mode of operation since time immemorial. Composers, who also tended to be performers with a particular instrumental expertise, would record their compositions on score sheets using standard musical notation. Even in the time of Mozart and other classical composers, the composer was managed by an agent who would arrange for performances for paying audiences. The income from performances provided the livelihood for both the composer and the agent, and the musical scores themselves were jealously guarded properties such that only the actual composer would be able to present the music in the manner it was intended to be presented. With copyright laws, musical scores became a commercial commodity and ownership of music was no longer the province only of composers.

Agents, patrons, and others could buy or otherwise acquire ownership of the copyright to a particular piece of music, giving them the sole right to say who could perform the copyrighted works, typically for a fee. Fast forward to the much more recent development of the recording industry and recorded music. In this industry, with a music production company, or recording label the company owns the copyright for the compositions it produces and delivers royalties to the label artists for their performance to the composer or writer of the particular work (the songwriter), as well as others involved in the production who are listed on the physical label of the recording medium distribution (e.g., 45-rpm singles, long-playing [LP] albums, extended play [EP] short albums, compact disc [CD], digital video disc [DVD] and Blu-ray discs). Others, such as background singers, studio musicians, and graphic artists, are generally paid a flat rate for their work on a particular production.

Those costs are recovered by the producers from commercial sales and licensing deals such as theme songs for TV shows and movies, and for use in advertisements. Promotion of the artists and commercial recordings is a very significant cost to the recording label, and that aspect is generally handled by agents, with costs being recovered from ticket sales as well. Taken altogether, the return to the artists and composers is historically low unless they become phenomenally successful commercially, perhaps forming their own recording labels and managing other artists. The Beatles, for example, began their commercial career with manager Brian Epstein under the Decca and Capitol record labels before forming their own Apple Records label. Many other contemporary artists, however, never achieve the level of success that would allow such a transition and remain instead as part of the recording industry establishment throughout their careers.

OVERVIEW

Overlooked by the industry are those artists and composers who either were not afforded the opportunity to sign with an established label, or who chose not to work in that milieu while creating their own songs. These are the independents, the musical equivalent of authors who choose to self-publish their books rather than accept small returns while the publisher garners the lion's share of the proceeds. For independent musicians and songwriters, this is where Bandcamp comes in.

Founded in 2008, Bandcamp is a non-label music storage and streaming service that can be used by independent artists and established recording artists. Bandcamp refers to itself as an online record store and music community where fans, discover, connect with, and directly support the artists they like. Users of Bandcamp are able to purchase musical recordings offered for sale as digital albums, individual music tracks or songs, vinyl record disks, CDs, and cassettes. T-shirts are also sold through Bandcamp. Of those sales, 82 percent of the proceeds is delivered directly to the artists rather than passing through the many hands reaching for a piece in a label company. Bandcamp directly retains about 10 percent of the proceeds from sales to support its operation, with the rest being used to pay various fees associated with licensed labeled pieces that are made available.

Bandcamp is the quintessential win-win-win situation for artists, users, and recording companies. For artists, the return on sales is immediate and substantial, especially for independent artists who have only to produce a quality recording of their work and upload it to Bandcamp for sale, with none of the ancillary costs associated with the traditional recording industry. For users, Bandcamp offers a vast amount of music that is simply not available from the traditional industry. This allows them to explore and find new music and artists that appeal to their tastes, whether as single songs or as album collections. For recording companies, Bandcamp provides a means to test market an artist's potential for success before offering to sign them on with the label and investing in the risks that come with promotion bringing great success. An independent artist who performs well on Bandcamp may, after all, be worth being offered a recording contract, though the artists may also choose to remain independent, in which case the recording company invests nothing and loses nothing. Recording companies are also able to provide the recordings they control through Bandcamp music sales, though this may seem somewhat counterintuitive.

In March of 2022, Bandcamp chief executive officer (CEO) and cofounder Ethan Diamond announced that the company had been acquired by Epic Games, a technology company best known for creating the video game *Fortnite*. Diamond confirmed that Bandcamp would continue to operate as usual but would also expand its offerings and make technological improvements following the acquisition.

—Richard M. Renneboog

Further Reading

"About Us." *Bandcamp*, 2023, bandcamp.com/about. Accessed 13 July 2023.

Arditi, David. *Streaming Culture: Subscription Services and the Unending Consumption of Culture*. Emerald Group, 2021.

Diamond, Ethan. "Bandcamp Is Joining Epic Games." *Bandcamp Updates*, 2 Mar. 2022, blog.bandcamp.com/2022/03/02/bandcamp-is-joining-epic. Accessed 13 July 2023.

Gordon, Steve. *The Future of the Music Business: How to Succeed with the New Digital Technologies*. Hal Leonard, 2011.

Oswald, Vanessa. *Indie Rock: Finding an Independent Voice*. Greenhaven Publishing LLC, 2018.

Ravens, Chal. "Good Vibrations: How Bandcamp Became the Heroes of Streaming." *The Guardian*, 25 June 2020, www.theguardian.com/music/2020/jun/25/bandcamp-music-streaming-ethan-diamond-online-royalties. Accessed 13 July 2023.

Riselvato, John. *Bandcamp for Artists and Music Labels: Marketing Tips to Sell More Music and Merch*. Amazon Digital Services LLC, 2020.

Rolston, Clyde Philip, Amy Macy, Tom Hutchison, and Paul Allen. *Record Label Marketing: How Music Companies Brand and Market Artists in the Digital Era*. CRC Press, 2015.

Zager, Michael. *Music Production: A Manual for Producers, Composers, Arrangers, and Students*. Rowman & Littlefield, 2021.

BIG DATA

ABSTRACT

In the cutting-edge world of data retrieval systems, big data is in-field shorthand that refers to the sheer mass of data produced daily by and within global computer networks at a pace that far exceeds the capacity of current databases and software programs to organize and process. Virtually any organization that gathers information—government, businesses, retail stores, services, hospitals, social media outlets—faces an enormous challenge in gathering information that is generated at massive rates every minute, reading that data into meaningful and coherent information, and then storing it efficiently and effectively. Given that the number of qualified computer software engineers savvy in this burgeoning field is far below industry needs, the strategies for directing and controlling big data represent a primary challenge to those who rely on computers for operating information.

BACKGROUND

The amount of information processed by computer networks globally every day is measured in terabytes (one trillion bytes) and petabytes (one quadrillion bytes). Such information includes general information, statistics, medical records, science and research data, phone records, business records, government records, and many other types of data. Data generation on that scale is measured four ways: (1) the variety of the information available (from cell phone records to air travel information, from social network inputs to stock trading); (2) the volume of information produced, measured per second; (3) the velocity at which the information is produced; and (4) its veracity (i.e., its meaningfulness). The ability to harness and direct that information represents an entirely new paradigm for virtually any organization from governments to businesses, from hospitals to school systems. The potential rewards are significant. Any interaction with computers—smartphones, e-readers, digital music players, tablets, laptops, internet searches—generates data about individuals.

OVERVIEW

Data about individuals is useful to businesses seeking to direct their resources most favorably, from accumulating customer profiles to directing product development and marketing. In the media and entertainment industries, for instance, companies analyze user data to glean insight into the media consumed by different demographics as well as user behavior. Data analytics is likewise a critical component of business operations within social media companies as well as in the field of advertising.

Beyond media and entertainment, big data could have numerous far-reaching research applications. Big data research could be directed toward the global monitoring of potential terrorist activities, given that chatter from such sources is nearly constant. Medical records could be synthesized and coordinated globally with and against available research data about treatments. Government records could be streamlined, and government services could be made far more efficient. Organization performance could be boosted, records could be made more accessible, and networks could communicate effectively across traditional boundaries. Information itself could become fluid and transparent.

Critics of big data research, however, point to a Big Brother scenario, in which intrusion into privacy could proceed unchecked. All computer activity could be monitored and theoretically stored by the government, by employers, by businesses—really by anyone with access to what would become unprecedentedly massive data reservoirs. In addition, such widespread harvesting of information creates new threats to intellectual property rights and copyright infringements. Gathering data, these critics point out, is not effectively analyzing it, thus stripping the data from the sociocultural context that provides its true meaning. People, they argue, are more than data points.

In recognition of the privacy ramifications of data collection and research, some governments introduced regulations governing the collection and use of individuals' data. One such regulation was the General Data Protection Regulation (GDPR), which was introduced in the European Union in 2016 and came into force two years later.

—Joseph Dewey

Further Reading

Cukier, Kenneth. *Big Data: A Revolution that Will Transform How We Live, Work, and Think*. Dolan, 2013.

Erl, Thomas, editor. *Big Data Fundamentals: Concepts, Drivers & Techniques*. Pearson, 2016.

Horner, Veronica. "How Big Data Analytics Is Changing the Media Industry." *SEI*, 5 Oct. 2020, www.sei.com/insights/article/how-big-data-analytics-is-changing-the-media-industry. Accessed 13 July 2023.

Kolb, Jason, and Jeremy Kolb. *The Big Data Revolution*. Applied Data Labs, 2013.

Pentland, Alex. "Saving Big Data from Itself." *Scientific American*, vol. 311, no. 2, 2014, pp. 65-67.

Schmidt, Eric, and Jared Cohen. *The New Digital Age: Reshaping the Future of People, Nations, and Business*. Knopf, 2013.

Smith, Michael D., and Rahul Telang. *Streaming, Sharing, Stealing: Big Data and the Future of Entertainment*. Reprint ed., MIT Press, 2017.

"What Is Big Data?" *Oracle*, www.oracle.com/big-data/what-is-big-data. Accessed 13 July 2023.

BOTS

ABSTRACT

The term "bot" refers to any autonomous software application that runs automated tasks over a network, often the internet. Typically performed at a much faster rate than that which would be possible for a human, the tasks are simple and repetitive. By some estimates, bots make up more than half the traffic on the internet. Bots fit broadly into four categories: social, commercial, malicious, and helpful—which can at times overlap. Most scholars also include automated personal assistants such as Amazon's Alexa, Apple's Siri, and Google Assistant in discussions of bots.

BACKGROUND

Some scholars trace the origin of bots to Alan Turing and the Turing Test. In 1964 Joseph Weizenbaum at the Massachusetts Institute of Technology (MIT) Artificial Intelligence Laboratory created the ELIZA, a social or chat bot programmed to respond to a number of keywords. Though incapable of "learning" through interaction alone, ELIZA caused several participants in the experiment to become emotionally attached to it during their "conversations." As Weizenbaum noted at the time, "extremely short exposures to a relatively simple computer program could induce powerful delusional thinking in quite normal people."

OVERVIEW

Modern chat bots are far more adaptive than ELIZA and employ natural language processing systems to relate keywords and patterns from a database and formulate their responses. Companies will often employ chatbots as a first layer of customer service on a website. Chatbots can also be routed through third-party platforms, such as WeChat or Facebook Messenger.

Another form of social bot is a fraudulent account on social media. This type of bot came to prominence during the 2016 American presidential election cycle. Often, these accounts feature profile images and details that make them appear as if they are real people. However, they interact on social media at a rate that no human possibly could. In a 2018 article for *Wired* magazine, Paris Martineau gave the example of a Twitter bot account that had tweeted more than 2,000 times in three days, averaging 660 retweets and seven original tweets per day.

This type of social bot is often referred to as a troll due to its programmed behavior, usually political. A well-programmed bot of this variety can be very difficult to discern from an actual person, even for social media companies themselves. Two prominent platforms for bots of this variety, Facebook and Twitter, both launched campaigns to rid themselves of bots following bad press in the aftermath of the 2016 presidential race. In 2022, Twitter leadership reported that bots represented less than 5 percent of the service's daily active users; however, other analyses, including a study conducted by the technology intelligence company Cyabra, suggested that that proportion could exceed 13 percent.

While politically motivated social bots are widely regarded as malicious, other social bots function as news aggregators and might retweet articles that feature certain keywords. Others are used to archive threads on social media or for parody, and some can be helpful or even commercial. Helpful or commer-

Photo via iStock/jittawit.21. [Used under license.]

cial bots, however, usually disclose that they are bots. Malicious social bots attempting to impersonate genuine accounts typically will not disclose their true nature.

Another variety of malicious bot is represented by "botnets," a portmanteau of "robot" and "internet." Botnets are composed of a number of internet-connected devices, each of which is running a particular bot application. Devices can be part of a botnet without their users knowing. The botnets can then be used to perform a variety of tasks, including distributed denial of service (DDoS) attacks, cryptocurrency mining, and data mining.

Computers compromised by a botnet are often referred to as zombie computers. Their users may have no idea that they are being employed for malicious purposes. Often, computers are infected via spam email or fraudulent downloads. They can also be infected by visiting an infected website, or by exploited vulnerabilities in a web browser. The decentralized nature of botnets makes them difficult to quarantine after a number of machines have been infected. Moreover, after zombie computers have reconnected to the botnet's "home," malicious software download packets often delete themselves, leaving little visible evidence of the botnet's presence on the zombie machine.

Legally, there is not much precedent for bots or botnets. Until 2016, many gaps existed in the Federal Rules of Criminal Procedure, creating substantial obstacles both for prosecuting the creators of botnets and also when attempting to de-infect zombie computers. Following the passage of the new regulation, which went into effect on December 1, 2016, investigators were allowed to bring a single warrant to search infected computers to one federal court rather than being required to craft identical warrants in up to ninety-four jurisdictions. Previously, individual warrants had to be issued for each computer, regardless of whether they were in the same jurisdiction, making the process slow and ineffectual.

The State of California passed a law that went into effect in July 2019 requiring chatbots to identify themselves as not being human. While most commercial chatbots already did so, the law's author, State Senator Robert Herzberg, said that the measure was particularly targeted at "deceptive commercial and political bots." Some legal scholars questioned whether this might constitute "compelled speech," either from the bot, its programmer, or the company that owns it. Moreover, as the law was at the state level, its overall efficacy remained uncertain. Companies that did not already have their bots disclose that they were not human would likely simply reprogram their bots to do so in order to do business within California, but malicious bots were unlikely to comply regardless of the legislation.

More legislation is likely to focus on bots, particularly as home assistants such as Amazon's Alexa, Google Assistant, and Apple's Siri become more prevalent. Questions have already been raised about the potential for Alexa to monitor its users in order to gain marketing information on them. In the early 2020s, however, users widely privileged the convenience of such bots over their potential negative side effects.

—J. N. Manuel

Further Reading

Dang, Sheila, Katie Paul, and Dawn Chmielewski. "Focus: Do Spam Bots Really Comprise Under 5% of Twitter Users? Elon Musk Wants to Know." *Reuters*, 13 May 2022, www.reuters.com/technology/do-spam-bots-really-comprise-under-5-twitter-users-elon-musk-wants-know-2022-05-13. Accessed 13 July 2023.

"Ensuring BotNets Are Not 'Too Big to Investigate.'" *US Department of Justice*, 22 Nov. 2016, www.justice.gov/archives/opa/blog/ensuring-botnets-are-not-too-big-investigate. Accessed 13 July 2023.

Gershgorn, Dave. "A California Law Now Means Chatbots Have to Disclose They're Not Human." *BotLaw*, *Quartz*, 3 Oct. 2018, qz.com/1409350/a-new-law-means-californias- bots-have-to-disclose-theyre-not-human. Accessed 13 July 2023.

Matineau, Paris. "What Is a Bot?" *Wired*, 16 Nov. 2018, www.wired.com/story/the-know-it-alls-what-is-a-bot. Accessed 13 July 2023.

Shah, H., K. Warwick, J. Vallverdú, and D. Wu. "Can Machines Talk? Comparison of Eliza with Modern Dialogue Systems." *Computers in Human Behavior*, vol. 58, 2016, pp. 278-95.

Stricklin, Kasey. *Social Media Bots: Laws, Regulations, and Platform Policies*. CNA, Sept. 2020, www.cna.org/reports/2020/10/DIM-2020-U-028193-Final.pdf. Accessed 13 July 2023.

Swaine, Jon. "Twitter Admits Far More Russian Bots Posted on Election Than It Had Disclosed." *The Guardian*, 19 Jan. 2018, www.theguardian.com/technology/2018/jan/19/twitter-admits-far-more-russian-bots-posted-on-election-than-it-had-disclosed. Accessed 13 July 2023.

Weizenbaum, J. *Computer Power and Human Reason: From Judgement to Calculation*. W. H. Freeman, 1976.

Browsers

ABSTRACT

A web browser is an application utilized to access data and information on the World Wide Web. Browsers may be used on personal computers as well as on mobile devices such as smartphones. In the third decade of the twenty-first century, popular browsers included Microsoft Edge, Google Chrome, Firefox, and Safari.

BACKGROUND

Following the advent of the publicly accessible World Wide Web in the early 1990s, a variety of browsers were developed to grant users access to the content available on the web. One of the most versatile browser applications of that era was Netscape Navigator, produced by Netscape Communications Corporation in 1994. As befit the graphics capabilities of the time, the on-screen appearance of Navigator was simple in design. Navigator was essentially limited to a hypertext markup language (HTML)-based display of text and still images. Within that limitation though, Navigator was a very capable and reliable browser with excellent operating speed. It also had one unique feature that no other browser application offered; its users could design and compose their own HTML-based web pages using native HTML code or with Navigator's built-in HTML coding functions and then view those pages before uploading them to a website.

Navigator also introduced a number of user-friendly features that made accessing the internet during the times of slow dial-up modem transfers almost tolerable for users. One such feature was the ability to view web pages as they were downloading rather than having to wait for the entire page to download before any of it could be viewed. Netscape also introduced the use of cookies, frames, proxy-auto configuration, and JavaScript, a kind of intermediate language that augmented mark-up languages by providing computational functionality that mark-up languages do not possess on their own.

Navigator, and its successor Netscape Communicator, were widely available at no cost through 2007 and subsequently continued to be found as legacy applications. They were written to run on a variety of operating platforms, including Windows, Linux, OS/2, Macintosh, and many of the UNIX variants. They looked and operated almost exactly the same on each operating system. Netscape's great success with Navigator and Communicator, however, ultimately lead to its downfall. Browsers developed to compete with them, chiefly Microsoft's Internet Explorer (IE), eventually displaced Netscape's browsers from their position as the leading browser applications of the time.

Given the power of Microsoft's marketing strategies, it should come as no surprise that the IE browser that came with every single personal computer (PC) operating on Microsoft Windows would displace Netscape's browsers at some point. Explorer was introduced in 1995 to compete with Netscape Navigator, and by 2003, IE held a 95 per-

Photo via iStock/lelepado. [Used under license.]

cent usage share of browser applications. In turn, IE decreased in popularity throughout the following years, as other competing browser applications had taken over. In 2015, Microsoft replaced IE with the browser Microsoft Edge.

OVERVIEW

In the third decade of the twenty-first century, numerous browser applications are available. While they all function in the same way, each one has its own unique appearance and built-in capabilities. Each browser has its own strengths, weaknesses, and configurability, and therefore its own usefulness. One common characteristic is the ability to open multiple web pages at one time and display those pages in separate tabs. This makes it easy for users to multitask, perform research, compare products, or otherwise maintain concurrent access to information.

Firefox. According to Mozilla.org, the producer of the browser application, Firefox is not just a browser, but a collection of products designed for users to promote safer and smarter browsing while online. The main feature of the Firefox family of products however, is the Firefox browser application itself. The browser is produced for desktop and laptop computers, mobile systems such as cell phones and tablets, and for enterprise computing systems and mainframes. In all iterations, the browser automatically blocks more than 2000 known data trackers, small third-party programs that monitor a user's online activity and returns a record of that activity to the data trackers' sources for different purposes.

Firefox was created in 2002, when Microsoft's IE 6 was the dominant browser in general use. The beta version of Firefox proved to be faster, more secure, and with better add-on features than IE 6. By 2009, Firefox had exceeded IE's share of browser usage, before both fell behind Google's Chrome browser in popularity. Originally part of a collection of utility programs called Mozilla Suite, Firefox (originally called Phoenix) was developed to be a stand-alone application in order to reduce what was thought of as Mozilla Suite's "bloated" code. The term refers to program code that requires so many coded instructions to run all of the program's features that it causes program functioning to bog down or become too slow to be useful. Firefox quickly proved itself to be a fast, efficient browser that is easy to use, with user-friendly usability and security features. While Firefox declined in popularity in the decades following its initial release, it remained the preferred browser for some users and in May of 2023 had a global market share of about 2.8 percent, according to *Statista*.

Google Chrome. Google's Chrome browser was first released in 2008 to compete with Firefox and quickly displaced both Firefox and IE as the browser of choice. By May of 2023, the browser had a global market share of 63 percent. Chrome is a fast, robust browser that is exceedingly simple to use due to its minimalist presentation. Initially, new tabs in Chrome displayed only a wallpaper screen image and the search bar/address bar called the "omnibox." Users type in a web address, search term, or keyword. If a search term or keyword is entered, Chrome uses the Google search engine to call up every instance of the word that it can locate. The minimalism of the new tab display was subsequently augmented to add a selection of topical links that appear each time a new tab is opened.

Chrome is readily configured and maintained by its users, with numerous preference settings and "housekeeping" utilities. Chrome also has a large library of extensions available as add-ons, allowing users to add features such as other languages, translation, mathematics, and others that are not part of the basic structure of Chrome's provided release. Many such extensions are written and produced by third-party developers and may not be entirely compatible with a particular version of Chrome.

Safari. Because Safari was developed to run on Mac computers rather than Windows PCs, it is a graphics-based browser and unique to the MacOS, iOS, and iPadOS operating systems. A Windows version was offered in 2007, but it was discontinued after several years. Safari was initially offered in 2003 and as of 2023 is in its sixteenth iteration. Its performance in the 2020 version was 50 percent faster than Google Chrome and consumed significantly less power than its competitors. In May 2023, Safari usage was second only to Google Chrome, with a global market share of about 20.9 percent. New versions of Safari were released every two years from 2003 to 2007 and were later released yearly.

Safari functions like all web browsers, the essential difference being that internet connections and data are displayed through graphics processing rather than the text-based processing of other computers. In graphics processing the hyperlinks to other web pages may be embedded within the digital code of graphics images, whereas in text-based processing the hyperlinks are merely associated with a string of text that typically identifies the uniform resource locator (URL) of the target of the hyperlink, or is perhaps just an underlined word or an instruction such as "click here." The hyperlinks in both text-based and graphics processing can also be associated with an image.

Chromium. As the name would suggest, Chromium, like Chrome, is a product of Google. Though not a browser per se, it is a free and open-source browser project representing the very large codebase used by Chrome. As such, Google does not provide an official stable version of Chromium. While the program codes within the Chromium codebase are the property of Google, it is available freely for use by other developers and has been utilized to produce other web browsers, including Opera, Microsoft Edge and Samsung Internet. Users of Chromium code are able to develop and enhance the code they use and add those new features to the Chromium codebase. In addition, they can use their features with the Chrome brand attached.

Microsoft Edge. The successor to IE, Microsoft Edge was first released in 2015. Much like IE, it was automatically provided alongside the Windows operating system. Versions of Edge were also released for other operating systems, including macOS. The browser underwent substantial updates in the years following its initial release, and in 2020, Microsoft released a new version of Edge that was based on Chromium. Edge ranked third among browsers globally as of May 2023, with a market share of about 5.3 percent.

Brave. Brave, launched in 2016, is one of the several web browsers that have been constructed using the Chromium codebase. Brave recaptures the minimalistic appearance of new tabs, displaying only a beautifully composed high-resolution photographic wallpaper background, an omnibox search bar, and a small number of quick access icons linked to as many web locations as the user has accessed frequently. The icons are unobtrusive and can be removed either manually or when the browser cache is cleared. Brave tabs also often display a promotion for cryptocurrency as the wallpaper image, which users may or may not find annoying. The promotion is due to the fact that Brave users are also provided the opportunity to garner a type of cryptocurrency called BAT that they can choose to pass on to developers, charitable organizations, keep for themselves, or just simply ignore. This is an interesting feature not provided by other web browsers.

Brave's main attractions are its built-in automatic blocking of ads and data trackers and its high user security capabilities. In normal use Brave runs the Chrome browser, but with enhanced user security features. Chrome itself gives the option of using a private window for browsing, as do other browsers, that does not log the browsing history of the user's current session. When the option to open a new private window is selected in Brave, however, the user is presented with a choice to use either Chrome's basic private window service or Brave's private window with Tor connectivity. With Tor connectivity, the user's IP address is completely masked as the browsing requests are passed through any number of other Tor server nodes. Accordingly, this may slow down browser response, and Brave warns that some websites may not function through Tor and so will not be accessible. The Tor connectivity option also requires that the Tor browser be resident on the user's device.

Opera. Opera is one of the oldest web browsers, having been developed in 1995 by its namesake company. Until 2005, Opera was available only as commercial software based on Opera's Presto engine. Opera was switched from Presto to Chromium in 2013 and continues to develop from the Chromium codebase. In 2016, the Opera brand was sold to a consortium of Chinese investors.

The Opera browser has many of the features common to Chromium-based browsers, such as ad blocking, tabbed browsing, private browsing, and others. A particular feature of interest in Opera is that an unlimited number of recently viewed web pages are shown as thumbnail views when a new tab is opened. Opera also has embedded links to social media messaging applications, including Facebook Messenger, WhatsApp, Telegram, Twitter, Instagram, and V Kontakte (VK). Opera is contracted to use the Google search engine as its default. Opera runs on Windows, Mac, Linux, iOS, and Android operating systems.

—Richard M. Renneboog and Jake D. Nicosia

Further Reading

"Global Market Share Held by Leading Internet Browsers from January 2012 to May 2023." *Statista*, May 2023, www.statista.com/statistics/268254/market-share-of-internet-browsers-worldwide-since-2009. Accessed 13 July 2023.

Gralla, Preston. *How the Internet Works*. 4th ed., Que Publishing, 1998.

"The History of Web Browsers." *Mozilla*, www.mozilla.org/en-US/firefox/browsers/browser-history. Accessed 13 July 2023.

Hofmann, Chris, Marcia Knous, and John V. Hedke. *Firefox and Thunderbird Garage*. Prentice Hall Professional Technical Reference, 2005.

Huddleston, Rob. *HTML, XHTML, and CSS: Your Visual Blueprint for Designing Effective Web Pages*. John Wiley & Sons, 2009.

La Counte, Scott. *The Ridiculously Simple Guide to Surfing the Internet with Google Chrome*. SL Editions, 2020.

LeJeune, Urban A., and Jeff Duntemann. *Netscape and HTML Explorer*. Coriolis Group Books, 1995.

Markelo, Steve. *Microsoft Edge: A Beginner's Guide to the Windows 10 Browser*. Conceptual Kings, 2015.

McDaniel, Adam. *HTML 5: Your Visual Blueprint for Designing Rich Web Pages and Applications*. John Wiley & Sons, 2011.

Mehta, Prateek. *Creating Google Chrome Extensions*. Springer Science + Business Media, 2016.

O'Leary, Timothy, Linda O'Leary, and Daniel O'Leary. *Computing Essentials 2023*. McGraw-Hill, 2022.

Stockson, Eric. *Brave Browser: Blockchain Internet Browsing Made Easy*. First Rank, 2019.

Wilton, Paul, and Jeremy McPeak. *Beginning JavaScript*. 5th ed., Wrox, 2015.

BUILDING INFORMATION MODELING

ABSTRACT

Building information modeling (BIM) is a field within architecture, design, and construction that focuses on creating, altering, and managing digital representations of buildings and other structures. BIM involves the use of modeling software and is considered an evolution of computer-aided design (CAD) as well as a superior alternative to two-dimensional (2D) CAD, which uses software to produce 2D models of physical structures.

Unlike traditional CAD, BIM models are parametric, meaning that each element of the design can be linked to each other element so that any change made to the design or plan affects all parts of the overall design. BIM models typically have an informational layer that can be used to integrate data about pricing and sourcing, maintenance, deterioration, thermal conservation, and energy usage. Data associated with a BIM model can also be integrated in such a way that the information changes automatically if the model is altered.

BACKGROUND

Building information modeling (BIM) is one of several computer-aided design models used by engineers, architects, and contractors working on construction and design projects. BIM is a form of computer design and is an evolution of the first digital design programs. The first computer-aided design (CAD) program was Sketchpad, developed in 1963 by Ivan Sutherland. The program used a light pen and a button interface to allow users to draw designs on a computer screen. When personal computers became a reality in the 1980s, the first practical digital design programs began to appear.

Charles Eastman was the first to propose creating a three-dimensional (3D) architecture system with linked data files. He first wrote about the concept in a 1974 paper and later received funding to begin working on the project. The Building Description System (BDS), a database that contained a library of building elements and was searchable by factors such as building material, cost, supplier, or subcontractor, was his first step.

Eastman went on to design the Graphical Language for Interactive Design (GLIDE) program in 1977, which contained many of the elements that became standard in later BIM programs. In 1982, the company Autodesk released its first version of AutoCAD software using two-dimensional (2D) design. The next major step in the development of BIM came from Budapest, Hungary, where physicist Gábor Bojár launched the ArchiCAD system in 1984. ArchiCAD used many of the same elements as Eastman's BDS system and was considered a major technical leap forward, though the lack of widespread personal computer use limited the adoption of the program and many architecture firms continued to use traditional CAD programs.

The personal computer explosion of the 1990s led to the development of a host of new BIM programs. Advancements in the complexity of computer processing and data storage allowed new versions of BIM software to incorporate more detailed data management systems, making BIM more practical for commercial use.

OVERVIEW

While architects and engineers differentiate between BIM and CAD, both are models for creating and modifying 3D data about physical structures. BIM software available by the early 2020s integrated drawing and visualization tools that are part of CAD programs but tended to include additional functionality that is not present in traditional CAD software.

One of the major advantages of BIM systems is the use of a parametric method for representing the relationship between objects. Traditional CAD systems tend to use explicit or coordinate-based models that treat each object as an individual, separate structure. By contrast, BIM systems are parametric, meaning that each element is modeled according to behaviors and attributes that are linked to every other element. If a designer changes the pitch, weight, thickness, or other qualities assigned to a roof, for instance, the other elements of the design automatically change as a result.

BIM systems also typically include data files associated with each component. In a BIM model, a wall will have an associated list that can show the materials needed for the project, the time needed to complete the component, the sourcing of materials, and a variety of other parameters. If a wall is supposed to have a door, the list connected to the wall will contain what designers call a "door schedule" that plans when the door is supposed to be installed within the wall. The data lists in BIM models are similarly linked to every related component in the model so that the lists automatically update whenever changes are made to any related element in the design.

While BIM systems are often described as 3D, some architects and designers see BIM as "n-dimensional," meaning that additional dimensions can be modeled within the program. In addition to the three physical dimensions, BIM models can also integrate and model other aspects of the project that include the time needed for completion of individual components as well as the entire project, cost and maintenance of the project, and even estimates of how use will affect the structure.

By the 2020s, architects and engineers often used BIM for projects with specific needs or goals. When constructing a concert venue, engineers, designers, and contractors can use BIM programs to model the acoustic properties of components, materials, and rooms within a structure. Similarly, BIM modeling can help estimate factors including thermal and electrical conservation, energy use, and the environmental impact of materials. All of this information and any changes can be shared with the entire team working on the project, keeping the planning and building process smooth and updated.

Despite significant advantages, the widespread adoption of BIM systems only began in the 2010s, due partly to the complexity of earlier BIM software and conservatism in the building industry that inspired loyalty to the AutoCAD methodology. One of the trends in BIM design is for contractors and designers to use BIM models to involve the consumer in the design process. Using a BIM model, a contractor can demonstrate alterations or options to a customer.

—Micah L. Issitt

Further Reading

"3D-4D Building Information Modeling." *US General Services Administration*, 6 Apr. 2022, www.gsa.gov/real-estate/design-construction/3d4d-building-information-modeling. Accessed 13 July 2023.

Borrmann, André, Markus König, Christian Koch, and Jakob Beetz, editors. *Building Information Modeling: Technology Foundations and Industry Practice*. Springer International Publishing, 2018.

Deutsch, Randy. *BIM and Integrated Design: Strategies for Architectural Practice*. Wiley, 2011.

Goldstein, Phil. "What Is Building Information Modeling Technology in Government?" *FedTech*, 16 June 2020, fedtechmagazine.com/article/2020/06/what-building-information-modeling-technology-government-perfcon. Accessed 13 July 2023.

Hardin, Brad, and Dave McCool. *BIM and Construction Management: Proven Tools, Methods, and Workflows*. 2nd ed., Wiley, 2015.

Kensek, Karen M., and Douglas E. Noble. *Building Information Modeling: BIM in Current and Future Practice*. Wiley, 2014.

Lu, Weisheng, and Chimay J. Anumba, editors. *Research Companion to Building Information Modeling*. Edward Elgar, 2022.

Quirk, Vanessa. "A Brief History of BIM." *ArchDaily*, 7 Dec. 2012, www.archdaily.com/302490/a-brief-history-of-bim. Accessed 13 July 2023.

Sacks, Rafael, Chuck Eastman, Ghang Lee, and Paul Teicholz. *BIM Handbook: A Guide to Building Information Modeling for Owners, Designers, Engineers, Contractors, and Facility Managers*. 3rd ed., Wiley, 2018.

C

Camera Operator/Videographer

ABSTRACT

Camera operators record video footage for use in television, film, and internet content. They work in enclosed studios and sets as well as in outdoor locations. Students aspiring to enter the profession should pursue studies in subjects such as visual arts, dramatic arts, and broadcast media.

BACKGROUND

Camera operators record video footage for use in television and film and on the internet. Camera operators work across all realms of visual media recording, including videography and cinematography. Professional camera operators are traditionally employed by companies specializing in visual media, including news media corporations, television networks, film production companies, and cable television stations. Camera operators are responsible for capturing events across a broad spectrum of subject matter, from live events such as news and sports competitions to interviews, concerts, wildlife, and documentary and feature films.

Camera-operation professionals work closely with production teams in order to successfully capture the overall visual scope and narrative focus of an event or project. Camera operators must be able to accomplish a variety of filming techniques as specified by project directors, writers, producers, and other creative and technical staff. The numerous interactions and extensive cooperation involved in camera operations requires keen interpersonal communication savvy in concert with extensive technical skills.

Professional camera operators are traditionally broken down into two distinct disciplines: studio operators and field operators.

OVERVIEW

While the traditional work environment for camera operators is an enclosed studio or set, many work in a variety of external locations and weather conditions. Many camera operators change locations from project to project, meaning that much of their time is spent on the road. Camera-operator crews who specialize in recording live sporting events travel from stadium to stadium to cover events. News camera operators travel to places throughout the world to cover breaking news stories. Similarly, documentary and wildlife camera operators may travel great distances depending on whether their focus is on a particular climate, animal, or natural habitat.

A sense of adaptability and willingness to try new projects is paramount for camera operators, particularly those who are just starting out in the field. Acquiring experience through participation in a variety of different projects can help build the creative and technical skill sets necessary for camera operation.

Camera operators enter the field from a variety of creative and dramatic-arts arenas. Many have a foundation of study in traditional arts, including photography, perspective, colorfield exploration, and three-dimensional design. Some enter videography through an interest in drama or narrative arts, including English, theater, and music performance, while others may be drawn to camera work through exposure to reportage, broadcasting, or news journalism. A wide variety of universities offer associate, undergraduate, and postgraduate study in video production, camera operation, and cinematography.

DUTIES AND RESPONSIBILITIES

The day-to-day duties of a camera operator begin with preproduction planning with other technical and creative members of a project's production staff. The scope of production meetings can vary from project to project.

Studio telecasts often entail obtaining footage from fixed camera locations and established camera settings. Other projects, such as live event coverage, feature-film shoots, and documentary filmmaking, can require extensive preparation and a variety of input from several different team members. In concert with production managers and other set technicians, camera operators must also make sure their equipment is in proper working order and secured prior to filming.

The shooting process itself also varies from project to project. While studio operators normally set shooting schedules with more traditional working hours, field camera operators can work during all times of the day, depending on what parameters a particular project entails. If a particular shot or visual component is not captured during initial shooting, camera operators are called upon to reshoot footage.

OCCUPATION SPECIALTIES

Cinematographers. Cinematographers film motion pictures. They usually have a team of camera operators

Photo via iStock/FabioFilzi. [Used under license.]

and assistants working under them. They determine the best angles and types of cameras to capture a shot. They may use stationary cameras that shoot whatever passes in front of them or use a camera mounted on a track and move around the action. Some operators sit on cranes and follow the action. Others carry the camera on their shoulder while they move around the action. Some cinematographers specialize in filming cartoons or special effects.

Studio Camera Operators. Studio camera operators work in a broadcast studio and videotape their subjects from a fixed position. There may be one or several cameras in use at a time. Operators normally follow directions that give the order of the shots. They often have time to practice camera movements before shooting begins. If they are shooting a live event, they must be able to adjust at a moment's notice and follow the instructions of the show's director.

Videographers. Videographers film or videotape private ceremonies or special events, such as weddings. They also may work with companies and make corporate documentaries on a variety of topics. Some videographers post short videos on websites for businesses. Most videographers edit their own material. Many videographers run their own business or do freelance work. They may submit bids, write contracts, and get permission to shoot on locations that may not be open to the public. They also get copyright protection for their work and keep financial records.

WORK ENVIRONMENT

Physical Environment. Immediate work environments of camera operators vary greatly by discipline. Studio operators work primarily in closed studio locations, while field operators work in a diverse array of locations, from neighborhood street corners to crowded arenas and natural wildlife habitats.

Human Environment. Strong collaborative skills are important for any camera operator, as he or she acts as the eyes for a particular creative or informative vision. Camera operators work in concert with production staff to deliver the most pertinent video that will advance a narrative outline and inform the viewing audience.

Technological Environment. Camera operators must possess a complex set of media-technology skills, ranging from extensive knowledge of video cameras and image-recording technology to basic knowledge of audio and lighting systems. Many camera operators are also versed in video-editing and special-effects software.

EDUCATION AND TRAINING

High School/Secondary. Students can prepare for a career in camera operation at the high-school level with courses in geometry, visual arts, introductory computer programming, dramatic arts, and broadcast media. In addition to a survey of major dramatic works and theatrical history, participation in the technical aspects of scholastic theatrical productions is also highly encouraged. Many camera operators attain a basic grasp of lighting, audio rigging, and visual display through participation in school plays or on student television networks. Summer study, volunteer work, and internship programs at local cable-access outlets or television news studios can also be a tremendous asset for high-school students eager to gain experience in videography.

College/Postsecondary. Individuals aspiring to a career as a camera operator have a wide variety of postsecondary education opportunity options available at the associate, bachelor, and graduate level. Introductory and certificate-level coursework in camera operation often entails a survey of the basics of the discipline, including routine equipment maintenance, tripod setup, and command of basic camera focus settings.

Undergraduate programs in cinematography cover not only the basics of camera operation but also the historical development of visual arts and the theoretical foundations of contemporary media. Undergraduates also study location recording and computer animation.

Graduate-level study of cinematography includes an in-depth survey of the history of television and film as well as advanced coursework on film theory. Individual conception and production of one or more films is required for students at the graduate level.

Adult Job Seekers. Depending on the nature of the programs they are involved in, studio camera operators may often have schedules that involve working in the early morning or late into the evening. Field camera operators are always on the move, either covering events throughout specific regions in tandem with other traveling journalists or traveling to locations far and wide for more specialized projects.

Professional Certification and Licensure. While no specific certification is required to work as a camera operator, affiliation with professional groups or associations, such as the Society of Camera Operators, can boost credentials and present opportunities for networking. Continuing coursework on emerging technologies also helps camera operators stay abreast of contemporary trends in visual media. Cinematographers in the motion-picture industry often obtain professional certification.

Additional Requirements. Patience, cooperation, and attention to detail are qualities that tremendously benefit camera operators across all realms of work. The physical capacity to manipulate large camera equipment is also desirable. Camera operators must possess a sound technical aptitude as well as the ability to adapt to new technologies quickly.

—John Pritchard

Further Reading

Brine, Kelly Gordon. *The Art of Cinematic Storytelling: A Visual Guide to Planning Shots, Cuts, and Transitions.* Oxford UP, 2020.

Brown, Blain. *The Basics of Filmmaking: Screenwriting, Producing, Directing, Cinematography, Audio, and Editing.* Routledge, 2020.

Schenk, Sonja, and Ben Long. *The Digital Filmmaking Handbook.* 7th ed., Foreing Films, 2021.

CAMERA TECHNOLOGIES

ABSTRACT

Camera technologies are concerned with the design, development, operation, and assessment of film and digital cameras. The field includes traditional photographic imaging as well as measuring and recording instruments in the fields of science, medicine, engineering, surveillance, and cartography. Camera technologies have widespread applications in almost every area of modern life, from digital video cameras used at banks to record information about customers to long-range space voyagers transmitting images back to scientists on Earth for study. Medical applications also have been developed to help surgeons "see" inside the human body and perform minimally invasive laparoscopic surgeries. Camera technologies continue to evolve and change rapidly as they integrate smaller digital technology for use in medicine, entertainment, industry, and science.

BACKGROUND

The term "camera" comes from *camera obscura*, an ancient mechanism for projecting images. Camera obscura was a pinhole device used to produce temporary images on flat surfaces in dark rooms. Camera obscura was improved during the sixteenth and seventeenth centuries, but it was not until 1727, when German physician Johann Heinrich Schulze accidentally created the first photosensitive compound, that the next leap forward in camera technologies occurred.

The early nineteenth century saw the application of Schulze's discovery to create development processes that translated temporarily viewed images to permanent images. By the mid-nineteenth century, direct positive images could be permanently affixed to glass (ambrotypes) or metal (tintypes or ferrotypes). In 1861, another development in camera technologies occurred when Scottish physicist James Clerk Maxwell demonstrated the first color photographs using a system of filters and slide projection called the "color separation method." During this same era, the application of photos to record current events and new areas of the country occurred as photographers captured images taken during the American Civil War and of the West.

The next innovation, by English physician Richard Leach Maddox in 1871, paved the way for a revolution in camera technologies with the development of the dry plate process—using an emulsion of gelatin and silver bromide on a glass plate. In the late nineteenth century, companies such as Kodak created commercially available cameras that used film to capture pictures. In 1907, the first commercial color film became available, and by 1914, the standard modern 24-by-36-millimeter (mm) frame and sprocketed 35 mm movie film were developed. In 1932, the idea of Technicolor for movies arose. In the Technicolor process, three black-and-white negatives were made in the same camera under different filters to create a colorful finished product. The rest of the twentieth century saw continued technical improvements and optimization of the quality and abilities of cameras and films, including development of multilayer color film, "instant" Polaroid film, underwater cameras, autofocus cameras, and the automatic diaphragm as well as introduction of the single-lens reflex (SLR) camera. The SLR camera changed photography by using a semiautomatic moving mirror system that permitted the photographer to see the exact image that would be captured by the film or digital imaging system.

In 1972, Texas Instruments was granted the patent for an all-electric camera. In 1982, camera technologies moved forward again with the introduction of Sony's still video camera. The ability to manipulate and transform images was then revolutionized with the 1990 release of Adobe Photoshop, which allowed nonprofessionals to transform photo images from their home computers. Also in the early 1990s, Kodak brought digital cameras to the general population by developing the photo compact disc (CD) system. The camera phone was introduced in 2000 in Japan. By 2003, more affordable digital cameras were made available and quickly became the dominant image-capturing format, so much so that Kodak ceased production of film cameras in 2004, with Nikon following suit with many of its film cameras as well, in 2006. With the advent of smartphones and tablets, digital photography took on an even greater emphasis in society as such technology enabled users to capture, edit, and share photographs directly from their personal mobile devices.

OVERVIEW

Camera technologies revolve around the science of capturing images. The field includes the design, development, operation, and assessment of both film and digital camera technologies. Film cameras use traditional photosensitive films to capture images by exposing the films to the right amount of light for the right length of time. Digital and video cameras use an array of photosensitive electronic devices or image sensors to capture images electronically and store them as a series of digital data.

Whether the camera is film, digital, or video, it typically consists of five basic parts: the body, the shutter and shutter-release button, the viewfinder, the camera lens complex, and the film or device that captures the image. The camera's body or casing is made of metal or high-grade plastic that holds the camera's other parts together and provides protection for the camera parts. The shutter captures the image, and the shutter-release button tells the camera to take the picture. The viewfinder is the window that shows the object or objects that compose the image that will be captured by the camera. The camera lens complex is made of several smaller pieces.

Camera lens complexes control and create different image effects, such as the ability to zoom, focus, and distortion correction, which combine to control the appearance of the image being captured. The optical lens is the curved piece of glass located on the outside of the lenses that focuses light into the camera and onto the film or digital sensor. The aperture is controlled by an aperture ring, which controls the size of the opening and determines how much light goes into the lens. The aperture size is measured by an f-stop number; the smaller the f-stop, the more light is let in. The final component of the camera is the photographic medium used to capture and store the image, film or digital.

HOW IT WORKS

The core element in camera technology is the capture of a desired image. The mechanism of capture varies depending on the type of camera used: film or digital. Film cameras capture images on a sheet of plastic coated with light-sensitive pigments (silver halide salts bonded by gelatin). The pigments contain variable crystal sizes that determine the sensitivity, contrast, and resolution of the film. When the film is sufficiently exposed to light or other electromagnetic radiation, it forms an image. The film is then developed using specialized chemical processes to create a visible image. Digital cameras use a special sensor or pixelated metal oxide semiconductors (photodiodes) made from silicon to convert the light that falls onto them into electrons in order to capture a desired image and store it in a memory device. The two most common types of sensors used in digital cameras are the charge-coupled device (CCD) and the complementary metal-oxide semiconductor (CMOS). A film recorded with a digital camcorder is captured in a similar way—saved as a series of frames rather than a single snapshot. In both traditional film and digital cameras, the image itself can be modified using different filters and lenses.

APPLICATIONS AND PRODUCTS

Astronomy and Physics. Camera technologies have expanded the ability of astronomers to explore the universe far beyond manned space exploration or early telescopes. High-powered telescopes and cameras placed in space have provided photographic evidence of rare astronomical events and features that have resulted in revisions of scientific theory. For example, photographs from the Hubble Space Telescope led to the discovery of dark energy, the hypothetical and unexplained force that seems to be drawing galaxies away from each other. Other unique types of cameras are powerful tools used to gain spectacular insights into universe phenomena such as star formation.

Medicine. Use of camera technologies has revolutionized medicine. Physicians can use cameras to evaluate and treat distant patients as far away as the astronauts on the space station. Closer to home, medical professionals can use tiny cameras and image-capturing devices in pill form to evaluate the inner workings of their patients' intestines. Many surgeries can be performed laparoscopically, where surgeons use tiny cameras on medical instruments that project images of their patients' bodies. The surgeons then use images and controls attached to consoles to guide their surgical implements within the patients' bodies precisely. A major benefit of surgeries using camera technologies is that they are less invasive—smaller incisions need to be made because manipulators using cameras can be extremely narrow. Surgeries can also be performed remotely using imaging-incorporated technology such as the da Vinci Surgical System to conduct operations such as prostatectomy, cardiac surgery, bariatric surgery, and various forms of neurosurgery.

Journalism. The adage that "a picture is worth a thousand words" summarizes the incredible impact of images in conveying current events, news, and history. Photojournalists using camera technologies add depth to news stories that cannot always be conveyed through words alone. The increased prevalence and decreased cost of digital cameras, particularly those found in smartphones, further facilitated a rise in citizen journalism during the first decades of the twenty-first century, enabling members of the public to capture powerful images of natural disasters, wars, protests, and other significant events.

Cartography. The field of cartography has been revolutionized through the use of camera technologies. The ability to take images of the geologic features of the earth from the ground or from space satellites has increased the accuracy and speed of mapmaking. In addition, companies such as Google have traveled the world to take street-level images of addresses that are then incorporated into maps and directions.

Engineering. Modern camera development has expanded the capabilities of engineering. Using computer technology and images, engineers can create models to predict a variety of outcomes from the impact of a head-on car collision to the minute changes in an electrical circuit. Modern images and specialized image-analysis systems can be used to enhance an observer's ability to make measurements from a large or complex set of images by improving accuracy, objectivity, or processing speed.

Entertainment. The development of camera technologies over time has provided entertainment options from the early moving pictures to specialized gaming systems. The use of advances in cameras can provide a more realistic or personalized experience. For example, some exercise- or dance-based video games incorporate motion-tracking cameras into their controls, while others, known as augmented reality (AR) games, combine camera technology and a variety of virtual elements to create an immersive gaming environment.

Surveillance and Security. The ability to use camera technologies to track and identify the movements of individuals or intruders often relies heavily on images. Photographic evidence can be used in court as proof of wrongdoing.

Military Operations. Satellite imaging using high-powered cameras has transformed war and peacetime military operations, as the images can provide real-time information on the movement of individuals, troops, and resources. Remotely piloted drone airplanes with cameras allow the military to explore locations that may be dangerous for humans to enter and provide needed information for military operations. The development of image-based technologies such as thermal and infrared cameras to detect changes in geologic features and topography are vital for planning military missions.

Exploration and Rescue. Sending robots and other unmanned machines with camera technologies to places too remote or dangerous for human beings to work in—outer space, great ocean depths, and disaster zones—is an important application. The space exploration robots *Voyager 1* and *2* have been traveling through space since 1977, sending back images of distant planets, their moons, and Earth itself. Images and cameras were also invaluable during the 2010 BP Deepwater Horizon oil spill, as underwater explorer robots equipped with cameras were able to send images of the damaged oil well as well as reports on how repairs were proceeding. Thermal-imaging cameras can also identify regions on fire even as they allow rescue robots to seek out injured humans trapped in fires or under rubble.

CAREERS AND COURSEWORK

There are many careers in a variety of industries that directly correlate to camera technologies, and entry-level requirements vary significantly by position. Given the wide spectrum of difference between the careers, a sampling of careers and coursework follows.

The motion picture and video industries provide career options such as cinematographers, camera operators, and gaffers, who work together to capture the scripted scenes on film and perform the actual shooting. Formal training is an asset to workers in film and television production, but experience, talent, creativity, and professionalism are likewise important factors in getting a job. In addition to colleges and technical schools, many independent centers, such as the American Film Institute (AFI), offer training programs on various aspects of filmmaking, such as screenwriting, editing, directing, and acting.

Another camera-technologies career option is photographer. Photographers produce and preserve images that paint a picture, tell a story, or record an event. They often specialize in areas such as portrait, commercial and industrial, scientific, news, or fine arts photography. Employers usually seek applicants with a "good eye," imagination, and creativity, as well as a good technical understanding of photography. Photojournalists or industrial or scientific photographers generally need a college degree. Freelance and portrait photographers need technical proficiency, gained through a degree, training program, or experience. Photography courses are offered by many universities, community colleges, and vocational and technical institutes. Basic courses in photography cover equipment, processes, and techniques. Learning good business and marketing skills is important, and some bachelor's degree programs offer courses that focus on those. Art schools offer useful training in photographic design and composition. A good way for a photographer to start out is to work as an assistant to an experienced photographer.

Still another option is a mapmaker or geographer who works with cameras and images to create maps. Photogrammetrists interpret the more detailed data from aircraft to produce maps. Most mapmakers work for architectural and engineering services companies, governments, and consulting firms. Remote-sensing specialists and photogrammetrists often have at least a bachelor's degree in geography or a related subject, such as surveying or civil engineering. Classes in statistics, geometry, and matrix algebra also are useful. Many remote-sensing specialists have degrees in the natural sciences, including forestry, biology, and geology. They often take courses in remote sensing or mapping while earning these degrees. Not everyone working in this field has a bachelor's degree, however. People who have an associate's degree or a certificate in remote sensing or photogrammetry usually begin as assistants and gain additional skills on the job. Taking high school or college-level classes in mapping, drafting, and science can also lead to assistant jobs. Some employers hire entry-level workers who do not have college training but do have an aptitude for math and visualizing in three dimensions.

SOCIAL CONTEXT AND FUTURE PROSPECTS

The ability to record and view the world will continue to change as new camera technologies allow individuals and companies to explore further. Future technological innovations in image capturing, size of cameras, storage of images, and viewing of images will change as the industry progresses.

Since the early days of camera technology, cameras have been used to bring the reality of different lives and social issues to the general population through books, exhibits, and newspapers. Early examples of this include Jacob Riis's 1890 publication *How the Other Half Lives*, which presented images of tenement life in New York City, and Lewis Hine's commission by the United States National Child Labor Committee to photograph children working in mills in 1909. A later example is the interest in the plight of Afghan women, which was stimulated in 1985 by photojournalist Steve McCurry's haunting photo of a young Afghan refugee named Sharbat Gula that appeared on the cover of *National Geographic*. In the first decades of the twenty-first century, camera technology helped to call public attention to income inequality, police brutality, and the hardships facing refugees, among other issues of international concern.

—*Dawn A. Laney*

Further Reading

Brine, Kelly Gordon. *The Art of Cinematic Storytelling: A Visual Guide to Planning Shots, Cuts, and Transitions*. Oxford UP, 2020.

Choi, Paul J., Rod J. Oskouian, and R. Shane Tubbs. "Telesurgery: Past, Present, and Future." *Cureus*, vol. 10, no. 5, 2018, www.cureus.com/articles/12751-telesurgery-past-present-and-future. Accessed 13 July 2023.

Gómez Cruz, Edgar, and Asko Lehmuskallio, editors. *Digital Photography and Everyday Life: Empirical Studies on Material Visual Practices*. Routledge, 2016.

Forbes, Ted. *Visually Speaking: Mastering Photography as a Visual Language*. Cider Mill Press, 2023.

Jenkinson, Mark. *Photography Careers: Finding Your True Path*. Focal Press, 2016.

Langford, Michael, et al. *Langford's Basic Photography*. 10th ed., Focal Press, 2015.

Long, Ben. *Complete Digital Photography*. CDP Press, 2018.

Pritchard, Michael. *A History of Photography in 50 Cameras*. Bloomsbury Visual Arts, 2014.

ChatGPT

ABSTRACT

ChatGPT is a chatbot powered by artificial intelligence (AI) technology. Created by the San Francisco, California-based technology company OpenAI, ChatGPT made global headlines following its November 2022 public release. ChatGPT's launch was widely characterized as an enormous success for AI technology in general and OpenAI in particular. The chatbot had attracted a reported 100 million active users by January 2023, approximately two months after its release, leading international media sources to describe it as the fastest-growing consumer software application in the history of computing.

Users interface with ChatGPT through a prompt-based format that mimics human dialogue. ChatGPT uses generative AI, which draws on training data to create original content from user prompts with high levels of realism. Users can add additional parameters to their input prompts, which allows ChatGPT to make corrections, revisions, and other alterations to its output.

BACKGROUND

AI technology aims to mimic the human mind's capacity to apply existing knowledge, integrate new knowledge, draw from situational observations, and learn from mistakes dynamically and in real time. In broad terms, AI works by combining large repositories of data with rapid, real-time forms of iterative processing and advanced computing algorithms. These features allow AI-powered systems to learn from established and emerging patterns and the characteristics found in their data sets. In computing, iterative processing describes updates to algorithm parameters, which functionally result in gradual performance improvements. Algorithms are complex computer codes, heavily modeled with mathematics, that deliver instructions that tell a computer or software application how to perform tasks or solve specific problems.

ChatGPT is a chatbot, which is a specific type of software designed to simulate interactive conversations with human users. The first chatbot in computing history, named ELIZA, was created at the Massachusetts Institute of Technology (MIT) in 1966 by the German American computer scientist Joseph Weizenbaum (1923-2008). ELIZA used techniques including substitution and pattern matching. Though simple by contemporary standards, ELIZA was novel and highly advanced for its time and was reportedly capable of convincing some users that they were conversing with another person.

In 1991, the Singaporean technology company Creative Labs released a program known as Doctor Sbaitso, which was compatible with computers running Microsoft's MS-DOS operating system. Doctor Sbaitso "conversed" with users in a format often likened to a psychologist talking to a patient and was the first chatbot to incorporate early AI technologies.

ALICE, also known as Alicebot, appeared in 1995 and was modeled on Weizenbaum's ELIZA program. An acronym for "artificial linguistic internet computer entity," ALICE used natural language processing and engaged in real-time conversations with users. ActiveBuddy's Smarterchild chatbot, released in 2001, had similar capabilities and became very popular in the early years of the commercial internet. According to *Technology* magazine, ActiveBuddy amassed more than 30 million followers on various instant messaging (IM) platforms that were popular at the time.

In 2010, Apple released its Siri digital personal assistant, marking a major advancement in chatbot technology. Siri drew on a natural language user interface (UI) when interacting with users, creating a more lifelike and realistic experience. Siri's natural language UI became the prototype for virtually all the successful chatbots that followed it, including ChatGPT.

OpenAI began developing ChatGPT in 2018, after the company released its initial generative pre-trained transformer (GPT) neural network machine learning model. ChatGPT initially used a version of the technology known as GPT-3.5, which OpenAI combined with AI-assisted natural language processing (NLP). NLP combines multiple deep machine-learning models with principles of computational linguistics to understand and interpret text-based prompts in ways that simulate human language processing. ChatGPT became the first chatbot in computing history to integrate GPT and NLP.

OVERVIEW

OpenAI released ChatGPT on November 30, 2022. The chatbot quickly became an international sensation for its advanced generative AI capabilities. At that time, ChatGPT created text-based output with high degrees of realism. Users could interface with ChatGPT to brainstorm ideas, learn about complex topics, get personalized product and service recommendations, explore emerging scientific research, translate written text into ninety-five supported languages, write and debug computer code, and generate customized written content, among many other applications. The platform's beta version was originally available free of charge, and OpenAI continued to make a basic free version of the platform available to users throughout the months that followed. In February 2023, OpenAI announced a paid subscription plan offering users unlimited access to ChatGPT and a wider range of features and services.

ChatGPT logo. Image via Wikimedia Commons. [Public domain.]

Despite drawing rave reviews from both technology insiders and users, ChatGPT also quickly ignited multiple controversies. While the platform displayed the capability to answer complex fact-based questions with high degrees of accuracy, it also gave some users error-filled answers, especially when prompted to draw inferences, explore nuances, analyze opinions, or make predictions. Educators around the world also expressed concern about ChatGPT's place in the classroom, as students almost immediately began using the software to write assignments and generate answers to homework questions. In January 2023, news agencies reported that ChatGPT passed a Master of Business Administration (MBA) examination issued by a professor at the prestigious Wharton School of the University of Pennsylvania. The incident highlighted that ChatGPT's platform had significant potential for educational misuse, which could have far-reaching consequences if students used it to gain credentials that they would not have been able to earn on their own merit. To address related concerns, education officials in New York City issued a blanket ban on ChatGPT use in school assignments shortly after the platform's initial release.

Another controversy centered on apparent biases in ChatGPT's responses to politicized questions, which according to some observers displayed a clear tendency toward anticonservative viewpoints. Screenshots of ChatGPT conversations that were shared on Twitter in February 2023 ignited the bias charges. The screenshots documented ChatGPT's contradictory and inconsistent responses to two similar prompts, one of which asked the software to write a positive poem about former US president Donald Trump (1946-) while another asked the system to do the same about President Joe Biden (1942-). In response to the Trump prompt, ChatGPT indicated that it could not create "partisan, biased, or political content," but when asked to write a positive poem about Biden, ChatGPT immediately produced a patriotic poem extolling the president's virtues. OpenAI's senior leadership publicly acknowledged the incident, characterizing it as a programming misstep and pledging to correct it in future releases.

Generative AI and ChatGPT also appeared poised to revolutionize how internet user content is created. Web developers and content creators quickly began using the platform as a research and writing aid for website copy, with some observers expressing con-

A 2023 Time *cover: "The AI Arms Race Is Changing Everything." Photo via Wikimedia Commons. [Public domain.]*

cern about the technology's potential to displace human workers while also reducing the usability and reliability of internet content. Technology insiders speculate that major search engines would likely add tools capable of detecting AI-generated content and promote human-written content when returning search results to users.

In March 2023, OpenAI released a new version of its generative pre-trained transformer learning model known as GPT-4. The GPT-4 model, which tech insiders described as being more capable of understanding nuanced and complex user inputs, was intended to power the next set of ChatGPT releases. GPT-4 could also accept inputs in both text- and image-based formats, which would allow the chatbot to interpret data from charts, graphs, and screenshots taken from scientific journals and academic papers. Reviews also indicated that GPT-4 made far fewer reasoning errors and factual mistakes, known in AI as "hallucinations," than the GPT-3.5 model used in the initial ChatGPT release. OpenAI quickly integrated GPT-4 into available versions of ChatGPT tool.

By January 2023, OpenAI had raised more than $11 billion across six rounds of venture capital funding, according to data published by *Crunchbase*. That month, the *Wall Street Journal* reported that OpenAI was involved in negotiations to complete a tender offering of shares that carried a company valuation of $29 billion. The sum would make OpenAI one of the most valuable technology startups in the United States, even though the company was earning relatively little income from revenues. However, the company planned to monetize ChatGPT through subscription services, with senior company leadership predicting that OpenAI would be earning $1 billion per year in revenues by 2024.

—*Jim Greene*

Further Reading

Dastin, Jeffrey, Krystal Hu, and Paresh Dave. "Exclusive: ChatGPT Owner OpenAI Projects $1 Billion Revenue by 2024." *Reuters*, 15 Dec. 2022, www.reuters.com/business/chatgpt-owner-openai-projects-1-billion-revenue-by-2024-sources-2022-12-15. Accessed 18 July 2023.

Hassenfeld, Noam. "Even the Scientists Who Build AI Can't Tell You How It Works." *Vox*, 15 July 2023, www.vox.com/unexplainable/2023/7/15/23793840/chat-gpt-ai-science-mystery-unexplainable-podcast. Accessed 18 July 2023.

Hughes, Alex. "ChatGPT: Everything You Need to Know about OpenAI's GPT-4 Tool." *BBC Science Focus*, 16 Mar. 2023, www.sciencefocus.com/future-technology/gpt-3. Accessed 18 July 2023.

Law, Marcus. "From ELIZA to ChatGPT: The Evolution of Chatbots Technology." *Technology*, 7 Dec. 2022, technologymagazine.com/articles/from-eliza-to-chatgpt-the-evolution-of-chatbots-technology. Accessed 18 July 2023.

Marr, Bernard. "The Best Examples of What You Can Do with ChatGPT." *Forbes*, 1 Mar. 2023, www.forbes.com/sites/bernardmarr/2023/03/01/the-best-examples-of-what-you-can-do-with-chatgpt. Accessed 18 July 2023.

Mayer, Grace. "OpenAI Cofounder Responds to Elon Musk's Criticism that ChatGPT Is Too 'Woke:' 'We Made a Mistake.'" *Business Insider*, 9 Mar. 2023, www.businessinsider.com/openai-cofounder-greg-brockman-responds-elon-musk-woke-criticism-chatgpt-2023-3?IR=T. Accessed 18 July 2023.

"OpenAI Funding." *Crunchbase*, 2023, www.crunchbase.com/organization/openai/company_financials. Accessed 18 July 2023.

Ortiz, Sabrina. "What Is GPT-4? Here's Everything You Need to Know." *ZDNet*, 15 Mar. 2023, www.zdnet.com/article/what-is-gpt-4-heres-everything-you-need-to-know. Accessed 18 July 2023.

Oxborrow, Ian. "What Is ChatGPT and Why Is It Controversial?" *National*, 24 Jan. 2023, www.thenationalnews.com/business/technology/2023/01/25/chatgpt-what-why-controversial. Accessed 18 July 2023.

Pocock, Kevin. "What Is ChatGPT? What Is It Used For?" *PC Guide*, 15 Mar. 2023, www.pcguide.com/apps/what-is-chat-gpt. Accessed 18 July 2023.

CINEMATOGRAPHER

ABSTRACT

Cinematographers guide the photographic style and look of a motion picture to ensure that it reflects the director's vision for the project. They work both on film sets and in production studios. Students aspiring to enter the profession should pursue studies in subjects such as film and photography.

BACKGROUND

Cinematographers, also known as directors of photography, define and help guide the photographic style or look of a motion picture. Cinematographers ensure that the director's vision for the film, such as its mood and appearance, is achieved. They receive guidance from directors on how photographic shots should be created and work with other set personnel to design and frame shots appropriately. Cinematographers have a strong knowledge of lighting, special effects, and other important pieces of filmmaking technology that are being used on the set. Many cinematographers are also specialized and only focus on areas such as special effects or location shots.

OVERVIEW

Cinematographers work on the movie set, directing the cameras in such a way that the best shot is framed and taken. Such sets are busy and complex, with different groups working together to render a scene. This work environment is often tense, particularly in light of budget concerns and production deadlines. Cinematographers also work in studio offices and production studios, where they coordinate with writers, directors, producers, and other key artistic and technical professionals in the filmmaking process. Cinematographers generally work long and erratic hours. Their work hours may vary based on the production deadlines and the amount of film direction with which they are charged. The work itself can be draining both physically and psychologically, particularly as it may call for multiple shots, angles, and camera mountings in order to achieve the best take.

Cinematographers are critical components of the filmmaking field. Cinematographers work closely with film directors and producers to make their artistic dreams a reality. They are also senior-level managers on the set and must be effective communicators, as they direct camera operators and many other production personnel to create the ideal shot. Cinematographers are exceptional students of film, having studied a wide range of past and present techniques and even developing innovative new approaches to filmmaking.

DUTIES AND RESPONSIBILITIES

As the head of a film's camera department, the cinematographer coordinates with the director and producer to determine the best action and blocking (the placement of actors and scene material) for the film. Based on the director's "shot list," the cinematographer determines how the cameras should be positioned, the type of lenses and filters to be used, and how the scene should be lit. After the scene is shot, the cinematographer ensures that the film is processed in accordance with the director's wishes.

In addition, the cinematographer acts as a type of set manager. The cinematographer or director of photography must coordinate the activities of gaffers (set electricians), lighting and audio equipment handlers, and camera operators. This managerial work is critical for ensuring that all film crew members operate according to the director's and/or producer's desires. Furthermore, the cinematographer is frequently at the creative heart of the production. He or she works with scriptwriters, set and costume designers, and even actors to ensure that shots are made according to specifications.

WORK ENVIRONMENT

Physical Environment. Cinematographers and directors of photography work primarily at movie studios and sets. These are complex locations with a wide range of working parts, departments, and individuals. Sets are often in large, enclosed, and ventilated studios and lots, or on location throughout the country and world. Depending on the set and the film needs, a cinematographer may work outdoors in a variety of weather conditions.

Human Environment. Cinematographers are senior-level managers, directing the actions of camera operators and equipment operators on a movie set. They also coordinate directly with other important figures on the set and in the studio, including directors, producers, set and costume designers, special effects crews, screenwriters, and actors.

Technological Environment. Cinematographers interact with many pieces of technical equipment while directing photography on the set. In addition to various cameras, lenses, and filters, they must work with lighting equipment and other set technologies. Off the set, they work with graphics software and related programs and systems. Cinematographers also must keep up with the changing filmmaking technology available on the market.

EDUCATION AND TRAINING

High School/Secondary. High school students are encouraged to take classes in photography, film, drama, and art. They must also study communications, computer science, and graphics. Interested high school students should also get involved in school audiovisual departments and clubs.

College/Postsecondary. Most cinematographers have postsecondary degrees from colleges or film schools. Many colleges offer bachelor's degrees in film studies and in fine arts, while a number of vocational and technical schools offer associate's degrees in specialized fields related to filmmaking. A large number of independent institutions, such as the American Film Institute (AFI), offer similar specialized training in cinematography. The majority of the most popular programs are located in cities with thriving film and broadcast industries, such as Los Angeles and New York.

Adult Job Seekers. Cinematographers attain their high-level jobs after gaining considerable experience in the film industry. Qualified adults who seek to become cinematographers should therefore be ready to work as a lighting specialist, camera operator, or similar role. This work experience helps build an aspiring cinematographer's qualifications and set-management skills. Cinematographers may also find opportunities by joining and networking through professional associations, such as the American Society of Cinematographers or the International Cinematographers Guild.

Professional Certification and Licensure. There is no licensure requirement for cinematographers. However, many individuals seek additional training and certification in cinematography from accredited universities. Such programs give job candidates highly valuable training that may enhance their job appeal.

Additional Requirements. Cinematographers should have strong artistic vision and capabilities. They must also demonstrate attention to detail and composition. As shooting a scene often requires multiple takes, angles, and camera mounts, a cinematographer should have patience and persistence. As set managers, they should be comfortable working with and directing people on the set. Finally, they must be able to meet the demands of producers and directors.

—*Michael Auerbach*

Further Reading

Brine, Kelly Gordon. *The Art of Cinematic Storytelling: A Visual Guide to Planning Shots, Cuts, and Transitions*. Oxford UP, 2020.

Brown, Blain. *Cinematography: Theory and Practice*. 4th ed., Routledge, 2022.

Hoser, Tania. *Introduction to Cinematography: Learning through Practice*. Routledge, 2018.

CLICKBAIT

ABSTRACT

The term "clickbait" refers to online content that features a headline designed to arouse curiosity, stir outrage, or otherwise encourage the reader to click on a link to a story. Forms of clickbait include curiosity bait, sharebait, and trickbait. Clickbait is typically regarded negatively, and websites known for using clickbait-style headlines often gain poor reputations.

BACKGROUND

In online journalism, clickbait refers to headlines on online content that entice readers to click on a link to a story. Online news and commentary websites rely on clickbait to drive up the number of clicks to their articles, thereby increasing traffic to their sites and attracting advertisers. Clickbait headlines are often provocative, written to arouse curiosity or stir outrage. The headlines attempt to lure readers through forward referencing, suspenseful wording, and the use of images. The term has taken on a negative connotation. Readers may feel manipulated after clicking to stories that overdramatize events, mischaracterize content, or advertise products.

Websites use clickbait to promote content that readers can share across social media platforms, generating instant clicks. The content may include news articles, think pieces, quizzes, or viral videos.

OVERVIEW

Clickbait exists in many different forms. Clickbait content that provides entertainment and is designed to be distributed on social media platforms is sometimes called "sharebait." Examples of sharebait include the many quizzes offered by the website *Buzzfeed*.

Other kinds of clickbait appeal to people's inquisitive nature; they grab readers' attention by using forward referencing, which includes phrases such as, "You won't believe what happens next." During the 2010s, the content-curating website *Upworthy* gained notice by using these types of headlines, which sometimes mixed in pronouns to add a mysterious slant. An example is the headline of the site's 2013 article "Calling Them 'Girls' Was Their First Mistake. What Happened Next Is a Great Moment in Fierceness." The practice, sometimes known as "curiosity bait," quickly spread to other sites. Curiosity bait can let down readers if "what happens next" in the story fails to surprise them or sensationalizes the news.

A form of clickbait called "trickbait" is used by websites that do not produce their own content to encourage people to click on their links. Trickbait often appears as a grid of stories, often called "sponsored content" or "promoted stories," at the bottom of legitimate news websites. The strategy is called trickbait because the sponsored stories can be mistaken for those of the news site. "Visual trickbait" refers to a story that appears in the "sponsored" grid accompanied by a photo that has no connection whatsoever to the article. The image misleads readers into clicking on the link.

As clickbait became increasingly common, readers and some media organizations criticized the practice, contributing to a growing backlash. Readers particularly expressed frustration when articles did

Fictional examples of "chumbox" style adverts, employing common clickbait tactics of using an information gap to encourage reader curiosity, and promising easy-to-read numbered lists. Photo by Lord Belbury, via Wikimedia Commons.

not deliver the content they promised or used clickbait headlines that were in poor taste. In response, Facebook announced in 2014 that it would crack down on stories fronted by clickbait headlines, while in 2015, Upworthy announced that it would stop aggregating clickbait content and instead produce original stories. Despite the widespread backlash against clickbait, the use of clickbait headlines and image thumbnails remained common on some websites, including video-sharing sites such as YouTube, well into the 2020s.

—*Rebecca Kivak*

Further Reading

Chowdhry, Amit. "Facebook Is Going to Suppress 'Click-Bait' Articles." *Forbes*, 26 Aug. 2014, www.forbes.com/sites/amitchowdhry/2014/08/26/facebook-is-going-to-suppress-click-bait-articles/#51f7bf8b47b0. Accessed 13 July 2023.

Closs, Wyatt. "Calling Them 'Girls' Was Their First Mistake: What Happened Next Is a Great Moment in Fierceness." *Upworthy*, 11 July 2013, www.upworthy.com/calling-them-girls-was-their-first-mistake-what-happens-next-is-a-great-moment-in-fierceness. Accessed 13 July 2023.

Council of Europe. *Journalism at Risk: Threats, Challenges and Perspectives*. Council of Europe, 2015.

Douglas, Nick. "Get Rid of YouTube's Terrible Clickbait Thumbnail Images." *Lifehacker*, 5 Aug. 2019, lifehacker.com/get-rid-of-youtubes-terrible-clickbait-thumbnail-images-1836969236. Accessed 13 July 2023.

Frampton, Ben. "Clickbait: The Changing Face of Online Journalism." *BBC News*, 14 Sept. 2015, www.bbc.com/news/uk-wales-34213693. Accessed 13 July 2023.

Hollows, Joanne. *Media Studies: A Complete Introduction*. Hodder & Stoughton, 2016.

Iglesias, Beatriz, and Josep Valor. "The Good and Bad Sides of Clickbait." *IESE*, 1 Mar. 2023, blog.iese.edu/the-media-industry/2023/03/01/the-good-and-bad-sides-of-clickbait. Accessed 13 July 2023.

Rubin, Victoria L. "Deception Detection and Rumor Debunking for Social Media." *The SAGE Handbook of Social Media Research Methods*, edited by Luke Sloan and Anabel Quan-Haase, Sage Publications, 2017, pp. 342-64.

Computer-Aided Design and Manufacturing

ABSTRACT

Computer-aided design and manufacturing is a method by which graphic artists, architects, and engineers design models for new products or structures within a computer's "virtual space." It is variously referred to by a host of different terms related to specific fields: computer-aided drafting (CAD), computer-aided manufacturing (CAM), computer-aided design and drafting (CADD), computer-aided drafting/computer-aided manufacturing (CAD/CAM), and computer-aided drafting/numerical control (CAD/NC). The generic term "computer-aided technologies" is, perhaps, the most useful in describing the computer technology as a whole. Used by professionals in many different disciplines, computer-aided technology allows designers in traditional fields such as architecture to render and edit two-dimensional (2D) and three-dimensional (3D) structures. Computer-aided technology has also revolutionized once highly specialized fields, such as video-game design and digital animation, into broad-spectrum fields that have applications ranging from the film industry to ergonomics.

BACKGROUND

Computer-aided design and manufacturing is the combination of two different computer-aided technologies: computer-aided design (CAD) and computer-aided manufacturing (CAM). Although the two are related by their mutual reliance on task-specific computer software and hardware, they differ in how involved an engineer or architect must be in the creative process. A CAD program is a three-dimensional (3D) modeling software package that enables a user to create and modify architectural or engineering diagrams on a computer. CAD software allows a designer to edit a project proposal or produce a model of the product in the virtual "space" of the computer screen. CAD simply adds a technological layer to what is still, essentially, a user-directed activity. CAM, on the other hand, is a related computer-aided technology that connects a CAD system to laboratory or machine-shop tools. There are many industrial applications for computer-directed manufacturing. One example of CAD/CAM might be when an automotive designer creates a 3D model of a proposed design using a CAD program, verifies the physical details of the design in a computer-aided engineering (CAE) program, and then constructs a physical model via CAD or CAE interface with CAM hardware (using industrial lasers to burn a particular design to exact specifications out of a block of paper). In this case, the automotive designer is highly involved with the CAD program but allows the CAD or CAE software to direct the activity of the CAM hardware.

CAD/CAM is, essentially, a modernization of the age-old process of developing an idea for a structure or a product by drawing sketches and sculpting models. Intrigued by the idea that machines could assist designers in the production of mathematically exact diagrams, Ivan Sutherland, a doctoral student at the Massachusetts Institute of Technology, proposed the development of a graphical user interface (GUI) in his 1963 dissertation, "Sketchpad: A Man-Machine Graphical Communication System." GUIs allow users to direct computer actions via manipulation of images rather than by text-based commands. They sometimes require the use of an input device, such as a mouse or a light pen, to interact with the GUI images, also known as icons. Although it was never commercially developed, Sketchpad was innovative in that a user could create digital lines on a screen and determine the constraints and parameters of the lines via buttons located along the side of the screen. In subsequent decades, the CAD software package employed the use of a mouse, light pen, or touch screen to input line and texture data through a GUI, which was then manipulated by computer commands into a virtual, or "soft," model. In the twenty-first century, CAM is more or less the mechanical-output aspect of CAD work, but in the early days, CAM was developed independently to speed up industrial production on assembly lines. Patrick J. Hanratty, regarded as the father of CAD/CAM, developed the first industrial CAM using numerical control (NC) in 1957. Numerical control—the automation of machining equipment by the use of numerical data stored on tape or punched on cards—had been in existence since the 1940s, as it was cheaper to program controls into an assembly machine than to employ an operator. Hanratty's PRONTO, a computer language intended to replace the punched-tape system (similar to punch cards in that the automaton motors reacted to the position of holes punched into a strip of tape) with analogue and digital commands, was the first commercial attempt to modernize numerical control and improve tolerances (limiting the range of variation between the specified dimensions of multiple machined objects).

OVERVIEW

Traditionally, product and structural designers created their designs on a drawing board with pencils, compasses, and T squares. The entire design process was a labor-intensive art that demanded not just the ability to visualize a proposed product or structure in exact detail but also extensive artistic abilities to render what, up until that point, had been only imagined. Changes to a design, whether due to a client's specific needs or the limitations of the materials needed for construction, frequently required the designer to "go back to the drawing board" and create an entirely new series of sketches. CAD/CAM was intended to speed up the design and model-making processes dramatically.

Computer-Aided Design (CAD). CAD programs function in the same way that many other types of computer software function: One loads the software into a computer containing an appropriate amount of memory, appropriate graphics card, and appropriate operating system to handle the mathematical calculations required by the 3D modeling process. There are quite literally hundreds of different CAD programs available for different applications. Some programs are primarily two-dimensional (2D) drawing applications, while others allow 3D drawing, shading, and rendering. Some are intended for use in construction management, incorporating project schedules, or for engineering, focused on structural design. There are a range of CAD programs of varying complexity for computer systems of different capabilities and operating systems.

Computer-Aided Manufacturing (CAM). CAM, on the other hand, tends to be a series of automated mechanical devices that interface with a CAD program installed in a computer system. The typical progress of a CAD/CAM system from design to soft model might involve an automotive designer using a light pen on a CAD system to sketch out the basic outlines of suggested improvements for an existing vehicle. Once the car's structure is defined, the designer can then command the CAD program to overlay a surface, or "skin," on top of the skeletal frame. Surface rendering is a handy tool within many CAD programs to play with the use of color or texture in a design. When the designer's conceptual drawings are complete, they are typically loaded into a computer-aided engineering program to ascertain the structural integrity and adherence to the requirements of the materials from which the product will ultimately be constructed. If the product design is successful, it is then transferred to a CAM program, where a product model, machined out of soft model board, can be constructed to demonstrate the product's features to the client more explicitly. Advertisements for CAD programs tend to emphasize how the software will enhance a designer's imagination and creativity. They stress the many different artistic tools that a designer can easily manipulate to create a virtual product that is as close as possible to the idealized conceptual design that previously existed only in the designer's imagination. On the other hand, CAM advertisements stress a connection of the CAM hardware to real-world applications and a variety of alterable structural materials ranging from blocks of paper to machinable steel. The two processes have fused over time, partly because of the complementary nature of their intended applications and partly because of clients' need to have a model that engages their own imaginations, but certain aspects of CAD (those technological elements most appropriate to the creation of virtual, rather than actual, worlds) may eventually allow the software package to reduce the need for soft models in terms of product development.

CAD in Video Games and Film Animation. CAD work intended for video games and film animation studios, unlike many other types of designs, may intentionally and necessarily violate one basic rule of engineering—that a design's form should never take precedence over its function. Industrial engineers also stress the consideration of ergonomics (the human body's requirements for ease, safety, and comfort when using a product) in the design process. In the case of the virtual engineer, however, the visual appeal of a video game or animation element may be a greater factor in how a design is packaged for a client. *Snow White and the Seven Dwarfs* was an animated film originally drawn between 1934 and 1937. Since animation was drawn by hand in those days—literally on drawing boards—with live models for the animated characters, the studio's founder, Walt Disney, looked for ways to increase the animated characters' visual appeal. One technique he developed was to draw human characters (in this case, Snow White herself) with a larger-than-standard size head, ostensibly to make her appear to have the proportions of a child and, thus, to be more appealing to film viewers. In order for the artists to draw their distorted heroine more easily, the actor playing Snow White, Marjorie Bell, was asked to wear a football helmet while acting out sections of the story's plot. Unfortunately for Disney, Bell complained so much about the helmet that it had to be removed after only a few minutes. In that instance, ergonomics took precedence over design. Later animation styles, influenced by designers' use of CAD animation software, were able to be even more exaggerated without the need for a live model in an uncomfortable outfit. For example, the character of Jessica Rabbit in Amblin Entertainment's *Who Framed Roger Rabbit?* (1988) has a face and body that were designed exclusively for visual appeal, rather than to conform to external world reality.

APPLICATIONS AND PRODUCTS

CAD and CAD/CAM have an extensive range and variety of applications.

Medicine. In the medical field, software that employs finite element methods (FEM; also called finite element analysis, or FEA), such as Abaqus FEA, allows designers to envision new developments in established medical devices—improving artificial heart valves, for example—or the creation of new medical devices. CAD software ideally allows a user to simulate real-world performance.

Aerospace. In aerospace engineering, programs can simulate the tensile strength and load capacity of a variety of structural materials as well as the effect of weather on performance and durability.

Mechanical Engineering. Mechanical engineering firms sometimes use Siemens Digital Industries Software for CAM or CAE functions in order to create and revise structural models quickly. Some forms of CAD have engineering applications that can anticipate a design's structural flaws to an even greater extent than an actual model might. One example of this application is SIMULIA, one of the simulation divisions of Dassault Systèmes that creates virtual versions of the traditional automobile safety tester—the crash-test dummy. In the real world, crash-test dummies are rather complex mechanical devices, but their virtual simulacra have been declared as effective in anticipating structural failure as the original models.

Animation. CAD programs have caused the rapid expansion and development of the animation industry—both in video games and in film animation. Models of living creatures, both animal and human, are generated through the use of 3D CAD software such as Massive (which was used by the graphic artists creating the battle sequences in Peter Jackson's *Lord of the Rings* film trilogy) and Autodesk Maya (considered an industry standard for animated character designs). Autodesk 3ds Max is one of the forerunner programs in the video-game industry, since it allows a great deal of flexibility in rendering surfaces. Certain video games, such as Will Wright's *Sims* series, as well as any number of massively multiplayer online role-playing games (MMORPGs) such as *World of Warcraft*, *Second Life*, and *EverQuest*, not only have CAD functions buried in the design of the game itself but also accept custom content designed by players as a fundamental part of individualization of a character or role. A player must design a living character to represent him or her in the video game, including clothing, objects carried, and weapons used, and the process of creation is easily as enjoyable as the actual game play.

Advances in computer-aided technology also enabled the creation of films such as *Avatar*, released by Twentieth Century Fox in 2009. Envisioned by director James Cameron as a sweeping epic in much the same vein as George Lucas's original *Star Wars* trilogy (1977-83), Avatar was originally scheduled to start filming in 1997. However, Cameron felt that computer-aided technology was not ready to create the detailed feature film he envisioned. Given that the film was also supposed to take advantage of newly developed 3D stereoscopic film techniques, Cameron did not start work on the computer-aided-technology aspects of his film until 2006, when he felt sure the technology was up to the task. Even with an extensive 3D software animation package at his command, the task of creating and populating an entire alien world was a daunting prospect. Most of the action of *Avatar* takes place on a distant planet, Pandora, which is intrinsically toxic to human beings. To get around this difficulty and successfully mine Pandora for a much-needed mineral, humans must use genetically created compatible bodies and interact with the native population, the Na'vi. Both the native creatures, who are 10 feet tall with blue skin, and the hybrid human-Na'vi "avatar" bodies were a challenge for animators to render realistically even using computer-aided technology. The full IMAX edition was released on December 18, 2009, having ultimately cost its creators somewhere between $280 million and $310 million. Fortunately for Cameron, the film ended up rewarding its creators with gross revenues of more than $2 billion and also won Academy Awards for art direction, cinematography, and visual effects. Cameron subsequently began work on several planned sequels, the first of which, *Avatar: The Way of Water*, premiered in 2022.

CAREERS AND COURSEWORK

Careers in computer-aided design and manufacturing tend to be affected most obviously by the intended purpose of the designer's work. There could be significant differences in the coursework required of a medical-device designer and that required of a film-animation specialist. Aside from coursework in rapid visualization, CAD operation, graphic illustration, digital type and image manipulation, digital photography, new media, and 3D modeling, those who wish to work with computer-aided technology should consider supplemental coursework in the field most relevant to their area of design. Various institutes have CAD/CAM labs that provide hands-on experience with computer numerical control (CNC) machines, CAD packages, computer-integrated manufacturing, and industrial robots.

SOCIAL CONTEXT AND FUTURE PROSPECTS

Computer-aided technologies, in general, are expected to keep developing and evolving. The possibilities of virtual reality are continuing to evolve as the need for structural and materials modeling increases. One common description of the future of computer-aided design and manufacturing is the term "exponential productivity." In other words, so much productive time in the fields of design has been wasted by the constant need to create by hand what can be better (and more quickly) produced by a computer. An architect or engineer might spend hours carefully working out the exact proportions of a given structure, more hours using a straight edge and compass to determine the exact measurements needed by the design, and even more time to clean up mistakes. On the other hand, computers can help that same architect draw straight lines, evaluate numerical formulas for proportions, and edit mistakes in a matter of minutes. When one considers the added benefits of a compatible computer-aided manufacturing program, one can further see the time saved in the quick and efficient production of a soft model.

—Julia M. Meyers

Further Reading

"Best General-Purpose CAD Software for Enterprise Businesses." *G2*, 2023, www.g2.com/categories/general-purpose-cad. Accessed 13 July 2023.

Eskilson, Stephen. *Digital Design: A History*. Princeton UP, 2023.

Heidmets, Rao. *The Animation Textbook*. CRC Press, 2022.

Kerlow, Isaac. *The Art of 3D Computer Animation and Effects*. 4th ed., Wiley, 2009.

Mund, Ben. "CAD/CAM Options for Increased Shop Productivity." *SME*, 27 Jan. 2021, www.sme.org/technologies/articles/2021/february/cadcam-options-for-increased-shop-productivity. Accessed 13 Jul. 2023.

Petty, Luke. "What Is Computer Aided Manufacturing? And How Is It Leading the Digital Site Revolution?" *Chapman Taylor*, 19 May 2016, www.chapmantaylor.com/en/insights/article/what-is-computer-aided-manufacturing-and-how-is-it-leading-the-digital-site/en. Accessed 13 July 2023.

Sarkar, Jayanta. *Computer Aided Design: A Conceptual Approach*. CRC Press, 2015.

Schell, Jesse. *The Art of Game Design: A Book of Lenses*. 2nd ed., CRC Press, 2015.

Schodek, Daniel, et al. *Digital Design and Manufacturing: CAD/CAM Applications in Architecture and Design*. Wiley, 2005.

COMPUTER ANIMATION

ABSTRACT

Computer animation is the process of creating animated projects for film, television, video games, or other media using any of a variety of specialized computer programs. As animation projects may range from short, simple clips to detailed and vibrant feature-length films, a wide variety of animation software is available, each addressing the particular needs of animators. The computer animation process includes several key steps, including modeling, adding lighting and special effects, and rendering, which are typically carried out by a team of animators.

BACKGROUND

Since the early twentieth century, the field of animation has been marked by frequent, rapid change. Innovation in the field has been far-reaching, filtering into film, television, advertising, video games, and other media. It was initially an experimental method and took decades to develop. Computer animation revitalized the film and television industries during the late twentieth and early twenty-first centuries, in many ways echoing the cultural influence that animation had decades before.

Prior to the advent of computer animation, most animated projects were created using a process that later became known as "traditional," or "cel," animation. In cel animation, the movement of characters, objects, and backgrounds was created frame by frame. Each frame was drawn by hand. This time-consuming and difficult process necessitated the creation of dozens of individual frames for each second of film.

As computer technology developed, computer researchers and animators began to experiment with creating short animations using computers. Throughout the 1960s, computers were used to create two-dimensional (2D) images. Ed Catmull, who later founded the studio Pixar in 1986, created a three-dimensional (3D) animation of his hand using a computer in 1972. This became the first 3D computer graphic to be used in a feature film when it appeared in *Futureworld* (1976). Early attempts at computer animation were found in live-action films. The 1986 film *Labyrinth*, for instance, notably features a computer-animated owl flying through its opening credits. As technology improved, computer animation became a major component of special effects in live-action media. While cel animation continued to be used in animated feature films, filmmakers began to include some computer-generated elements in such works. The 1991 Walt Disney Studios film *Beauty and the Beast*, for instance, featured a ballroom in one scene that was largely created using a computer.

In 1995, the release of the first feature-length computer-animated film marked a turning point in the field of animation. That film, *Toy Story*, was created by Pixar, a pioneer in computer animation. Over the following decades, Pixar and other studios, including Disney (which acquired Pixar in 2006) and DreamWorks, produced numerous computer-animated films. Computer animation became a common process for creating animated television shows as well as video games, advertisements, music videos, and other media.

In the early twenty-first century, computer animation also began to be used to create simulated environments accessed through virtual reality equipment such as headsets. Much of the computer-animated content created during this time featured characters

Photo via iStock/AndreyPopov. [Used under license.]

and surroundings that appeared 3D. However, some animators opted to create 2D animations that more closely resemble traditionally animated works in style.

OVERVIEW

From designing the original animation model to creating algorithms that control the movement of fluids, hair, and other complex systems, computer software has drastically changed the art of animation. Through software that can manipulate polygons, a face can be rendered and further manipulated to create a number of images much more efficiently than with hand-drawn illustrations. Thus, the detail of the imaging is increased, while the time needed to develop a full animation is reduced.

Creating a feature-length computer-animated production is a complex and time-intensive process that is carried out by a large team of animators, working with other film-industry professionals. When creating a 3D computer-animated project, the animation team typically begins by drawing storyboards. Storyboards are small sketches that serve as a rough draft of the proposed scenes.

Next, animators transform 2D character designs into 3D models using animation software. They use animation variables (avars) to control the ways in which the 3D characters move, assigning possible directions of movement to various points on the characters' bodies. The number of avars used and the areas they control can vary widely. The 2006 Pixar film *Cars* reportedly used several hundred avars to control the characters' mouths alone. Using such variables gives animated characters a greater range of motion and often more realistic expressions and gestures. After the characters and objects are modeled and animated, they are combined with backgrounds as well as lighting and special effects. All of the elements are then combined to transform the 3D models into a 2D image or film. This process is known as 3D rendering.

Animating a 2D computer-animated work is somewhat different from its 3D counterpart, in that it does not rely on 3D modeling. Instead, it typically features the use of multiple layers, each of which contains different individual elements. This method of animating typically features keyframing. In this procedure, animators define the first and last frames in an animated sequence and allow the computer to fill in the movement in between. This process, which in traditionally animated films was a laborious task done by hand, is often known as "inbetweening," or "tweening."

TOOLS

Various animation programs are available to animators, each with its own strengths and weaknesses. Some animation software, such as Maya and Cinema 4D, are geared toward 3D animation. Others, such as Adobe Animate (previously known as Adobe Flash), are better suited to 2D animation. Adobe Animate has commonly been used to produce 2D cartoons for television, as it is considered a quick and low-cost means of creating such content. Major animation studios typically use proprietary animation software, thus ensuring that their specific needs are met.

In addition to animation software, the process of computer animation relies heavily on hardware, as many steps in the process can be taxing for the systems in use. Rendering, for example, often demands a sizable amount of processing power. As such, many studios make use of render farms, which are large, powerful computer systems devoted to that task.

—Joy Crelin

Further Reading

"Computer Animator." *Art Career Project*, 15 July 2021, theartcareerproject.com/careers/computer-animation. Accessed 13 July 2023.

Glawion, Alex. "Building the Best PC for 3D Animation (2023 Guide)." *CGDirector*, 3 Jan. 2023, www.cgdirector.com/best-computer-for-animation. Accessed 13 July 2023.

Heidmets, Rao. *The Animation Textbook*. CRC Press, 2022.

Highfield, Roger. "Fast Forward to Cartoon Reality." *Telegraph*, 13 June 2006, www.telegraph.co.uk/technology/3346141/Fast-forward-to-cartoon-reality.html. Accessed 13 July 2023.

"Our Story." *Pixar*, 2023, www.pixar.com/our-story-pixar. Accessed 13 July 2023.

Parent, Rick. *Computer Animation: Algorithms and Techniques*. 3rd ed., Elsevier, 2012.

Sito, Tom. *Moving Innovation: A History of Computer Animation*. MIT Press, 2013.

Winder, Catherine, and Zahra Dowlatabadi. *Producing Animation*. Focal, 2011.

Computer Languages, Compilers, and Tools

ABSTRACT

Computer languages are used to provide the instructions for computers and other digital devices based on formal protocols. Low-level languages, or machine code, were initially written using the binary digits needed by the computer hardware, but since the 1960s, languages have evolved from early procedural languages to object-oriented high-level languages, which are more similar to English. There are many of these high-level languages, with their own unique capabilities and limitations, and most require some type of compiler or other intermediate translator to communicate with the computer hardware. Computer languages underlie the multitude of programs in use in the twenty-first century, from word-processing and design programs to data-analysis and simulation software. The popularity of the internet has created the need to develop numerous applications and tools designed to share data across the internet.

BACKGROUND

Early computers such as Electronic Numerical Integrator and Computer (ENIAC), the first general-purpose computer, were based on the use of switches that could be turned on or off. Thus, the binary digits of 0 and 1 were used to write machine code. In addition to being tedious for a programmer, the code had to be rewritten if used on a different type of machine, and it certainly could not be used to transmit data across the internet, where different computers all over the world require access to the same code.

Assembly language evolved by using mnemonics (alphabetic abbreviations) for code instead of the binary digits. Because these alphabetic abbreviations of assembly language no longer used the binary digits, additional programs were developed to act as intermediaries between the human programmers writing the code and the computer itself. These additional programs were called compilers, and this process was initially known as compiling the code. This compilation process was still machine and vendor dependent, however, meaning, for example, that there were several types of compilers that were used to compile code written in one language. This was expensive and made communication of computer applications difficult.

The evolution of computer languages from the 1950s has accompanied technological advances that have allowed languages to become increasingly powerful, yet easier for programmers to use. FORTRAN and COBOL languages led the way for programmers to develop scientific and business application programs, respectively, and were dependent on a command-line user interface, which required a user to type in a command to complete a specific task. Several other languages were developed, including Basic, Pascal, PL/I, Ada, Lisp, Prolog, and Smalltalk, but each of these had limited versatility and various problems. The C and C++ languages of the 1970s and 1980s, respectively, emerged as the most useful and powerful languages and are still in use. These were followed by development tools written in the Java and Visual Basic languages, including integrated development environments with editors, designers, debuggers, and compilers all built into a single software package.

OVERVIEW

The traditional process of using a computer language to write a program has generally involved the initial design of the program using a flowchart based

Image via iStock/matejmo. [Used under license.]

on the purpose and desired output of the program, followed by typing the actual instructions for the computer (the code) into a file using a text editor, and then saving this code in a file (the source code file). A text editor is used because it does not have the formatting features of a word processor. An intermediate tool called a "compiler" then has been used to convert this source code into a format that can be run (executed) by a computer.

However, in the 2010s, a tool much faster and more efficient than compilers, called "interpreters," gained prominence and replaced most compilers. Larger, more complex programs have evolved that have required an additional step to link external files. This process is called "linking," and it joins the main, executable program created by the compiler to other necessary programs. Finally, the executable program is run and its output is displayed on the computer monitor, printed, or saved to another digital file. If errors are found, the process of debugging is followed to go back through the code to make corrections.

HOW IT WORKS

BIOS and Operating System. The programs within the BIOS are the first and last programs to execute whenever a computer device is turned on or off. These programs interact directly with the operating system (OS). The early mainframe computers that were used in the 1960s and 1970s depended on several different operating systems, most of which are no longer in usage, except for UNIX and disk operating system (DOS). DOS was used on the initial microcomputers of the 1980s and early 1990s, and it is still used for certain command-line specific instructions.

Graphical User Interfaces (GUIs). Microsoft dominates the PC market with its many updated operating systems, which feature user-friendly GUIs. These operating systems consist of computer programs and software that act as the management system for all of the computer's resources, including the various applications most taken for granted, such as Word (for documents), Excel (for mathematical and spreadsheet operations), and Access (for database functions). Each of these applications is a program itself, and there are many more that are also available.

Since the 1980s, many programming innovations increasingly have been built to involve the client-server model, with less emphasis on large mainframes and more emphasis on the GUIs for smaller microcomputers and handheld devices that allow consumers to have deep color displays with high resolution and voice and sound capabilities. However, these initial GUIs on client computers required additional upgrades and maintenance to be able to interact effectively with servers.

World Wide Web (WWW). The creation of the World Wide Web provided new means of accessing information, and the widespread use of the internet led to the creation of new programming languages and tools. The browser was developed to allow an end user (client) to be able to access web information, and hypertext markup language (HTML) was developed to display web pages. Because the client computer was manufactured by many different companies, the Java language was developed to include applets, which are mini-programs embedded into web pages that could be displayed on any type of client computer. This was made possible by a special type of compiler-like tool called the "Java Virtual Machine," which translated byte code.

APPLICATIONS AND PRODUCTS

FORTRAN and COBOL. FORTRAN (sometimes written as Fortran) was developed by a team of programmers at IBM and was first released in 1957 to be used primarily for highly numerical and scientific applications. It derived its name from formula translation. Initially, it used punched cards for input, because the text editors were not available in the

1950s. It has evolved but still continues to be used primarily in many engineering and scientific programs. Several updated versions have been released, including the 2023 revision Fortran 2023. FORTRAN77, released in 1980, had the most significant language improvements. Common Business-Oriented Language, or COBOL was released in 1959 with the goal of being used primarily for tracking retail sales, payroll, inventory control, and many other accounting-related activities.

C and C++. The C computer language was the predecessor to the C++ language. Programs written in C were procedural and based on the usage of functions, which are small programming units. As programs grew in complexity, more functions were added to a C program. The problem was that eventually it became necessary to redesign the entire program, because trying to connect all of the functions was too difficult. C++ was created in the 1980s based on the idea of objects grouped into classes as the building blocks of the programs. Object-oriented programming made developing complex programs much more efficient.

Microsoft.NET. In June 2000, Microsoft introduced a suite of languages and tools named Microsoft.NET along with its new language called "Visual C#." Later known simply as .NET or the .NET framework, Microsoft.NET is a software infrastructure that consists of many programs that allow a user to write programs for a range of new applications such as server components and web applications by using new tools. Although programs written in Java can be run on any machine, as long as the entire program is written in Java, .NET allows various programs to be run on the Windows OS. Additional advantages of .NET involve its use of Visual C#. Visual C# provides services to help web pages already in existence, and C# can be integrated with the Visual Basic and Visual C++ languages, which facilitate the work of web programmers by allowing them to update existing web applications, rather than having to rewrite them.

The .NET framework uses a common type system (CTS) tool to compile programs written in a variety of languages into an intermediate language. This common intermediate language (CIL) can then be compiled to a common language runtime (CLR). The result is that the .NET programming environment promotes interoperability to allow programs originally written in different languages to be executed on a variety of operating systems and computer devices. This interoperability is crucial for sharing data and communication across the internet.

SOCIAL CONTEXT AND FUTURE PROSPECTS

The internet continues to bring the world together at a rapid pace, which has both positive and negative ramifications. Consumers have much easier access to many services, such as online education and telemedicine, and can use free search tools to locate doctors, learn more about any topic, comparison shop and purchase, and immediately access software, movies, pictures, and music. However, along with this increase in electronic commerce involving credit card purchases, bank accounts, and additional financial transactions has been the increase of cybercrime. Thousands of dollars are lost each year to various internet scams and the accessing of private financial information by hackers. Some programmers even use computer languages to produce viruses and other destructive programs for purely malicious purposes, which have a negative impact on computer security. Such phenomena have given rise to the development of improved security features within computer languages and tools.

—Jeanne L. Kuhler

Further Reading

Das, Sumitabha. *Your UNIX/Linux: The Ultimate Guide*. 3rd ed., McGraw-Hill, 2012.

Dhillon, Gupreet. "Dimensions of Power and IS Implementation." *Information and Management*, vol. 41, no. 5, 2004, pp. 635-44.

Horstmann, Cay. *Big Java: Early Objects*. 7th ed., John Wiley & Sons, 2018.

Lee, Kent D. *Foundations of Programming Languages*. 2nd ed., Springer, 2017.

O'Leary, Timothy, Linda O'Leary, and Daniel O'Leary. *Computing Essentials 2023*. McGraw-Hill, 2022.

Petzold, Charles. *Code: The Hidden Language of Computer Hardware and Software*. 2nd ed., Microsoft Press, 2022.

Scott, Michael L. *Programming Language Pragmatics*. 4th ed., Morgan Kaufmann, 2016.

Sellers, Audrey. "Top Programming Languages of 2023." *Coding Dojo*, 3 Feb. 2023, www.codingdojo.com/blog/top-programming-languages. Accessed 13 July 2023.

COMPUTER MEMORY AND STORAGE

ABSTRACT

There are different types of memory inside a computer, including temporary memory, read-only memory (ROM), random-access memory (RAM), and programmable read-only memory (PROM). Computer storage technologies include hard drives and cloud storage. Memory and storage are used for different purposes. Fast, temporary memory such as RAM is used to make quick calculations, while hard drives are used for long-term storage of programs and files. Without memory and storage, computers would not be able to function in any meaningful capacity.

BACKGROUND

In their earliest days, computers were strictly mechanical devices. They used punch cards for memory and output. These machines were developed for utility rather than for the multitude of tasks for which modern computers are designed. They were primarily used for complex calculations.

Alan Turing, a famous computer scientist, is credited with the idea for the first multipurpose computer. In the 1930s, J. V. Atanasoff created the first computer that contained no gears, cams, belts, or shafts. Atanasoff and his team then designed the first computer with functioning, nonmechanical memory devices. Primitive when compared to today's devices, Atanasoff's creation allowed the computer to solve up to twenty-nine equations simultaneously.

The next major leap in computing power was the usage of vacuum tubes. In 1994, professors John Mauchly and J. Presper Eckert built the first tube-powered electronic calculator. This is commonly considered the first digital computer. It was a massive machine, taking up the entirety of a large room. They soon began to market this computer to governments and businesses. However, tube computers became obsolete in 1947 with the invention of the transistor.

Ten years later, the transistor was used by Robert Noyce and Jack Kilby to create the first computer chip. This spurred the development of the first devices recognizable as modern computers. Computers took another leap forward with the graphical user interface (GUI), which projects options as images on a screen instead of requiring users to learn to code. Computers advanced further with the inventions of random-access memory (RAM) in 1970 and floppy disks in 1971. Floppy disks were a form of permanent storage used in the early days of computers. They could easily be transferred from one computer to another, making them ideal for transporting information. Floppy disks were made obsolete by CD-ROMs, which are small plastic discs that store large amounts of information.

OVERVIEW

Computer memory and storage are measured in binary digits, called bits. One bit is an extremely small amount of information. Eight bits is the equivalent of one byte. 1,024 bytes is called a kilobyte (KB); 1,024 KB makes a megabyte (MB); 1,024 MB makes a gigabyte (GB); and 1,024 GB makes a terabyte (TB). Over time, the cost of large amounts of computer memory and storage has drastically fallen.

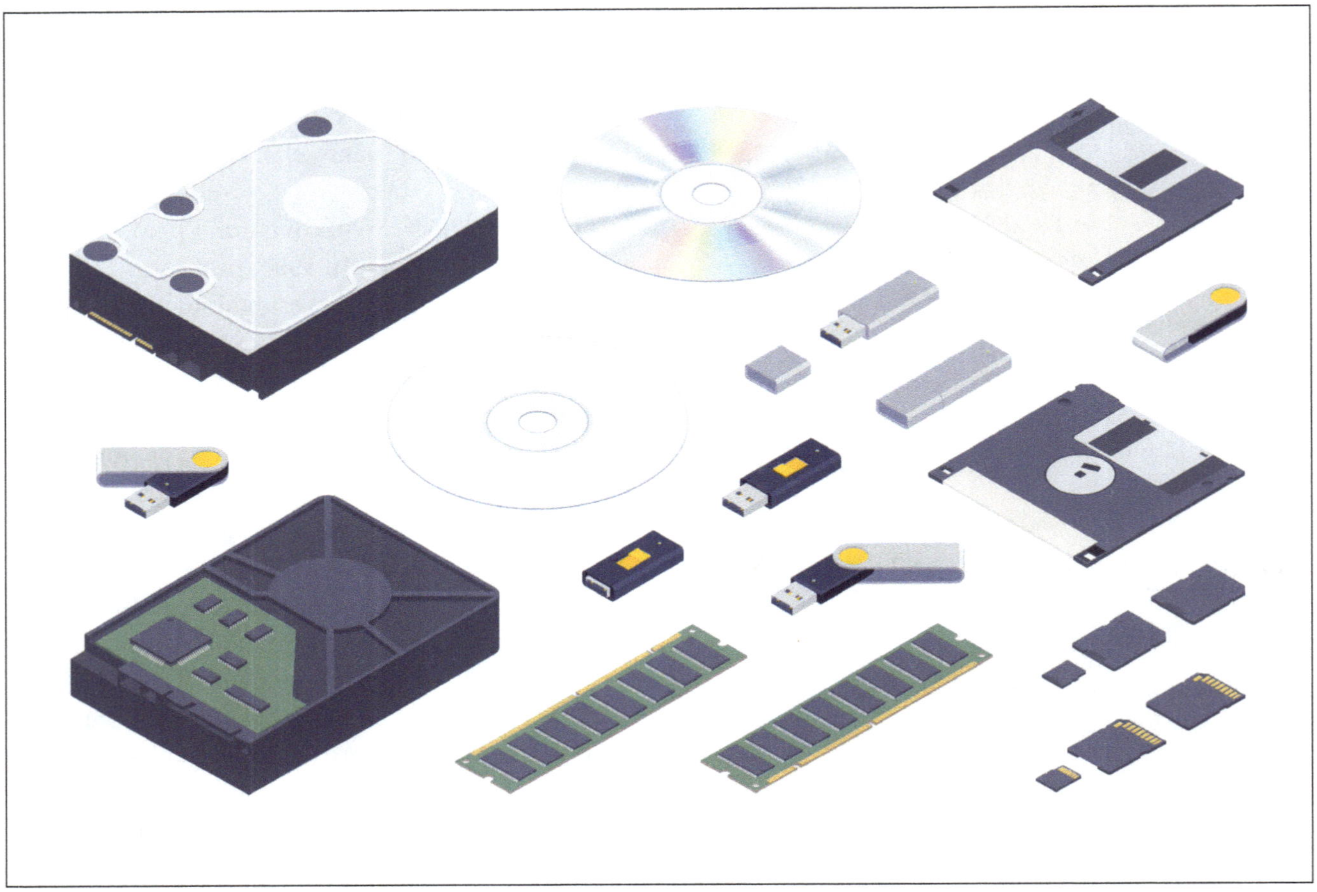

Image via iStock/Andrew_Rybalko. [Used under license.]

However, the amount of memory and storage required by computers has also drastically increased.

Computers contain and utilize several types of memory and storage.

Temporary Memory. The most common type contained in a computer is temporary memory, which is designed to hold information for only a short period. Most of a computer's temporary memory is RAM. RAM is designed to quickly write and erase information. It performs calculations, runs scripts, and enacts most of the computer's functions. A computer with more RAM can perform more functions at once, making it more powerful and more capable of running resource-intensive programs, including programs used to edit audiovisual media or create digital art.

Permanent Storage. Permanent storage may refer to several devices. In most cases, information is stored on the computer's hard disk drive (HDD). Some hard disk drives use a spinning disk and an actuator arm. The actuator arm writes to the spinning disk by rearranging electrons. In this scenario, the entire inside of the HHD is located inside an airtight seal. These hard drives can be found in many sizes. However, HHDs of the twenty-first century are often found in capacities of hundreds of gigabytes to terabytes.

Many high-quality computer manufacturers have begun replacing HHDs with solid-state drives (SSDs). Solid-state drives contain no moving parts. In most instances, these drives can read and write data much faster than HHDs. Because they have no moving parts, SSDs are also much quieter than hard disk drives. However, they are also more expensive than HHDs. For this reason, if a device needs large quantities of storage, it may be more cost-effective to use HHDs. However, if the device needs to be able to access data quickly, be durable, or be compact in size, manufacturers may use an SSD.

External Storage. Some computers utilize external forms of storage. These are drives located outside the device. If it is connected to the device by a universal serial bus (USB) cable or other interface, it is called an "external hard drive." External hard drives are easily transferable from one device to another, making them useful for quickly moving large media files. They may also be used to back up large amounts of important files, protecting them from computer malfunction or viruses.

If the external storage is accessed through the internet, it is referred to as cloud storage. Many services offer large amounts of external storage for purchase. This may be used for server backups, media storage, or any number of other applications. Cloud storage allows users to expand the storage capacity of their machines without making any physical alterations to the computers. The cloud also features many of the same benefits as an external hard drive. Files can easily be relocated to a new machine in the event of a hardware or software failure and can likewise be shared among those collaborating on a group project, such as an engineering design. Additionally, renting space from a cloud storage service may be cheaper than purchasing and installing additional physical storage devices.

—*Tyler Biscontini*

Further Reading

Brant, Tom. "SSD vs. HDD: What's the Difference?" *PCMag*, 26 Aug. 2022, www.pcmag.com/news/ssd-vs-hdd-whats-the-difference. Accessed 13 July 2023.

"Data Measurement Chart." *University of Florida*, www.wu.ece.ufl.edu/links/dataRate/DataMeasurementChart.html. Accessed 13 July 2023.

"Hard Drive." *Computer Hope*, 18 Oct. 2022, www.computerhope.com/jargon/h/harddriv.htm. Accessed 13 July 2023.

"How Computers Work: The CPU and Memory." *University of Rhode Island*, homepage.cs.uri.edu/faculty/wolfe/book/Readings/Reading04.htm. Accessed 13 July 2023.

O'Leary, Timothy, Linda O'Leary, and Daniel O'Leary. *Computing Essentials 2023*. McGraw-Hill, 2022.

"Storage vs. Memory." *PCMag*, www.pcmag.com/encyclopedia/term/storage-vs-memory. Accessed 13 July 2023.

"Timeline of Computer History." *Computer History Museum*, www.computerhistory.org/timeline/computers. Accessed 13 July 2023.

Computer Modeling

ABSTRACT

Computer modeling is the process of designing a representation of a particular system of interacting or interdependent parts in order to study its behavior. Models that have been implemented and executed as computer programs are called "computer simulations."

BACKGROUND

A computer model is a programmed representation of a system that is meant to mimic the behavior of the system. A wide range of disciplines, including meteorology, physics, astronomy, biology, and economics, use computer models to analyze different types of systems. When the program representing the system is executed by a computer, it is called a "simulation."

One of the first large-scale computer models was developed during the Manhattan Project by scientists designing and building the first atomic bomb. Early computer models produced output in the form of tables or matrices that were difficult to analyze. It was later discovered that humans can see data trends more easily if the data is presented visually. For example, humans find it easier to analyze the output of a storm-system simulation if it is presented as graphic symbols on a map rather than as a table of meteorological data. Thus, simulations that produced graphic outputs were developed.

Computer models are used when a system is too complex or hard to study using a physical model. For example, it would be difficult to create a physical model representing the gravitational effects of planets and moons on each other and on other objects in space, although the National Aeronautics and Space Administration (NASA) has done exactly that for the Voyager space probes.

There are several different types of models. Static models simulate a system at rest, such as a building design. Dynamic models simulate a system that changes over time. A dynamic model could be used to simulate the effects of changing ocean temperatures on the speed of ocean currents throughout the year. A continuous model simulates a system that changes constantly, while a discrete model simulates a system that changes only at specific times. Some models contain both discrete and continuous elements. A farming model might simulate the effects of both weather patterns, which constantly change, and pesticide spraying, which occurs at specified times.

OVERVIEW

To create a computer model, one must first determine the boundaries of the system being modeled and what aspect of the system is being studied. For example, if the model is of the solar system, it might be used to study the potential effect on the orbits of the existing planets if another planet were to enter the solar system.

To create such a model, a computer programmer would develop a series of algorithms that contain the equations and other instructions needed to replicate the operation of the system. Variables are used to represent the input data needed. Examples of variables that might be used for the solar system model include the mass, diameter, and trajectory of the theoretical new planet. The values that define the system, and thus how the variables affect each other, are the parameters of the system. The parameters control the outputs of the simulation when it is run. Different values can be used to test different scenarios related to the system and problem being studied. Example parameters for the solar system model might include the orbits of the known planets, their distance from the sun, and the equations that relate an object's mass to its gravity. Certain parameters can be changed to test different scenarios each time a simulation is run. Because parameters are not always constant, they can be difficult to distinguish from variables at times.

Computer models have provided great benefits to society. They help scientists explore the universe, understand the earth, cure diseases, and discover and test new theories. They help engineers design buildings, transportation systems, power systems, and other items that affect everyday life. With the development of more powerful computer systems, computer models will remain an important mechanism for understanding the world and improving the human condition.

—Maura Valentino

Further Reading

Abbasov, Iftikhar B., editor. *Computer Modeling in the Aerospace Industry*. Wiley/Scrivener Publishing, 2020.

"Computational Modeling." *National Institute of Biomedical Imaging and Bioengineering*, May 2020, www.nibib.nih.gov/science-education/science-topics/computational-modeling. Accessed 13 July 2023.

Edwards, Paul N. *A Vast Machine: Computer Models, Climate Data, and the Politics of Global Warming*. MIT Press, 2010.

Kojiæ, Miloš, et al. *Computer Modeling in Bioengineering: Theoretical Background, Examples and Software*. Wiley, 2008.

Law, Averill M. *Simulation Modeling and Analysis*. 6th ed., McGraw-Hill, 2023.

Mityushev, Vladimir, Wojciech Nawalaniec, and Natalia Rylko. *Introduction to Mathematical Modeling and Computer Simulations*. CRC Press, 2018.

Morrison, Foster. *The Art of Modeling Dynamic Systems: Forecasting for Chaos, Randomness, and Determinism*. 1991. Dover, 2008.

Seidl, Martina, et al. *UML@Classroom: An Introduction to Object-Oriented Modeling*. Springer, 2015.

Computer Networks

ABSTRACT

Computer networks consist of the hardware and software needed to support communications and the exchange of data between computing devices. The computing devices connected by computer networks include large servers, business workstations, home computers, and a wide array of smart mobile devices. The most popular computer network application is email, followed by exchanging audio and video files. Computer networks provide an infrastructure for the internet, which in turn provides support for the World Wide Web.

BACKGROUND

A computer network is a collection of computer devices that are connected in order to facilitate the sharing of resources and information. Underlying the computer network is a communications network that establishes the basic connectivity of the computing devices. This communications network is often a wired system but can include radio and satellite paths as well. Devices used on the network include large computers, used to store files and execute applications; workstations, used to execute small applications and interact with the servers; home computers, connected to the network through an internet service provider (ISP); and mobile devices, connected to the network by radio wave transmissions. Middleware is the software that operates on top of the communications network to provide a software layer for developers to add high-level applications, such as a search engine, to the basic network.

The scientists who developed the first computers in the 1950s recognized the advantage of connecting computing devices. Teletype machines were in common use at that time, and many of the early computers were "networked" with these teletype machines over wired networks. By 1960, American Telephone and Telegraph (AT&T) had developed the first modem to allow terminal access to mainframes, and in 1964, IBM and American Airlines introduced the Semi-Automated Business Research Environment (SABRE) networked airline reservation system.

The Defense Department created the Advanced Research Projects Agency Network (ARPANET) in 1966 to connect its research laboratories with college researchers. The early experience of ARPANET led the government to recognize the importance of being able to connect different networks so they could interoperate. One of the first efforts to promote interoperability was the addition of packet switching to ARPANET in 1969. In 1974, Robert Kahn and Vinton Cerf published a paper on packet-switching networks that defined the transmission control protocol/internet protocol (TCP/IP), and in 1980, the US government required all computers it purchased to support TCP/IP. When Microsoft added a TCP/IP stack to its Windows 95 operating system, TCP/IP became the standard wide area network in the world.

The development of the microcomputer led to the need to connect these devices to themselves and the wide area networks. In 1980, the Institute of Electrical and Electronics Engineers (IEEE) 802 standard was announced. It has provided most of the connectivity to networks since that time, although many wireless computing devices can connect with Bluetooth.

The high-level applications are what most people using the network see as the network. Some of the most important computer network applications provide communications for users. Earlier examples of this are email, instant messaging, chat rooms, and videoconferencing. Later examples of communications software are the multitude of social networks, such as Facebook and Twitter. Other high-level applications allow users to share files. One of the oldest and still quite popular file-sharing programs is the file transfer protocol (FTP) program. Another way to use computer networks is to share computing power. The Telnet program (terminal emulation program for TCP/IP networks) allowed one to use an application on a remote mainframe in the early days of networking, and twenty-first-century web services allow one to run an application on a mobile device while getting most of the functionality from a remote server.

OVERVIEW

Computer networks consist of the hardware needed for networking, such as computers and routers; the software that provides the basic connectivity, such as the operating system; and middleware and applications, the programs that allow users to use the network. In understanding how these components work together, it is useful to look at the basic connectivity of the wide area network and contrast that to the way computers access the wide area network.

Wide Area Networks. A wide area network is one that generally covers a large geographic area and in which each pair of computers connect via a single line. The first wide area networks consisted of a number of connected mainframes with attached terminals. By connecting the mainframes, a user on one system could access applications on every networked computer. IBM developed the Systems Network Architecture (SNA), which was a popular early wide area network in the United States. The X.25 packet switching protocol standard provided a common architecture for the early wide area networks in Europe. Later, as all computers provided support for the TCP/IP protocol, it became possible for different computer networks to work together as a single network. In the twenty-first century, any device on a single network, whether attached as a terminal by X.25 or as a part of a local area network (LAN), can access applications and files on any other net-

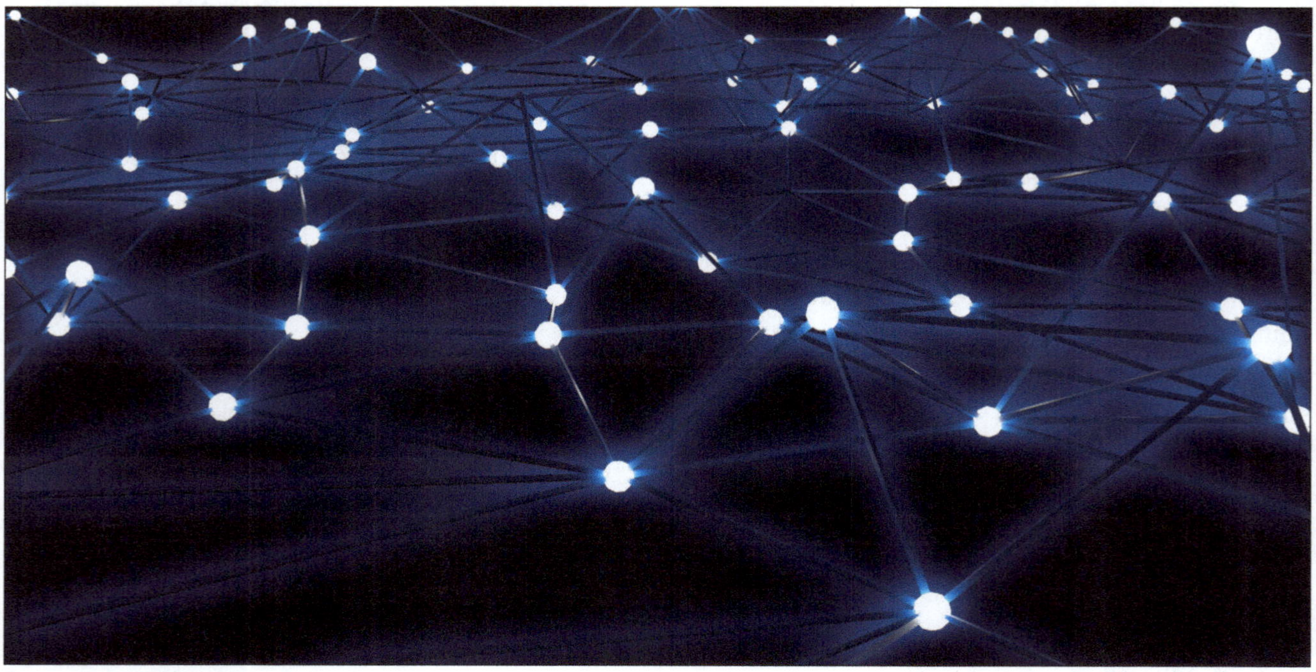

Image via iStock/photoman. [Used under license.]

work on the internet. This complete connectivity allows browsers on any type of mobile device to access content over the World Wide Web, because it runs transparently over the internet.

Routing. A key to making the internet operate efficiently is the large number of intermediate devices that route IP packets from the computer making a connection to the computer receiving the data. These devices are called routers, and they are the heart of the internet. When a message is sent from a computer, it is decomposed into IP packets, each having the IP address of the receiver. Then each packet is forwarded to a border router of the sender. From the border router, the packet is routed through a set of intermediate routers, using an algorithm-like best path, and delivered to the border router of the receiver, and then the receiver. Once all the packets have arrived, the message is reassembled and used by the receiver.

Local Area Networks. A local area network (LAN) is a collection of computers, servers, printers, and the like that are logically connected over a shared media. A media access control protocol operates over the LAN to allow any two devices to communicate at the link level as if they were directly connected. There are a number of LAN architectures, but Ethernet is the most popular LAN protocol for connecting workplace computers on a LAN to other networks. Ethernet is also usually the final step in connecting home computers to other networks using regular, cable, and Asymmetric Digital Subscriber Line (ASDL) modems. The original coaxial cable Ethernet, developed by Robert Metcalfe in 1973, has been largely supplanted by the less expensive and easier to use twisted-pair networks. The IEEE 802.11 wireless LAN is a wireless protocol that is almost as popular as Ethernet.

Wireless Networks. A laptop computer's initial connection to a network is often through a wireless device. The most popular wireless network standard is the IEEE 802.11, which provides a reliable and secure connection to other networks by using an access point, a radio transmitter/receiver that usually includes border-routing capabilities. The Bluetooth network is another wireless network that is often used for peer-to-peer connection of cameras and cell phones to a computer and then through the computer to other networks. Cell phones also provide network connectivity of computing devices to other networks.

APPLICATIONS AND PRODUCTS

Computer networks have many applications. Some improve the operation of networks, while others provide services to people who use the network. The World Wide Web supplies many exciting new applications, and technologies such as teleconferencing and cloud computing have revolutionized computing. The adoption of such technologies was spurred on by the COVID-19 pandemic, during which many workers and schoolchildren relied on these technologies from their homes.

Network Infrastructure. Computer networks are complex, and many applications have been developed to improve their operation and management. A typical example of the software developed to manage networks is Cisco's Internetwork Operating System (IOS) software, an operating system for routers that also provides full support for the routing functions.

Communications and Social Networking. Communications programs are among the most popular computer applications in use. Early file-sharing applications, such as FTP, retain their popularity, and later protocols such as BitTorrent, which enables peer-to-peer sharing of very large files, and file hosting services such as Dropbox, which provides cloud storage and online backup services, are widely used. Teleconferencing through platforms such as Zoom is used to create virtual workplaces, and Voice over Internet Protocol (VoIP) allows businesses and home users to use the internet for digital telephone service.

The earliest computer network communications application, email, is still the largest and most successful network application. One accesses email services through an email client. The client can be a stand-alone program, such as Microsoft Outlook, or it can be web-based (webmail), accessed using a browser such as Safari, Microsoft Edge, or Google Chrome. The email server can be a proprietary system such as HCL Notes or one of the many webmail programs, such as Gmail, Outlook.com, or Yahoo! Mail.

One of the most important types of applications for computer networks is social networking sites. Facebook is by far the most widely used online social network, claiming more than 2.9 billion monthly active users as of April 2023, although services such as Twitter, TikTok, and YouTube are also very popular. Facebook was conceived by Mark Zuckerberg in 2004 while he was still a student at Harvard University. Facebook members can set up a profile that they can share with other members. The site supports a message service, event scheduling service, and news feed. By setting up a "friend" relationship with other members, one can quickly rekindle old friendships and gain new acquaintances. Although Zuckerberg's initial goal was to create an interactive environment for individuals, Facebook has fast become a way to promote businesses, organizations, and even politicians. Facebook does not charge a subscription fee and has a business model based on selling customer information and advertising as well as working with partners. Facebook's parent company, Meta, also owns several other social media services, including Instagram and WhatsApp, which each boasted about 2 billion monthly active users as of April 2023.

Cloud Computing. Cloud computing refers to the hosting of data and applications on remote servers and networks ("the cloud") instead of on local computers. Among other things, this makes the data and software services accessible from multiple locations, such as both home and office computers, as well as from mobile devices. This has benefits for data accessibility as well as businesses' approach to their technology needs, as cloud computing essentially allows them to outsource some of their information technology (both hardware and software) for an ongoing service fee rather than up-front capital costs. For individuals, it allows ready access to one's files and programs anywhere one has access to the internet. Platforms include Microsoft Azure, Amazon Web Services, and Google Cloud, among others. Communication services such as Slack and Microsoft Teams are further examples of cloud computing that gained popularity among businesses during the COVID-19 pandemic.

Next to email, the most common application is word processing, and Microsoft Word is the dominant word processor. Microsoft offers online versions of Word and its other Microsoft Office software through a service called Microsoft 365. Google offers similar cloud-based services, such as Google Workspace. As more and more applications are produced to run in the cloud, it is predicted that this will become the dominant form of computing, replacing the desktop computers of the early twenty-first century.

CAREERS AND COURSEWORK

A major in computer science is the traditional way to prepare for a computer networking job. Students first need courses in ethics, mathematics, and physics to form the basis for a computer science degree. Then they take courses in computer hardware and software. Those getting a computer science degree often take jobs developing network software or managing a network.

A major in information systems is another way to prepare for a computer networking job. Students must take courses in mathematics and business as a background for this degree in addition to courses on information systems development. Those getting degrees in computer information systems often take jobs as network managers, especially in small businesses.

In addition to the traditional academic programs that prepare someone for a computer networking job, a large number of professional training programs result in certification. Novell was the first company to develop a certification program for its NetWare network operating system. The Cisco Certified Network Professional (CCNP) program is a large certification program that covers a variety of topics related to networking, including network design.

SOCIAL CONTEXT AND FUTURE PROSPECTS

The development of computer networks and network applications has often resulted in some challenging legal issues. File exchange programs are very popular but can have real copyright issues. Napster developed a successful peer-to-peer file exchange program but was forced to close after only two years of operation by a court decision in 2001. Legislation has played an important role in the development of computer networks in the United States. The initial deregulation of the communications system and the breakup of AT&T in 1982 reduced the cost of the base communications system for computer networks and thus greatly increased the number of networks in operation. Opening up NSFNET for commercial use in 1991 made the internet possible.

Networking and access to the web have been increasingly important to business development and improved social networking. The emergence of cloud computing as an easy way to do computing has been just as transformative, especially as evidenced during the COVID-19 pandemic that began in 2020. Although there are still security and privacy issues to be solved, people are increasingly using mobile devices to access a wide variety of web services through the cloud. Using just a smartphone, people are able to communicate with friends and associates, transact business, compose letters, pay bills, read books, and watch films.

—George M. Whitson III

Further Reading

Comer, Douglas. *Computer Networks and Internets*. Pearson, 2015.

_____. *The Internet Book: Everything You Need to Know about Computer Networking and How the Internet Works*. 5th ed., Chapman and Hall/CRC, 2018.

Dordal, Peter L. "An Introduction to Computer Networks." *Loyola University Chicago*, 2015, intronetworks.cs.luc.edu/current/html. Accessed 13 July 2023.

FitzGerald, Jerry, Alan Dennis, and Alexandra Durcikova. *Business Data Communications and Networking*. 14th ed., Wiley, 2020.

Forouzan, Behrouz. *Data Communications and Networking*. 5th ed. McGraw-Hill, 2013.

"Global Social Media Statistics." *Datareportal*, Apr. 2023, datareportal.com/social-media-users. Accessed 13 July 2023.

Griffith, Eric. "What Is Cloud Computing?" *PCMag*, 15 Feb. 2022, www.pcmag.com/news/what-is-cloud-computing. Accessed 13 July 2023.

O'Leary, Mike. *Cyber Operations: Building, Defending, and Attacking Modern Computer Networks*. 2nd ed., Apress, 2019.

O'Leary, Timothy, Linda O'Leary, and Daniel O'Leary. *Computing Essentials 2023*. McGraw-Hill, 2022.

Stallings, William. *Data and Computer Communications*. 10th ed., Prentice Hall, 2014.

COMPUTER SOFTWARE

ABSTRACT

Computer software refers to one or more programs that determine how a computer operates. Software works along with a computer's hardware, or physical parts, to make the computer function efficiently for various tasks. There are two main kinds of computer software: system software and application software. System software is mainly used by the computer to help its parts communicate and cooperate. Application software is generally added to allow humans to perform specific tasks with the computer. Together, these kinds of software make computers valuable machines for work and entertainment alike.

BACKGROUND

Computers are electronic machines that can accept and interpret different forms of data and perform various operations using the data. People use computers for a great variety of purposes ranging from complex scientific research to entertainment. After several generations of development, computers have taken on a wide range of forms in the modern world. People today may use large computers on desktops or smaller laptop computers. Notebook computers and relatively tiny mobile versions become more popular each year, and embedded computers are critical parts of most automobiles and many appliances.

Despite the great variety in modern computers, they share some important similarities. One of the main similar features of all computers is that they use hardware and software. Hardware is a term referring to the physical parts of a computer. These parts include easily identifiable external features such as the central processing unit (CPU), keyboard, and mouse or touchpad. They also include smaller, internal features such as hard disks, or devices that allow the computer to store and access information. Hardware is important because it creates a basis upon which people may add software that will make the computer perform specific tasks.

OVERVIEW

Computer software is a much broader category than computer hardware. Software refers to all the programs that manufacturers or users may install on a computer. Software may be used by the computer, by the user, or by both to organize and use files and other data for specific tasks.

The most important software on a computer, and the first program that should be installed, is called the operating system (OS). The operating system serves as a wide-reaching interface, or a digital workstation through which users can install, organize, and operate other, smaller programs. It also helps the computer share its memory and other resources among its various tasks, handle input and output functions, and store new data.

For most computer users, the operating system is the first thing visible once the computer loads after it is turned on. Microsoft Windows is one of the most common operating systems in use, but some types of computers and computer hardware may work better with other operating systems. Some of the main competitors to Windows include macOS, ChromeOS, iOS, Linux, and Android.

Although the operating system is the most important software, there are dozens of other kinds of software available to modern computer users for both personal and professional tasks. This software can generally be divided into two main categories based on their types, the work they do, and the fields in which they are most commonly used. These main categories are system software and application software.

SYSTEM SOFTWARE

System software includes any kind of program that controls other programs or the computer's systems. These programs are meant to function closely with the hardware and are useful for making the computer and all of its other programs run more efficiently. The operating system is one prime example of system software. Another important example is a compiler, a program that translates the codes, or computer "language," used by the programs and hardware.

Other types of system software include programming tools such as loaders, linkers, and assemblers, which also gather and translate data. There are also drivers, or programs that assist hardware in performing its necessary tasks. All of these kinds of system software help the various features of a computer communicate and cooperate. Some also allow users to diagnose and solve problems with, or maximize the efficiency of, the computer or its programs.

APPLICATION SOFTWARE

The other main category of software is application software. Whereas system software is mostly used to help the computer function properly, application software is generally more focused on the needs of the user. This category includes all programs that are added to a computer to perform tasks for people using the computer. These tasks can be for productivity, for fun, or both, and may be general in nature or meant for specific goals.

Hundreds of examples of application software are available to computer users, each designed to meet specific user needs. Most of this software falls into broad categories. For instance, multimedia software lets computer users view, listen to, or even create visual and audio files. Spreadsheet and database software helps users store, sort, and analyze large amounts of information. Word processing software lets users create various kinds of documents, and presentation software helps users display their creations for others. Internet browsers, gaming programs, and communication tools are also popular types of application software.

—Mark Dziak

Further Reading

Campbell-Kelly, Martin, William F. Aspray, Jeffrey R. Yost, Honghong Tinn, and Gerardo Con Díaz. *Computer: A History of the Information Machine*. 4th ed., Routledge, 2023.

"Different Types of Software." *Introduction to IT English*, web2.uvcs.uvic.ca/elc/sample/ite/u01/u1_1_03.html. Accessed 13 July 2023.

O'Leary, Timothy, Linda O'Leary, and Daniel O'Leary. *Computing Essentials 2023*. McGraw-Hill, 2022.

Patterson, David A., and John L. Hennessy. *Computer Organization and Design: The Hardware/Software Interface*. 6th ed., Morgan Kaufmann, 2020.

Rosencrance, Linda. "Software." *Tech Target*, Mar. 2021, www.techtarget.com/searchapparchitecture/definition/software. Accessed 13 July 2023.

Conceptual Art

ABSTRACT

Conceptual art was a movement in which the concept behind a work of art took precedence over the work itself. Some of the art created during this movement, which took place during the 1960s and 1970s, is considered highly unorthodox. Conceptual art, also called conceptualism, often features everyday materials and "found objects" and incorporates text, photography, maps, and videos.

BACKGROUND

Conceptual art stemmed from earlier art movements such as Dada and Surrealism, which frequently featured people and objects in unusual and imaginative ways or presented unexpected objects as artworks. Perhaps the most notable example was Marcel Duchamp's *Fountain*, which is a photograph of a urinal with the name "R. Mutt" written on it. The term "conceptual art" in reference to the movement grew out of a series of articles and exhibits beginning in the early 1960s. In his 1967 essay "Paragraphs on Conceptual Art," Sol LeWitt explains, "When an artist uses a conceptual form of art, it means that all of the planning and decisions are made beforehand and the execution is a perfunctory affair." In other words, an artist spends a considerable amount of time thinking about an idea to present with an artwork but makes the final product unremarkable in appearance.

Early examples of conceptual art include *Twentysix Gasoline Stations* (1962), a book of photographs by Ed Ruscha, in which the artist presents simple photos of twenty-six gas stations in Oklahoma. *Art as Idea* (1965) by Joseph Kosuth features printed dictionary definitions of common words in newspapers or as photographs. Kosuth's *One and Three Chairs* (1965) shows a picture of a wooden chair hung on a wall, a wooden chair against the wall, and the definition of a chair also hung on the wall. An early pillar of conceptual art was John Baldessari's "word paintings," one of which features the words "I will not make any more boring art" written in cursive script in rows from the top of the canvas to the bottom. Other word paintings were created by professional sign painters hired by Baldessari. Most of these were simple statements on canvas about art or aesthetics. The intention of these works may be to taunt viewers, challenging their ideas about what constitutes a work of art.

OVERVIEW

The Art & Language group presented many of its ideas about conceptual art in a journal called *Art Language*. In its exhibition *Air/ Show Air Conditioning* (1966-77), the group simply presented a series of statements and diagrams—the diagrams gave instructions for the installation of air conditioning, while the texts addressed both air conditioning and art itself in an attempt to relate the two. Another significant conceptual art exhibition was organized by Seth Siegelaub. The show's title, *January 1-31 1969*, indicated when the exhibit would take place; its subtitle, "0 Objects, 0 Painters, 4 Artists ... 32 Works, 1 Exhibition, 2000 Catalogues," was similarly descriptive. Strangely, one of the works in the show was not even there—it consisted of radio waves sent by transmitters planted behind the wall of the exhibition space. The frequency of these waves was stated on a plaque.

Other conceptual artists included Joseph Beuys, John Latham, and Michael Craig-Martin. Beuys wrapped a piano in felt to illustrate human's inwardness and inability to communicate. Latham threw a party that grew into a piece of performance art—the guests were asked to take pages from a book by art critic Clement Greenberg, chew them, and spit them out. The pages were then mashed, fermented, and placed in a test tube to be brought back to the library from which the book was checked out. In one of his works, Craig-Martin cut several boxes in half and exchanged their parts.

As with many art movements, the conceptual art movement's timeline is difficult to pin down. However, *Six Years*, a definitive book by art critic Lucy Lippard, places the movement between 1966 and 1972. Many conceptual artists emerged during this period—all of whom strove to create art that communicated a thought or a group of thoughts instead of art that was pleasant to look at. Conceptual art was an intense movement that took place during a period of social and cultural upheaval. Conceptual art is controversial. Those who think the movement was beneficial to art history see it as a kind of rebellion against commercialism. Others think the art created during this movement is boring and pretentious. Still others doubt whether conceptual art should be considered art at all.

Some contemporary artists carry on the tradition of conceptual art. Artist Marina Abramovic staged a performance during her exhibit *The Artist Is Present* at the Museum of Modern Art in 2010 in which she simply sat at a table and stared at visitors. Abramovic sat silently and perfectly still for 750 hours as various museum visitors sat across from her—some of them moved to tears by the experience. In addition, critics have characterized some artists who incorporate computer technology into their work as influenced by the conceptual and postconceptual art movements.

—*Max Winter*

Further Reading

"Conceptual Art." *Tate*, www.tate.org.uk/learn/online-resources/glossary/c/conceptual-art#introduction. Accessed 13 July 2023.

Greenwood, Elizabeth. "Wait, Why Did That Woman Sit in the MoMA for 750 Hours?" *The Atlantic*, 2 July 2012, www.theatlantic.com/entertainment/archive/2012/07/wait-why-did-that-woman-sit-in-the-moma-for-750-hours/259069. Accessed 13 July 2022.

Hoy, Meredith. *From Point to Pixel: A Genealogy of Digital Aesthetics*. Dartmouth College P, 2017.

"Joseph Beuys." *Walker*, walkerart.org/collections/artists/joseph-beuys. Accessed 13 July 2023.

"Post-Conceptualism and Digital Art: The Future of Art?" *US Advisory*, www.ugoscalia.com/post/post-conceptualism-and-digital-art-the-future-of-art. Accessed 13 July 2023.

Samuels, Miranda. "Tragedy of the Non-Visible: Aesthetic and Ontological Differences in Early Computer versus Conceptual Art." *Leonardo*, vol. 56, no. 2, 2023, pp. 159-63.

Schellekens, Elizabeth. "Conceptual Art." *Stanford Encyclopedia of Philosophy*, 21 Aug. 2022, plato.stanford.edu/archives/fall2022/entries/conceptual-art. Accessed 13 July 2023.

COPYWRITER

ABSTRACT

Copywriters create the written content used in print advertisements, websites, brochures, commercials, and other marketing materials. They may be employed by specific companies or marketing agencies or may work on a freelance basis. Students aspiring to enter the profession should take courses in subjects such as English, marketing, and journalism.

BACKGROUND

Copywriters work within the communication and information sectors. They research and prepare the written words that accompany advertising, promotional, and marketing materials. These include brochures, print advertising, press releases, scripts for television and radio commercials, websites, direct mail pieces, and any other communications that call for an ability to write engaging and persuasive content.

Copywriters are often employed by marketing and advertising agencies, but they may also work as independent freelancers. Copywriting is a highly collaborative role, which usually demands working with a team of creative colleagues and supervisors. Copywriters interact with other communication specialists, such as marketers, brand strategists, advertising executives, public relations executives, graphic designers, art directors, multimedia technicians, and editors. They are also likely to work with clients and business specialists across a broad range of industries, topics, products, and services.

OVERVIEW

Copywriters work in an office environment, although the role is sufficiently flexible to accommodate working from any place where computer and telecommunication technologies are readily available. Freelance copywriters may work remotely, from a home office or other setting.

Full-time copywriters generally work for marketing and advertising agencies or within other communication outlets. A full-time employee may expect to work forty hours per week during normal office hours. He or she may be required to work longer hours, as needed.

Copywriting attracts graduates and professionals who have a strong grounding in written communications. This occupation suits people with an interest in writing and a flair for creative expression and engaging an audience. Copywriters must be able to write fluently on a broad range of topics and to manage multiple projects concurrently. They usually have a strong interest in trends and markets.

In addition to having excellent writing skills, copywriters must possess strong research and analytical abilities. They must have advanced oral communication and collaboration skills and the ability to produce creative work under pressure. Copywriting demands good organizational, prioritization, and time management skills. Copywriters must also be

Image via iStock/VectorUp. [Used under license.]

able to respond positively to constructive criticism and feedback about the work they produce.

DUTIES AND RESPONSIBILITIES

The copywriter's day is characterized by periods of independent and collaborative work. As a member of a creative team, copywriters meet with their colleagues, supervisors, and clients on a daily basis to brainstorm and present ideas as well as develop, analyze, and critique creative strategies and solve problems. Copywriters spend solitary time researching, drafting advertising and promotional materials, and revising content as needed. Research may entail activities like referring to consumer surveys or conducting interviews. The solitary aspects of the occupation demand high levels of self-discipline and self-motivation. This is especially important for freelance copywriters who spend much of their time working alone and managing their own workloads.

The copywriter's daily output includes written content and concepts for brochures, print advertising, billboards, press releases, scripts for television and radio commercials, websites, direct mail pieces, and other marketing materials. The copywriter may find that the projects and assignments they work on are subject to tight timeframes and strict deadlines. It is a daily challenge for copywriters to produce high quality creative work under pressure.

Copywriters may be expected to contribute within their workplaces more widely. This may include some administrative duties, such as tracking project hours.

WORK ENVIRONMENT

Physical Environment. Office settings predominate. Copywriters in full-time employment generally work for small to large marketing and advertising firms or in other corporate contexts. Freelance copywriters often work from home.

Human Environment. Copywriting demands strong collaborative skills. Copywriters interact with other communication and creative specialists, such as marketers, brand strategists, advertising executives, public relations executives, graphic designers, art directors, multimedia technicians, and editors. They are also likely to work with clients and business specialists across a broad range of industries.

Technological Environment. Copywriters use technologies that range from telephone, email, and the internet to word processing software. Copywriters may also be expected to work with web content management systems and blogging software. Advanced computing skills and an understanding of multimedia, social media, and emerging media technologies are considered an advantage.

EDUCATION AND TRAINING

High School/Secondary. High school students can best prepare for a career in copywriting by taking courses in English literature, language and composition, social studies, journalism, and business communications. The creative nature of copywriting may be explored through art and graphic design; the business aspects through business studies, accounting, and entrepreneurship; and the technology aspects through computer literacy. Courses such as history and sociology can also prepare the student for synthesizing research into written materials. Psychology and cultural studies may provide an understanding of group and individual responses to written and visual messaging. Extracurricular school activities that involve writing can also provide students with an opportunity to develop their writing skills and learn from others prior to graduation. Such activities might include entering writing competitions and writing for school newspapers or club newsletters.

College/Postsecondary. Graduates from a diverse range of disciplines can become copywriters. The most common pathway to copywriting is by obtaining an undergraduate degree in communications, advertising, marketing, or journalism. Coursework in the social sciences can also be helpful preparation for writing persuasive content. A large number of colleges and universities offer courses in copywriting. There are also an increasing number of certified and noncertified copywriting programs, seminars, and workshops offered by private companies, academies, and professional associations.

Postsecondary students interested in a career in copywriting are also encouraged to become involved in extracurricular, club, or volunteer roles where they can develop their writing skills and begin to build a writing portfolio.

Adult Job Seekers. Adults seeking a career transition into copywriting are advised to develop copywriting experience and a portfolio through part-time or volunteer work for a charity, nonprofit organization, local club, or association. Writing a personal blog is another way to build a portfolio while also developing the daily discipline of writing.

An increasing number of copywriting opportunities are advertised on nontraditional job sites. The proliferation of web-based freelance writing sites allows people with little or no previous copywriting experience to upload their own articles for sale or bid for writing projects.

Taking writing courses may assist with networking and portfolio development. Self-guided learners may also benefit from reading "how to" books about copywriting or by undertaking short courses, seminars, or workshops.

Professional Certification and Licensure. There are no required professional certifications or licenses for copywriting. Practical experience generally outweighs formal qualifications in this occupation. Some professional associations provide certifications in the field. The American Marketing Association offers the Professional Certified Marketer (PCM) certification, which requires the completion of an exam. Consult credible professional associations within the field and follow professional debate as to the relevancy and value of any certification program.

Additional Requirements. The most important attribute for prospective copywriters is a love of writing combined with interest in a broad range of topics. In addition to advanced writing skills, copywriters must possess strong social and cultural awareness, an understanding of trends and markets, and the ability to write about products and services in emotionally engaging terms. A portfolio of writing samples is essential.

—*Kylie Grimshaw Hughes*

Further Reading

Bly, Robert W. *The Copywriter's Handbook: A Step-by-Step Guide to Writing Copy That Sells.* 4th ed., St. Martin's Griffin, 2020.

Kuehn, Scott A., and James Andrew Lingwall. *The Basics of Media Writing: A Strategic Approach.* CQ Press, 2017.

Lingwood, Gyles, and Mark Shaw. *Copywriting: Successful Writing for Design, Advertising, and Marketing.* 3rd ed., Laurence King, 2022.

CRAFT/FINE ARTIST

ABSTRACT

Craft and fine artists create original works of art for sale and exhibition. They use a variety of materials and create a wide range of works, from paintings to sculptures to jewelry. Students aspiring to enter the profession may benefit from formal instruction in a given craft or art form as well as classes in business.

BACKGROUND

Craft and fine artists use a variety of materials and techniques to create original works of art for sale and exhibition. Craft artists create objects, such as pottery, glassware, and textiles, that are designed to be functional. Fine artists, including painters, sculptors, and illustrators, create pieces of art more for aesthetics than for function. Artists may employ the use of technology such as graphic-design software, three-dimensional (3D) printing, and artificial intelligence (AI) in the creation of their art.

OVERVIEW

Many artists work in fine- or commercial-art studios located in office buildings, warehouses, or lofts. Others work in private studios in their homes. Some artists share studio space, where they also may exhibit their work. Artists may also choose to work outdoors, depending upon their medium and artistic intent.

Most craft and fine artists work full-time, although part-time and variable schedules are also common. Many hold another job in addition to their work as an artist. During busy periods, artists may work additional hours to meet deadlines. Those who are self-employed usually determine their own schedules, although those employed in forensics may be on call.

Professional craft and fine artists are creative individuals by nature and have a desire to not only share their creations, but to turn their passion into a source of income. Over half of all craft and fine artists are self-employed, meaning these individuals must have a drive to succeed in a crowded and highly entrepreneurial field. In many cases, they should expect to work another job to make a living wage.

DUTIES AND RESPONSIBILITIES

Artists create objects that are beautiful, thought provoking, and sometimes shocking. They often strive to communicate ideas or feelings through their art.

Craft artists work with many different materials, including ceramics, glass, textiles, wood, metal, and paper. They use these materials to create unique pieces of art, such as pottery, quilts, stained glass, furniture, jewelry, and clothing. Many craft artists also use fine art techniques—for example, painting, sketching, and printing—to add finishing touches to their products.

Fine artists typically display their work in museums, in commercial or nonprofit art galleries, at craft fairs, in corporate collections, on the internet, and in private homes. Some of their artwork may be commissioned (requested by a client), but most is sold by the artist or through private art galleries or dealers. The artist, gallery, and dealer together decide in advance how much of the proceeds from the sale each will keep.

Most craft and fine artists spend their time and effort selling their artwork to potential customers and building a reputation. This can include in-person showings and online sales through marketplaces de-

Photo via iStock/visualspace. [Used under license.]

signed specifically for selling one's own art. In addition to selling their artwork, many artists have at least one other job to support their craft or art careers.

Some artists work in museums or art galleries as art directors or as archivists, curators, or museum workers, planning and setting up exhibits. Others teach craft or art classes or conduct workshops in schools or in their own studios. Those in forensics may choose their disciplines, such as composite art, image modification (such as age progression), and postmortem and facial reconstruction.

OCCUPATION SPECIALTIES

Cartoonist. Cartoonists create simplified or exaggerated drawings to visually convey political, advertising, comic, or sports concepts. Some cartoonists work with others who create the idea or story and write captions. Others create plots and write captions themselves. Most cartoonists have humorous, critical, or dramatic talent, in addition to drawing skills.

Ceramic Artist. Ceramic artists shape, form, and mold artworks out of clay, often using a potter's wheel and other tools. They glaze and fire pieces in kilns, which are large, special furnaces that dry and harden the clay.

Digital Artist. Digital artists use design and production software to create interactive art online. The digital imagery may then be transferred to paper or some other form of printmaking or made available directly on web-accessible devices. Digital artists may also use AI in the creation of their art, including the use of robotic systems or algorithms, although that practice is a controversial one.

Fiber Artist. Fiber artists use fabric, yarn, or other natural and synthetic materials to weave, knit, crochet, or sew textile art. They may use a loom to weave fabric, needles to knit or crochet yarn, or a sewing machine to join pieces of fabric for quilts or other handicrafts.

Fine-Art Painter. Fine-art painters paint landscapes, portraits, and other subjects in a variety of styles, ranging from realistic to abstract. They may work in a variety of media, such as watercolors, oil paints, and acrylics.

Furniture Maker. Furniture makers cut, sand, join, and finish wood and other materials to make handcrafted furnishings.

Glass Artist. Glass artists process glass in a variety of ways—such as by blowing, shaping, staining, or joining it—to create artistic pieces. Some processes require the use of kilns, ovens, and other equipment and tools that bend glass at high temperatures. These workers also decorate glass objects, such as by etching or painting.

Illustrator. Illustrators create pictures for books, magazines, and other publications and for commercial products, such as textiles, wrapping paper, stationery, greeting cards, and calendars. Illustrators increasingly use computers in their work. They might draw in pen or pencil and then scan the image, using software to add color, or they might use a special pen to draw images directly onto the computer.

Jewelry Artist. Jewelry artists use metals, stones, beads, and other materials to make objects for personal adornment, such as earrings or necklaces.

Medical/Scientific Illustrator. Medical and scientific illustrators combine drawing skills with knowledge of biology or other sciences. Medical illustrators work with computers or with pen and paper to create images, 3D models, and animations of human anatomy and surgical procedures. Scientific illustrators draw animal and plant life, atomic and molecular structures, and geologic and planetary formations. These illustrations are used in medical and scientific publications and in audiovisual presentations for teaching purposes. Some medical and scientific illustrators work for lawyers, producing exhibits for court cases.

Printmaker. Printmakers create images on a silk screen, woodblock, lithography stone, metal etching plate, or other types of matrices. A hand printing press then creates the final work of art, inking and transferring the matrix to a piece of paper.

Public Artist. Public artists create large paintings, sculptures, and displays called "installations" that are meant to be seen in open spaces. These works are typically displayed in parks, museum grounds, train stations, and other public areas.

Sculptor. Sculptors design and shape 3D works of art, either by molding and joining materials such as clay, glass, plastic, and metal or by cutting and carving forms from a block of plaster, wood, or stone. Some sculptors combine various materials to create mixed-media installations, such as by incorporating light, sound, and motion into their work.

Sketch Artist. Sketch artists are a type of illustrator who often use pencil, charcoal, or pastels to create likenesses of subjects. Their sketches are used by law enforcement agencies to help identify suspects, by the news media to show courtroom scenes, and by individual customers for their own enjoyment.

Tattoo Artist. Tattoo artists use stencils and draw by hand to create original images and text on skin. With specialized needles, these artists use a variety of styles and colors based on their clients' preferences.

Video Artist. Video artists record avant-garde, moving imagery that is typically shown in a loop in art galleries, museums, or performance spaces. These artists sometimes use multiple monitors or create unusual spaces for the video to be shown.

WORK ENVIRONMENT

Physical Environment. Work settings for artists vary depending upon the person and the type of art being created. Artists may also desire to have particular surroundings in order to inspire and stimulate creativity, or to allow them to focus on their work.

Artists' studios are usually well lit and ventilated. However, artists may be exposed to fumes from glue, paint, ink, and other materials. They may also have to deal with dust or other residue from filings, splattered paint, or spilled cleaning and other fluids. Artists often wear protective gear, such as breathing masks and goggles, to remain safe from exposure to harmful materials. Ceramic and glass artists must use caution in working with materials that may break into sharp pieces and in using equipment that can get extremely hot, such as kilns. Artists and related workers have one of the highest rates of injuries and illnesses of all occupations.

Human Environment. Depending upon the person and nature of their art, human interaction during working hours may be limited. However, artists commonly share their studios with like-minded individuals to help cover rental and other costs and possibly to share ideas and techniques. Artists also must be comfortable interacting with the public when selling their art, be it in person or online, and must be able to cope with criticism of their work. When selling or showing art in public, artists must also interact with gallery owners.

Technological Environment. Technological requirements vary widely in the field of craft and fine art. Much depends upon the type of art being created and the medium being used. Most artists work with their hands and must be familiar with the tools and techniques required to accomplish what they envision. A painter must know the benefits and limitations of certain canvases, paints, and brushes, for example, while a sculptor must understand the properties of the material being used to sculpt. Painters may also become skilled at framing. Artists increasingly also use computer technology, including advances in areas such as AI and 3D printing, in novel ways to create new forms of art. Generative adversarial networks (GANs) are sometimes used by technologically based artists, whereby a generator is used to create images, and a discriminator chooses which images are considered successful. Also popular are nonfungible tokens (NFTs), which are unique units of data that can take the form of photos, videos, and audio and are stored on a blockchain, similar to cryptocurrency.

EDUCATION AND TRAINING

High School/Secondary. High school art classes can teach prospective craft artists the basic drawing skills they need. Craft and fine artists improve their skills through practice and repetition. English classes are also important for improving communication skills. High school students interested in pursuing art as a business should also take fundamental business classes and be comfortable with practical math. Computer courses are also an asset for students interested in pursuing any form of computer-based art.

College/Postsecondary. Most fine artists pursue postsecondary education to improve their skills and job prospects. A formal educational credential is typically not needed to be a craft artist. However, it is difficult to gain adequate artistic skills without some formal education. Computer courses may be desirable for artists interested in using cutting-edge technology in their work, although many of these skills can be learned at home.

A number of colleges and universities offer bachelor's and master's degrees in subjects related to fine arts. In addition to studio art and art history, postsecondary programs may include core subjects, such as English, marketing, social science, and natural science. Independent schools of art and design also offer postsecondary education programs, which can lead to a certificate in an art-related specialty or to an associate, bachelor's, or master's degree in fine arts.

The National Association of Schools of Art and Design (NASAD) accredits more than three hundred postsecondary institutions with programs in art and design. Most of these schools award a degree in art.

Education gives artists an opportunity to develop their portfolio, which is a collection of an artist's work that demonstrates his or her styles and abilities. Portfolios are essential, because art directors, clients, and others look at them when deciding whether to hire an artist or to buy the artist's work. In addition to compiling a physical portfolio, many artists choose to create a portfolio online.

Adult Job Seekers. Adults who are looking to turn their artistic passion into a business should become familiar with art currently on the market in their chosen niche and start to build an online presence. Doing so can simultaneously help curate a portfolio, which is vital so an artist can show examples of their work to prospective buyers. Individuals seeking to become forensic artists should approach local law enforcement agencies or look for related job postings. Adults can also attend continuing education courses, such as night school, to improve techniques or learn new methods.

Professional Certification and Licensure. Those who want to teach fine arts at public elementary or secondary schools usually must have a teaching certificate in addition to a bachelor's degree.

Additional Requirements. Craft and fine artists improve their skills through practice and repetition. They can train in several ways other than, or in addition to, formal schooling. Craft and fine artists may train with simpler projects before attempting something more ambitious.

Some artists learn on the job from more experienced artists. Others attend noncredit classes or workshops or take private lessons, which may be offered in artists' studios or at community colleges, art centers, galleries, museums, or other art-related institutions.

—*Stuart Paterson*

Further Reading

Costello, Vic. *Multimedia Foundations: Core Concepts for Digital Design*. 3rd ed., Focal Press, 2023.

DeWitte, Debra J., M. Kathryn Shields, and Ralph M. Larmann. *Gateways to Art: Understanding the Visual Arts*. 4th ed., Thames & Hudson, 2023.

Paul, Christiane. *Digital Art*. 3rd ed., Thames & Hudson, 2015.

CREATIVE COMMONS

ABSTRACT

Creative Commons (CC) is an international nonprofit organization that provides easy access to creative works that others can legally build upon and share through its range of copyright licenses known as Creative Commons licenses. These licenses, which are completely free of charge, allow the makers of various creative works to easily communicate with others which rights to their works they reserve and which they voluntarily waive. In other words, CC licenses are a standardized way for creators of all types to grant permission for the public to legally use their work within the structure of established copyright law. Since it was originally founded in 2001, CC has worked to overcome the legal obstacles involved with copyright law to make it easier for people around the world to share information and creativity. CC licenses enjoy widespread use, notably serving as the default license of Wikipedia and covering millions of Flickr images and YouTube videos.

BACKGROUND

Creative Commons (CC) was primarily created to simplify the sometimes-complicated concepts of copyright and copyright law. Copyright is the legal right of owners of intellectual property to reproduce their creations as they see fit. Put another way, copyright means that the creators of intellectual property and those to whom they grant permission are the only people allowed to make copies of the work in question. Copyright law, which varies from country to country, is the legal framework that governs how copyright works.

The basic idea of copyright is relatively straightforward. When someone uses significant mental effort to create an original product, that product is automatically viewed as intellectual property. To ensure that the creator gets appropriate credit and compensation for their work, it is necessary for their work to be protected from unauthorized duplication. Many things can be considered intellectual property, including books, films, music, and art. Copyright law is the mechanism through which intellectual property is protected.

Copyright law defines original works as those created entirely through independent thought and without any reliance on duplication. A work produced in this manner is referred to as an original work of authorship (OWA). When an OWA is created, it automatically becomes copyrighted. This means that no one else has the right to use or duplicate the work without the creator's consent. It is

Image via iStock/YuriyVlasenko. [Used under license.]

important to note, however, that the concept of copyright applies only to tangible works. Intangible creations like ideas, names, and logos may instead be protected through trademarks or patents.

The specifics of US copyright law clearly define the rights enjoyed by the holders of copyrights. According to the law, the creator of a copyrighted work has the exclusive right to reproduce the work, create new works derived from the original, publicly distribute copies of the work, display or perform the work publicly in various ways as applicable, and authorize others to exercise these same rights in relation to the work. Further, copyright law establishes that in the case of works created after January 1, 1978, legal copyright protection lasts throughout the lifetime of the OWA's creator, plus an additional seventy years.

OVERVIEW

As an organization, Creative Commons provides an alternative to traditional copyright law that makes it easier for the creators of OWAs to allow others to use and share their works. It accomplishes this feat by offering a collection of CC licenses that afford creators the ability to decide for themselves how others are permitted to use their work. By releasing their work under the CC license of their choice, creators clearly define what the public can and cannot do with the work in question. From the other side of the equation, the use of CC licenses gives members of the public the ability to freely use a copyrighted work without having to seek permission from the creator unless they want to do so in a way that is not covered by the specific CC license in play. On top of all that, CC licenses are also advantageous because they all allow the open use of covered works for educational purposes. This affords teachers and students the opportunity to duplicate, share, and in some cases even modify CC works without having to seek permission.

The six distinct types of CC licenses place different restrictions on how a work may be used, but all share some common characteristics. All CC licenses allow users to copy a work, distribute a work, display or perform a work, communicate a work, and format shift verbatim copies of a work. Beyond that, each type of CC license varies in its degree of permissiveness. The six types of CC licenses include: CC BY (Attribution), CC BY-SA (Attribution-ShareAlike), CC BY-NC (Attribution-NonCommercial), CC BY-NC-SA (Attribution-NonCommercial-ShareAlike), CC BY-ND (Attribution-NoDerivs), and CC BY-NC-ND (Attribution-NonCommercial-NoDerivs).

Each license is denoted by special symbols that appear on or near a licensed work. CC BY, the most permissive CC license, allows users to copy, distribute, change, and build upon an existing work in virtually any way as long as the work is properly attributed to the original creator. It also allows for commercial use. The CC BY-SA license allows users to do all of the same things that a CC BY license does, except that it requires users who change or build upon an existing work to license the modified materials under the same terms. CC BY-SA licenses allow for commercial uses as well. CC BY-NC licenses are the same as CC BY licenses, but the works they protect can only be used for noncommercial purposes. CC BY-NC-SA licenses are the same as CC BY-SA licenses, except that they prohibit commercial uses. CC BY-ND licenses allow users to copy and distribute a work in any medium or format, but only in unaltered form. Commercial use is permitted. CC BY-NC-ND licenses are the same as CC BY-ND, except commercial uses are not permitted.

CC's work goes well beyond issuing licenses. The organization also provides helpful public domain tools, works with other institutions and governments to create and implement open licensing protocols, develops technologies that make it easier to find and use open license materials, offers a course for those who wish to become open license experts, and more.

—Jack Lasky

Further Reading

"About the Licenses." *Creative Commons*, 2023, creativecommons.org/licenses. Accessed 13 July 2023.

Bailey, Jonathan. "What Is Creative Commons Anyway?" *Plagiarism.org*, 24 July 2018, www.plagiarism.org/blog/2018/07/24/what-is-creative-commons-anyway. Accessed 13 July 2023

Kenton, Will. "Copyright." *Investopedia*, 10 June 2023, www.investopedia.com/terms/c/copyright.asp. Accessed 13 July 2023.

"What Is Copyright?" *Copyright.gov*, www.copyright.gov/what-is-copyright. Accessed 13 July 2023.

"What Is a Creative Commons License?" *Copyright Alliance*, 2023, copyrightalliance.org/faqs/what-is-creative-commons-license. Accessed 13 July 2023.

Crowdfunding

ABSTRACT

Crowdfunding is a method by which an individual or group raises money for a project, cause, or business by receiving donations from a large group of individuals, often by means of many small donations from people who have no relationship to each other beyond interest in the project. Crowdfunding began largely as a way for musicians, artists, and other creative people to gain support for their projects and relies heavily on the Internet to reach potential funders. Crowdfunding is also used to raise money for philanthropic causes and to raise investment capital for new businesses.

BACKGROUND

Despite its burgeoning popularity in the internet era, crowdfunding is not a new concept. For centuries artists and publishers have sought subscriptions to finance creative projects. In the United States, notable examples include the public solicitation of contributions to pay for the pedestal for the Statue of Liberty in 1884 and the 1933 public contributions to build a swimming pool in the White House so that President Franklin Roosevelt, who was disabled as the result of polio, could enjoy water therapy.

There are two primary models for crowdfunding: donation-based crowdfunding and investment crowdfunding. In donation-based crowdfunding, donors contribute money for a project, such as the creation of a film or book, and in turn receive gifts such as a personal note from the author, a coffee mug or T-shirt advertising the project, or a copy of the product created (e.g., a copy of the book or film). This model is similar to the fundraising method used on National Public Radio (NPR) and PBS television, where people make a charitable contribution to their local station or to a particular program they enjoy and may elect to receive some branded item in return. In investment crowdfunding, businesses sell debt or equity stakes to investors, and those investors will share in the company's profits, if there are any.

OVERVIEW

In the first decades of the twenty-first century, crowdfunding initiatives were primarily carried out on the internet. The 2000s and 2010s saw the establishment of several prominent online crowdfunding platforms, including Indiegogo, founded in 2008; Kickstarter, founded in 2009; and GoFundMe, founded in 2010. Popular crowdfunding sites also included AngelList, founded in 2010, which enabled startup companies to raise money from investors.

Each site had its own set of rules. For instance, Kickstarter hosted campaigns for creative projects and did not permit fundraising for charity purposes, while GoFundMe permitted fundraising for most kinds of project and also allowed users to raise funds for personal needs such as medical expenses. In addition, some sites, including Kickstarter, required that project creators set a goal (amount of money to be raised) and a deadline; if the goal was not met by the deadline, no money was collected from the donors who pledged it. Other sites, such as Indiegogo, allowed campaign creators to keep the donations from a partially funded campaign.

The field of crowdfunding grew rapidly throughout the early twenty-first century. According to Chance Barnett, writing for *Forbes* in August 2013, in 2012 the crowdfunding industry raised $2.7 billion across more than one million individual campaigns. In May 2014, a report by the Crowdfunding Centre estimated that, worldwide, more than $60,000 were being raised via crowdfunding every hour, and 442 crowdfunding campaigns were being initiated each day. In 2022, the website *Fundera* reported, nearly 6.5 million crowdfunding campaigns were conducted worldwide, with campaigns conducted in North America generating more than $17 billion in total.

—*Sarah E. Boslaugh*

Further Reading

Barnett, Chance. "Top 10 Crowdfunding Sites for Fundraising." *Forbes*, 8 May 2013, www.forbes.com/sites/chancebarnett/2013/05/08/top-10-crowdfunding-sites-for-fundraising/?sh=4dd914103850. Accessed 13 July 2023.

Clifford, Catherine. "Crowdfunding Generates More Than $60,000 an Hour." *Entrepreneur*, 19 May 2014, www.entrepreneur.com/starting-a-business/crowdfunding-generates-more-than-60000-an-hour/234051. Accessed 13 July 2023.

Cunningham, William Michael. *The JOBS Act: Crowdfunding for Small Businesses and Startups*. Springer, 2012.

Di Pietro, Francesca. *Crowdfunding for Entrepreneurs: Developing Strategic Advantage through Entrepreneurial Finance*. Routledge, 2020.

Dredge, Stuart. "Kickstarter's Biggest Hits: Why Crowdfunding Now Sets the Trends." *Guardian*, 17 Apr. 2014, www.theguardian.com/technology/2014/apr/17/kickstarter-crowdfunding-technology-film-games. Accessed 13 July 2023.

Dushnitsky, Gary. "What If Crowdfunding Becomes the Leading Source of Finance for Entrepreneurs or Growing Companies?" *Forbes*, 20 May 2015, www.forbes.com/sites/lbsbusinessstrategyreview/2015/05/20/what-if-crowdfunding-becomes-the-leading-source-of-finance-for-entrepreneurs-or-growing-companies/?sh=248b28aa4faa. Accessed 13 July 2023.

Lawton, Kevin, and Dan Marom. *The Crowdfunding Revolution: How to Raise Venture Capital Using Social Media*. McGraw. 2012.

Schneiderhan, Erik, and Martin Lukk. *GoFailMe: The Unfulfilled Promise of Digital Crowdfunding*. Stanford UP, 2023.

Shepherd, Maddie. "Crowdfunding Statistics: Market Size and Growth." *Fundera*, 23 Jan. 2023, www.fundera.com/resources/crowdfunding-statistics. Accessed 13 July 2023.

Trigonis, John T. *Crowdfunding for Filmmakers: The Way to a Successful Film Campaign*. Michael Wiese Productions, 2013.

Crowdsourcing

ABSTRACT

Crowdsourcing can be thought of a specialized variant of outsourcing, one where specific content is solicited from a large, and often anonymous, online community. Individuals respond to a call for information by providing small amounts of data. The entities that sponsor crowdsourcing include corporations, academic institutions, governments, and individuals. Contributors to crowdsourcing projects may be paid professionals or volunteers merely interested in sharing their knowledge.

BACKGROUND

The term *crowdsourcing* made its first print appearance in the June 2006 issue of *Wired* magazine, and by 2013 it was an established way of conducting business on the Internet. Although crowdsourcing is a creation of the internet era, information-gathering projects with considerable numbers of participants have existed for decades. Crowdsourcing served as the means for the creation of the Oxford English

Dictionary, was responsible for new methods for determining a ship's location in the Longitude Prize contest in Great Britain, and became a way to collect genealogical data for members of the Church of Jesus Christ of Latter-day Saints.

The Internet Movie Database (IMDB) and Wikipedia are two of the most successful examples of crowdsourced internet resources. Another well-received crowdsourcing project was the New York Public Library's "What's on the Menu," which began in 2011 and sought to transcribe more than 45,000 historical restaurant menus from the 1840s to the 2010s. In 2011 the California Digital Newspaper Collection utilized crowdsourcing to correct the text of digitized newspapers, some of which predate California's founding in 1850. The Katrina People Finder Project was developed in the wake of Hurricane Katrina in 2005; more than ninety thousand entries were added in an effort to keep track of the victims of the natural disaster. Carnegie Mellon University's CAPTCHA system, established in 2000, is a type of implicit crowdsourcing in which users identify characters as part of a logon sequence. User responses, in turn, are used to determinate text that optical character recognition (OCR) software cannot, thereby aiding the digitization of historical text documents.

OVERVIEW

The appeal of crowdsourcing is based on its relatively inexpensive cost and its ability to undertake problems too complex to solve with an organization's existing staff, as well as the idea that harnessing the knowledge of multiple individuals, who possess different knowledge and experiences, produces better, more accurate data. While individuals who participate in crowdsourcing are typically demographically diverse, many have academic training or significant professional experience relevant to the project. Some critics decry the quality and accuracy of crowdsourcing projects, but by the 2010s, the practice had become commonplace across a wide array of industries and endeavors.

Specialized forms of crowdsourcing allow the efforts of multiple users to achieve numerous types of distinct goals. Crowd creation is used frequently to solve complicated scientific problems. Crowdfunding involves collecting small sums of money from numerous contributors to meet fundraising goals for a start-up company or an artistic endeavor. Crowd searching leverages large networks of internet and smartphone users to ascertain the location of a missing item or individual. Crowd voting makes use of communal judgment to grade, sort, and categorize web-based content; it is the most common type of crowdsourcing and is prominently used on websites such as Reddit.

—Leon James Bynum

Further Reading

Botto, Richard. *Crowdsourcing for Filmmakers: Indie Film and the Power of the Crowd*. Routledge, 2017.

Brabham, Daren C. *Crowdsourcing*. MIT Press, 2013.

Estelles-Miguel, Sofia, Ignacio Gil-Pechuán, and Fernando J. Garrigos-Simon. *Advances in Crowdsourcing*. Springer, 2015.

Hargrave, Marshall. "Crowdsourcing: Definition, How It Works, Types, and Examples." *Investopedia*, 20 Nov. 2022, www.investopedia.com/terms/c/crowdsourcing.asp. Accessed 13 July 2023.

Howe, Jeff. *Crowdsourcing: Why the Power of the Crowd Is Driving the Future of Business*. Crown, 2009.

Palmer, Daniel. *Photography and Collaboration: From Conceptual Art to Crowdsourcing*. Routledge, 2017.

Sloane, Paul, editor. *A Guide to Open Innovation and Crowdsourcing: Advice from Leading Experts*. Kogan, 2012.

Suie, Daniel, Sara Elwood, and Michael Goodchild, editors. *Crowdsourcing Geographic Knowledge: Volunteered Geographic Information (VGI) in Theory and Practice*. Springer, 2013.

Surowiecki, James. *The Wisdom of Crowds: Why the Many Are Smarter than the Few and How Collective Wisdom Shapes Business, Economies, Societies, and Nations*. New York: Doubleday, 2004.

Tsai, Wei-Tek, et al., editors. *Crowdsourcing: Cloud-Based Software Development*. Springer, 2015.

Cryptography

ABSTRACT

Early cryptography focused on ensuring that written messages could be sent to their intended recipients without being intercepted and read by other parties. This was achieved through various encryption techniques. Encryption is based on a simple principle: the message is transformed in such a way that it becomes unreadable. The encrypted message is then transmitted to the recipient, who reads it by transforming (decrypting) it back into its original form. The development of modern computer systems in the 1950s changed the world of cryptography in major ways, and further technological advancements rendered cryptography increasingly essential to business, government, and society throughout the late twentieth and early twenty-first centuries.

BACKGROUND

The word "cryptography" comes from the Greek words *kryptos* ("hidden," "secret") and *graphein* ("writing"). Early forms of encryption were based on ciphers. A cipher encrypts a message by altering the characters that comprise the message. The original message is called the "plaintext," while the encrypted message is the "ciphertext." Anyone who knows the rules of the cipher can decrypt the ciphertext, but it remains unreadable to anyone else.

Among the earliest ciphers used were the transposition cipher and the substitution cipher. A transposition cipher encrypts messages by changing the order of the letters in the message using a well-defined scheme. One of the simplest transposition ciphers involves reversing the order of letters in each word. When encrypted in this fashion, the message "MEET ME IN THE PARK" becomes "TEEM EM NI EHT KRAP." More complicated transposition ciphers might involve writing out the message in a particular orientation (such as in stacked rows) and then reading the individual letters in a different orientation (such as successive columns).

A substitution cipher encodes messages by substituting certain letters in the message for other letters. One well-known early substitution cipher is the Caesar cipher, named after Julius Caesar. This cipher encodes messages by replacing each letter with a letter that is a specified number of positions to its right or left in the alphabet. For example, Caesar is reported to have used a left shift of three places when encrypting his messages.

Early ciphers were relatively easy to break, given enough time and a working knowledge of statistics. By the 1920s, electromechanical cipher machines called "rotor machines" were creating complex ciphers that posed a greater challenge. The best-known example of a rotor machine was the Enigma machine, used by the German military in World War II. Soon after, the development of modern computer systems in the 1950s would change the world of cryptography in major ways.

OVERVIEW

With the introduction of digital computers, the focus of cryptography shifted from just written language to any data that could be expressed in binary format. The encryption of binary data is accomplished through the use of keys. A key is a string of data that determines the output of a cryptographic algorithm. While there are many different types of cryptographic algorithms, they are usually divided into two categories. Symmetric-key cryptography uses a single key to both encrypt and decrypt the data. Public-key cryptography, also called "asymmetric-key cryptography," uses two keys, one public and one private. Usually, the public key is used to encrypt the data, and the private key is used to decrypt it.

When using symmetric-key cryptography, both the sender and the recipient of the encrypted message must have access to the same key. This key must be exchanged between parties using a secure channel, or else it may be compromised. Public-key cryptography does not require such an exchange.

This is one reason that public-key cryptography is considered more secure.

Another cryptographic technique developed for use with computers is the digital signature. A digital signature is used to confirm the identity of the sender of a digital message and to ensure that no one has tampered with its contents. Digital signatures use public-key encryption. First, a hash function is used to compute a unique value based on the data contained in the message. This unique value is called a "message digest," or just "digest." The signer's private key is then used to encrypt the digest. The combination of the digest and the private key creates the signature. To verify the digital signature, the recipient uses the signer's public key to decrypt the digest. The same hash function is then applied to the data in the message. If the new digest matches the decrypted digest, the message is intact.

WHY IS CRYPTOGRAPHY IMPORTANT?

The ability to secure communications against interception and decryption has long been an important part of military and international affairs. In the modern age, the development of new computer-based methods of encryption has had a major impact on many areas of society, including law enforcement, international affairs, military strategy, business, and media. It has also led to widespread debate over how to balance the privacy rights of organizations and individuals with the needs of law enforcement and government agencies. Businesses, governments, and consumers must deal with the challenges of securing digital communications for commerce and banking on a daily basis. The impact of cryptography on society is likely to increase as computers grow more powerful, cryptographic techniques improve, and digital technologies become ever more important.

—Maura Valentino

Further Reading

Hoffstein, Jeffrey, Jill Pipher, and Joseph H. Silverman. *An Introduction to Mathematical Cryptography*. 2nd ed., Springer, 2014.

"How Cryptography and Web3 Can Help Restore Trust in Digital Media." *Stanford Engineering*, 17 June 2022, engineering.stanford.edu/magazine/how-cryptography-and-web3-can-help-restore-trust-digital-media. Accessed 13 July 2023.

Katz, Jonathan, and Yehuda Lindell. *Introduction to Modern Cryptography*. 3rd ed., Chapman and Hall, 2020.

Menezes, Alfred J., Paul C. van Oorschot, and Scott A. Vanstone. *Handbook of Applied Cryptography*. CRC Press, 1996.

Niederreiter, Harald, and Chaoping Xing. *Algebraic Geometry in Coding Theory and Cryptography*. Princeton UP, 2009.

Paar, Christof, Jan Pelzl, and Tim Güneysu. *Understanding Cryptography: A Textbook for Students and Practitioners*. Springer Cham, 2023.

CURATOR/ARCHIVIST

ABSTRACT

Curators and archivists collect, appraise, organize, and preserve a vast array of physical and digital items of historical, cultural, and scientific interest. They are employed by museums, libraries, universities, cultural centers, and other institutions. Students aspiring to enter the profession must often pursue graduate studies in library science, archival studies, or museum studies.

BACKGROUND

Archivists and curators are preservationists of human culture and history and the natural world. They collect, appraise, organize, and preserve documents, artwork, specimens, ephemera, films, and many other objects for historical and educational purposes. Archivists usually handle documents and records that are of historical value. Curators are more likely to manage cultural or biological items, such as artwork or nature collections.

OVERVIEW

Archivists work in libraries, government depositories, universities, and historical museums, while curators are more often employed in art museums, zoos, nature centers, and other cultural or scientific institutions. Each typically divides the workweek between independent projects and interaction with other staff and outsiders, such as dealers, researchers, and the public.

People interested in archivist or curator positions value the contributions of humans or the natural world and realize their importance in research. They are scholars who possess good organizational skills and a knack for handling irreplaceable items that are often fragile and extremely valuable. They need to be both detail-oriented and aware of larger cultural, scientific, and/or historical contexts. Other important traits include critical thinking, leadership ability, oral and written communication skills, and a high level of integrity.

DUTIES AND RESPONSIBILITIES

Archivists and curators build on their institution's collections by purchasing items or receiving them as gifts, often the result of bequests. A collection donated by a celebrated author might consist of boxes of unpublished manuscripts and drafts, personal correspondence, publishing contracts, and other printed matter. A collection obtained from a philatelist might include rare postal stamps, philatelic books and journals, microscopes, antique magnifying glasses, and other materials.

The archivist or curator is usually responsible for deciding what items to keep based on physical condition, financial, historical, and cultural value, and relevance to the institution's mission or purpose. While assessing each item, he or she authenticates its provenance (date and origin) and researches the item for any additional relevant information. The archivist or curator also determines how best to preserve and store items. For example, special cabinets may have to be ordered or an item may be given to a conservator for repairs.

Next, the archivist or curator catalogues or classifies items in a database so scholars can access the information. These databases also allow archivists or curators to keep track of their collections, provide reference service, and plan exhibits. Many different classification systems are used, although the most common one in the United States is the Library of Congress Classification System. Some items may be given a taxonomic classification as well as a call number.

Curators and archivists have other tasks in addition to their preservation work. Curators and archivists often write articles, grant proposals, and annual reports. Depending on their work environment, they may give tours and presentations to the public. Curators and archivists may also take care of other administrative duties or oversee assistants who handle some of these responsibilities, or they may do everything themselves.

OCCUPATION SPECIALTIES

Museum Technicians. Museum technicians prepare specimens for museum collections and exhibits. They preserve and restore specimens by reassembling fragmented pieces and creating substitute pieces.

Art Conservators. Art conservators coordinate the examination, repair, and conservation of art objects.

Historic-Site Administrators. Historic-site administrators manage the overall operations of a historic structure or site.

Museum Registrars. Museum registrars maintain records of the condition and location of objects in museum collections and oversee the movement of objects to other locations.

WORK ENVIRONMENT

Physical Environment. Archivists and curators tend to work at least part of the time in climate-controlled storage facilities. They may have to wear gloves or masks to protect items from human contamination. They sometimes deal with dust, mold, and insect infestations. Fieldwork may include visits to off-site locations such as auctions, schools, and private residences.

Human Environment. Archivists and curators usually report to a director and may supervise assistants, volunteers, or interns. In some cases, the curator is the director and reports to a board of administrators. Archivists and curators also interact with clerical staff and fellow preservation professionals, such as librarians, conservators, or museum technicians. They also work with researchers and other members of the public who use their facilities.

Technological Environment. Archivists and curators rely heavily on computers for research, database management, file sharing, and communication. They also use a variety of digitization equipment for preservation purposes, including digital photography and video cameras. Microscopes are often used for detail work. In many cases, they must be familiar with radio-frequency identifications (RFIDs) and other inventory control and antitheft systems. Some archivists and curators specialize in born-digital materials, such as digital art and multimedia content, digital manuscripts, and digital records. As such, they must have a strong understanding of data-transfer technologies and the principles of digital preservation.

EDUCATION AND TRAINING

High School/Secondary. Archivist and curator positions require advanced education. A strong college preparatory program with electives in the areas of professional interest will provide the best foundation for postsecondary studies. History courses are especially important for aspiring archivists and curators. Students interested in becoming a curator of art should take art history and appreciation courses. Botany, zoology, and other natural sciences are important for curators of natural history. Students should also consider volunteering or working part-time in a library, museum, or other similar institution.

College/Postsecondary. A bachelor's degree in history, art history, botany, political science, or other relevant discipline, with additional coursework in archival or museum studies, is the minimum requirement; however, most positions require a master's degree or doctorate in the specialized discipline or a master's degree in library science, archival studies, or museum studies. Business and public administration courses may also be useful. An internship or other work experience in a related institution is typically required for employment. Continuing education courses are expected as part of ongoing professional training.

Adult Job Seekers. Adults who have experience working at a relevant institution, researching a particular type of collection, or writing grant proposals or fundraising have an advantage over inexperienced graduates, as maturity and experience are often desired in addition to education.

Advancement is highly dependent upon the size of the institution. In larger institutions, advancement usually takes the form of increasing responsibility, such as a supervisory or directorial position. In government positions, one can move into higher pay grades with proper experience and education. Consulting is also an option for experienced professionals.

Professional Certification and Licensure. Licensing is typically not necessary for archivists and curators, although some employers may require certification by a professional organization, such as the Academy of Certified Archivists (ACA). A master's degree and archival experience are necessary before one can take the ACA written exam for certification. Those interested in becoming certified should consult credible professional associations within the field and follow professional debate as to the relevancy and value of any certification program.

Additional Requirements. Physical strength is needed to lift heavy boxes or other items, and good eyesight is needed for detail work. Membership in professional archivist or curator associations may provide access to networking opportunities and professional development programs.

—*Sally Driscoll*

FURTHER READER

Arp, Charlie. *Archival Basics: A Practical Manual for Working with Historical Collections*. Rowman & Littlefield, 2019.

Dekker, Annet, editor. *Curating Digital Art: From Presenting and Collecting Digital Art to Networked Co-curation*. Valiz, 2021.

Müller, Katja. *Digital Archives and Collections: Creating Online Access to Cultural Heritage*. Berghahn, 2021.

D

Data Mining

ABSTRACT

Advances in technology in the latter half of the twentieth century led to the accumulation of massive data sets in government, business, industry, and various sciences. Extracting useful information from these large-scale data sets required new mathematical and statistical methods to model data, account for error, and handle issues like missing data values and different variable scales or measures. Data mining uses tools from statistics, machine learning, computer science, and mathematics to extract information from data, especially from large databases.

BACKGROUND

Data mining has roots in three areas: classical statistics, artificial intelligence, and machine learning. In the late 1980s and early 1990s, companies that owned large databases of customer information, in particular credit card banks, wanted to explore the potential for learning more about their customers through their transactions. The term "data mining" had been used by statisticians since the 1960s as a pejorative term to describe the undisciplined exploration of data. It was also called "data dredging" and "fishing." However, in the 1990s, researchers and practitioners from the field of machine learning began successfully applying their algorithms to these large databases in order to discover patterns that enable businesses to make better decisions and to develop hypotheses for future investigations.

Partly to avoid the negative connotations of the term "data mining," researchers coined the term "knowledge discovery in databases" (KDD) to describe the entire process of finding useful patterns in databases, from the collection and preparation of the data to the end product of communicating the results of the analyses to others. This term gained popularity in the machine learning and artificial intelligence (AI) fields, but the term "data mining" is still used by statisticians. Those who use the term "KDD" refer to data mining as only the specific part of the KDD process where algorithms are applied to the data. The broader interpretation will be used in this discussion.

Software programs to implement data mining emerged in the 1990s and continue to evolve today. There are open-source programs and many commercial programs that offer easy-to-use graphical user interfaces (GUIs), which can facilitate the spread of data mining practice throughout an organization.

OVERVIEW

The concepts involved in data mining are drawn from many mathematical fields, such as fuzzy sets, developed by mathematician and computer scientist Lotfi Zadeh, and genetic algorithms, based on the work of mathematicians such as Nils Barricelli. Because of the massive amounts of data processed, data mining relies heavily on computers, and mathematicians contribute to the development of new algorithms and hardware systems. For example, the Gfarm Grid File System was developed in the early twenty-first century to facilitate high-performance petascale-level computing and data mining.

The specific types of tasks that data mining addresses are typically broken into four types: predictive modeling (classification, regression), segmentation (data clustering), summarization, and visualization.

Predictive modeling is the building of models for a response variable for the main purpose of predicting the value of that response under new—or future—values of the predictor variables. Predictive modeling problems, in turn, are further broken into classification problems or regression problems, depending on the nature of the response variable being predicted. If the response variable is categorical (e.g., whether a customer will switch telephone providers at the end of a subscription period or will stay with his or her current company), the problem is called a "classification." If the response is quantitative (e.g., the amount a customer will spend with the company in the next year), the problem is a "regression problem." The term "regression" is used for these problems even when techniques other than regression are used to produce the predictions. Because there is a clear response variable, predictive modeling problems are also called "supervised problems" in machine learning. Sometimes there is no response variable to predict, but an analyst may want to divide customers into segments based on a variety of variables. These segments may be meaningful to the analyst, but there is no response variable to predict in order to evaluate the accuracy of the segmentation. Such problems with no specified response variable are known as "unsupervised learning problems."

Summarization describes any numerical summaries of variables that are not necessarily used to model a response. For example, an analyst may want to examine the average age, income, and credit scores of a large batch of potential new customers without wanting to predict other behaviors. Any use of graphical displays for this purpose, especially those involving many variables at the same time, is called "visualization."

ALGORITHMS

Data mining uses a variety of algorithms (computer code) based on mathematical equations to build models that describe the relationship between the response variable and a set of predictor variables. The algorithms are taken from statistics and machine learning literature, including such classical statistical techniques as linear regression and logistic regression and time series analysis, as well as more recently developed techniques like classification and regression trees (ID3 or C4.5 in machine learning), neural networks, naïve Bayes, K-nearest neighbor techniques, and support vector machines.

One of the challenges of data mining is to choose which algorithm to use in a particular application. Unlike the practice in classical statistics, the data miner often builds multiple models on the same data set, using a new set of data (called the "test set") to evaluate which model performs best.

Recent advances in data mining combine models into ensembles in an effort to collect the benefits of the constituent models. The two main ensemble methods are known as "bootstrap aggregation" (bagging) and "boosting." Both methods build many (possibly hundreds or even thousands of) models on resampled versions of the same data set and take a (usually weighted) average (in the case of regression) or a majority vote (in the case of classification) to combine the models. The claim is that ensemble methods produce models with both less variance and less bias than individual models in a wide variety of applications. This is an area of extensive research in data mining.

APPLICATIONS

Data mining techniques are being applied everywhere there are large data sets. A number of important application areas include the following.

Customer Relationship Management (CRM). Credit card banks formed one of the first groups of companies to use large transactional databases in an attempt to predict and understand patterns of customer behavior. Models help banks understand acquisition, retention, and cross-selling opportunities.

Image via iStock/mathisworks. [Used under license.]

Risk and Collection Analytics. Predicting both who is most likely to default on loans and which type of collection strategy is likely to be successful is crucial to banks.

Direct Marketing. Knowing which customers are most likely to respond to direct marketing could save companies billions of dollars a year in junk mail and other related costs.

Fraud Detection. Models to identify fraudulent transactions are used by banks and a variety of government agencies including state comptroller's offices and the Internal Revenue Service (IRS).

Terrorist Detection. Data mining has been used by various government agencies in an attempt to help identify terrorist activity—although concerns of confidentiality have accompanied these uses.

Genomics and Proteomics. Researchers use data mining techniques in an attempt to associate specific genes and proteins with diseases and other biological activity. This field is also known as "bioinformatics."

Health Care. Data mining is increasingly used to study efficiencies in physician decisions, pharmaceutical prescriptions, diagnostic results, and other health-care outcomes.

CONCERNS AND CONTROVERSIES

Privacy issues are some of the main concerns of the public with respect to data mining. In fact, some kinds of data mining and discovery are illegal. There are federal and state privacy laws that protect the information of individuals. Nearly every website, credit card company, and other information collecting organization has a publicly available privacy policy. Social networking sites, such as Facebook, have been criticized for sharing and selling information about subscribers for data mining purposes. In health care, the Health Insurance Portability and Accountability Act of 1996 (HIPAA) was enacted to help protect individuals' health information from being shared without their knowledge. Other regulations dealing with data privacy include the General Data Protection Regulation (GDPR), which was introduced in the European Union in 2016 and came into effect two years later.

—*Richard De Veaux*

Further Reading

Berry, M. A. J., and G. Linoff. *Data Mining Techniques for Marketing, Sales and Customer Support.* Wiley, 1997.

Plotkin, David. *Data Stewardship: An Actionable Guide to Effective Data Management and Data Governance.* 2nd ed., Academic Press, 2020.

Wiggins, Chris, and Matthew L. Jones. *How Data Happened: A History from the Age of Reason to the Age of Algorithms.* Norton, 2023.

Witten, Ian H., Eibe Frank, Mark A. Hall, and Christopher J. Pal. *Data Mining: Practical Machine Learning Tools and Techniques.* 4th ed., Morgan Kaufmann, 2016.

Data Protection

ABSTRACT

Originating in the 1980 Organisation for Economic Co-operation and Development (OECD) Privacy Guidelines, data protection regimes typically include a usually wide definition of "personal information" and place a heavy emphasis on the knowledge of and/or consent to collection of personal information by the data subject. Globally, there are two broad approaches taken to protect personal information: a sectoral approach in which various legislative regimes create standards and rules in discrete areas of the economy and an alternative approach in which the collection, use, and disclosure of personal information are regulated under a single data protection regime.

BACKGROUND

The origin of comprehensive data protection regimes may be found in the 1980 Organisation for Economic Co-operation and Development (OECD) Privacy Guidelines, the core of which was the adoption of eight "fair information practice principles" (FIPPS) intended to give direction to the collection, use, and disclosure of personal information. Though having jointly developed the OECD Guidelines, the United States and Europe went in two different directions regarding their applicability in the years that followed.

The United States saw them as a framework of useful guidelines that could be freely adopted by industry if they so choose, while in Europe they were gradually strengthened and made directly enforceable. This strengthening culminated in the passage of the European Union (EU) Data Protection Directive. In one form or another, an enforceable version of the FIPPS can be found in all other modern data protection regimes, such as Canada's Personal Information & Protection of Electronic Documents Act, Hong Kong's Personal Data (Privacy) Ordinance, South Africa's Protection of Personal Information Act, and so on.

Because of this common ideological heritage, data protection regimes typically share some general features. These include a usually wide definition of "personal information" (such as any information that can lead to an identifiable individual) and of activities that count as "data processing" (both manual and automatic). Data protection regimes place a heavy emphasis on the knowledge of and/or consent to collection of personal information by the data subject and a strict limitation on the purposes to which collected information can be put. In other words, an individual must generally be informed at the time of collection of her or his information as to the purpose of the collection, and any new purpose must receive new consent.

OVERVIEW

Globally, there are two broad approaches taken to protect personal information. One may be generally described as sectoral in nature: There are various legislative regimes that create standards and rules in discrete areas of the economy. Beyond those areas, privacy protection is left purely to the free market: If people desire privacy, then they can (in theory) pay for it. Different jurisdictions regulate different areas, and at different levels of protection.

The United States, for instance, has legislated privacy protections in various areas—including, but not limited to, telecommunications, health information, credit reporting, and websites aimed at children—but there is little consistency as to the kind of privacy protection offered in each area. Proponents of a sectoral or free-market approach argue that excessive privacy laws impose costs on business and are therefore a threat to technological innovation and economic growth.

But the sectoral model is a global outlier. Critics argue that it leaves gaps in the law and creates confusing inconsistencies. Leaving privacy protections to the free market in those gaps is ineffective since there are great disparities in bargaining power between individuals and large organizations that seek to profit off their personal information. As a result, the approach adopted by much of the rest of the world is to legislate a single data protection regime applicable to all organizations that seek to collect, use, or disclose personal information. Broad state involvement through data protection regimes is justified because privacy is understood as a human right connected to individuality, dignity, and autonomy; while there may be economic costs to robust privacy rights, they are justified because of the values at play.

THEORY AND PRACTICE

Generally speaking, under a data protection regime, *all* organizations that collect, use, or disclose personal information will be subject to the same rules; this is in clear contrast to the sectoral approach. Some jurisdictions, however, may choose to have two separate regimes for public and private organizations. Likewise, while all data protection regimes grant exemptions for data processing or collection in specified areas such as journalism, statistical research, or public security, only some may grant exemptions to all noncommercial or personal use of information. Finally, while the definition of "personal information" tends to be broad, some regimes treat certain classes of information as particularly sensitive (e.g., medical information, financial records, and political opinions) and thus deserving of additional protections, while others do not.

Data protection regimes are also typically supervised by an independent commissioner, though the exact power of that commissioner may vary. Canada's Privacy Commissioner, for instance, lacks the power to issue fines directly to violators of the relevant law, while under the EU General Data Protection Regulation (adopted in 2016 and implemented in 2018), national data protection authorities are granted the power to issue sanctions to organizations found to be in noncompliance. Beyond sanctioning, other powers that may vary across regimes include the possibility of an independent lawsuit absent public complaint, compelling reporting of data breaches, strength of independence from government, and so on.

Further, shared features most data protection regimes include a right of individual access to information held about them, and the right to seek erasure or correction of information where it is incorrect. However, differing interpretations of this basic principle (drawn from the FIPPS) can be found in the recent discussion over the right to be forgotten—whether individuals can request online sources to remove information about them. Interpretation of the Data Protection Directive by the European Court of Justice has found such a right to exist.

In *Google Spain v. Gonzalez* (2014), the court found that existing provisions that granted the data subject the right to erasure of irrelevant data could support the right to be forgotten and required the well-known search engine to delete links to a news

Image via iStock/lemono. [Used under license.]

story regarding the complainant's financial difficulties from some fifteen years prior. The General Data Protection Regulation explicitly recognizes and expands the application of this right. In other jurisdictions, however, the prevailing argument is that as long as the information is true, forcing its deletion is incompatible with the right to free expression.

Indeed, the debate over the right to be forgotten is indicative of a new set of challenges that data protection regimes face. The FIPPS were developed in the 1970s in response to a particular set of privacy issues relevant at the time. The twenty-first century, however, is, an age of "always-on" internet-connected devices that are carried voluntarily and an economy the lifeblood of which is information. There are thus disputes about whether data protection regimes can be adapted (or need to be adapted) to respond to these changes.

—*Stuart A. Hargreaves*

Further Reading

Bygrave, Lee A. "The Place of Privacy in Data Protection Laws." *University of New South Wales Law Journal*, vol. 24, no. 1, 2001, pp. 277-83.

_____. *Data Privacy Law, an International Perspective.* Oxford UP, 2014.

Cate, Fred. "The Changing Face of Privacy Protections in the European Union and the United States." *Indiana Law Review*, vol. 33, 1999, pp. 173-232.

Jarmul, Katharine. *Practical Data Privacy: Enhancing Privacy and Security in Data*. O'Reilly Media, 2023.

Koops, Bert-Jaap. "The Trouble with European Data Protection Law." *International Data Privacy Law*, vol. 4, no. 4, 2014, pp. 250-61.

Levin, Avner, and Mary Jo Nicholson. "Privacy Law in the United States, the EU, and Canada: The Allure of the Middle Ground." *University of Ottawa Law & Technology Journal*, vol. 2, no. 2, 2005, pp. 357-95.

Mantlero, Alessandro. "The EU Proposal for a General Data Protection Regulation and the Roots of the 'Right to Be Forgotten.'" *Computer Law & Security Review*, vol. 29, no. 3, June 2013, pp. 229-35.

Quinn, Brendan. *Data Protection Implementation Guide: A Legal, Risk and Technology Framework for the GDPR*. Wolters Kluwer, 2021.

Shoenberger, Allen. "Privacy Wars." *Indiana International & Comparative Law Review*, vol. 17, 2007, pp. 355-93.

Whitman, James Q. "Two Western Cultures of Privacy: Dignity versus Liberty." *Yale Law Journal*, vol. 113, no. 6, Apr. 2004, pp. 1151-221.

DEBUGGING

ABSTRACT

Debugging is the process of identifying and addressing errors, known as "bugs," in computer systems. It is an essential step in the development of all kinds of programs, from consumer programs such as web browsers and video games to the complex systems used in transportation and infrastructure. Debugging can be carried out through a number of methods depending on the nature of the computer system in question.

BACKGROUND

Debugging is the process of testing software or other computer systems, noting any errors that occur, and finding the cause of those errors. Errors, or "bugs," in a computer program can seriously affect the program's operations or even prevent it from functioning altogether. The goal of debugging is to get rid of the bugs that have been identified. This should ensure the smooth and error-free operation of the computer program or system.

Computer programs consist of long strings of specialized code that tell the computer what to do and how to do it. Computer code must use specific vocabulary and structures to function properly. As such code is written by human programmers, there is always the possibility of human error, which is the cause of many common bugs. Perhaps the most common bugs are syntax errors. These are the result of small mistakes, such as typos, in a program's code. In some cases, a bug may occur because the programmer neglected to include a key element in the code or structured it incorrectly. For example, the code could include a command instructing the computer to begin a specific process but lack the corresponding command to end it.

Bugs fall into one of several categories, based on when and how they affect the program. Compilation errors prevent the program from running. Run-time errors, meanwhile, occur as the program is running. Logic errors, in which flaws in the program's logic produce unintended results, are a particularly common form of bug. Such errors come about when a program's code is syntactically correct but does not make logical sense. For instance, a string of code with flawed logic may cause the program to become caught in an unintended loop. This can cause it to become completely unresponsive, or freeze. In other cases, a logic error might result when a program's code instructs the computer to divide a numerical value by zero, a mathematically impossible task.

OVERVIEW

Bugs may interfere with a program's ability to perform its core functions or even to run. Not all bugs are related to a program's core functions, and some programs may be usable despite the errors they contain. However, ease of use is an important factor that many people consider when deciding which program to use. It is therefore in the best interest of software creators to ensure that their programs are as free of errors as possible. In addition to testing a program or other computer system in house prior to releasing them to the public, many software companies collect reports of bugs from users following its release. This is often done through transfers of collected data commonly referred to as memory dumps. They can then address such errors through updates known as software patches.

While bugs are an inconvenience in consumer computer programs, in more specialized computer systems, they can have far more serious consequences. In areas such as transportation, infrastructure, and finance, errors in syntax and logic can place lives and livelihoods at risk. Perhaps the most prominent example of such a bug was the so-called Y2K bug. This bug was projected to affect numerous computer systems beginning on January 1, 2000. The problem would have resulted from existing practices related to the way dates were written in computer programs. However, it was largely averted through the work of programmers who updated the affected programs to prevent that issue. As the example of the far-reaching Y2K bug shows, the world's growing reliance on computers in all areas of society has made thorough debugging even more important.

IDENTIFYING AND ADDRESSING BUGS

The means of debugging vary based on the nature of the computer program or system in question. However, in most cases bugs may be identified and addressed through the same general process. When a bug first appears, the programmer or tester must first attempt to reproduce the bug in order to identify the results of the error and the conditions under which they occur. Next, the programmer must attempt to determine which part of the program's code is causing the error to occur. As programs can be quite complex, the programmer must simplify this process by eliminating as much irrelevant data as possible. Once the faulty segment of code has been found, the programmer must identify and cor-

Image via iStock/LumerB. [Used under license.]

rect the specific problem that is causing the bug. If the cause is a typo or a syntax error, the programmer may simply need to make a small correction. If there is a logic error, the programmer may need to rewrite a portion of the code so that the program operates logically.

DEBUGGING IN PRACTICE

Programmers use a wide range of tools to debug programs. As programs are frequently complex and lengthy, automating portions of the debugging process is often essential. Automated debugging programs, or "debuggers," search through code line by line for syntax errors or faulty logic that could cause bugs. A technique known as delta debugging provides an automated means of filtering out irrelevant information when the programmer is looking for the root cause of a bug.

Different types of programs or systems often require different debugging tools. An in-circuit emulator is used when the computer system being tested is an embedded system (i.e., one located within a larger system) and cannot otherwise be accessed. A form of debugging known as integration testing is often used when a program consists of numerous components. After each component is tested and debugged on its own, they are linked together and tested as a unit. This ensures that the different components function correctly when working together.

—*Joy Crelin*

Further Reading

Burger, Arnold S. *Debugging Embedded and Real-Time Systems*. Newnes, 2020.

Foote, Steven. *Learning to Program*. Pearson, 2015.

Lettnin, Djones, and Markus Winterholer, editors. *Embedded Software Verification and Debugging*. Springer-Verlag New York, 2017.

McCauley, Renée, et al. "Debugging: A Review of the Literature from an Educational Perspective." *Computer Science Education*, vol. 18, no. 2, 2008, pp. 67-92.

Myers, Glenford J., Tom Badgett, and Corey Sandler. *The Art of Software Testing*. 3rd ed., Wiley, 2012.

O'Leary, Timothy, Linda O'Leary, and Daniel O'Leary. *Computing Essentials 2023*. McGraw-Hill, 2022.

Patt, Yale, and Sanjay Patel. *Introduction to Computing Systems: From Bits & Gates to C/C++ & Beyond*. 3rd ed., McGraw-Hill, 2019.

"What Is Debugging?" *AWS*, 2023, aws.amazon.com/what-is/debugging. Accessed 13 July 2023.

Zeller, Andreas. *Why Programs Fail: A Guide to Systematic Debugging*. Kaufmann, 2009.

Deepfake

ABSTRACT

The term "deepfake" refers to emerging technology that allows computer users to create fabricated but highly convincing sounds, static images, and moving pictures. Deepfake technologies are assisted by advanced artificial intelligence (AI), mostly from a class of AI known as "generative adversarial networks" (GANs). Using sophisticated algorithms, GANs manipulate user-supplied input to generate sounds, images, and videos, resulting in strikingly realistic simulated content. The word "deepfake" was derived from the words "deep learning" and "fake."

BACKGROUND

The invention of deepfake technology is generally credited to Ian Goodfellow, a machine-learning expert who created his first generative adversarial network (GAN)-powered deepfakes in 2014 while he was a PhD student at the University of Montreal. Goodfellow went on to work for several major companies carrying out research within the artificial intelligence (AI) space, including Google, OpenAI, Apple, and the Google subsidiary DeepMind.

Deepfake technology allows computer users to create fabricated but highly convincing sounds, static images, and moving pictures. It relies on algorithms, which are systematized sequences of programming instructions that tell a computer how to handle a

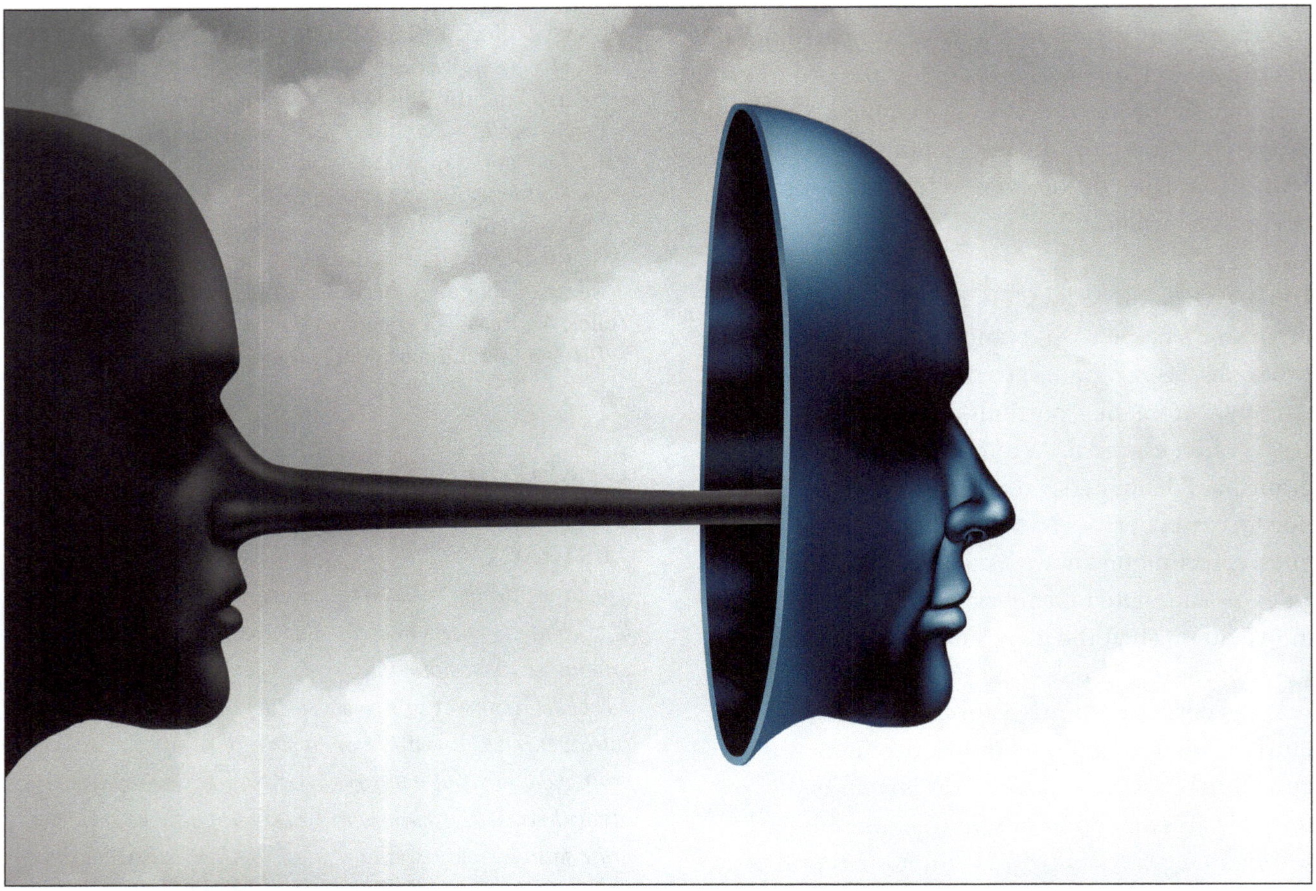

Image via iStock/wildpixel. [Used under license.]

complex task. In particular, it uses advanced AI-powered GAN processes that push the limits of conventional algorithms. Most algorithms are focused on simply sorting or classifying data, while GANs use multiple algorithms that try to trick each other into categorizing a manufactured sound, image, or video as real. Specifically, they function by simultaneously adopting the roles of generator and discriminator, where the generator is responsible for drawing on user input to manufacture a fake sound clip, image, or video clip and the "discriminator" is responsible for comparing the simulated content against the authentic input. GANs are capable of testing simulated content against millions of evaluative parameters very quickly, ultimately allowing users to generate fake sounds, images, and videos realistic enough to convince viewers of their authenticity.

One definitive aspect of deepfake technology is that it does not require much initial input to create a believable result. A *Popular Mechanics* article from August 2019 noted that GANs only need a few images to generate output that appears genuine to the untrained eye.

OVERVIEW

Deepfake software is openly available for download on the internet, enabling any user with the requisite computer skills to use it to produce fake audio, image, and video content. According to a *BBC News* report published in October 2019, which drew on data supplied by the cybersecurity firm Deeptrace, approximately 15,000 deepfake videos were online by that point, marking a sharp rise from the nearly 8,000 the firm counted in December 2018. Observers and analysts believe that amateur computing hobbyists are responsible for a large majority of existing deepfake content and emphasize that deepfake production is a worldwide phenomenon. The BBC report also noted that Deeptrace's analysis found 96 percent of deepfakes to be pornographic in nature, with most simulated videos superimposing the likenesses of famous actresses onto the bodies of adult performers.

Deeptrace's report also addressed claims that deepfake videos were used in recent political campaigns in Malaysia and Gabon. According to the firm, allegations that deepfake videos were used to influence voters in both countries did not withstand scrutiny and can be dismissed as false. Deeptrace did note that deepfake videos had the potential to be weaponized for political purposes. However, as of the report's October 2019 publication date, the firm's expert analysts believed the most pressing threat came from the technology's potential misuse as a cybercriminal and cyberbullying tactic.

A different analysis, conducted in February 2019 by a collaborative group known as Witness Media Lab, reported that existing deepfake technologies required a significant level of specialized knowledge to use effectively. However, Witness Media Lab researchers also stated that the end-user landscape was changing quickly, with increasingly advanced deepfake technologies requiring less and less user skill. Witness Media Lab's conclusions matched the Deeptrace analysis, with both organizations agreeing that the production of simulated, personalized, and highly sophisticated fake content represented a pressing threat to individual users, and particularly to girls and women who faced the risk of having their likenesses imposed on explicit adult videos for the purposes of phishing, extortion, harassment, and cyberbullying.

Deepfakes continued to proliferate throughout the next several years, with the number of deepfake videos online exceeding 85,000 by the end of 2020, according to a 2021 report by the firm Sensity. While some security experts predicted that deepfakes would be widely used to create political propaganda and false news stories in an effort to influence the results of the 2020 US presidential election, deepfake technology did not make a significant impact on the 2020 presidential campaigns or on the election's outcome. However, deepfake technology remained concerning in the early 2020s due to its frequent use in the creation of explicit videos as well as in online scams. One prominent group of scammers became known for its livestreamed deepfake videos featuring technology executive Elon Musk, which the group used to entice victims into sending cryptocurrency to the perpetrators. Though present on other websites as well, the scam became particularly common on YouTube, with scammers gaining unauthorized access to popular YouTube accounts in order to reach large audiences and make their fake livestreams appear more legitimate.

IMPACT

During their early years of existence, persuasive deepfakes were largely confined to sounds, static images, and a genre of content commonly known as "talking head videos," which depicted immobile individuals speaking. However, as the underlying technologies continued to advance and improve, experts voiced concerns that deepfake technology heralded the impending arrival of a dangerous virtual landscape. Its potential criminal and political applications were particularly worrisome to many observers. The consensus among experts was that deepfake technology was likely to introduce unprecedented complications to the problem of "fake news" through artificially manufactured but believable video clips featuring politicians and other high-profile public figures.

With authorities on the topic agreeing that the technology poses new dangers in the virtual environment, many experts believed that deepfakes would soon become a central part of political disinformation campaigns and cyberwarfare. In response to such concerns, US government agencies and other cybersecurity stakeholders worked to develop technology capable of recognizing deepfakes in what some observers described as a kind of virtual arms race meant to limit or prevent deepfakes from exerting a disruptive or damaging influence on society.

—Jim Greene

Further Reading

Cellan-Jones, Rory. "Deepfake Videos 'Double in Nine Months.'" *BBC News*, 7 Oct. 2019, www.bbc.com/news/technology-49961089. Accessed 13 July 2023.

Chen, Angela. "Three Threats Posed by Deepfakes That Technology Won't Solve." *MIT Technology Review*, 2 Oct. 2019, www.technologyreview.com/s/614446/deepfake-technology-detection-disinformation-harassment-revenge-porn-law. Accessed 13 July 2023.

Gregory, Sam, and Eric French. "How Do We Work Together to Detect AI-Manipulated Media?" *Witness Media Lab*, 2019, lab.witness.org/projects/osint-digital-forensics. Accessed 13 July 2023.

Libby, Kristina. "This Bill Hader Deepfake Video Is Amazing: It's Also Terrifying for Our Future." *Popular Mechanics*, 13 Aug. 2019, www.popularmechanics.com/technology/security/a28691128/deepfake-technology. Accessed 13 July 2023.

Petkauskas, Vilius. "Report: Number of Expert-Crafted Video Deepfakes Double Every Six Months." *Cybernews*, 28 Sept. 2021, cybernews.com/privacy/report-number-of-expert-crafted-video-deepfakes-double-every-six-months. Accessed 13 July 2023.

Porup, J. M. "How and Why Deepfake Videos Work—And What Is at Risk." *CSO*, 10 Apr. 2019, www.csoonline.com/article/3293002/deepfake-videos-how-and-why-they-work.html. Accessed 13 July 2023.

Shao, Grace. "What 'Deepfakes' Are and How They May Be Dangerous." *CNBC*, 13 Oct. 2019, www.cnbc.com/2019/10/14/what-is-deepfake-and-how-it-might-be-dangerous.html. Accessed 13 July 2023.

Simonite, Tom. "Prepare for the Deepfake Era of Web Video." *Wired*, 6 Oct. 2019, www.wired.com/story/prepare-deepfake-era-web-video. Accessed 13 July 2023.

Tidy, Joe. "YouTube Accused of Not Tackling Musk Bitcoin Scam Streams." *BBC*, 10 June 2022, www.bbc.com/news/technology-61749120. Accessed 13 July 2023.

"What Is a Deepfake?" *Economist*, 7 Aug. 2019, www.economist.com/the-economist-explains/2019/08/07/what-is-a-deepfake. Accessed 13 July 2023.

DESIGN

ABSTRACT

Design is the development of a plan or process to achieve or create something. Designs include objects, events, and many other elements of daily life—anything that is created by humankind. Successful design is appealing to its intended audience for its appearance and usefulness. However, it must consider practical matters, such as materials and engineering concerns. An attractive and innovative structure, for example, must be constructed to meet safety standards. Examples of design arenas include business and organization strategies; structures, such as buildings and bridges; websites; publications; interiors; vehicles; landscapes; tools, ranging from paper clips to robotic arms; clothing and footwear; and product packaging. While early designs were likely achieved through trial and error—people simply used what they had and adopted changes that seemed to work better—effort and expertise may be required for modern design efforts. An organization may employ a number of teams to research, provide input and expertise, and develop various aspects of a design project.

BACKGROUND

Archaeologists are unsure when early humans developed creativity. The earliest tools were likely spears and digging sticks made of wood, bone, and other materials. Humans used stone tools that did not vary for about 1.6 million years. The tools they crafted were designed with a purpose—to help them survive by performing tasks better than they could with their bare hands—even though tool design remained static.

Between forty thousand and ninety thousand years ago, during the Upper Paleolithic period, humans began painting cave walls and creating items with shell beads. They carved flutes and figures. Stone tools also began to evolve. This development indicates that as the human brain became larger, humans developed the ability to form ideas and design and create with a purpose.

Anthropologists call the human ability to take an idea and further develop and refine it "cultural ratcheting." Design frequently involves looking at old ideas and finding new solutions. Anthropologists believe that as society developed and people lived in larger groups, the chances of individuals within the group having the ability to improve tools and processes increased.

Some aspects of design changed the course of human history. In agriculture, for example, early humans used primitive tools such as digging sticks and stone axes to cultivate crops. Eventually, a simple plow was created to dig more efficiently. As animals were domesticated, humans altered the plow to use animals to pull it. The development of metal tools allowed people to design a better tool for carving into soil. All of these design changes improved human living conditions. Humans who successfully raised crops had more and better food, which allowed them to develop their brains and continue to advance ideas.

While humans were changing physical aspects of their lives, they were also developing communication. Early pictographs likely developed from gestures and sounds people used to communicate. Humans designed writing systems to convey messages.

While early design was largely practical, eventually people became concerned about aesthetics. Early shelters gave way to decorative structures. Simple body coverings were replaced by clothing the wearer deemed attractive. By the third millennium BCE, civil design was evident in urban planning, such as the prehistoric city of Mohenjo Daro in Pakistan. By the late eighteenth century CE, design was recognized as a practice.

OVERVIEW

Design is an element of all human creations and plans. It encompasses practical and aesthetic concerns. In many cases, design begins with a need and desire to address the need. Designers may require expertise to help define the problem and desired goal.

In the visual arts, elements of design include color, direction, form, line, point, shape, space, texture, and value. Visual art design also includes design principles, such as balance, contrast, dominance, emphasis, gradation, harmony, movement, pattern, perspective, proportion, repetition, rhythm, unity, and variety. A skilled artist employs a variety of techniques to achieve an effect through drawing, painting, sculpting, carving, or molding. Many of these elements and principles are useful in other fields. For example, someone designing the interior of a car should consider placement of instruments from both an aesthetic perspective and a practical consideration. While the driver must be able to view the speedometer with ease, the designer may choose fonts or other elements to make the display visually pleasing.

Civil, transportation, and architectural design often involve engineering. Engineering addresses the practical elements of design. Computer-aided design (CAD) is increasingly central to the design process, allowing designers to create two-dimensional (2D) and three-dimensional (3D) models of projects. CAD also offers the benefit of simulating situations to test a design and evaluate its performance. Civil and transportation designers draft and scale drawings and maps, such as those used for the construction of bridges and roads. Civil engineers also design cities and other elements of human civilization. Designs take into consideration the community and human experiences, for example, by integrating green spaces and wide, walkable sidewalks into an urban environment. Access to highways and rivers should be designed for the benefit of citizens and industry. Civil engineers must also consider future expansion in a design and plan where and how growth will occur.

Universal design, or the design of items to be accessible to everyone, arose late in the twentieth century. The movement seeks to address communications, environments, and products to ensure no one is barred from access because of a disability or health condition. Universal design is prominent in building design and redesign. It is also important to infrastructure, such as sidewalks and streets to ensure all individuals can safely travel.

Business model design or strategy involves evaluation of a market and goals. Managers must consider both short-term and long-term goals. In designing a business model, one should consider the value of what the business has to offer; test it in the market; seek feedback and advice; and plan, execute, and evaluate a limited rollout.

Most businesses and organizations design a corporate structure. This may include departments, which have specific goals and responsibilities and may work with other departments on the same level or may function in a hierarchy. A small business may need only a few departments, such as finance, human resources, and production. The management and executive teams must be defined and their responsibilities and authority clearly stated. Job descriptions at all levels should include their places in the hierarchy of the organization and expectations for employees. A well-designed organizational structure helps members perform at their best.

—Josephine Campbell

Further Reading

Bi, Zhuming, and Xiaoqin Wang. *Computer Aided Design and Manufacturing*. Wiley-ASME Press, 2020.

Costello, Vic. *Multimedia Foundations: Core Concepts for Digital Design*. 3rd ed., Focal Press, 2023.

DiChristina, Mariette. "How Human Creativity Arose." *Scientific American*, 1 Mar. 2013, www.scientificamerican.com/article/how-human-creativity-arose. Accessed 13 July 2023.

Eskilson, Stephen. *Digital Design: A History*. Princeton UP, 2023.

"History of Universal Design." *Centre for Excellence in Universal Design*, 2020, universaldesign.ie/what-is-universal-design/history-of-ud. Accessed 13 July 2023.

Leurs, Bas, and Isobel Roberts. "What Do We Mean by Design?" *Design Council*, 9 Nov. 2018, www.designcouncil.org.uk/news-opinion/what-do-we-mean-design. Accessed 13 July 2023.

Margolin, Victor. *World History of Design: Two-Volume Set*. Bloomsbury, 2015.

Pringle, Heather. "The Origins of Creativity." *Scientific American*, Oct. 2016, www.scientificamerican.com/article/the-origins-of-creativity. Accessed 13 July 2023.

"What Is Design?" *University of Illinois at Chicago College of Architecture, Design, and the Arts*, design.uic.edu/what-is-design. Accessed 13 July 2023.

Design Thinking

ABSTRACT

Design thinking is a thought process that depends on examining all sides of an issue from both a practical and a creative perspective before deciding what course of action is most likely to achieve the desired goal. Design thinking, which is a major tool within the business world, is a form of solution-focused thinking. The major aspects of design thinking are understanding the practical and emotional needs of a client, using prototypes or physical models to explore possible ways of achieving goals, and being willing to try different paths even though they may result in failure. Unlike the scientific method of thinking, which is based on a thorough analysis of a problem and involves both observation and experimentation, design thinking begins with a specific goal in mind and then investigates all possible paths that travel toward that goal. With design thinking, it is often necessary to redefine terms, repeat steps, and reexamine initial ideas before achieving the ultimate goal.

BACKGROUND

Design thinking arose out of the need to modernize existing businesses and government agencies as advances in technology continued to change the traditional ways that these entities carried out their activities. Choices were often restricted by financial parameters, and any solution was useless if it did not accomplish the client's desired goals. Thus, technical feasibility, economic viability, and user desirability became the chief tenets of design thinking. The groundwork for design thinking was established during and immediately after World War II as scientists, social scientists, and other researchers struggled to meet the needs of the government and military by developing new strategies such as operational research methods, which depended on thorough analysis of an issue, and management decision-making techniques that used spreadsheets to examine different aspects of an issue.

The concept behind design thinking was introduced by Herbert A. Simon, an American political scientist and economist, in *The Sciences of the Artificial* (1969). Simon identified three critical steps to engaging in decision-making: first, thoroughly investigate the issue to discover all relevant information; second, analyze the information and identify all possible courses of action; and finally, make a choice about which path should be taken. Simon was awarded the Nobel Prize in Economic Sciences in 1978 for his work. The concept of design thinking received a boost in 1973 with the publication of Robert H. McKim's *Experiences in Visual Thinking*, in which he outlines seeing, imaging, and idea sketching as steps to identifying and examining possible solutions to problems.

In the 1980s, L. Bruce Archer, a British engineer, began expanding on ways to engage in "designerly" thinking, noting that it was vastly different from methods used by scientists and scholars. Archer is credited with coining the term "design thinking." Over the course of the decade, other engineers, including those in Germany and Japan, adopted design thinking. Influential studies in the field were also produced by architect Bryan Lawson in *How Designers Think* (1980) and by Peter Rowe, a professor of design and architecture at Harvard University, in *Design Thinking* (1987). The popularity of design thinking led to an upsurge in the number of design journals being published. By the following decade, most university business schools were offering classes in design thinking, and the concept spread into other fields.

In the early twenty-first century, the business press further popularized design thinking, offering articles that suggested that it was an important element in creative thinking. Design thinking also became increasingly popular with engineering schools.

OVERVIEW

Experts in design thinking usually agree that there are four rules that govern the process. First, one must maintain the emphasis on the needs of humans rather than simply engaging in technical experimentation. If choices are made without recognizing those needs, then the process of design thinking fails. Second, ambiguity is used to guide the search for possible solutions, preventing the designer from becoming limited by narrow ways of thinking. Third, design thinking must always involve redesign, which calls for retaining traditional tenets of design while employing new tools and innovative ways of thinking. The final rule of design thinking emphasizes the need to use tangible methods of improving communication as the design thinking process is carried out.

Many designers have discovered that design thinking is ideal for dealing with problems that have not been well defined. In cases where existing technology is incapable of finding a solution to a problem, design thinking may be used to clarify the issue and identify what needs to be done to deal with the issue at hand. One of the chief purposes of design thinking is to use visual thinking to clarify problems that may not be fully defined.

Design thinking may employ such strategies as synthesis, divergent thinking, analysis, and convergent thinking in order to arrive at a solution. Divergent thinking is the process of brainstorming all possible solutions to a problem or all options in a given situation, while convergent thinking applies rules and logic to a limited set of options to determine which is best. With synthesis and divergent thinking, the designer may offer a number of ideas while considering a single aspect of the situation. After doing this, analysis and convergent thinking can be used to reach a decision about what should be done.

While the phrases "design methods" and "design processes" are often used interchangeably, there are distinct differences in their meanings. The former refers to the techniques and strategies used within the field. Those techniques include interviewing interested parties, gathering data on product use, creating physical models, and analyzing the issue at hand. Design process, on the other hand, is more involved with determining how a project should be carried out and involves such actions as allotting tasks and identifying what equipment might be needed.

Design thinking is always concerned with examining all possible solutions to a problem. Solutions are possible only when they are based on the needs of stakeholders involved in the project. By examining all aspects of the situation, it is possible to arrive at creative solutions that address client needs while allowing a designer the freedom to explore and create.

—Elizabeth Rholetter Purdy and Elena Popan

Further Reading

Archer, L. Bruce. "Whatever Became of Design Methodology?" *Design Studies*, vol. 1, no. 1, 1979, pp. 17-20.

Ben Manmoud-Jouini, Sihem, Christophe Midler, and Philippe Silberzahn. "Contributions of Design Thinking to Project Management in an Innovation Context." *Project Management Journal*, vol. 47, no. 2, 2016, pp. 144-56.

Dijksterhuis, Eva, and Gilbert Silvius. "The Design Thinking Approach to Projects." *Project Management Development—Practices and Perspectives*, vol. 5, no. 6, 2016, pp. 67-81.

Fulwood, Nathan. "Opinion: Design Thinking in Publishing and Media." *What's New in Publishing*, 22 June 2021, whatsnewinpublishing.com/opinion-design-thinking-in-publishing-and-media. Accessed 13 July 2023.

Kolko, Jon. "Design Thinking Comes of Age." *Harvard Business Review*, Sept. 2015, hbr.org/2015/09/design-thinking-comes-of-age. Accessed 13 July 2023.

Kumar, Kaushik, Divya Zindani, and J. Paulo Davim. *Design Thinking to Digital Thinking*. Springer Nature Switzerland, 2020.

Leifer, Larry J., Christoph Meinel, and Hasso Plattner. *Design Thinking: Understand—Improve—Apply*. Springer, 2011.

Lewrick, Michael, Patrick Link, and Larry Leifer. *The Design Thinking Toolbox*. Wiley, 2020.

Liedtka, Jeanne. "Why Design Thinking Works." *Harvard Business Review*, Sept. 2018, hbr.org/2018/09/why-design-thinking-works. Accessed 13 July 2023.

Liggett, Susan, Rae Earnshaw, and Jill Townsley. *Creativity in Art, Design and Technology*. Springer Cham, 2023.

Retna, Kala S. "Thinking about 'Design Thinking': A Study of Teacher Experiences." *Asia Pacific Journal of Education*, vol. 36, suppl. 1, 2016, pp. 5-19.

Rowe, Peter G. *Design Thinking*. MIT Press, 1987.

Shearer, Allan W. "Abduction to Argument: A Framework of Design Thinking." *Landscape Journal*, vol. 34, no. 2, 2015, pp. 127-38.

Desktop and E-publishing

ABSTRACT

Desktop publishing is the process of producing a publication on a home or work computer. Electronic publishing (e-publishing) often employs desktop publishing, especially by individuals or independent publishers. The rise of desktop and e-publishing in the twenty-first century created new opportunities for writers and independent publishing companies. However, it also introduced a number of challenges for writers, publishers, readers, and researchers.

BACKGROUND

Desktop publishing is the process of producing a book, report, or other publication on a home or work computer. E-publishing, short for electronic publishing, often employs desktop publishing, especially by individuals or independent publishers, but it also refers to the publication of e-books by traditional publishers. E-publishing and Amazon's dominant role in distribution have put pressure on traditional publishers to adapt to a changing market in which authors can exercise control over when, how, and for what cost their works are released. These changes have also put pressure on writers who choose to publish their work independently, as they must become experts in elements of publishing typically overseen and paid for by traditional publishers, such as cover and page design; the hiring of editors, proofreaders, indexers, and other professional contractors; and marketing and distribution. Additionally, desktop and e-publishing have forced readers to evaluate publications in a new way. Open-source journals, for example, give researchers access to a vast number of scientific and academic articles, but the reader must assess a publication for quality or peer-review in the absence of an editorial process that ensures a paper has been thoroughly reviewed.

Desktop and e-publishing are fascinating to many writers because they allow an author's voice to be expressed in a direct, more-or-less unfiltered way. Publications that are released from large presses often undergo considerable editing and revision, especially if the author is not highly skilled as a writer but possesses information the publisher is interested in publishing. Even skilled and experienced authors receive varying degrees of editorial guidance in the areas of craft and story development before a publisher will consent to publish a manuscript and may find their work ultimately rejected if the publisher remains unsatisfied with it. While such revisions are generally approved by the author, they do tend to smooth over characteristics of an author's style, word choice, and, at times, opinion.

While desktop and e-publishing are often considered together, other nontraditional forms of publishing occurred long before e-publishing was possible. Access to photocopying machines made it possible for individuals and collectives to publish local newspapers, fliers, and magazines from home or by using photocopiers located in schools, libraries, grocery stores, and other community centers or local businesses. Small magazines, often called zines, were produced by individuals or small groups to express political opinions, engage in popular culture and music, and/or distribute art. Traditionally, most zines had low distribution numbers, often less than one thousand copies, and were either sold in local bookstores or given away for free. The advent of desktop and e-publishing facilitated the publication and distribution of zines and similar publications, in part because the internet made it easier to buy and sell zines in both physical and electronic forms. To encourage future research about this influential form of desktop publishing, many libraries began collecting zines and worked to develop means of storing and filing such publications.

OVERVIEW

The emergence of computer-based desktop publishing and e-publishing have enabled writers to produce professional-looking publications. Access to personal computers and the development of word-processing and desktop publishing software rapidly expanded the ability of writers to publish their own work. For many writers, access to personal computers and publishing software occurred before the internet was available or was fast enough to allow consistent online publication. This means that there was a period in which publications produced on personal computers were printed at home, or sent to print shops, and then distributed. As the internet developed, many of those publications moved to online forums, blogs, and journals. The emergence of technology that encouraged portability, such as laptops, tablets, smartphones and dedicated e-book readers such as the Kindle, further encouraged electronic publication.

As a result of all of these changes and the number of publications emerging each year, the number of questions that scholars and researchers need to address before they analyze a publication has expanded. Some scholars are concerned with variables regarding the type of publication software that is used, the platforms which publication occurs on, the ways in which the publication is edited, how distribution occurs, and in what ways writers, editors, and graphic designers are paid for their work.

The emergence of desktop and e-publishing has caused a good deal of skepticism among scholars and professional journalists. The primary concern is that academic and journalistic works have not been properly vetted through a process known as peer review. Most professional news outlets and academic publications go through a process of peer review in which the work read and evaluated by experts in the subject of the publication (in the case of journalism, traditionally that person is an editor).

In the case of scholarly journals, experts evaluate the publication to ensure that the information is accurately presented, that research has been ethically conducted, and that the manuscript advances research and scholarship within the subject. Often, this process is done anonymously though blind-peer review in which the reviewers do not know the identity of the writer, and therefore their judgments are not clouded by their like or dislike of the writer. This process takes a long time, especially if the reviewers suggest revisions and then want to see the manuscript again before they agree to publish the text. The amount of time that this takes, and the selectivity of reviewers means that some scholars have difficulty publishing their work as quickly as they would like. They may be pressured because they have findings that they would like to be released immediately, or because their annual review requires that they have published a number of articles or books each year. These pressures have made desktop and e-publishing attractive to academics, although the quality of such publishing is often dubious.

Problems have emerged as the number of online journals rapidly expands. While older journals are attached to universities or publishing houses with long reputations, some new journals are organized by untraceable offices. These publications may charge academics to publish their work, whereas older journals typically publish research without a charge to the author. These "pay-to-publish" publications are sometimes labeled as predatory journals because they prey on academics that need to quickly publish to keep or secure their job.

OTHER ISSUES

Communications scholars and researchers are interested in the ways that desktop and e-publishing are affecting reading and communication in multiple public spheres. For example, do desktop and e-publishing make it possible for readers to find a wider diversity of texts, or are readers continuing to focus on texts published by traditional publishing houses but now available in electronic formats?

Many textbooks have been made available in electronic formats, which enable students to carry all of their textbooks on a tablet, computer, or phone. This certainly makes packing for a day of classes easier, but academics are still researching if these electronic texts are as effective as paper publications. They ask if using electronic texts changes the ways that students take notes, study, and pay attention to their assignments. Because these texts are so new, researchers are still working to determine what questions should be asked and how to measure student learning via electronic texts. It is clear that electronic texts will continue to exist for a long time. Further, there are many advantages to desktop and e-publishing. Therefore, communication researchers are searching for ways to take advantage of these new tools and capabilities while ensuring that the public is still able to evaluate the information presented to them, and that academics are able to determine if the information they are reviewing has been properly reviewed.

—Allison Hahn

Further Reading

Al-Khatib, Aceil, and Jaime A. Teixeira da Silva. "Stings, Hoaxes and Irony Breach the Trust Inherent in Scientific Publishing." *Publishing Research Quarterly*, vol. 32, no. 3, 2016, pp. 208-19.

Clark, Jocalyn, and Richard Smith. "Firm Action Needed on Predatory Journals." *BMJ*, no. 350, 2015, p. h210.

Costello, Vic. *Multimedia Foundations: Core Concepts for Digital Design*. 3rd ed., Focal Press, 2023.

Cox, Debbie. "Developing and Raising Awareness of the Zine Collections at the British Library." *Art Libraries Journal*, vol. 43, no. 2, 2018, pp. 77-81.

Henrie, Curtis R., Lisa R. Halverson, and Charles R. Graham. "Measuring Student Engagement in Technology-Mediated Learning: A Review." *Computers & Education*, vol. 90, 2015, pp. 36-53.

Luther, Jason, Frank Farmer, and Steve Parks. "Special Issue Editors' Introduction: The Past, Present, and Future of Self-Publishing: Voices, Genres, Publics." *Community Literacy Journal*, vol. 12, 2017, pp. 1-4.

Whitington, John. *How Computers Make Books: From Graphics Rendering, Search Algorithms, and Functional Programming to Indexing and Typesetting*. Manning, 2023.

DESKTOP PUBLISHER

ABSTRACT

Desktop publishers arrange and prepare design layouts for materials intended for publication, including books, newspapers, and magazines. They use digital word-processing and design software in the course of their work and may also create and modify digital graphics. Students aspiring to enter the profession should study subjects such as journalism, composition, art, and drafting.

BACKGROUND

Desktop publishers use digital word-processing and design software to arrange and prepare design layouts for newspapers, books, magazines, and other materials for print or online publication. They may also work as graphic designers, creating and modifying graphics for publication. Desktop publishers typically work closely with editors, designers, graphic artists, and computer programmers in the design of web-based or printed publications. Depending on their area of employment, desktop publishers may spend time editing and writing text for publication in addition to supervising layout and design. Desktop publishing is utilized by a variety of companies, from traditional print publishers to the many businesses that produce web-based publications for advertising and informational purposes.

OVERVIEW

Desktop publishers work closely with writers, editors, and designers to complete projects for publication. Those who work on web-based publications may work with web designers or computer programmers to create websites for online publication. They may also work with graphic designers to create and refine fonts, images, and other graphics. Desktop publishing can be performed in a standard office, using a desktop or laptop computer. Some desktop-publishing specialists may be able to complete part of their work off-site in a home office.

Desktop publishing attracts individuals interested in graphic design, printing, and technical artistry. Many desktop publishers have backgrounds in art or writing. A strong knowledge of both subjects is helpful to anyone interested in the field. Digital design for web publications has become an important facet in the field of publishing, and demand has increased for desktop-publishing specialists with experience and expertise in web design, digital art, and computer programming.

DUTIES AND RESPONSIBILITIES

The printing process is typically divided into three stages: prepress, printing, and postpress activities. Desktop publishers are generally considered part of the prepress team, and they may also spend time working with professionals during the printing process.

Desktop publishers usually work full time during regular business hours, though overtime or alternate schedules are sometimes necessary during seasonal peak periods or when working on time-sensitive projects. For example, newspapers may require night workers to prepare for early-morning publication. Holidays and special events may also require employees to work extra hours. In some cases, desktop publishers work part-time hours or during varying schedules depending on the deadlines for upcoming projects. Some desktop publishers are able to work from a home office, but this requires frequent communication and collaboration with other individuals involved in various parts of a project.

Photo via iStock/Zorica Nastasic. [Used under license.]

During a typical day, a desktop publisher communicates with writers, editors, project managers, and designers about ongoing and upcoming projects. They also gather pictures or artwork from various sources and then arrange the images into a cohesive and attractive layout for publication. Desktop publishers write, edit, proofread, and arrange text and work with various typefaces. Desktop publishers also merge files from design and publishing programs and therefore need to have extensive knowledge of a variety of digital design, writing, editing, and image-processing programs.

WORK ENVIRONMENT

Physical Environment. Offices and workspaces for desktop publishers tend to be quiet, indoor environments featuring desks and worktables for laying out print samples. Most desktop publishers work at a desk or worktable with a computer, and they also spend time in a printing room or print shop checking on projects.

In some cases, desktop publishers work in offices that are attached to industrial printing presses, as is the case with many large-scale magazines, newspapers, and book publishers. Desktop publishers do not typically work in a printing room environment but may visit the printing press to examine printed examples of their projects to ensure that the digital layout transitions well to the printed copy.

Human Environment. Desktop publishers are part of a creative team and work closely with writers, editors, and various design professionals to complete projects. In some cases, companies hire desktop publishers to work on interoffice or professional publications specific to their industry, which often requires publishing professionals to work closely with industry specialists. A pharmaceutical company interested in publishing printed or web-based information about a specific medication, for instance, would expect a desktop publisher to work closely with their in-house sales team as well as with physicians and pharmacists to ensure the information was accurate and clear. In some small companies, there is less interaction between groups since desktop publishers tend to work independently and are responsible for the writing, editing, and graphic design duties in addition to organizing layouts for publication.

Technological Environment. Desktop publishers use specific digital publishing software and other computer programs to create and manage their projects. The digital design software programs produced by Adobe Inc. have become one of the industry standards in this field, but it is important for desktop publishing professionals to have experience in a variety of digital composition and graphic design software programs. Desktop publishing professionals must also understand computer operation and technical maintenance. Continued education is necessary in order to stay abreast of software changes or new programs relevant to the field. Many of the skills and technologies employed by desktop publishers are made use of in the creation of digital publications.

EDUCATION AND TRAINING

High School/Secondary. High school students can prepare for a career in desktop publishing by taking classes in journalism, composition, art, and drafting. Some high schools offer graphic design programs and train students in the use of printing and publishing software. In addition, desktop publishers should be knowledgeable about grammar, spelling, and editing procedures, as they typically must check and revise their work prior to printing. Basic English skills developed through a high school education are essential for those pursuing work in the publishing field.

College/Postsecondary. Community colleges and technical institutes sometimes offer courses and programs in the field of desktop publishing. Some desktop publishers hold associate's degrees in journalism or graphic design. Individuals can also enter the publishing field by pursuing a bachelor's degree or higher certification in graphic design, graphic arts, or a specific print-oriented program. Journalism and technical writing programs are viable avenues into desktop publishing when supplemented with courses in computers, computer software, and graphic design.

Adult Job Seekers. Adults interested in pursuing desktop publishing as a career are advised to seek out community college or technical certification programs in their area. Individuals with a background in manual or computer design may have a competitive advantage in transitioning to the desktop publishing field.

Professional Certification and Licensure. Desktop publishers are not legally required to hold a professional certificate or license; however, some employers may require workers to undergo specific training. Some technical institutes, community colleges, and universities offer certification courses in the software programs used in desktop publishing as well as training programs in other aspects of computer-aided programming and design. Desktop publishers typically receive on-the-job training while working under experienced professionals. Some independent desktop publishing specialists work as freelancers or independent contractors and work for several different clients.

Additional Requirements. Desktop publishers must possess attention to detail and the interest and ability to make aesthetic judgments that combine artistic sensibility with technical precision. Because exact measurements are needed when developing and creating projects, desktop publishers also benefit from experience with basic mathematics and geometry.

—Micah L. Issitt

Further Reading

Costello, Vic. *Multimedia Foundations: Core Concepts for Digital Design*. 3rd ed., Focal Press, 2023.

Hardy, David Leicester. *Introduction to Digital Media Design: Transferable Hacks, Skills, and Tricks*. Bloomsbury Visual Arts, 2022.

Whitington, John. *How Computers Make Books: From Graphics Rendering, Search Algorithms, and Functional Programming to Indexing and Typesetting*. Manning, 2023.

DIGITAL ART

ABSTRACT

Digital art is art created using computer technology. It includes digital sculpture, digital cinema, digital fine art, digital illustration, and digital architecture. Digital art may utilize any amount of digital technology during its creation or in its presentation. Digital art changed the way art is created and provided new ways for people to experience art. Sound, light, and video can all be incorporated into installations. Digital art can also be much simpler, requiring only a television or computer screen, and can easily be transmitted electronically. This transition has helped many artists bypass traditional representation and gallery presence; some use crowdsourcing to fund their art and may use social media to share their work. Early digital artists were interested in the juncture of art and technology. The medium was not only a way to manipulate and create art but also often a means of social commentary.

BACKGROUND

Digital art can trace its roots back to the dawn of the information age in the 1950s. The development and availability of new technology, including television and computer technology, eventually drew the interest of members of the art community. The digital art movement is often considered to have begun in 1965, when some artists began experimenting with computers.

As more technology emerged and advanced, two artists and two engineers decided to form a collective to foster connections between art and technology. Engineers Billy Klüver and Fred Waldhauer

Irrational Geometrics *digital art installation 2008 by Pascal Dombis. Photo via Wikimedia Commons. [Public domain.]*

and artists Robert Rauschenberg and Robert Whitman formed Experiments in Art and Technology (EAT). The collaborators incorporated a variety of technologies into their art, including video and audio projection. While not all of the EAT installations and performances qualified as digital art, the collective fostered further development through its openness to new technology. The boundaries between art and science were swept away.

Klüver was employed by the Bell Labs scientific research company in New Jersey. A computer graphics specialist who worked with Klüver at Bell Labs created the first piece of widely recognized digital art in 1966. Kenneth Knowlton used a photograph of a young woman to create an image composed of computer pixels. His *Young Nude* bridged the gap between historical art traditions and the twentieth century. The traditional nude figure was brought into the information age and rendered in pixels.

Far less traditional digital art projects followed. In 1969, for example, Allan Kaprow created *Hello*. This was not a static work; rather, it was an experience involving multiple people interacting through television monitors. Other artists used early computers, television, and radio to create works and comment on the increasing connectivity technology was allowing. Some artists saw these societal changes as beneficial, while others seemed to warn of encroachment on personal privacy.

The first known artist to use an algorithm to make art was German artist and mathematician Frieder Nake. In 1965, he measured early twentieth-century artist Paul Klee's *High Roads and Byroads* (1929), using the proportion and the relationship between the horizontal and vertical lines to create an algorithm. Nake also introduced random variables so the computer would have to make choices. Computers did not have screens that could display the images, so Nake used a pen plotter, a mechanical device that draws a pen across paper under the computer's control. The resulting work, from which he created prints, is called *Hommage a Paul Klee, 13/9/65 Nr.2*.

Major advancements during the 1970s further inspired artists. The Apple II computer, for example, was the first personal computer that could be used to create color graphics. The modem, developed in 1979, opened up data transfer possibilities. During the 1980s, computer animation developed rapidly, and artists began using design and animation software to create their own works.

One pioneer in digital media development was the visual effects company Industrial Light and Magic (ILM). Based in San Francisco, California, the company was founded by director George Lucas in 1975 to work on the visual effects for his groundbreaking 1977 film, *Star Wars*. ILM assembled a small team of artists, engineers, and scientists in its computer division during the 1970s. This team broke barriers in digital media, finding unexpected new ways to alter audio, graphics, and video.

Digital art advanced further with the widespread availability of the internet that developed during the 1990s. Artists began using the internet to create and disseminate works. One such work, *My Boyfriend Came Back from the War* (1996), was created by Olia Lialina and made use of hyperlinks, which viewers clicked through to reveal the artist's narrative. While this was an example of digital art, many critics began to view internet art (or net.art) as a new and separate artistic movement. In the decades that followed, the internet took on increasing importance as a venue for the display and distribution of digital art, although traditional art venues such as museums and galleries remained important, particularly to artists creating interactive or immersive digital installations.

OVERVIEW

Many art historians divide digital art into two categories. Object-oriented artworks are those that use digital technologies as tools to create otherwise traditional forms of art, such as prints, paintings, photographs, and sculptures. Process-oriented visuals are those that are created, distributed, and stored digitally—without the technology, they are inaccessible or essentially do not exist.

Object-oriented artworks may include digital painting, digital photography, digital collage, and digital sculpture. The latter may be designed using a computer and turned into a physical object. However, if it remains a three-dimensional (3D) digital image, it is likely classified as a process-oriented visual. Digital installations, including immersive works, may be both object-oriented and process-oriented artworks.

Some artists use programming languages to create works. These may involve an artist setting up a concept, creating the necessary code, and sitting back to see the images the computer generates. In some cases, visitors influence the art by their presence, bringing the viewer in as a collaborator.

The digital sphere readily lends itself to mixed-media work. Computers can combine still and moving images, alter sound and light, incorporate snippets of photographs, and much more. Typical process-oriented objects include internet and networked art. These works explore the workings of networks and computing structures. Works created for internet distribution are categorized as process-oriented objects.

—*Josephine Campbell*

Further Reading

Danae Human Intelligence. "What Is Digital Art? Definition and Scope of the New Media." *Digital Art Weekly*, 31 Aug. 2018, medium.com/digital-art-weekly-by-danae-hi/what-is-digital-art-definition-and-scope-of-the-new-media-f645058cfd78. Accessed 13 July 2023.

"Digital Art." *Art Story*, 2019, www.theartstory.org/movement-digital-art-history-and-concepts.htm#beginnings_header. Accessed 13 July 2023.

Eskilson, Stephen. *Digital Design: A History*. Princeton UP, 2023.

Mon, Sierra. "Becoming a Concept Artist for a Hollywood Film." *ArtStation*, 17 July 2017, magazine.artstation.com/2017/07/concept-artist-hollywood-film. Accessed 13 July 2023.

Moritz, William. "Digital Harmony: The Life of John Whitney, Computer Animation Pioneer." *Animation World Magazine*, Aug. 1997, www.awn.com/mag/issue2.5/2.5pages/2.5moritzwhitney.html. Accessed 13 July 2023.

Prust, Z. A., and Peggy B. Deal. *Graphic Communications: Digital Design and Print Essentials*. 7th ed., Goodheart-Willcox, 2022.

"Summary of Digital Art." *Art Story*, 2019, www.theartstory.org/movement-digital-art.htm. Accessed 13 July 2023.

Digital Divide

ABSTRACT

The term "digital divide" refers to the division between individuals able to access computer and internet technology—typically in their homes—and those unable to do so. Such a lack of access can occur for several reasons. These may include the absence of reliable infrastructure, such as broadband internet, in a particular geographical area; the cost of internet service; or a lack of computer ownership. In some cases, an individual may have partial access to internet and computing technologies but nevertheless struggle to complete certain tasks. For example, an individual may be able to access the internet using a mobile device such as a smartphone but may experience spotty connectivity or be unable to use websites not optimized for mobile use.

BACKGROUND

Information and communication technology (ICT) has pervaded almost every aspect of society, from dating and the labor market to governance and shopping. However, not every member of society has been successfully integrated into the information society. The increasing isolation of those suffering from this digital inequality continues to deepen the divide. While information technology (IT) persists in conquering ever more aspects of social life, the new form of inequality is affecting the social and economic prospects of those left without ICT knowledge or access. Without access to digital information or resources, these people face a digital divide that sociologists are working to overcome.

The term digital divide rose to fame in the mid-nineties, predominantly due to its use by Vice President Al Gore in a 1996 speech. Originally, it referred to the distribution of personal computers in US households. With the turn of the century, however, the question of internet access became a crucial aspect of the debate in the United States and around the world.

OVERVIEW

In a global perspective, the difference between developed and developing nations in regard to IT access is immense. While in developed countries it is perfectly normal for more than half of households to own computers, in many developing nations it may be nearly impossible to find more than two people out of a hundred who own a computer. This disparity became the topic of two World Summit on the Information Society conferences hosted by the United Nations in Geneva and Tunis (2003, 2005) as well as a series of further related events. The continuing aim of these conferences, whose reports can be found online, is to find workable solutions for bridging the international digital divide. In the same spirit, a group of faculty members of the Massachusetts Institute of Technology (MIT) began an initiative called "One Laptop per Child," which built inexpensive, rugged, energy-efficient laptop computers and distributed them to children in countries such as Uruguay, Rwanda, and Mongolia. These computers enabled teachers worldwide to better educate children about technology and their world.

But the term "digital divide" does not just describe a global disparity. Within developed nations, too, access to IT can be unequal, and this inequality's effects are becoming increasingly grave. Within the United States itself, attempts have been made to bridge the gap in internet access, including trying to regulate the companies supplying this access. In 2011, one of the nation's largest internet providers, Comcast, released a product called "Internet Essentials" as part of a Federal Communications Commission (FCC) mandate issued upon the company's acquisition of NBC Universal. Internet Essentials, priced at $10 per month, was designed to allow low-income families the ability to purchase an affordable subscription. While many argued that the company did not make a genuine effort to reach out and market this product and initially included too many restrictions, leading to a minimum impact, Comcast did loosen one of the eligibility requirements and ran a special back-to-school promotion for the program in 2014. As IT continues to pervade everyday life and the job market, those without access are left further and further behind.

CAUSES OF THE DIGITAL DIVIDE

The causes behind the national digital divide are manifold. Poverty and social class are issues that come into play even within the most developed nations. These issues can be described in terms of access to cultural capital or symbolic capital, a theoretical conception originally formulated by French sociologist Pierre Bourdieu. According to Bordieu's model, members of a lower social class have little or no opportunity to acquire the traits, habits, or information necessary to accomplish a rise in status, income, class, or livelihood. In the worst cases, a lack of information, which is increasingly available only online, would bar these classes from informed participation in civic life and democracy.

In the case of the digital divide, a lack of cultural capital would make it much harder for children born into low socioeconomic classes to gain the knowledge necessary to command IT. Possessing this command, according to the Bourdieu theory, is a necessary form of capital if one is to be a part of modern society, which increasingly relies on the use of technology such as email and internet videophones. From arranging a date to securing job information to handling client agendas, the demands of everyday private and professional life require not only the capability to use IT but also the ability to do so with ease.

From the point of view of those already fluent in the use of this technology, it may seem very simple to be able to handle technology. But even something as simple as reading a website and finding its significant content requires a thorough initiation and consistent access.

MEDIA MULTITASKING

The requirements have reached a level designated as media multitasking. Successfully performing a variety of simultaneous tasks, such as coordinating activities on a cell phone while simultaneously surfing the internet, requires a great deal of prior learning. Further, the learning process for such habitual routines is fairly time-consuming and gets harder with the increased age of the learner.

Therefore, one must consider factors beyond mere class status when considering the digital divide, such as those of age and generation. Elderly people often have difficulty adjusting to swift changes in technology. The two major problems they are facing, even when they are presented with access, are their own fear or resentment and a form of technological illiteracy. For those who weren't educated into this cultural technique at an early age, the learning process can be long and difficult.

SYMBOLIC-CAPITAL THEORY

The symbolic-capital theory applies in these cases, too. Consider the following analogy: buying into a fledgling market requires only a small amount of start-up capital, but once a market has gained momentum, it becomes much more difficult and requires an even higher margin of investment to buy into it. It is the same with IT. If a person has not grown up within the development of the technology, then more effort and more symbolic capital will be required of them in order to obtain even a moderate amount of technological knowledge.

In other words, those on the losing side of digital inequality will face a widening gap between themselves and the technologically literate as well as an increasingly steep learning curve as they try to catch up.

This effect is visible in the impact that internet use has on US workers. Just demonstrating the ability to use internet technology, whether at work or at home, has a significant positive impact on workers' earnings and chances on the job market. The ability to use the internet, symbolized, for example, by possessing one's own email address, can not only serve as an indicator to employers that a worker is computer literate and therefore more employable but also enable workers to find better job opportunities.

In sum, it is necessary to shrink the digital divide, for knowledge of ICT must now be considered a condition of participation not only in the labor market, but also within the political and social spheres in the forms of access to certain markets as well as government services. However, ICT must not be seen as only a mechanism of exclusion. ICT can also be a means to empower those who have been excluded. Many citizens, previously barred from certain forms of political and social participation, or, because of age or disability, dependent on others when it comes to fulfilling certain administrative requirements, now have a chance at participation and independence through technology in the form of e-government. Even if this goal is not equally realized across all US states, efforts are being made all over the United States to ensure that more people have easy access to government information and services.

APPLICATIONS

Because spreading information about consumer products becomes more difficult in a globalizing market, consumer empowerment has become increasingly important. Take the example of health care: a number of cheap generic medical products, for example, are only available through internet sources. Studies have shown that internet access and use can be positively correlated with personal health care, and the promotion of personal health is negatively affected by digital inequality. On the other hand, though, access and use can be linked to better health as well as increased opportunities for social and economic participation. Enabling access to e-health is therefore a crucial factor to reduce the effects of digital inequality.

One group of actors in particular is struggling with the new digital age and represents a special dimension of the digital divide: the print media. Newspapers, magazines, and journals have in the past been perceived as the agents that upheld a system of critical control over political and commercial developments. But, faced with the free-floating information of the web, print media have a hard time adapting. With every major paper now maintaining its own website, readers are beginning to question whether buying a paper at the newsstand is truly an efficient way of accessing up-to-date information.

Further, with dwindling sales and subscribers, the number of advertisers willing to financially support quality publishing is also decreasing.

The effect on journalism has been fairly grave, most importantly in the dimension of the quality. The quality of information (including contextualization, critical perspectives, reliability, validity, and thoroughness of research) is one of the least explicitly discussed, yet most important, factors in the functioning of modern economies and political systems. To some commentators, iconic publications such as the *New York Times* and *New Republic* have represented the shift from organized, in-depth editorial content to providing more space for images in conjunction with coverage of popular trends and even trending online games such as Wordle. The quality of information itself also contributes to the digital divide, since those who have access to or can afford high quality information can more easily increase their activities and gains.

BRIDGING THE DIGITAL DIVIDE

On the global scale, bridging the digital divide will be an effort that is wrought with conflict. One must factor in questions of gender, ethnicity, race, and religion, all of which are addressed differently from one nation to another. In some countries, internet access is regulated by the state, and content considered harmful, whether for political or ideological reasons, is banned. Governments often fear the internet's potential for aiding democratic and revolutionary movements in spreading their ideas and creating a coherent power base. This fear was proven to be warranted when dissenters used the internet to organize and report on massive protests during the Arab Spring uprisings that began in late 2010 and early 2011.

But at the same time, the progress of IT itself makes it much harder to bridge this divide. More and more intricate internet content demands not only better hardware, but also ever faster broadband connections. Delivering these to every region in the world is an enormous challenge.

Several economic theorists subscribing to rather simple accounts of the digital divide often hold to theories of access, claiming that merely providing everyone with technology such as personal computers and broadband connections will eventually solve the problem of exclusion. However, studies have shown that just providing technological access is not enough to guarantee inclusion. Even in the nineteenth century, simply being literate was not sufficient for active civic, social, or political participation. In the same regard, owning a computer with internet access is unlikely to solve the problems of poverty either globally or nationally. Additionally, the problems caused by the digital divide on the one hand, and by IT itself on the other—for example, information overload, invalid information distribution, identity theft, etc.—lead to a transformation of the political structure of modern societies themselves.

THE DIGITAL DIVIDE TODAY—AND TOMORROW (POSTPANDEMIC)

By 2019 the United States faced significant disparities in broadband adoption, often along socioeconomic, racial, and ethnic lines. According to a 2019 survey by the Pew Research Center, 94 percent of US adults living in a household with an annual income of more than $100,000 had broadband internet at home, and 94 percent also owned a desktop or laptop computer. However, only 56 percent of adults living in households making $30,000 or less per year—defined by Pew as lower-income households—had home broadband access, and only 54 percent owned a desktop or laptop computer. About 26 percent of lower-income households relied on smartphones as their sole means of accessing the internet at home. Further, while 79 percent of white adults had broadband internet at home, only 66 percent of black adults and 61 percent of Hispanic adults did. Geographical disparities also persisted. The 2019 Pew survey found that only 63 percent of rural adults had broadband internet at home, as compared to 75 percent of urban adults and 79 percent of suburban adults.

The issue of the digital divide took on particular significance beginning in early 2020, when the COVID-19 pandemic prompted many businesses to shift to telecommuting and schools to transition to virtual-learning programs. Those shifts, carried out quickly and suddenly, highlighted the challenges facing those without reliable access to computer technology or the internet. A Pew Research Center survey found in April 2020 that 53 percent of US adults described the internet as "essential" during the pandemic, yet for many, that essential tool was effectively out of reach.

In addition, the shift to virtual learning for many schoolchildren underscored the ways in which the digital divide interfered with students' ability to attend online classes and complete assignments. Surveying parents whose children's schools had shifted to virtual learning by April 2020, Pew found that 36 percent of lower-income parents answered that it was very or somewhat likely that their children would be unable to complete work because they had no access to a computer at home. In addition, 40 percent of lower-income parents stated that it was very or somewhat likely that the children would need to use public internet services, such as public wireless hotspots, because they had no reliable internet service at home.

Because of the inequities that the COVID-19 pandemic illuminated, government bodies throughout the United States introduced a number of new initiatives designed to close the digital divide, many of them focusing on increasing access to broadband internet. During late 2020 and early 2021, the federal government passed multiple COVID-relief bills that, along with other forms of financial aid, funded programs such as the Emergency Broadband Benefit, a new FCC subsidy program that took effect in May 2021. Under that temporary program, most eligible households would receive a broadband subsidy of up to $50 per month. Households located on tribal lands could receive up to $75 per month. This benefit expanded upon the existing Lifeline program, which, prior to the pandemic, had offered monthly broadband subsidies of $9.25 to most participants.

The FCC also announced the launch of the Emergency Connectivity Fund Program. That $7.17 billion initiative would provide schools and libraries with the funds to pay for computers and mobile devices, networking equipment, and internet services that could be used by students and others who needed such technology.

In addition to the initiatives carried out at the federal level, states such as Texas and Alabama introduced new programs designed to supply devices, internet access, or both to students in need. Cities likewise introduced new programs. Phoenix, Arizona, for example, began a pilot program that would bring internet service to apartment complexes where local students lived, while Philadelphia, Pennsylvania, and Chicago, Illinois, offered free internet service to public-school students. School districts throughout the country worked to address the lack of computers in some homes, in many cases lending devices such as laptops and tablets to students engaged in virtual learning.

—Alexander Stingl and Joy Crelin

Further Reading

Anderson, J. "Fighting to Bridge the Digital Divide." *Social Policy*, vol. 44, 2014, pp. 56-57.

Anderson, Monica, and Madhumitha Kumar. "Digital Divide Persists Even as Lower-Income Americans Make Gains in Tech Adoption." *Pew Research Center*, 22 June 2021, www.pewresearch.org/fact-tank/2019/05/07/digital-divide-persists-even-as-lower-income-americans-make-gains-in-tech-adoption. Accessed 13 July 2023.

Beck, U. *Power in the Global Age: A New Global Political Economy*. Polity Press, 2006.
Bourdieu, P. "The Forms of Capital." *Handbook of Theory and Research in the Sociology of Education*, edited by J. C. Richardson, Greenwood Press, 1986, pp. 241-58.
Castells, M. *The Internet Galaxy: Reflections on the Internet, Business and Society*. Oxford UP, 2001.
Chen, W. "The Implications of Social Capital for the Digital Divides in America." *Information Society*, vol. 29, 2013, pp. 13-25.
Epstein, D., E. C. Nisbet, and T. Gillespie. "Who's Responsible for the Digital Divide? Public Perceptions and Policy Implications." *Information Society*, vol. 27, 2011, pp. 92-104.
Fung, B. "Comcast Is Expanding Its $10-A-Month Internet Program for the Poor." *Washington Post*, 4 Aug. 2014, www.washingtonpost.com/blogs/the-switch/wp/2014/08/04/comcast-is-expanding-its-10-a-month-internet-program-for-the-poor. Accessed 13 July 2023.
Goldstein, Phil. "How Cities Are Forging Partnerships to Close the Digital Divide." *StateTech*, 10 Dec. 2020, statetechmagazine.com/article/2020/12/how-cities-are-forging-partnerships-close-digital-divide. Accessed 13 July 2023.
Habermas, J. *The Structural Transformation of the Public Sphere*. MIT Press, 1991.
Hollander, P. "Popular Culture, the New York Times and the New Republic." *Society*, vol. 51, 2014, pp. 288-96.
Kong, S. C. "A Curriculum Framework for Implementing Information Technology in School Education to Foster Information Literacy." *Computers & Education*, vol. 51, 2008, pp. 129-41.
Lash, S., and J. Urry. *Economies of Signs and Space*. SAGE, 1994.
Logue, D., and M. Edwards. "Across the Digital Divide." *Stanford Social Innovation Review*, vol. 11, 2013, pp. 66-71.
Masic, I., and E. Suljevic. "An Overview of E-Health Implementation in Countries, Members of the European Union." *Acta Informatica Medica*, vol. 15, 2007, pp. 242-45.
Matusitz, J., and G. M. Breen. "E-health: A New Kind of Telemedicine." *Social Work in Public Health*, vol. 23, 2007, pp. 95-113.
Pick, J. B., and R. Azari. "Global Digital Divide: Influence of Socioeconomic, Governmental, and Accessibility Factors on Information Technology." *Information Technology for Development*, vol. 14, 2008, pp. 91-115.
Romm, Tony. "Millions of Low-Income Americans Will Receive Internet Access Rebates under New $7 Billion Broadband Stimulus Plan." *Washington Post*, 22 Dec. 2020, www.washingtonpost.com/technology/2020/12/22/internet-rebate-coronavirus-stimulus. Accessed 13 July 2023.
Sanz, E., and G. Turlea. "Downloading Inclusion: A Statistical Analysis of Young People's Digital Communication Inequalities." *Innovation: The European Journal of Social Sciences*, vol. 25, 2012, pp. 337-53.
Sullivan, Emily Tate. "The Digital Divide Has Narrowed, but 12 Million Students Are Still Disconnected." *EdSurge*, 27 Jan. 2021, www.edsurge.com/news/2021-01-27-the-digital-divide-has-narrowed-but-12-million-students-are-still-disconnected. Accessed 13 July 2023.
Vogels, Emily A., et al. "53% of Americans Say the Internet Has Been Essential during the COVID-19 Outbreak." *Pew Research Center*, 30 Apr. 2020, www.pewresearch.org/internet/2020/04/30/53-of-americans-say-the-internet-has-been-essential-during-the-covid-19-outbreak. Accessed 13 July 2023.
Wei, L., and D. Hindman. "Does the Digital Divide Matter More? Comparing the Effects of New Media and Old Media Use on the Education-Based Knowledge Gap." *Mass Communication & Society*, vol. 14, 2011, pp. 216-35.

Digital Libraries and Artifacts

ABSTRACT

Digital libraries organize, store, and preserve digitized and born-digital materials. They extend the reach of traditional libraries and provide students, researchers, and members of the public with access to a wide range of documents and artifacts.

BACKGROUND

In the early 1990s, when traditional library systems were first galvanized by the sweeping possibilities of digitalizing centuries of paper artifacts and documents into computer databases, library scientists believed a global network of digitalized materials was just around the corner. This grand digital library would be accessible to individual users anytime on any computer, thus rendering obsolete the notion of a library as a public building servicing a given community.

However, that bold vision proved to be a significant challenge to bring to fruition. Computer software experts and library scientists charged with devising the systems, system links, and databases and amassing the archives to create digital libraries realized that reaching their goal would involve overcoming a myriad of complexities that still remained unresolved in the early twenty-first century.

OVERVIEW

Not every available archive of information, documents, specialized publications, and data is a digital library. Instead, much like traditional libraries, digital libraries stress organization of the materials in addition to the traditional functions of long-term storage and preservation.

Digital libraries offer an important strategy for extending the reach of traditional libraries, and public and university libraries routinely subscribe to global databases for patron use in addition to participating in ongoing projects to convert centuries of print material into digital format. Although library science is well on its way to catching up to the possibilities of digital collections, distinctions have been established between materials that have to be digitalized—a process that is relatively quick, cheap, and applicable to basically to any publication from before 1990—and materials that are "born digital."

Those charged with developing the templates for digital libraries stress the virtual system's need to organize the ever-growing body of materials to permit efficient and transparent access to users worldwide, given that there will be no central digital library, simply links that connect databases and archives around the world. Such a system would require well-trained professionals to move users smoothly through an often-intimidating network of information.

The development of digital libraries poses a number of significant challenges, including designing the superstructure of an architecturally sound interrelated network of systems, rewriting existing international copyright laws for reproduction and distribution of materials, and generating metadata—that is, data that describes and can be used to catalog primary materials, much like traditional card catalogs or indexes.

In addition, libraries must keep up with the ever-expanding body of data that includes not only traditional print materials, such as books and periodicals, but also films, music, government records, and scientific and research data. Coordinating a theoretically unlimited number of digital libraries, each storing and organizing a specific area of available materials, and putting that network within reach of users ably assisted by digital librarians is the challenge that remains for library scientists and computer engineers.

—Joseph Dewey and Laura Nicosia

Further Reading

Calhoun, Karen. *Exploring Digital Libraries: Foundations, Practice, Prospects*. Facet, 2018.

Corrado, Edward M., and Heather Moulaison Sandy. *Digital Preservation for Libraries, Archives, and Museums*. 2nd ed., Rowman & Littlefield, 2017.

Gartner, Richard. *Metadata in the Digital Library: Building an Integrated Strategy with XML*. Facet, 2021.

Lanagan, James, and Alan Smeaton. "Video Digital Libraries: Contributive and Decentralized." *International Journal on Digital Libraries*, vol. 12, no. 4, 2012, pp. 159-78.

Mizruchi, Susan L., editor. *Libraries and Archives in the Digital Age*. Palgrave Macmillan, 2020.

Xie, Iris, and Krystyna Matusiak. *Discover Digital Libraries: Theory and Practice*. Elsevier, 2016.

Digital Photography

ABSTRACT

Throughout most of history, recording a scene or producing a detailed image was a laborious, time-intensive process. Artists created paintings and etchings from sketches. The invention of the camera allowed people without these artistic talents to record images, and the development of digital cameras further democratized the fields of photography and photo editing. Digital cameras have many advantages over traditional film cameras.

BACKGROUND

The first cameras were large, cumbersome tools that required the use of glass or metal plates, chemicals, and other equipment. In 1884, George Eastman introduced the technology of film to replace the large plates. Sixteen years later, in 1900, the company he founded, Kodak, would allow everyone to take pictures with a simple yet effective camera, the Brownie.

Digital photography was the next step in allowing the general public to produce photographic images with ease. In 1975, Cromemco released the Cyclops, a digital camera available commercially; however, much of the technology used was parlayed into other products. Though development had begun much earlier, with prototypes surfacing in the 1970s, the first digital camera widely available on the market was the Dycam Model 1, released in 1990 by Swiss tech company Logitech.

Photo via iStock/Dejan Ilic. [Used under license.]

Digital photography increased in popularity throughout the next several decades, becoming the predominant form of photography among the public by the 2020s. That increase in popularity was facilitated in part by improvements to the cameras found in mobile devices such as smartphones. The rise of digital photography was also linked to the development of photo-sharing websites and social media, through which individuals could easily share photographs they had taken.

OVERVIEW

Digital cameras use electronic photodetectors to capture images. Although the science of light and optics (lenses) is largely the same as with traditional film cameras, the image is stored electronically, without the use of light-sensitive film. Electronic images are stored either as RGB color space—where tints of red, green, and blue can be combined to various degrees to form any of all possible colors—or raw data, sometimes also called digital negatives, which store all photographic metadata, including light, color, color saturation, white balance, contrast, and sharpness. The advantages of a camera that can store raw images is that the digital negatives contain significantly more information to produce higher quality and more accurate photographs during photo editing. In either case, the images are stored on solid-state digital memory devices.

The technical (rather than aesthetic) quality of a digital Image depends on several factors. One is pixel count: the number of red, green, or blue image points within a given space. Modern cameras record images that contain millions of pixels (1 megapixel = 1,000,000 pixels). Manufacturers and vendors commonly call attention to megapixels when advertising cameras. Although these numbers can help the consumer compare cameras, the number of pixels along either axis of the photograph (rather than the entire area) is a much more accurate factor in determining the quality of a photograph, as each increase in linear resolution results in the square of total resolution.

Digital cameras have many advantages over traditional film cameras. First, they allow the photographer to take many photos easily, without relying on the limitations of the number of exposures on a roll of film. Digital cameras also allow the photographer to see an image immediately and decide whether it is necessary to reshoot a scene. Finally, digital photography allows the photographer to edit photographs in a photo-editing program such as Adobe Photoshop or Lightroom, bringing out the dynamic range of colors and shades in the photograph to reveal greater detail. Some photographers also use photo editing to make changes in the light, hue, and texture of a photograph for aesthetic purposes.

In the world of photography, there are purists who prefer the warmth of traditional film (and traditional photo developing), just as in the music world there are audiophiles who prefer the warmth and dynamic range of vinyl recordings. Digital photography will never replace traditional analog photography, but the two can easily coexist. The camera industry continues to evolve, as advances in camera technologies continue.

—Daniel L. Berek

Further Reading

"History of Photography." *PBS*, www.pbs.org/wgbh/americanexperience/features/eastman-history-photography. Accessed 13 July 2023.

Long, Ben. *Complete Digital Photography*. 9th ed., CDP Press, 2018.

Peterson, Bryan. *Understanding Exposure: How to Shoot Great Photographs with Any Camera*. 4th ed., Amphoto Books, 2016.

Taylor, David. *Digital Photography Complete Course*. DK, 2015.

Digital Storytelling

ABSTRACT

Digital storytelling refers to the tendency of individuals to narrate aspects of their daily lives with the use of new media technology such as podcasts, videos, blogs, and social networks. It has also been described as multimedia creations that combine several digital aspects to construct a narrative. These tools through which people share their lives with others are growing in number, diversity, and sophistication. Although the expression may sound like an oxymoron, with "storytelling" recalling oral traditions and "digital" evoking modern technology, digital storytelling constitutes the postmodern adaptation of the ancient art of narrating stories. It is expanding both in terms of practitioners and audience and has also acquired a relevant status in all levels of education due to its engaging and interactive characteristics.

BACKGROUND

Digital storytelling is made possible in large part by the accessibility and user-friendliness of new media hardware, such as digital cameras, digital voice recorders, and software such as Windows Movie Maker and Apple iMovie. Through these means, people who are not necessarily professional storytellers produce accounts of their lives or of things that have happened to them. These accounts are then shared with larger communities on the internet via video-sharing websites such as YouTube and Vimeo, as well as other means of electronic distribution such as social media and apps such as Instagram and TikTok.

In digital storytelling, the ancient art of constructing a narrative is given a new angle through technology that allows storytellers to use multimedia techniques to intertwine words and digitized stills, moving images, music, and animation. Narrative techniques are therefore blended with technological resources to construct digital stories. This gives the storytelling a new form, as these narratives are often nonchronological, nonsequential, and much shorter than conventional narratives. In addition, new media technology allows for greater interaction with the audience.

OVERVIEW

Because of such qualities, digital storytelling has attracted increasing numbers of authors and audiences. Evidence of this is the growing popularity of the Center for Digital Storytelling (CDS) in Berkeley, California. Founded in 1988, the center has worked to develop digital storytelling techniques and competencies that appeal to hundreds of organizations and thousands of people all over the world.

The interaction that digital storytelling affords makes it particularly well suited for the education sector as well as for the service sector, where stories have been used to carry out programs in the areas of prevention, inclusion, welfare, and health services.

Moreover, teachers and instructors can also use digital stories to generate interest and engagement for their digital-native students. Digital stories can appeal to diverse learning styles, allowing instructors to present abstract or conceptual information in a more understandable way.

Digital storytelling allows students to become creators in the classroom. They acquire technological skills, synthesize information, and write for an authentic audience with a meaningful or entertaining message. The benefits of digital storytelling are certainly verifiable by the passion with which students approach this kind of composition process.

—Luca Prono and Laura Nicosia

Further Reading

Alexander, Bryan. *The New Digital Storytelling: Creating Narratives with New Media*. ABC-CLIO, 2011.

Frazel, Midge. *Digital Storytelling Guide for Educators*. ISTE, 2010.

Gitner, Seth. *Multimedia Storytelling for Digital Communicators in a Multiplatform World*. 2nd ed., Routledge, 2022.

Hartley, John, and Kelly McWilliam, editors. *Story Circle: Digital Storytelling around the World*. Blackwell, 2009.

Lambert, Joe, and Brooke Hessler. *Digital Storytelling: Capturing Lives, Creating Community*. 5th ed., Routledge, 2018.

Miller, Carolyn Handler. *Digital Storytelling: A Creator's Guide to Interactive Entertainment*. 4th ed., CRC Press, 2019.

Miller, Lisa C. *Make Me a Story: Teaching Writing through Digital Storytelling*. Stenhouse, 2010.

Ohler, Jason B. *Digital Storytelling in the Classroom: New Media Pathways to Literacy, Learning, and Creativity*. 2nd ed., Sage, 2013.

Sylvester, Ruth, and Wendy-lou Greenidge. "Digital Storytelling: Extending the Potential for Struggling Writers." *Reading Teacher*, vol. 63, no. 4, 2009, pp. 284-95.

Digital Watermarking

ABSTRACT

Digital watermarking protects shared or distributed intellectual property by placing an additional signal within the file. This signal can be used to inform users of the copyright owner's identity and to authenticate the source of digital data. Digital watermarks may be visible or hidden.

BACKGROUND

A paper watermark is an image embedded within another image or a piece of paper. It can be seen by shining light on the image. Watermarks are used on banknotes, passports, and other types of paper documents to verify their authenticity. Similarly, digital watermarking involves embedding data within a digital signal in order to verify the authenticity of the signal or identify its owners. Digital watermarking was invented in the late 1980s or early 1990s. It uses techniques that are also used in steganography (the concealment of messages, files, or other types of data within images, audio, video, or even text).

OVERVIEW

Digital watermarking is a technique that embeds digital media files with a hidden digital code. These hidden codes can be used to record copyright data, track copying or alteration of a file, or prevent alteration or unauthorized efforts to copy a copyrighted file. Digital watermarking is therefore commonly used for copyright-protected music, video, and software downloads. Governments and banks also rely on it to ensure that sensitive documents and currency are protected from counterfeiting and fraud.

Most digital watermarks are not detectable without an algorithm that can search for the signal embedded in the carrier signal. In order for a carrier signal to be watermarked, it must be tolerant of noise. Noise-tolerant signals are generally strong signals that resist degradation or unwanted modulation. Typically, digital watermarks are embedded in data by using an algorithm to encode the original signal with a hidden signal. The embedding may be performed using either public- or private-key encryption, depending on the level of security required.

QUALITIES OF DIGITAL WATERMARKS

One way to classify digital watermarks is by capacity, which measures how long and complex a watermarking signal is. The simplest type is 1-bit watermarking. This is used to encode a simple message that is meant only to be detected or not (a binary result of 1 or 0). In contrast, multibit watermarking embeds multiple bits of data in the original signal. Multibit systems may be more resistant to attack, as an attacker will not know how or where the watermark has been inserted.

Watermarks may also be classified as either robust or fragile. Robust watermarks resist most types of modification and therefore remain within the signal after any alterations, such as compression or cropping. These watermarks are often used to embed copyright information, as any copies of the file will also carry the watermark. Fragile watermarks cannot be detected if the signal is modified and are therefore used to determine if data has been altered.

In some cases, a digital watermark is designed so that users can easily detect it in the file. For instance, a video watermark may be a visible logo or text hovering onscreen during playback. In most cases, however, digital watermarks are hidden signals that can only be detected using an algorithm to retrieve the watermarking code. Reversible data hiding refers to cases in which the embedding of a watermark can be reversed by an algorithm to recover the original file.

APPLICATIONS OF DIGITAL WATERMARKING

A primary function of digital watermarking is to protect copyrighted digital content. Audio and video players may search for a digital watermark contained in a copyrighted file and only play or copy the file if it contains the watermark. This essentially verifies that the content is legally owned.

Certain types of programs, known colloquially as crippleware, use visible digital watermarks to ensure that they are legally purchased after an initial free evaluation period. Programs used to produce digital media files, such as image- or video-editing software, can often be downloaded for free so users can try them out first. To encourage users to purchase the full program, these trial versions will output images or videos containing a visible watermark. Only when the program has been registered or a product key has been entered will this watermark be removed.

Some creators of digital content use digital watermarking to embed their content with their identity and copyright information. They use robust watermarks so that even altered copies of the file will retain them. This allows a content owner to claim their work even if it has been altered by another user. In some cases, watermarked data can be configured so that any copies can be traced back to individual users or distributors. This function can be useful for tracing illegal distribution of copyrighted material. It can also help investigations into the unauthorized leaking of sensitive or proprietary files.

Digital watermarking has also been used to create hidden watermarks on product packaging. This is intended to make it easier for point-of-sale equipment to find and scan tracking codes on a product. Digital watermarking is also increasingly being used alongside or instead of regular watermarking to help prevent the counterfeiting of important identification papers, such as driver's licenses and passports.

—Micah L. Issitt

Further Reading

Bas, Patrick, et al. *Watermarking Security*. Springer Singapore, 2016.

Chao, Loretta. "Tech Partnership Looks beyond the Bar Code with Digital Watermarks." *Wall Street Journal*, 12 Jan. 2016, www.wsj.com/articles/tech-partnership-looks-beyond-the-bar-code-with-digital-watermarks-1452623450. Accessed 13 July 2023.

Giri, Kaiser J., Shabir Ahmad Parah, Rumaan Bashir, and Khan Muhammad, editors. *Multimedia Security: Algorithm Development, Analysis and Applications*. Springer Nature Singapore, 2021.

Gupta, Siddarth, and Vagesh Porwal. "Recent Digital Watermarking Approaches, Protecting Multimedia Data Ownership." *Advances in Computer Science*, vol. 4, no. 2, 2015, pp. 21-30.

Nematollahi, Mohammad Ali, Chalee Vorakulpipat, and Hamurabi Gamboa Rosales. *Digital Watermarking: Techniques and Trends*. Springer Singapore, 2017.

Patel, Ruchika, and Parth Bhatt. "A Review Paper on Digital Watermarking and Its Techniques." *International Journal of Computer Applications*, vol. 110, no. 1, 2015, pp. 10-13.

"Quick Facts about the Digital Watermarking Alliance." *Digital Watermarking Alliance*, digitalwatermarkingalliance.org/about/quick-facts. Accessed 13 July 2023.

Su, Qingtang. *Color Image Watermarking*. Tsinghua UP/Walter de Gruyter, 2017.

Director/Producer

ABSTRACT

Directors plan, coordinate, and manage the creative aspects of a film, television, radio, or theatrical production. Producers plan, coordinate, and manage the business side of a production. Students aspiring to enter either profession should study subjects such as film, theater, business, and event planning.

BACKGROUND

Directors and producers oversee all aspects of a film, television, radio, or theatrical production. Directors plan, coordinate, and manage the creative aspects of the production, including interpreting scripts, casting talent, approving artistic designs, and directing the work of actors, cinematographers, set designers, wardrobe designers, and other members of the cast and crew. Producers plan, coordinate, and manage the business side of a production, which includes raising money, approving and developing the script, and performing any related administrative tasks. In most cases, directors and producers must both report to the executive producer (usually the person or entity who finances the project), who must approve all final decisions.

OVERVIEW

Like actors, directors and producers must be willing to work an irregular schedule with long hours and evening and weekend work, punctuated by frequent periods of unemployment. Productions may last from one day to several months, and during that time, directors and producers are expected to be on call and available to solve problems that arise before, during, and after a production has finished. They may also be away from home, or "on location," for extended periods. The irregular hours and intense competition in these occupations can result in stress, fatigue, and frustration. Most directors and producers must work day jobs or other employment unrelated to entertainment.

Prospective directors should be highly creative, confident, and possess a strong desire to tell stories. They must be extremely organized, be natural leaders, and understand all aspects of coordinating a theatrical or film production, including the role that each cast and crew member plays in the successful completion of a production. Prospective producers should be detail-oriented people who have a desire to take on both small and large tasks. Producers should enjoy planning, coordinating, and organizing an event from start to finish and should be willing to handle and resolve any issues that arise.

DUTIES AND RESPONSIBILITIES

There are many different styles of directing films and plays, just as there are many different styles of acting. Directors are ultimately responsible for the appearance, stylistic and emotional tone, and aesthetic organization of a dramatic production. A film studio or independent producer normally hires a director through the director's agent or manager. Before production begins, a director auditions and chooses actors, holds rehearsals, and prepares the cast for production. He or she also consults with set designers, choreographers, cinematographers, music supervisors, and other creative personnel to plan and develop a successful production. During production, a director guides and oversees the entire creative execution of a project, often with help from assistant directors and production assistants. Once production is finished, a director oversees any postproduction responsibilities, such as video and sound editing, graphic design, and music selection.

Producers are responsible for handling the business aspects of a production. They secure funds, set budget limitations, coordinate schedules, and ensure smooth management of the whole project. Producers also work with directors to approve their decisions regarding talent, locations, and other creative

Photo via iStock/selimaksan. [Used under license.]

choices, as well as to ensure that deadlines are met and money is spent according to financier instructions. Larger productions usually require the services of associate or line producers to assist the producer with his or her duties.

OCCUPATION SPECIALTIES

Stage Directors. Stage directors interpret scripts, direct technicians, and conduct rehearsals to create stage presentations.

Motion Picture Directors. Motion picture directors read and interpret scripts, conduct rehearsals, and direct the activities of cast and technical crews for motion picture films.

Television Directors. Television directors interpret scripts, conduct rehearsals, and direct television programs.

Radio Directors. Radio directors direct radio rehearsals and broadcasts.

Casting Directors. Casting directors audition and interview performers for specific parts.

Motion Picture Producers. Motion picture producers initiate and manage all the business needs of a motion picture production.

WORK ENVIRONMENT

Physical Environment. Most directors and producers work on set during the production of a theatrical project. Set locations vary greatly and may be indoors or outdoors in any weather conditions. Some productions are held in different locations across the country or around the world. Before production begins (during "preproduction") and after a production finishes (during "postproduction"), directors and producers may work from an office or home studio.

Human Environment. Directors and producers constantly interact with other cast and crew members. Their coworkers typically include executive producers, actors, production staff, set designers, costume and makeup personnel, and assistants. Producers regularly work with external vendors, such as caterers, insurance representatives, and establishment owners.

Technological Environment. Directors and producers employ a wide variety of tools and equipment to assist them in the completion of their daily tasks. Directors use video cameras, lighting and sound equipment, two-way radios, cell phones, audiovisual editing equipment and software, and the internet. Producers use schedules, budgets, contracts, email and the internet, laptops, cell phones, and other devices.

EDUCATION AND TRAINING

High School/Secondary. High school students who wish to become directors or producers should have an inherent interest in the dramatic arts and should foster that interest by pursuing academic study in English literature, theater, public speaking, communications, and cinema. They should also learn as much as they can about management, business, and event planning. Involvement in school groups or extracurricular activities, such as drama clubs, plays, musical productions, dance performances, film clubs, and photography clubs, can provide a solid background in the arts. They should also enroll in a basic acting class to become familiar with the fundamentals of acting, dramatic literature, and theater production.

College/Postsecondary. Although an undergraduate degree is not necessarily required for one to become a director or producer, many people consider it essential to have received some formal training at the postsecondary level. Many universities and colleges offer bachelor's degree programs in the dramatic arts. Some directors find it beneficial to have studied directing, filmmaking, writing, acting, designing, radio broadcasting, film history, or public speech at the college level. Producers can benefit by taking undergraduate business courses in marketing, public relations, management, and finance.

After obtaining a bachelor's degree, some directors and producers earn a master of fine arts (MFA) degree in directing, producing, acting, or screenwriting. Some conservatories, such as the American Film Institute (AFI) in Los Angeles, offer MFA programs that teach students the practical skills needed to start a career in filmmaking. Often, students are required to complete a thesis film as part of their coursework, designed to simulate a large-scale production. Producers and directors must raise money, find talent, and promote their thesis films.

Adult Job Seekers. Prospective directors and producers possess varying levels of experience. Those who attend conservatories often make valuable connections with faculty and other students, which eventually lead to production work. Others become apprentices, interns, or assistants for established directors or producers. Some job seekers begin by taking other employment positions in the entertainment industry and working their way up to director or producer positions through networking and industry contacts.

Many directors and producers are members of professional organizations, such as the Producers Guild of America and the Directors Guild of America, which protect the rights of the producers and provide networking opportunities.

Professional Certification and Licensure. Directors and producers are not required to receive any kind of professional certification or licensure in dramatic production. There is no official training for producers, but many directors train or take classes in directing and cinematography.

Additional Requirements. Directing and producing are highly competitive fields, and few people are able to achieve financial stability through these occupations. Candidates must be able to handle criticism well, demonstrate emotional and physical stamina, and remain incredibly driven to succeed. Being talented is not enough to make one successful in these fields—directors and producers must not give up easily, especially after experiencing rejection. They should be self-promoters who are passionate about their work and use every opportunity to meet potential investors, employers, and talent. Long hours and demanding or difficult employers or work conditions are common in these occupations.

—Briana Nadeau

Further Reading

Brine, Kelly Gordon. *The Art of Cinematic Storytelling: A Visual Guide to Planning Shots, Cuts, and Transitions.* Oxford UP, 2020.

Brown, Blain. *The Basics of Filmmaking: Screenwriting, Producing, Directing, Cinematography, Audio, and Editing.* Routledge, 2020.

Rabiger, Michael, and Mick Hurbis-Cherrier. *Directing: Film Techniques and Aesthetics.* Routledge, 2020.

DVDs

ABSTRACT

Digital video discs (DVDs) are a form of optical disc storage primarily used by consumers to watch films and television programs. They arrived in the mid-1990s and soon overtook the supremacy of videocassettes for home viewing. By the 2020s, however, DVDs were quickly becoming obsolete as consumers increasingly favored the Blu-ray disc format, cloud storage, and streaming media.

BACKGROUND

In the 1970s and 1980s, the video home system (VHS) videocassette recorder (VCR) revolutionized home entertainment by allowing viewers to videotape television programs and watch them at their leisure and to rent or buy films on videotape. This viewing flexibility was undercut for some because the picture and sound quality of videocassettes was generally inferior to what could be seen and heard on television and especially in movie theaters.

Two years after its introduction in 1976, VHS had a challenger in the laser disc, which offered much sharper images, offering 425 lines of horizontal resolution in contrast to VHS's 240. With Criterion's special edition release of *Citizen Kane* (1941) in 1985, which provided supplemental material about the film (generally referred to as bonus materials, or extras), another advantage was introduced. Laser discs never caught on with the public, however, because of their cumbersome size (comparable to long-playing records), their susceptibility to damage, their high cost, and their inability to record television programs.

Consumers waited patiently while a cheaper, more adaptable technology was developed. By 1993, Philips and Sony had developed the smaller MultiMedia Compact Disc, while Toshiba had created its Super Density Disc. Philips and Sony, wanting to avoid a format war, eventually decided to proceed with the specifications of the Toshiba disc, and the DVD was born. Though the industry initially insisted that DVD stood for "digital versatile disc," most consumers believed it meant "digital video disc," and the nomenclature was changed.

Digital video discs and DVD players entered the US market in late 1997. At first, only recent popular films were offered, with older and more esoteric films becoming available late in the 1990s. Some consumers were hesitant at first because the first DVDs were not recordable, but those who had been reluctant to buy laser discs quickly embraced the cheaper, smaller (the same size as audio compact discs), and easier-to-store discs. These buyers found that the picture, with the same resolution as laser discs, was vastly superior to that of videocassettes, and when connected to the proper equipment (receiver, speakers), multichannel audio was possible. As a result, Pioneer, the main champion of laser discs, abandoned that format in June 1999.

OVERVIEW

Though DVDs were initially slow to catch on with the general public, matters soon changed as more and more consumers heard about the improved audiovisual quality and saw it demonstrated in electronics stores. The entertainment industry had assumed that DVDs would slowly supplement or perhaps even replace VHS, but experts underestimated the public by assuming that consumers would continue to rent much more than they purchased. Though some had developed VHS collections of their favorite films, the bulky tapes took up considerable room. Digital video discs occupied much less space, were at first only slightly costlier than VHS cassettes, and were generally more durable (though prone to scratching), inviting multiple viewings. They also did not require rewinding as VHS tapes did.

Contributing to this early success was the inclusion of extras. Following the lead of laser discs, the makers of DVDs offered commentaries by filmmakers and film scholars. Viewers could watch films either with or without commentaries, which explained the production details of the films and, with older films, placed them in a historical context. Consumers wanted more and more extras, and soon many films included bonus material such as on-set interviews, documentaries about the making of the films, and deleted scenes.

Digital video discs also offered viewers the option, for many films made since the mid-1950s, of watch-

Photo via iStock/kyoshino. [Used under license.]

ing them in a full-screen format—which filled every inch of screen space but cut off the corners of the images—or in the film's original wide-screen format, meaning that the tops, bottoms, and sides of images were no longer excised. This pleased many cinephiles who had lamented the formatting changes made to classic films upon VHS release. As television screens became larger, the wide-screen format option became more important.

In 1998, 2 million DVD players were sold in the United States, three thousand titles were available, and 9.3 million DVDs were sold. In 1999, 4 million players were sold, 6,300 titles were available, and 20 million DVDs were sold. Digital video discs and DVD players had an advantage that VHS and VCRs did not: the internet. By the late 1990s, consumers had discovered the ease of purchasing online, finding that players and DVDs could be found at lower prices than in stores. By the end of 1999, four titles had sold over a million copies each: *Austin Powers: The Spy Who Shagged Me* (1999), *The Matrix* (1999), *Saving Private Ryan* (1998), and *Titanic* (1997).

Internet commerce and DVDs have been linked since the introduction of the format. While videocassettes had been rented from stores such as the Blockbuster chain, DVDs could be rented online from Netflix starting in April 1998. When the company introduced a flat monthly fee for unlimited rentals of DVDs delivered quickly by mail in September 1999, online renting of DVDs took off.

IMPACT

Because DVDs were introduced relatively late in the 1990s, their full impact did not become clear until the early twenty-first century. Once the initial consumer excitement over improved picture, sound, and storage subsided, more and more titles were demanded, especially classic and obscure films, along with increased extras. Many enthusiastic DVD collectors built libraries of hundreds or thousands of titles. Digital video disc producers were pleasantly shocked by the number of people wanting to own complete seasons of their favorite television shows. What began as an alternative to VHS eventually changed the US population's viewing habits as much as that earlier innovation. As a result, VHS itself slowly faded, with the last mass-market VHS title issued in 2006.

However, DVDs themselves soon faced challenges from newer technologies. Blu-ray Discs, which use blue lasers instead of the red lasers of DVDs, were officially introduced in 2006 and allow much higher resolution images and a greater amount of data. Cooperation from companies such as Sony and Philips again led to a standardization of formats, beating out Toshiba's HD DVD competitor. Ultrahigh-definition Blu-ray technology later followed. These discs and players competed with DVDs, and retailers promoted Blu-ray releases of films with new special features in order to attract consumers who may already have purchased the DVD version. Though Blu-ray technology was initially expensive compared to DVDs, and therefore less popular, it eventually ate into the market share of DVDs. Perhaps even more importantly, the 2010s also saw consumers increasingly turn to online streaming services for video content. Netflix pivoted from its DVD-mailing business to focus on streaming films and television shows, for example, and sales of DVDs declined.

—Michael Adams

Further Reading

Barlow, Aaron. *The DVD Revolution: Movies, Culture, and Technology*. Praeger, 2005.

Bennett, James, and Tom Brown, editors. *Film and Television after DVD*. Routledge, 2008.

Schenk, Sonja, and Ben Long. *The Digital Filmmaking Handbook*. 7th ed., Foreing Films, 2021.

Stokel-Walker, Chris. "Why Do DVDs Still Exist?" *Wired*, 27 Nov. 2021, www.wired.com/story/why-do-dvds-exist. Accessed 13 July 2023.

Taylor, Jim. *Everything You Ever Wanted to Know About DVD: The Official DVD FAQ*. McGraw-Hill, 2004.

Wroot, Jonathan, and Andy Willis, editors. *DVD, Blu-ray and Beyond: Navigating Formats and Platforms within Media Consumption*. Palgrave Macmillan, 2017.

E

E-LEARNING

ABSTRACT

E-learning leverages communications and multimedia technology to broaden the learning environment for students. The internet has had a significant impact on the ways in which students learn and teachers teach; e-learning, or electronic learning, extends the learning environment into this virtual realm. E-learning takes place in an online, computer-based environment and covers a broad range of teaching techniques and practices. These include online instructional presentations, interactive lessons, and computer-supported in-class presentations.

BACKGROUND

Also known as online learning or virtual education, e-learning is commonly used in public high schools or at colleges and universities. It may use of a variety of electronic media, including, but not limited to, text, streaming video, instant messaging, document sharing software, Blackboard learning environments, webcams, blogging, and streaming video. E-learning's key benefits include broader access to education by a wide range of students; collaborative interaction between students, peers, and teachers; development of technological skills and knowledge; and independent study.

E-learning is not synonymous with distance or mobile learning. Instead, e-learning refers solely to the use of online, computer-based technologies to complete an area of instruction. E-learning can take place in or out of a classroom and can be led either by a teacher in real time or completed at a pace set by the student, although many e-learning courses include some hybrid version of those two scenarios. The virtual learning environment is typically collaborative and often incorporates a blog or wiki entries to facilitate this relationship.

OVERVIEW

Researchers suggest that the success of e-learning is compounded when the information presented is coherent in design and does not include extraneous material. Educators suggest that when using an e-learning environment with their students, instructors should still employ good pedagogical practices and techniques as a foundation upon which to build. Additionally, many educators cite the interactions between a student and teacher and between students and their peers as crucial for the successful completion of an e-course.

E-learning is also at the forefront of college and university education structures, with a significant number of university curriculums offering courses that take place solely in an online, e-learning environment. These courses, called massive open online courses, or MOOCs, are often available to anyone in the world without charge. Additionally, other universities have taken e-learning a step further to build entirely mobile programs of study, establishing online higher-education programs where students can complete a bachelor's or master's degree program entirely in an e-learning format.

While the popularity of e-learning and distance education programs was already growing during the first decades of the twenty-first century, the emergence of the COVID-19 pandemic in early 2020 resulted in a dramatic shift toward distance learning in the United States, particularly at the postsecondary level. According to data reported by the National

Photo via iStock/Prostock-Studio. [Used under license.]

Center for Education Statistics (NCES), 6 million US undergraduates were enrolled in at least one distance learning course in the fall of 2019, but by the fall of 2020, that number had risen to 11.8 million. In addition, more than half of the undergraduates who took at least one distance learning course in the fall of 2020 were exclusively enrolled in distance courses. In light of such trends, e-learning technologies and methods took on newfound importance at many colleges and universities in the United States during that period.

E-learning is also being implemented by many companies and organizations to keep their employees up to date on various topics and to ensure compliance with the organization's policies and practices, offering training seminars in a completely online environment. This frees up resources by placing the burden of learning on the student, alleviating some oversight and human resources costs associated with training workforces.

—*Laura L. Lundin*

Further Reading

Berge, Zane L., and Lin Muilenburg. *Handbook of Mobile Learning*. Routledge, 2013.

Clark, Ruth Colvin, and Richard E. Mayer. *E-Learning and the Science of Instruction: Proven Guidelines for Consumers and Designers of Multimedia Learning*. 5th ed., Wiley, 2023.

"Distance Learning." *IES NCES*, 2022, nces.ed.gov/fastfacts/display.asp?id=80. Accessed 13 July 2023.

Donally, Jaime. *The Immersive Classroom: Create Customized Learning Experiences with AR/VR*. International Society for Technology in Education, 2021.

Garrison, D. Randy. *E-Learning in the 21st Century: A Community of Inquiry Framework for Research and Practice*. 3rd ed., Routledge, 2016.

Hill, Janette R., Llyan Song, and Richard E. West. "Social Learning Theory and Web-Based Learning Environments: A Review of Research and Discussion Implications." *American Journal of Distance Education*, vol. 23, no. 2, 2009, pp. 88-103.

Mayer, Richard E., and Logan Fiorella, editors. *The Cambridge Handbook of Multimedia Learning*. 3rd ed., Cambridge UP, 2022.

Wang, Chih-Hsuan, David M. Shannon, and Margaret E. Ross. "Students' Characteristics, Self-Regulated Learning, Technology Self-Efficacy, and Course Outcomes in Online Learning." *Distance Education*, vol. 34, no. 3, 2013, pp. 302-23.

ELECTRONIC COMMERCE TECHNOLOGY

ABSTRACT

Electronic commerce (e-commerce) is the buying, selling, and transfer of products and services using the internet. It offers enormous advantages for supply-chain management and the coordination of distribution channels, in addition to being convenient for consumers. Convenience is a big selling point; in 2021, global retail e-commerce sales reached about $5.2 trillion, according to the statistics platform Statista. A great amount of technology goes into supporting the online platforms used for e-commerce, facilitating financial transactions, and coordinating the logistics of inventory and shipping.

BACKGROUND

E-commerce refers to the communications between customers, vendors, and business partners over the internet. The term e-business incorporates the additional activities carried out within a business using intranets, such as communications related to production management and product development. Many also view e-business as referring to collaborations with partners and e-learning organizations.

In addition to online purchases of goods and services, e-commerce involves bill payments, online banking, e-wallets, smart cards, and digital cash. E-commerce depends on secure connections to the internet. Many precautions to ensure security are necessary to maintain successful e-commerce, including public-key cryptography, Secure Sockets Layer (SSL) and Transport Layer Security (TLS) protocols, digital signatures, digital certificates, firewalls, and antivirus programs. New technology is constantly being developed to protect against computer worms and viruses and other cyberattacks.

In 1969, the Advanced Research Projects Agency (now the Defense Advanced Research Projects Agency, or DARPA) of the US Department of Defense proposed a method to link together the computers at several universities to share computational data via networks. This network became known as ARPANET, the precursor of the internet. As a result, electronic mail (email) was developed, along with protocols for sending information over phone lines in packets. The protocols for the transmission of these packets of data came to be known as transmission control protocol (TCP) and internet protocol (IP). Together, these two protocols, known as TCP/IP, are still in use and are responsible for the efficient communication conducted through the network of networks referred to as the internet.

Oxford University graduate Tim Berners-Lee (b. 1955) initially created the World Wide Web in 1989 while working at CERN, the European Organization for Nuclear Research, and made it available to the public in 1991. During this time, Cisco Systems was growing to become the first company to produce the broad range of hardware products that allowed ordinary individuals to access the internet. In 1993, Marc Andreessen (b. 1971) and Eric Bina (b. 1964), employees at the National Center for Supercomputing Applications (NCSA), created Mosaic, the first web browser that supported clickable buttons and links and allowed users to view text and images on the same page. New software and programming-language developments rapidly followed, allowing ordinary consumers easy access to the internet. As a result, companies saw the opportunity to gain customers, resulting in the creation of online businesses, including Amazon.com in 1994, eBay in 1995, and PayPal and Priceline in 1998.

OVERVIEW

The engineers who developed ARPANET created digital packets to use for transmitting data via packet switching. The theory was that it would be faster and cheaper to transmit digital data using small packets that could be sent, or routed, to their destination in the most efficient way possible, even if the original message had to be split up into smaller packets that were then joined back together at their destination. In order to accomplish the packaging of data and transmission via the best routes, the engineers who developed ARPANET also developed TCP. The first and most common access method to the internet was through the wiring that has transmitted telephone calls. However, wireless internet connections can now be made much faster from many locations, even through cellular phones. The economy has become dependent on digital communication, and the companies that sell the most goods and services have a strong online presence. A great deal of planning goes into the maintenance of an effective website.

E-Commerce Business Establishment. After first developing a practical business plan, the process for establishing an online business that will be able to successfully compete for sales can be overwhelming. One way to start an e-business is by using a turnkey solution, which is essentially a prepackaged type of software specifically for a new business. An alternative is to use the services of an internet incubator, which is a company that specializes in e-business development. An internet incubator typically obtains

Photo via iStock/gorodenkoff. [Used under license.]

ownership of at least 50 percent of the business and may also enlist funding help from venture capitalists to get started. Web-hosting companies sell space on a web server to customers, maintain enough storage space for the website, and provide support services. A domain name for the website must be chosen and registered. This domain name is to be used in the URL (uniform resource locator) for the website. The URL, or website's internet address, consists of three parts: the host name, which is shown by the "www" for World Wide Web; the domain name, which is usually the name of the company; and lastly, the top-level domain (TLD), which describes the type of organization that owns the domain name, such as ".com" for a commercial organization or ".gov" for a government organization. An initial public offering (IPO) of stock to assist with funding usually follows for enterprises that achieve a certain level of success.

Design of Markets and Mechanisms of Transactions. Initially the primary e-commerce activities were business-to-business (B2B) transactions. These activities quickly expanded to include sales to consumers via electronic retailing (e-tailing), often called business-to-consumer (B2C). Since the late 1990s, e-commerce has expanded to include consumer-to-consumer (C2C) websites, including eBay, and consumer-to-business (C2B) services such as Priceline, where several hotels will compete for the purchase dollars of consumers. Each of these types of transactions can be completed within the general structure of one of the many different types of e-commerce models to generate revenue.

Automated Negotiation and Peer-to-Peer Distribution Systems. Auction models allow an internet user to assume the role of a buyer using either the reverse-auction model, where the buyer sets a price and sellers have to compete to beat that price, or the reverse-price model, where the seller sets the minimum price that will be accepted. Auction sites such as eBay update listings, feature items, facilitate payments, and earn submission and commission fees, but they leave the process of delivery up to the actual buyers.

Dynamic-pricing models include the name-your-price companies, such as Priceline, that use a shopping bot to collect bids from customers and deliver these bids to the providers of services to see if they are accepted. A shopping bot is a computer program that searches through vast amounts of information, then collects, summarizes, and reports the information. This is one example of the use of intelligent agents, or software programs that have been designed to gather information, by e-businesses. Priceline's immense success is due in part to its use of this technology; however, the company phased out its name-your-price tool for flights in 2016 and for car rentals in 2018.

Network Resource Allocation: Electronic Data Interchange (EDI). Portal models present a variety of news, weather, sports, and shopping all on one web page, allowing a visitor to see an overview and then choose to obtain more in-depth information. Vertical portals are specific for a single item, while horizontal portals function as search engines with access to a large range of items. Storefront models require a product line to be accessible online via the merchant server so that customers can select items from the database of products and collect them in the order-processing technology called a shopping cart. Businesses use EDI as a standardized protocol for communication to monitor daily inventory, shipments, and payments. Standardized forms for invoices and purchase orders are routinely accessible via the use of extensible markup language, or XML. Companies such as TIBCO Software and Commerce One (which was later acquired by Perfect Commerce and in turn by Proactis) were created to help companies move their businesses to the web via B2B techniques. The transition of traditional brick-and-mortar stores to click-and-mortar stores has helped decrease lead time and has caused an increase of just-in-time (JIT) inventory management. JIT inventory management allows e-businesses to save money because the companies do not overbuy goods and create an inventory surplus that they then have to worry about storing and selling, which decreases overhead costs.

APPLICATIONS AND PRODUCTS

E-commerce has spawned a variety of applications and products that both facilitate e-businesses and enhance people's lives.

Consumer Products. Smartphones and tablet computers have been among the most popular consumer digital purchases. The proliferation of mobile devices, including laptop computers and smartphones, boosts the e-commerce industry by giving consumers the means to remain almost constantly connected to the internet and therefore more likely to make online purchases. The rapid expansion of the market for application software, or apps, in the 2010s presented a major development in internet retailing, including the controversial practice of in-app purchases. In the 2020s, developments such as augmented reality (AR) technology benefits e-commerce, especially fashion and décor industries, as it provides a virtual interaction with products before purchases. Other tools include artificial intelligence (AI) and progressive web applications (PWAs). AI helps e-commerce companies personalize customer experiences and enables them to offer benefits such as visual search, which allows customers to submit pictures of products they are interested in.

Both contact and contactless "smart cards," which resemble credit cards, have been developed to store much more information (banking, retail, identification, health care) on a microprocessor embedded in the card. Examples of smart cards include cards used to make contactless fare payments in public-transportation systems throughout the United States, such as the Massachusetts Bay Transportation Authority's CharlieCard. Smart cards are more secure than credit cards because they are encrypted and password protected.

Security Applications and Products. Companies have been created to help merchants accept credit-card payments online, which are called card-not-present (CNP) transactions. These companies, such as Stripe and Square, offer services to facilitate the authentication and authorization processes through SSL or TLS protocols to minimize fraud. Additional security features include firewalls, encryption, and antivirus software. A number of companies, including many major credit-card companies, have introduced digital wallets that allow customers to save their shipment address and payment information securely in an online database so that purchases can be made with one click of the mouse, instead of having to reenter the same information each time. In 1999, the Electronic Commerce Modeling Language (ECML) emerged as the protocol for digital-wallet usage by merchants. PayPal can be used to transfer payments between consumers securely by simply creating an account using an email address and a credit card or checking account, which is then used to pay for goods and services. PayPal is especially secure because credit-card information is checked before the transaction actually begins. This allows for payment to take place in real time, minimizing the opportunity for fraud. Other applications, such as Venmo, have made person-to-person payments by phone extremely easy, and services such as Apple Pay and Google Pay can be used to make secure online payments in addition to offline contactless payments.

Applications Using Wireless Transactions. Mobile e-commerce, or m-commerce, has become increasingly important due to significant advances in wireless technology. The fourth generation of mobile technology, called 4G, offers a maximum speed of 150 megabits per second (Mbps), while successor 4G Long Term Evolution Advanced (LTE-A) offers a maximum speed of between 300 Mbps and 1,000

Mbsp, or one gigabit per second (Gbps). The fifth generation of mobile technology, 5G, is significantly faster, with a maximum speed of between one and ten Gbps. With the increased popularity of apps, such as those offered by Apple's iOS App Store and Android's Google Play, m-commerce has become an important segment of the e-commerce market.

CAREERS AND COURSEWORK

Universities such as the University of California in Los Angeles and Arizona State University offer undergraduate courses in digital and integrated marketing communication. Job titles, career paths, and salaries within the field of e-commerce vary a great deal. Some of the typical job titles include website developer, web designer, database administrator, web analyst, website manager, and programmer. Jobs related to web content require skills in web-development tools and software languages, including hypertext markup language (HTML), XML, Java, and JavaScript. Database administrators focus less on these web tools and languages and more on database-related tools, such as structured query language (SQL), Microsoft Access, and Oracle Database. Knowledge of computer networks and operating systems is also very helpful. Aspirants can work as e-commerce specialists, online sales consultants, customer service specialists, and operations logistics.

SOCIAL CONTEXT AND FUTURE PROSPECTS

Due in part to the easy accessibility of goods and services via the internet, online businesses have become increasingly competitive, with more made-to-order goods being produced by companies of all sizes and corresponding decreases in the costs associated with the maintenance of a large inventory. Since so much more data is exchanged digitally via stock trades, mortgages, purchases of consumer goods, payment of bills, and banking transactions, it is conceivable that eventually digital cash and smart cards could replace traditional cash completely. Because consumers enjoy the comparison shopping among goods offered by companies all over the world, as well as the twenty-four-hour-a-day, seven-days-a-week convenience of online shopping, sales at traditional brick-and-mortar stores will no doubt continue to decline, or at least migrate toward those goods that consumers prefer not to purchase online.

Online retailers such as Amazon continue to experiment with methods for more efficient supply chain and logistics systems; one notable project has involved the use of drones to deliver packages to consumers after an online purchase. Other shipping innovations have included Amazon's partnership with the US Postal Service (USPS) to provide package delivery on Sundays and the creation of local centers where consumers can pick up purchased items rather than wait for local delivery.

Despite continuing growth, e-commerce does face important challenges. Security threats such as hacking, identity theft, and other types of cyber theft have become problems, which has only increased the need for better security tools, such as digital certificates and digital signatures. Increased security needs will continue to fuel the ever-expanding internet security industry.

Beginning in 2020, the coronavirus (COVID-19) pandemic negatively affected global economic activities and industries alike due to lockdowns imposed by governments. The only sector that saw growth during this period was e-commerce, as consumers and businesses used this channel to buy and sell products and services online. Experts believed that the impact of the pandemic on the growth of the e-commerce sector would be long-term due to increased digitalization.

—Jeanne L. Kuhler

Further Reading

Castro, Elizabeth, and Bruce Hyslop. *HTML and CSS*. 8th ed., Peachpit Press, 2014.

Chaffey, Dave, Tanya Hemphill, and David Edmundson-Bird. *Digital Business and E-Commerce Management*. 7th ed., Pearson Education, 2019.

Chevalier, Stephanie. "Global Retail E-Commerce Sales 2014-2026." *Statista*, 21 Sept. 2022, www.statista.com/statistics/379046/worldwide-retail-e-commerce-sales. Accessed 13 July 2023.

Cramer-Flood, Ethan. "Global Ecommerce Update 2021: Worldwide Ecommerce Will Approach $5 Trillion This Year." *eMarketer, Business Insider Intelligence*, Jan. 2021, www.emarketer.com/content/global-ecommerce-update-2021. Accessed 13 July 2023.

Jansen, Mark, and Paula Beaton. "5G vs. 4G: How Does the Newest Network Improve on the Last?" *DigitalTrends*, 22 Apr. 2022, www.digitaltrends.com/mobile/5g-vs-4g. Accessed 13 July 2023.

Laudon, Kenneth C., and Carol Guercio Traver. *E-Commerce 2019*. 15th ed., Pearson Education, 2020.

Sherif, Mostafa Hashem. *Protocols for Secure Electronic Commerce*. 3rd ed., CRC Press, 2018.

Turban, Efraim, Carol Pollard, and Gregory Wood. *Information Technology for Management: On-Demand Strategies for Performance, Growth and Sustainability*. 11th ed., John Wiley & Sons, 2018.

Umar, Amjad. "IT Infrastructure to Enable Next Generation Enterprises." *Information Systems Frontiers*, vol. 7, no. 3, 2005, pp. 217-56.

Wei, June. *Mobile Electronic Commerce: Foundations, Development, and Applications*. CRC Press, 2015.

ELECTRONIC MUSIC

ABSTRACT

Electronic music is music produced and heard through electronic means. Though based in part on the musical innovations of earlier decades, electronic music experienced a substantial increase in popularity during the 1990s, creating many new genres and subgenres of music. It remained popular in the decades the followed, during which it influenced the evolution of mainstream pop, hip-hop, and rock music.

BACKGROUND

The development of electronic music was made possible in large part by major advances in music technology that took place throughout the twentieth century, beginning with the invention of the triode vacuum tube in 1906. In 1920, Russian engineer Léon Theremin invented the instrument that bears his name, which was patented in 1928 and went on to become iconic as a wholly electronic instrument. The theremin consists of two loop antennae that control oscillators; one regulates amplitude (volume) and the other regulates frequency (pitch). The antennae sense movements of the performer's hands without actual physical contact to create changes in the electronic signal and the resulting eerie music is amplified and transmitted through speakers.

The 1930s saw further developments, including the invention of recording with magnetic tape, which would allow many new techniques such as sampling and sound effects, and the digital coding method called pulse code modulation (PCM). Meanwhile musicians began to experiment with electronic means of amplifying acoustic instruments in order to achieve greater volume in band settings. The earliest experiments with electric guitars started in the early 1930s, with the first recording of a magnetic pickup made in 1938. Electric guitars would go on to influence popular music heavily.

One instrument closely associated with electronic music is the synthesizer, the first successful programmable example of which, the RCA Mark II, was completed in 1959 at the Columbia-Princeton Electronic Music Center. Engineering physicist Robert Moog revived interest in the theremin in 1961 and applied some of its principles to the invention of the Moog synthesizer, introduced in 1964. His invention used electronic circuit boards to alter the sound of music produced by a keyboard and was the first compact and portable synthesizer. It quickly caught on with the advent of rock and funk music styles and was used in concerts by groups such as the Beatles and the Rolling Stones in the late 1960s. Moog is acknowledged to be an American pioneer in electronic music.

An artist in his studio making music with a midi controller. Photo via iStock/alvarez. [Used under license.]

Many other designs followed, including widespread use of digital technology with the increasing popularity of synthesizers in the 1980s. One of the most influential devices was the Yamaha DX7 digital synthesizer that used frequency modulation (FM) to generate sounds; it was released in 1983, the same year that the musical instrument digital interface (MIDI) standard for connecting electronic instruments and computers was established. The wide variety of sounds available with advanced synthesizers greatly impacted music across all genres, and as the cost of such devices decreased, they became easily available to all types of musicians.

OVERVIEW

Electronic music developed over the course of several decades, and a number of influential artists experimented with electronic and computer-based music technologies throughout the 1970s and 1980s, particularly in countries such as Japan, the United Kingdom, and Germany. However, electronic music gained extensive momentum in the United States during the 1990s, a decade in which developments in music coincided with technological shifts and cultural trends that proved to be compatible with electronic artists' musical sensibilities. Advances in electronic music equipment, in particular, generated a turn in the American music industry toward increasingly accessible and easy-to-use synthesizers and computers, which allowed individual musicians to create music in the domestic confines of their house or apartment. Consequently, a number of dance music artists, or deejays, emerged from major urban centers where they found an audience who embraced their music in all-night dance clubs.

House music, coming out of Chicago, and techno music, originating in Detroit, spread quickly into other US cities during the early 1990s, combining with other genres like European Trance music and British trip-hop to form new subgenres such as acid techno, anthem house, and epic trance, all of which contributed to and became associated with the growing underground rave scene.

Two major electronic music genres, however, remained detached from rave culture. Krautrock combined psychedelic rock with avant-garde electronics to create a dark, mechanical sound more appealing to listeners at home than audiences at concerts or nightclubs. The audience, then, was relatively small in the United States, but German Krautrock groups like Kraftwerk, who describe their own music as robot pop, had far-reaching influence on more popular musicians of the latter 1990s such as New York deejay Moby. The other major genre not formally associated with rave culture, ambient music, is often characterized by what it lacks, frequently leaving out lyrics, beats, and the typical song structure of a general rhythm leading to a lyrical or melodic refrain. Begun by British musician and producer Brian Eno, ambient music, like Krautrock, had a limited audience but widespread influence on other genres, like trance, and other musicians, finding its most popular expression, again, in Moby.

Although electronic dance music in the latter 1990s became associated with rave culture and the drug use associated with raves, this did not deter some music artists from going mainstream. The English electronica group The Prodigy found mainstream success in the United States with *Fat of the Land* (1997), which included bestselling singles "Breathe" and "Firestarter," both of which were made into music videos that aired on MTV that year. Moby was the only American electronic artist to gain consistent mainstream success in the United States during that period. His 1999 platinum album, *Play*, met with success on the radio and on MTV, and all eighteen songs on the album were licensed out for commercials, television shows, and movies, earning him criticism from techno purists and praise from mainstream critics. Nevertheless, during the late 1990s, Moby's name was synonymous with electronic music in the United States.

As a genre in itself, electronic music gained a larger audience in the 1990s, though, with a few exceptions, it generally went unacknowledged by mainstream critics and news media. However, the same sounds that had fueled clubgoers in urban American dance clubs were readily adopted by popular musicians and producers as musical background for singers' lyrics and vocal melodies. Particularly influential in the spread of electronic music into the pop mainstream, C+C Music Factory, an American group formed by two former deejays, Robert Clivills and David Cole, released their incredibly successful debut album in 1990, *Gonna Make You Sweat (Everybody Dance Now)*, which combined the sounds of Chicago house music with R&B female vocals and deep-voiced male rapping. Though the contribution from dance genres to the genres of R&B, rap, and hip-hop varied in the 1990s, C+C Music Factory set a standard of combining electronically produced rhythms with soulful vocals and rap lyrics in all three categories.

Marking a general turn in electronic and popular music, the deejays and producers of dance music also became crucial participants in the production of mainstream popular music. For example, two British dance music producers, Neal Slateford and Nick Batt, remixed American alternative musician Suzanne Vega's 1981 a cappella song, "Toms Diner," and the remix became a hit upon its release in 1990. Beyond the remixing and sampling that became common among dance music deejays in the 1990s, pop singer Madonna collaborated with dance and ambient musician William Orbit for her 1998 album, *Ray of Light*, which won three Grammy Awards in 1999. Deejays such as Orbit, who also worked with rock musicians Seal, Sting, Peter Gabriel, and Prince, were often behind the scenes as collaborators and producers of some of the most popular songs of the decade.

Electronic music remained popular throughout the first decade of the twenty-first century, with artists such as French duo Daft Punk proving particularly influential. Subsequent decades saw the rise of multiple prominent electronic genres, including dubstep and electronic dance music (EDM). EDM exerted a strong influence on mainstream pop, and a number of deejays and producers working within that genre—including David Guetta, Diplo, Calvin Harris, Avicii, and others—received widespread notice in their own right. Dubstep, popularized in the United States by artists such as Skrillex, likewise helped to shape the sound of many hit singles.

In addition to pop, genres such as rock and hip-hop likewise displayed strong electronic influences throughout that period. Electronic music and musicmaking techniques also contributed to the remixing and sampling cultures of the era as well as the development of "microgenres" such as vaporwave and hyperpop.

—Bethany Thivierge, Lawrence Schwegler, and Joy Crelin

Further Reading

"Electronica Music Guide: The History and Sound of Electronica." *MasterClass*, 14 Sept. 2021, www.masterclass.com/articles/electronica-music-guide. Accessed 13 July 2023.

Gilbert, Jeremy, and Ewan Pearson. *Discographies: Dance Music, Culture, and the Politics of Sound*. Routledge, 1999.

Holmes, Thomas B. *Electronic and Experimental Music: Pioneers in Technology and Composition*. Routledge, 2002.

Manning, Peter. *Electronic and Computer Music*. Oxford UP, 2004.

Payling, Dave. *Electronic Visual Music: The Elements of Audiovisual Creativity*. Focal Press, 2023.

Shapiro, Peter, editor. *Modulations: A History of Electronic Music Throbbing Words on Sound*. Caipirinha Productions, 2000.

Encryption

ABSTRACT

Encryption is a process in which data is translated into code that can only by read by a person with the correct encryption key. It focuses on protecting data content rather than preventing unauthorized interception. Encryption is essential in intelligence and national security and is also common in commercial applications. Various software programs are available that allow users to encrypt personal data and digital messages.

BACKGROUND

The study of different encryption techniques is called cryptography. The original, unencrypted data is called the "plaintext." Encryption uses an algorithm called a "cipher" to convert plaintext into ciphertext. The ciphertext can then be deciphered by using another algorithm known as the decryption key or cipher key.

A key is a string of characters applied to the plaintext to convert it to ciphertext, or vice versa. Depending on the keys used, encryption may be either symmetric or asymmetric. Symmetric-key encryption uses the same key for both encoding and decoding. The key used to encode and decode the data must be kept secret, as anyone with access to the key can translate the ciphertext into plaintext. The oldest known cryptography systems used alphanumeric substitution algorithms, which are a type of symmetric encryption. Symmetric-key algorithms are simple to create but vulnerable to interception.

In asymmetric-key encryption, the sender and receiver use different but related keys. First, the receiver uses an algorithm to generate two keys, one to encrypt the data and another to decrypt it. The encryption key, also called the "public key," is made available to anyone who wishes to send the receiver a message. (For this reason, asymmetric-key encryption is also known as "public-key encryption.") The decryption key, or private key, remains known only to the receiver. It is also possible to encrypt data using the private key and decrypt it using the public key. However, the same key cannot be used to both encrypt and decrypt.

Asymmetric-key encryption works because the mathematical algorithms used to create the public and private keys are so complex that it is computationally impractical determine the private key based on the public key. This complexity also means that asymmetric encryption is slower and requires more processing power. First developed in the 1970s, asymmetric encryption is the standard form of encryption used to protect internet data transmission.

OVERVIEW

Authentication is the process of verifying the identity of a sender or the authenticity of the data sent. A common method of authentication is a hashing algorithm, which translates a string of data into a fixed-length number sequence known as a "hash value." This value can be reverted to the original data using the same algorithm. The mathematical complexity of hashing algorithms makes it extremely difficult to decrypt hashed data without knowing the exact algorithm used. For example, a 128-bit hashing algorithm can generate 2^{128} different possible hash values.

In order to authenticate sent data, such as a message, the sender may first convert the data into a hash value. This value, also called a message digest, may then be encrypted using a private key unique to the sender. This creates a digital signature that verifies the authenticity of the message and the identity of the sender. The original unhashed message is then encrypted using the public key that corresponds to the receiver's private key. Both the privately encrypted digest and the publicly encrypted message are sent to the receiver, who decrypts the

Image via iStock/JuSun. [Used under license.]

original message using their private key and decrypts the message digest using the sender's public key. The receiver then hashes the original message using the same algorithm as the sender. If the message is authentic, the decrypted digest and the new digest should match.

ENCRYPTION SYSTEMS IN PRACTICE

One of the most commonly used encryption programs is Pretty Good Privacy (PGP). It was developed in 1991 and combines symmetric- and asymmetric-key encryption. The original message is encrypted using a unique one-time-only private key called a "session key." The session key is then encrypted using the receiver's public key, so that it can only be decrypted using the receiver's private key. This encrypted key is sent to the receiver along with the encrypted message. The receiver uses their private key to decrypt the session key, which can then be used to decrypt the message. For added security and authentication, PGP also uses a digital signature system that compares the decrypted message against a message digest. The PGP system is one of the standards in personal and corporate security and is highly resistant to attack. The data security company Symantec acquired PGP in 2010 and subsequently incorporated the software into many of its encryption programs. The relevant division of Symantec was in turn acquired by the company Broadcom in 2019.

Encryption can be based on either hardware or software. Most modern encryption systems are based on software programs that can be installed on a system to protect data contained in or produced by a variety of other programs. Encryption based on hardware is less vulnerable to outside attack. Some hardware devices, such as self-encrypting drives (SEDs), come with built-in hardware encryption and are useful for high-security data. However, hardware encryption is less flexible and can be prohibitively costly to implement on a wide scale. Essentially, software encryption tends to be more flexible and widely usable, while hardware encryption is more secure and may be more efficient for high-security systems.

—Micah L. Issitt

Further Reading

Delfs, Hans, and Helmut Knebl. *Introduction to Cryptography: Principles and Applications*. 3rd ed., Springer, 2015.

History of Cryptography: An Easy to Understand History of Cryptography. Thawte, 2013.

Holden, Joshua. *The Mathematics of Secrets: Cryptography from Caesar Ciphers to Digital Encryption*. Princeton UP, 2018.

"How to: Use PGP for Windows." *Surveillance Self-Defense*, 17 June 2018, ssd.eff.org/module/how-use-pgp-windows. Accessed 13 July 2023.

Levy, Stephen. "Almost 50 Years into the Crypto Wars, Encryption's Opponents Are Still Wrong." *Wired*, 21 July 2023, www.wired.com/story/plaintext-50-years-into-the-crypto-wars-encryptions-opponents-are-still-wrong. Accessed 22 July 2023.

EXTENSIBLE MARKUP LANGUAGE

ABSTRACT

Extensible markup language (XML) is a programming language used to categorize and describe data. As a type of markup language, XML uses tags or rules for tags to add information about data. Tags used in XML coding help describe the data inside the tags. XML does not rely on any particular software or hardware. It also is written in plain text, so it can be shared easily among different programs and devices.

BACKGROUND

The World Wide Web Consortium (W3C)—a group of volunteers and paid employees who create standards for the web—wanted to create a new type of markup language that was that useful on the internet. The group decided to base the new language on another type of markup language called standard generalized markup language (SGML). Standard generalized markup language is made up of elements, and these elements usually consist of a start tag, content, and an end tag (e.g., <firstname>Jose</firstname>). The group from W3C structured XML in much the same way.

Members of W3C developed the first draft of XML in 1996. They distributed coding instructions in a twenty-five-page manual and released it to the public. The W3C recommended the use of XML in 1998. Soon, XML became an important markup language that many people used for different purposes. Some people used XML on the web, and others used it in electronic publishing, in databases, and in other types of documents. The W3C working group intended to create a markup language for the internet, but they created a markup language capable of being used in many different situations.

Throughout the history of XML, groups at W3C and other individuals and groups around the world have worked to further develop the language. The W3C publishes its work so others can use XML and make changes and enhancements to the language. Multiple conferences about XML, its standards, and its uses are held around the world each year, proving the language's continued popularity.

OVERVIEW

Even though XML is a markup language, it is different from some other markup languages because it was created to label and categorize data rather than display data. One of the most popular markup languages is hypertext markup language (HTML). HTML adds information about the format and the appearance of text. XML, however, adds information about the data included inside the tags. The goal of HTML is to display text in a certain way, while the goal of XML is to describe data.

XML and HTML have another important difference: XML tags are not predefined, but HTML tags are. A person using HTML has to memorize or look up specific tags to use to have data display in a particular way. XML users, however, can develop unique tags to describe data. One other difference between XML and HTML is that XML is more rigid. HMTL is a fairly flexible language and can tolerate some coding errors. XML, however, has to be formatted very specifically because errors can be introduced easily.

Image via Wikimedia Commons. [Public domain.]

USES OF XML

One of the main uses of XML is to categorize data. XML can be used on the web to classify information. The idea of data categorization is especially important on the Semantic Web, which is the web of data. XML also can be used to separate data from other elements on a web page. For example, a programmer can use XML to keep the data of a web page separate from the HTML coding on that page. This separation allows the programmer to update the data without updating the HTML.

Another important use of XML is data transport. At times, data has to be shared between two different programs that are incompatible with one another. Because XML does not rely on a specific type of software, it is compatible with many different programs. Sometimes XML can help two incompatible programs share data with each other. In a similar way, data sometimes has to be shared when a platform (either software or hardware) is being upgraded. These upgrades can be time consuming because of the large amount of data that needs to be shared. However, XML is useful because it can make transferring the data much simpler.

ADDING XML TO DOCUMENTS

One way people use XML is to add markup to documents. Often in documents using XML, the first line of code is called the declaration line. The declaration line states that XML is being used and identifies the version of XML being used (e.g., <?xml version='1.0' encoding='UTF-8'?>). The line after the declaration is a line that describes the root element. This information tells what type of document is being coded (e.g., <note>). The lines after the root element contain the content of the document and can be coded to show the function of each piece of data (e.g., <heading>, <body>, <conclusion>). The document then ends with a tag that shows the text is complete (e.g., </note>).

People using XML have to remember some important rules about the language. If any opening tag is used, a corresponding closing tag must also be used (e.g., <firstname> and </firstname> or <note> and </note>). Other markup languages, such as HTML, do not always require closing tags, but XML does. Along the same lines, XML tags have to be nested properly. That means the elements opened first must be closed last. For example, <i><b>word</i></b> is nested incorrectly, but <i><b>word</b></i> is nested correctly. Another factor people using XML have to remember is that the language is case sensitive, so the tag <Note> and the tag <note> are not the same.

—Elizabeth Mohn

Further Reading

"A Brief SGML Tutorial." *W3C*, www.w3.org/TR/WD-html40-970708/intro/sgmltut.html. Accessed 13 July 2023.

Campbell-Kelly, Martin, William F. Aspray, Jeffrey R. Yost, Honghong Tinn, and Gerardo Con Díaz. *Computer: A History of the Information Machine*. 4th ed., Routledge, 2023.

"Development History." *W3C*, 6 Jan. 2003, www.w3.org/XML/hist2002. Accessed 13 July 2023.

"Extensible Markup Language (XML)." *W3C*, 11 Oct. 2016, www.w3.org/XML. Accessed 13 July 2023.

Frain, Ben. *Design with HTML5 and CSS*. 4th ed., Packt, 2022.

"XML." *PC Mag*, www.pcmag.com/encyclopedia/term/55048/xml. Accessed 13 July 2023.

"XML Tutorial." *W3Schools*, www.w3schools.com/xml. Accessed 13 July 2023.

F

Facebook

ABSTRACT

The world's largest online social networking service, Facebook enables users to connect and share with other people, referred to on the site as "friends." With its billions of active users located throughout the world, Facebook has brought people together in unprecedented ways and provided innovative sharing and communication tools. Nevertheless, the site and its policies have also raised concerns about privacy and safety.

BACKGROUND

When Facebook was first released in 2004, several predecessors and early competitors already existed. Friendster, a social network for finding people with similar interests (with the possible intention of dating), was founded in 2002 by Jonathan Abrams and Rob Pazornik. The year 2003 saw the launch of Myspace, a media-heavy site that let users customize their profile pages. LinkedIn, a social networking site for business professionals, was also launched in 2003.

In addition to more general social networking sites, campus-specific social networks were founded at many colleges and universities in the early 2000s. The first was Club Nexus, launched at Stanford University in 2001, but Columbia University and several other schools also offered similar services, many created and managed by students. There were also several social networks aimed at college students in general, including the Daily Jolt and Collegester.com. Despite Facebook's late start compared to its competitors, the site ultimately gained the largest user base of all social networking sites, becoming one of the highest-trafficked sites of the decade.

Harvard University student Mark Zuckerberg developed two websites in 2003 that would set the stage for the service that became Facebook. CourseMatch, which helped students determine their courses based on what others were taking, and Facemash, which allowed people to compare images of fellow students, were immediate hits and proved that college students were ready to take their socializing online.

Based on the popularity of these two sites and Zuckerberg's initial experiments with image-commenting capabilities, the first version of Facebook was launched on February 4, 2004, at TheFaceBook.com, allowing students to create profiles and connect with other students. Within a day

Image via Wikimedia Commons. [Public domain.]

of the launch, between twelve hundred and fifteen hundred people had signed up for the service.

After approximately one month of exclusive access for Harvard students, Facebook opened up to Stanford, Columbia, and Yale universities. By June, the site had extended its access to users at forty different colleges and universities.

The early popularity of Facebook stemmed from a number of factors, including widespread internet access, which was initially greater on college campuses than in the general population, and a focused target demographic of college students who constantly socialized, had numerous social connections, and were willing to share personal information online with others. Additionally, the site was frequently updated by its users, prompting nearly 60 percent of Facebook's users to visit the site daily and some 20 percent of users to visit the site more than five times a day.

Facebook grew quickly. Zuckerberg and cofounder Moskovitz moved to California's Silicon Valley and received their first funding—$500,000 from PayPal cofounder Peter Thiel—in June 2004. The site reached 1 million users by November 2004, narrowly missing a goal set by Thiel of 1.5 million users for the year. In Palo Alto, Zuckerberg connected with Sean Parker, founder of the online music-sharing service Napster, who served as Facebook's president for a short, but influential, time.

Following Thiel's initial investment, Facebook continued to grow, attracting investment interest from venture capital companies. Zuckerberg turned down several acquisition offers, some from large and profitable companies such as Viacom, which wanted to combine the service with MTV.com. Interest in Facebook was sparked by its huge user base, rapid growth, high levels of repeat usage, and attractive demographic of users: young college students. Even the *Washington Post* displayed interest in investing in Facebook, but the venture capital firm Accel Partners outbid the news outlet by offering $12.7 million in funding in exchange for a 15 percent stake in the company. The investment would be crucial in acquiring the staff to develop the features and partnerships that would ultimately help Facebook become the dominant social network of the 2000s.

OVERVIEW

Facebook added many capabilities throughout its evolution, but early core Facebook features included the user's profile, wall, status update, and notifications.

Facebook user profiles include basic personal information, such as name, age, and gender; background information, such as hometown or schools attended; interests and hobbies; and other customizable information. Unlike many other social networking sites, Facebook does not make it easy to find people through shared interests; the emphasis is on maintaining existing connections, not creating new ones. A user's wall is a profile section where other users can post greetings, links, or photos. A status update is a message that a user can post to Facebook to be displayed to their friends on the site.

Users can receive notifications, get "tagged" in photos, send and receive private messages, and chat with friends/acquaintances. Other ways to communicate on Facebook include groups and events. Users can create groups for general communication around a particular topic or cause, or they can schedule events such as birthday parties or protests. The vast range of Facebook events and groups shows the flexibility and power of the platform, which has often been used for purposes of political organization and activism.

FACEBOOK PLATFORM AND FACEBOOK CONNECT

By May 2007, 1 percent of all internet time was spent on Facebook. In the same month, Facebook launched the Facebook Platform at its first annual

F8 Developers Conference. The platform enabled software developers to create applications that integrated with the service, such as online games that could be played within Facebook. This move accelerated the company's rise, as its users began to download more and more applications that prompted them to log in to Facebook multiple times a day. By November 2007, more than 8,000 apps had been developed on the platform, a number that would rise to 33,000 by July 2008.

The success of the platform showed that Facebook's primary value proposition was not its features but was instead the connections between its users. The platform also offers a social graph, which features the connections between users and their shared interests. It then offered access to the social graph to other software developers, a move that had nothing to do with risking its own success but rather with ensuring it: by selling its social data to other companies and developers, Facebook no longer needed to develop multiple types of functionality for the site. Instead, it could serve as the platform for myriad applications developed by others.

In late 2008, Facebook released Facebook Connect, a secure single sign-on system that allowed people to access any website using their Facebook login. This not only was convenient for its users but also enabled the social graph to track users' activity on other websites.

ADVERTISING, BUSINESS MODEL, AND CORPORATE CULTURE

By the fall of 2007, Facebook had 50 million users. Microsoft invested $240 million in the company as part of an advertising deal, taking a 1.6 percent stake in Facebook. As the company continued to grow, Facebook hired Google executive Sheryl Sandberg in 2008 to serve as chief operating officer (COO). The company was first cash-flow positive in 2009, due mostly to advertising.

Outside Facebook HQ, 2017. Photo by Minette Lontsie, via Wikimedia Commons.

Facebook represented a revolution in advertising, because it provided not only a platform to broadcast product messages, but it also generated marketing and usage data to inform the successful targeting of those messages. For example, Facebook made it possible to target an advertisement about a running shoe sale in San Francisco only to users who live in the area and have indicated interest in running. This type of market targeting was unprecedented and prompted significant experimentation by other advertisers. However, many Facebook users eventually expressed discomfort with the degree to which the information they share on the service could be used to gear specific advertisements to them. This lack of user control and transparency factored heavily into widespread criticisms of Facebook in the late 2010s, especially in relation to the numerous scandals that emerged during that time.

PRIVACY, CRITICISM, AND BLOCKING

In addition to providing user data for advertisement targeting, Facebook has been widely criticized for perceived privacy infringements. Although users are typically able to limit the content they post to be visible only to friends on the site, these settings often default to sharing information publicly. This not only aggravates some users, but it has allegedly resulted in serious situations of people losing their jobs, not being hired for a job, or not being accepted to a college because of content posted on Facebook.

In December 2009, Facebook changed all profiles to be public—meaning visible to anyone, even nonusers and nonfriends—by default, requiring users to change their individual settings if they wanted to keep their information private. Because many users preferred to keep their information visible to friends only, the switch angered many users and caused some to stop using the service. Several organizations joined a US Federal Trade Commission (FTC) complaint against Facebook for its lack of privacy protections.

Facebook's ability to open up connections between people who have fallen out of touch and to facilitate communication between such parties also proved to have negative aspects. For example, the platform has been blamed for playing a role in divorces, with spouses discovering information such as private messages between users. A December 2009 study found that 20 to 33 percent of all divorce filings in the United Kingdom cited Facebook as a reason for the separation, reflecting a potential downside of widespread sharing. Other researchers and advocates turned attention to the social media platform's role in bullying and other damaging behavior. Meanwhile, some individuals and organizations began to suggest that users can form an addiction to social media, with potentially harmful mental and even physical effects.

Other concerns about Facebook have come from various sources. The site has been blocked in many countries, including China and Iran in 2009, due to concerns about its role as a tool for political activism in opposition to parties in power.

INITIAL PUBLIC OFFERING

In July 2010, Facebook announced that it had 500 million users, making it the world's largest online social network. By the end of 2011, based on page views, Facebook was second only to Google as the most-used website in the United States. Then, on May 17, 2012, after years of rebuffing acquisition offers and urgings to go public, Facebook staged an initial public offering (IPO) on the NASDAQ stock exchange, raising some $16 billion by selling more than 400 million shares at $38 a share in what was the third largest IPO in US history.

FAKE NEWS, MISINFORMATION, AND OTHER CONTENT CONCERNS

In August 2016, Facebook fired their team of editors responsible for their trending news section. This decision was made following accusations that the trending articles had a liberal political bias. The human editors were replaced with algorithms that determined what the biggest trending stories were. This turned out to not be an effective alternative, however. The algorithms allowed fake news stories to be at the top of Facebook's trending stories. These fake news stories would have previously been caught by editors.

Surveys in the mid- to late-2010s showed that increasing numbers of Facebook users got their news through posts on the social media site. In the aftermath of the 2016 US presidential election, Facebook was the subject of great criticism for allowing fake news stories, particularly ones that had to do with the election, to proliferate unchecked. In mid-November 2016, Zuckerberg announced that the company would not allow fake news sites to advertise and asked users for their help in catching fake news articles. However, media reports suggested the company's initiatives had little immediate success. Indeed, criticisms grew as investigations exposed Russian interference in the 2016 election. Many commentators suggested that fake news stories, often created by or with support from Russian agents seeking to support Republican presidential candidate Donald Trump and undermine Democratic candidate Hillary Clinton, may have had a considerable effect on the controversial election.

In early 2017, Zuckerberg announced a shift in the company's focus from connecting friends and family to creating a social infrastructure that would connect a global community. In an over five-thousand-word letter posted to his own Facebook page on February 16, Zuckerberg announced revised company goals of globally promoting peace and understanding as well as prosperity and freedom. He stated that through the use and implementation of artificial intelligence (AI) technology, Facebook would be helping to make the world safer by removing fake news stories, blocking terrorist propaganda, and helping those suffering from mental illness or bullying.

Despite these efforts, high-profile criticism of Facebook's treatment of fake news, conspiracy theories, hate speech, and similar content continued. For example, many mainstream media outlets considered the temporary suspension of the personal account of Alex Jones, head of the notoriously inaccurate and inflammatory right-wing site InfoWars, in July 2018 to be an inadequate response to the damage caused by his discourse. Facebook did subsequently ban several prominent InfoWars-related pages over violations of its hate speech policy. In May 2019, the company officially banned Nation of Islam leader Louis Farrakhan, Jones, and a handful of prominent right-wing extremists and white supremacists from the site.

The year 2020, which saw the worldwide spread of the coronavirus disease 2019 (COVID-19) pandemic as well as a highly contentious US presidential election, was also marked by criticism of the amount of harmful misinformation, disinformation, and hateful or even inflammatory speech communicated through social media platforms, particularly Twitter and Facebook. While Facebook, along with other platforms, began making efforts in late 2020 to make changes to aspects such as algorithms and protocols to help limit or prevent the spread of misinformation, disinformation, and inflammatory posts, many critics argued that the company had not done enough quickly enough, as President Trump himself as well as his supporters and general conspiracy theorists had long been posting content that included baseless or false claims regarding the election process as well as COVID-19.

After a large group of Trump supporters considered to have been persuaded by Trump's rhetoric as recently as a rally held by the outgoing president that same day stormed the US Capitol building on January 6, 2021, to disrupt Congress's certification of Joe Biden's election win, Facebook promptly announced that it had taken Trump's Instagram and Facebook accounts down. By June, in response to its independent Oversight Board, the company announced that it had issued a formal two-year suspension of Trump's accounts; as Facebook also indicated that it was working on protocols to be more transparently judicious, regardless of the account holder, of its content, Trump argued that it was unfair censorship.

CONTROVERSY OVER FACEBOOK LIVE

After Facebook's livestreaming feature, Facebook Live, was made available for public use in January 2016 (having previously been restricted to verified VIP users), the company began drawing criticism for allowing users to stream videos of criminal activity, violence, and even deaths. Facebook's control over and reaction time with regard to such content was limited, as it relied primarily on AI algorithms and user reports to moderate user content. Mathew Ingram reported for *Fortune* magazine that "in some cases, the company has taken swift action to remove the videos, but in others it has chosen to leave them up with a warning about the disturbing content."

Videos that were allowed to remain with a warning included one in which Chicago resident Antonio Perkins was killed in a drive-by shooting and another in which Minneapolis resident Philando Castile was fatally shot by police while his girlfriend (who filmed and livestreamed the encounter) and her young daughter remained in their pulled-over car. Other violent and otherwise disturbing videos were removed with varying reaction times.

On April 16, 2017, Cleveland resident Robert Godwin Sr. was shot and killed in a video that the killer livestreamed on Facebook. The killer had also posted an earlier video announcing his intention to commit the murder, and he posted one afterward in which he confessed to the crime. Facebook was criticized for leaving the videos up for around two hours, during which time they went viral across multiple social media platforms. Later the same month, a man in Thailand posted videos that showed him killing his eleven-month-old daughter and then hanging himself. The videos remained on Facebook for around twenty-four hours until they were removed. Both cases renewed questions about Facebook's responsibility for the content of its users and how the company could do more to prevent such tragedies. Statements by Facebook personnel disavowed the killers' actions and acknowledged the need for the site to do better in preventing such content from being seen.

CAMBRIDGE ANALYTICA, THE "UGLY TRUTH" MEMO, AND OTHER SCANDALS

With considerable negative publicity already surrounding Facebook over livestreamed killings, the proliferation of fake news and its role in US and world politics, and other issues, additional scandals involving the social media giant emerged in late 2017 and early 2018. Content from the leaked documents known as the "Paradise Papers" revealed in November 2017 indicated parties linked to the Russian government had made considerable investments in Facebook (and Twitter). One of the individuals involved in the investments, Yuri Milner, was noted to have connections to Jared Kushner, son-in-law of President Donald Trump and an aide in his administration. The revelation sparked further concerns over Facebook's political influence and manipulation by those seeking to sway public opinion.

A larger scandal, however, unfolded beginning in March 2018 when it was revealed that tens of millions of Facebook users' personal data had been sold to the British political analysis firm Cambridge Analytica, which had worked on the Trump campaign. Much of the private user data was harvested through apps, with users often unaware. Facebook initially did not acknowledge the data breach, and then suggested the data had been recovered when in fact Cambridge Analytica still had access to it. Facebook also reportedly attempted to stop news of the story from being published. When details of the episode broke, it garnered significant public backlash against Facebook as well as attention from lawmakers in both the United States and Great Britain, who wanted explanations for the data leak and its political ramifications.

In July 2019, it was announced that Facebook had agreed to pay a $5 billion settlement with the FTC reached in response to the company's mishandling of user privacy. As part of the settlement, the company also acquiesced to instituting new oversight policies, particularly setting up an independent privacy committee on Facebook's board to limit Zuckerberg's control over decisions related to the issue. Criticism followed over whether this regulation was strict enough. Later that year, on November 17, the banking data for about 29,000 Facebook employees—which was stored on unencrypted hard drives—was stolen. The hard drives were not reported to be missing until November 20, and Facebook did not inform employees about the data breach until December 13.

CORPORATE NAME CHANGE

While the Facebook site and app had continued to remain one of the most popular among social media platforms, in October 2021 its parent company announced that it was changing its longtime name of Facebook and rebranding as Meta. Zuckerberg argued that the corporation had extended its reach and conceptual goals so far past just social networking platforms that it did not make sense to continue operating under the name, considered wholly unrepresentative, of just one of its apps. Facebook the app, which would keep its name, was one part, its company's chief executive officer (CEO) stressed, of a concentrated effort toward creating a "metaverse" in which the technologies of augmented reality and virtual worlds would be combined. This development was widely discussed in the media, with some commentators noting that the move likely had a role in the company's desire to distance itself from past scandals, particularly after highly publicized accusations of a conscious disregard for protecting users' safety had been made by former employees that same month.

Following its corporate rebrand, Meta continued to operate Facebook in addition to pursuing a variety of other ventures related to social media, virtual reality, and artificial intelligence. In July of 2023, the company launched a new social networking service called Threads, with which Meta sought to compete with the increasingly embattled social network Twitter. Threads proved popular upon its debut, amassing 100 million users during its first five days of operation.

IMPACT

In becoming one of the largest social networks in the world, Facebook revolutionized online communications, human connections, activism, and advertising. The site has helped people maintain connections with others in unprecedented ways and has even been used to mobilize grassroots political efforts. However, it has also exposed significant challenges that come with the great influence of social media, including issues of privacy, free speech, and even impacts on physical and mental health. As the social network continues to evolve, it will undoubtedly have further new applications (and implications) for corporations, advertisers, governments, law enforcement, and individuals.

—Craig Belanger and Kerry Skemp

Further Reading

Almasy, Steve. "Thailand Baby Killing: Facebook Removes Video." *CNN*, 26 Apr. 2017, www.cnn.com/2017/04/25/asia/thailand-baby-killed-facebook-live-trnd. Accessed 13 July 2023.

Chokshi, Niraj. "Facebook Helped Drive a Voter Registration Surge, Election Officials Say." *New York Times*, 12 Oct. 2016, www.nytimes.com/2016/10/13/us/politics/facebook-helped-drive-a-voter-registration-surge-election-officials-say.html. Accessed 13 July 2023.

Collins, Ben. "Facebook to Restrict Livestream Feature after Christchurch Attack." *NBC News*, 14 May 2019, www.nbcnews.com/tech/tech-news/facebook-restrict-livestream-feature-after-christchurch-attack-n1005741. Accessed 13 July 2023.

Dwoskin, Elizabeth, and Craig Timberg. "Facebook Wanted 'Visceral' Live Video: It's Getting Live-Streaming Killers and Suicides." *Washington Post*, 17 Apr. 2017, www.washingtonpost.com/business/technology/facebook-wanted-visceral-live-video-its-getting-suicides-and-live-streaming-killers/2017/04/17/a6705662-239c-11e7-a1b3-faff0034e2de_story.html. Accessed 13 July 2023.

Feiner, Lauren, and Salvador Rodriguez. "FTC Slaps Facebook with Record $5 Billion Fine, Orders Privacy Oversight." *CNBC*, 24 July 2019, www.cnbc.com/2019/07/24/facebook-to-pay-5-billion-for-privacy-lapses-ftc-announces.html. Accessed 13 July 2023.

Griffin, Andrew. "Facebook Live: Site Adds Huge New Update to Feature That Lets People Stream in Their Timeline." *Independent*, 6 Apr. 2016, www.independent.co.uk/life-style/gadgets-and-tech/news/facebook-live-site-adds-huge-new-update-to-feature-that-lets-people-stream-in-their-timeline-a6971566.html. Accessed 13 July 2023.

Hern, Alex. "Mark Zuckerberg's Letter Annotated: What He Said and What He Didn't." *Guardian*, 17 Feb. 2017, www.theguardian.com/technology/ng-interactive/2017/feb/17/mark-zuckerberg-facebook-letter-annotated-what-he-said-what-he-didnt. Accessed 13 July 2023.

Herrman, John. "What Happens When Facebook Goes the Way of Myspace?" *New York Times*, 12 Dec. 2018, www.nytimes.com/2018/12/12/magazine/what-happens-when-facebook-goes-the-way-of-myspace.html. Accessed 13 July 2023.

Hill, Kashmir. "Facebook's Mark Zuckerberg: 'We've Made a Bunch of Mistakes.'" *Forbes*, 29 Nov. 2011, www.forbes.com/sites/kashmirhill/2011/11/29/facebooks-mark-zuckerberg-weve-made-a-bunch-of-mistakes. Accessed 13 July 2023.

Ingram, Mathew. "Facebook Killing Another Example of Live Video Feature's Dark Side." *Fortune*, 17 Apr. 2017, fortune.com/2017/04/17/facebook-killing. Accessed 13 July 2023.

Isaac, Mike. "Facebook Renames Itself Meta." *New York Times*, 28 Oct. 2021, www.nytimes.com/2021/10/28/technology/facebook-meta-name-change.html. Accessed 13 July 2023.

Kirkpatrick, David. *The Facebook Effect: The Inside Story of the Company That Is Connecting the World*. Simon & Schuster, 2010.

Klein, Ezra. "Mark Zuckerberg on Facebook's Hardest Year, and What Comes Next." *Vox*, 2 Apr. 2018, www.vox.com/2018/4/2/17185052/mark-zuckerberg-facebook-interview-fake-news-bots-cambridge. Accessed 13 July 2023.

Lacy, Sarah. *Once You're Lucky, Twice You're Good: The Rebirth of Silicon Valley and the Rise of Web 2.0*. Gotham Books, 2008.

Lincoln, Sian, and Brady Robards. "10 Years of Facebook." *New Media & Society*, vol. 16, no. 7, 2014, pp. 1047-50.

Mezrich, Ben. *The Accidental Billionaires: The Founding of Facebook, a Tale of Sex, Money, Genius and Betrayal*. Doubleday, 2009.

Moscaritolo, Angela, and Chloe Albanesius. "Zuckerberg's Vision for Facebook: A Global Community Backed by AI." *PC Magazine*, 16 Feb. 2017, www.pcmag.com/news/351809/zuckerbergs-vision-for-facebook-a-global-community-backed. Accessed 13 July 2023.

"Most Popular Social Networks Worldwide as of January 2023, Ranked by Number of Monthly Active Users." *Statista*, Jan. 2023, www.statista.com/statistics/272014/global-social-networks-ranked-by-number-of-users. Accessed 13 July 2023.

Newitz, Annalee. "Facebook Fires Human Editors, Algorithm Immediately Posts Fake News." *Ars Technica*, 29 Aug. 2016, arstechnica.com/business/2016/08/facebook-fires-human-editors-algorithm-immediately-posts-fake-news. Accessed 13 July 2023.

Olson, Parmy. "Facebook Wants Users to Help It Weed Out Fake News." *Forbes*, 6 Dec. 2016, www.forbes.com/sites/parmyolson/2016/12/06/facebook-users-fake-news. Accessed 13 July 2023.

Osnos, Evan. "Can Mark Zuckerberg Fix Facebook Before It Breaks Democracy?" *New Yorker*, 17 Sept. 2018, www.newyorker.com/magazine/2018/09/17/can-mark-zuckerberg-fix-facebook-before-it-breaks-democracy. Accessed 13 July 2023.

Paul, Katie, and Akash Sriram. "Meta's Twitter Rival Threads Surges to 100 Million Users Faster Than ChatGPT." *Reuters*, 10 July 2023, www.reuters.com/technology/metas-twitter-rival-threads-hits-100-mln-users-record-five-days-2023-07-10. Accessed 13 July 2023.

Rosenberg, Matthew, and Sheera Frenkel. "Facebook's Role in Data Misuse Sets of Storms on Two Continents." *New York Times*, 18 Mar. 2018, www.nytimes.com/2018/03/18/us/cambridge-analytica-facebook-privacy-data.html. Accessed 13 July 2023.

Stinson, Liz. "Facebook Reactions, the Totally Redesigned Like Button, Is Here." *Wired*, 24 Feb. 2016, www.wired.com/2016/02/facebook-reactions-totally-redesigned-like-button. Accessed 13 July 2023.

Sulleyman, Aatif. "Facebook Live Killings: Why the Criticism Has Been Harsh." *Independent*, 27 Apr. 2017, www.independent.co.uk/life-style/gadgets-and-tech/features/facebook-live-killings-ai-artificial-intelligence-not-blame-fatalities-murders-us-steve-stephens-a7706056.html. Accessed 13 July 2023.

Ziobro, Paul, and Jeff Horwitz. "Facebook Suspends Donald Trump for at Least Two Years." *Wall Street Journal*, 4 June 2021, www.wsj.com/articles/facebook-suspends-donald-trump-for-two-years-11622825480. Accessed 13 July 2023.

G

Game Theory

ABSTRACT

Game theory is a tool that has come to be used to explain and predict decision making in a variety of fields. It is used to explain and predict both human and nonhuman behavior as well as the behavior of larger entities such as nation states. Game theory is often quite complex mathematically, especially in new fields such as evolutionary game theory.

BACKGROUND

Game theory is a means of modeling individual decision making when the decisions are interdependent and when the actors are aware of this interdependence. Individual actors (which may be defined as people, firms, and nations) are assumed to maximize their own utility—that is, to act in ways that will provide the greatest benefits to them. Games take a variety of forms such as cooperative or noncooperative; they may be played one time only or repeated either indefinitely or with a finite ending point, and they may involve two or more players. Game theory is used by strategists in a variety of settings.

The mathematical theory of games was developed by the mathematician John von Neumann and the economist Oskar Morgenstern, who published *The Theory of Games and Economic Behavior* in 1944. Various scholars have contributed to the development of game theory, and in the process, it has become useful for scholars and practitioners in a variety of fields, although economics and finance are the disciplines most commonly associated with game theory.

Some of the seminal work in game theory was done by the mathematician John Nash in the early 1950s, with later scholars building on his work. Nash along with John Harsanyi and Reinhard Selten received a Nobel Prize in Economic Sciences in 1994 for their work in game theory. The political scientist Thomas Schelling received a Nobel Prize in Economic Sciences in 2005 for his work in predicting the outcomes of international conflicts.

The underlying mathematical assumptions found in much of game theory are often difficult to understand. However, the explanations drawn from game theory are often intuitive, and people can engage in

John von Neumann. Photo via Los Alamos National Laboratory, via Wikimedia Commons.

decision making based on game theory without understanding the underlying mathematics. The formal approach to decision making that is part of game theory enables decision makers to clarify their options and enhance their ability to maximize their utility.

Classical game theory has quite rigid assumptions that govern the decision process. Some of these assumptions are so rigid that critics have argued that game theory has few applications beyond controlled settings.

Later game theorists developed approaches to strategy that have made game theory applicable in many decision-making applications. Advances such as evolutionary game theory take human learning into account, and behavioral game theory factors emotion into decision making in such a way that it can be modeled and predictions can be made.

OVERVIEW

Game theory is the application of mathematical reasoning to decision making to provide quantitative estimates of the utilities of game players. Implicit in the decision-making process is the assumption that all actors act in a self-interested fashion so as to maximize their own utilities. Rationality, rather than altruism or cooperation, is a governing principle in much of game theory. Actors are expected to cooperate only when doing so benefits their own self-interest.

Noncooperative Games. Most games are viewed as noncooperative in that the players act only in their own self-interest and will not cooperate, even when doing so might lead to a superior outcome. A well-known noncooperative game is the prisoners' dilemma. In this game, the police have arrested Moe and Joe, two small-time criminals, for burglary. The police have enough information to convict the two men for possession of burglar's tools, a crime that carries a sentence of five years in prison. They want to convict the two men for burglary, which carries a ten-year sentence, but this requires at least one man to confess and testify against the other. It would be in each man's best interest to remain silent, as this will result in only a five-year sentence for each. The police separate Moe and Joe. Officers tell Moe that if he confesses, he will receive a reduced sentence (three years), but if Joe confesses first, Moe will receive a harsher sentence (twelve years). At the same time, other officers give Joe the same options. The police rely on the self-interest of each criminal to lead him to confess, which will result in ten-year sentences for both men. If the various payoffs are examined, cooperation (mutual silence), which means two five-year sentences, is the most rewarding overall. However, if the two criminals act rationally, each will assume that the other will confess and therefore each will confess. Because each man cannot make sure that the other will cooperate, each will act in a self-interested fashion and end up worse off than if he had cooperated with the other by remaining silent.

As some game theorists such as Robert Axelrod have demonstrated, it is often advantageous for players to cooperate so that both can achieve higher payoffs. In reality, as the prisoners' dilemma demonstrates, players who act rationally often achieve an undesirable outcome.

Sequential Games. Many games follow a sequence in which one actor takes an action and the other reacts. The first actor then reacts, and so the game proceeds. A good way to think of this process is to consider a chess game in which each move is countered by the other player, but each player is trying to think ahead so as to anticipate his or her opponent's future moves. Sequential games are often used to describe the decision process of nations, which may lead to war if wrong reactions occur.

Sequential games can be diagramed using a tree that lists the payoffs at each step, or a computer program can be used to describe the moves. Some sequential games are multiplayer games that can become quite complicated to sort out. Work in game theory suggests that equilibrium points change at each stage of the game, creating what are called sequential equilibria.

Simultaneous Games. In some games, players make their decisions at the same time instead of reacting to the actions of the other players. In this case, they may be trying to anticipate the actions of the other player so as to achieve an advantage. The prisoners' dilemma is an example of a simultaneous game. Although rational players may sometimes cooperate in a sequential game, they will not cooperate in a simultaneous game.

APPLICATIONS AND PRODUCTS

Game theory is most commonly used in economics, finance, business, and politics, although its applications have spread to biology and other fields.

Economics, Finance, and Business. Game theory was first developed to explain economic decision making, and it is widely used by economists, financial analysts, and individuals. For example, a firm may want to analyze the impact of various options for responding to the introduction of a new product by a competitor. Its strategists might devise a payoff matrix that encompasses market responses to the competitor's product and to the product that the company introduces to counter its competitor. Alternatively, a firm might prepare a payoff matrix as part of decision process concerning entry into a new market.

Game theory cannot be used to predict the stock market. However, some game theorists have devised models to explain investor response to such events as an increase in the interest rate by the Federal Reserve Board. At the international level, there are various games that can be used to explain and predict the responses of governments to the imposition of various regulations such as tariffs on imports. Because of the large number of variables involved, this sort of modeling is quite complex and still in its infancy.

Some businesses might be tempted to develop a game theoretic response (perhaps using the prisoners' dilemma) to the actions of workers. For example, a company can develop a game theoretic response to worker demands for increased wages that will enable the company to maximize its utility in the negotiating process. Even if a game does not play out exactly as modeled, a company gains by clarifying its objectives in the development of a formal payoff matrix. Companies can also develop an approach to hiring new employees or dealing with suppliers that draws on game theory to specify goals and strategies to be adopted.

Politics. Some of the most interesting work in game theory has occurred in explaining developments in international affairs, enabling countries to make better decisions in the future. Game theory is often used to explain the decision process of the administration of President John F. Kennedy during the Cuban Missile Crisis. Other game theoretic explanations have been developed to examine a country's decision to start a war, as in the Arab-Israeli conflicts. Game theorists are able to test their formal game results against what actually occurred in these cases so as to enhance their models' ability to predict.

Other game theorists have devised game theoretic models that explain legislative decision making. Much of the legislative process can be captured by game theoretic models that consider the step-by-step process of legislation. In this process, members from one party propose legislation and the other party responds, then the first part often responds, and the responses continue until the legislation is passed or defeated. Most of these models are academic and do not seem to govern legislative decision making, at least not explicitly.

Biology. Some evolutionary biologists have used game theory to describe the evolutionary pattern of some species. One relationship that is often described in game theoretic terms is the coevolution of predators and prey in a particular area. In this case, biologists do not describe conscious responses but rather situations in which a decline in a prey species affects predators or an increase in predators leads to a decline in prey and a subsequent decline in predators. Biologists use this relationship (called the hawk-dove game by the biologist John Maynard Smith) to show how species will evolve so as to better fit an evolutionary niche, such as the development of coloration that enables prey to better conceal itself from predators.

Social Media, Marketing, and Entertainment. Some researchers study users' engagement with social media services through the lens of game theory, while others use game theory as a means of better understanding the effects of social media marketing. In addition, game theory can influence the design of games themselves, providing insights into the potential behavior of players as well as the fairness of a given game.

IMPACT ON INDUSTRY

Much initial work in game theory was done by academics and was not readily adopted by practitioners. Over time, business and political leaders have come to understand that the application of game theoretic principles could possibly enhance their decision making. For example, in the 1950s, organizations such as the RAND Corporation developed game theoretic models that created different scenarios for how the United States should respond to the potential of nuclear war with the Soviet Union. These models were quite cumbersome and their predictability was questioned because of the rigidity of the models and the difficulties of running extensive computer simulations. More powerful computers and the evolution of game theory (including behavioral game theory and the recognition of sequential equilibria) have enabled game theorists to develop quite sophisticated simulation models, some of which can be run on laptop computers, although some international relations and financial models still require supercomputers.

Simple Interpersonal Models. Some businesses have applied the formal approach to decision making used in game theory to labor relations or in developing pay scales. Many of these simple models can be worked out quickly so that executives can make quick decisions. Executives trained in game theory can follow a somewhat formal approach that enables them to evaluate the impact of their decisions in a clear fashion. Of course, employees can also counter using game theoretic logic. The use of ad hoc game theory is becoming commonplace in many businesses.

Game Theoretic Simulations. Assisted by powerful computers that can combine massive amounts of data in multitudes of scenarios, decision makers use game theory in making major decisions that affect market share, profit, and the long-term success of firms. Some financial institutions use game theory to consider the impact of financial instruments such as derivatives. Analysts in such governmental agencies as the Congressional Budget Office use game theoretic simulations to study the potential impact of legislation such as quotas or tariffs or of providing assistance to various domestic industries. In these cases, game theoretic approaches are combined with econometric models to examine the effect of various possible courses of action. National security analysts use various games to predict the responses of foreign countries to moves made by the United States. In one case, analysts used theory to examine responses that US policy makers might follow in responding to an attack with a biological weapon.

Improved Decision Making. Because nations and businesses do not act in isolation, they are starting to use game theory to predict the responses of other nations and businesses. These models are complex, but they still follow the same principles found in simple games such as the prisoners' dilemma. The development of evolutionary game theory has enabled game theorists to better predict the outcomes of games with repeated rounds that have multiple equilibria, as have the use of concepts from psychology.

CAREERS AND COURSEWORK

Game theory can be used by individuals to help them to make better decisions, but it is also used to develop business or political strategy, to predict aspects of financial markets, and to describe evolutionary processes in biology. Many people practice game theory, often without knowing it. Formal game theory, driven by extensive mathematical modeling and computer applications, is used in industry and government to help guide decision making. Only a few mathematicians are full-time game theorists; most game theorists are people who use game theory to enhance decisions or to describe decision processes.

Although game theory is a field of applied mathematics, many graduate programs in social sciences require students to familiarize themselves with at least the basics. Students planning on advanced work in game theory can benefit from courses in statistics and formal logic, as well as any computer programming and any coursework that enhances their ability to deal with quantitative material. As behavioral game theory develops, gaining knowledge of psychology (and possibly neuroscience) will become important for some applications. Many game theorists, however, are self-taught.

SOCIAL CONTEXT AND FUTURE PROSPECTS

Game theory is an evolving field that can become esoteric and divorced from the realities of practical decision making. It can also be an intuitive, essentially nonmathematical approach to enhancing decision making. Both formal and informal game theoretic approaches are likely to be used in business and government in the future. However, most observers agree that sophisticated actors who are aware of the principles of game theory are likely to prevail over those who follow an ad hoc approach to decision making.

Game theory is not a perfect guide to decision making. At times, it has led to overly simplistic approaches that are derived from a narrow view of utility. With the introduction of newer conceptual frameworks, game theory has become less rigid and better able to model human decision making. In the future, sophisticated computer simulations based on game theory are likely to be used for corporate and governmental decision making.

—John M. Theilmann

Further Reading

Abdey, James. *Business Analytics: Applied Modelling and Prediction*. Sage Publications, 2023.

Anderson, Eric. *Social Media Marketing. Game Theory and the Emergence of Collaboration*. Springer, 2010.

Binmore, K. G. *Game Theory: A Very Short Introduction*. Oxford UP, 2007.

Bueno de Mesquita, Bruce. *The Predictioneer's Game: Using the Logic of Brazen Self-Interest to See and Shape the Future*. Random House, 2009.

Fisher, Len. *Rock, Paper, Scissors: Game Theory in Everyday Life*. Basic Books, 2008.

Gintis, Herbert. *Game Theory Evolving: A Problem-Centered Introduction to Modeling Strategic Behavior*. 2nd ed., Princeton UP, 2009.

Miller, James. *Game Theory at Work: How to Use Game Theory to Outthink and Outmaneuver Your Competition*. McGraw-Hill, 2003.

Tyler, Dustin. "How to Use Game Theory in Video Game Development." *Game Designing*, 10 June 2023, www.gamedesigning.org/learn/game-theory. Accessed 13 July 2023.

GAMIFICATION

ABSTRACT

The term "gamification" refers to the use of video-game logic and psychology in real-world environments. Gamification assumes that productivity and efficiency can be enhanced by creating artificial narratives and a game feel around otherwise routine endeavors, thus increasing engagement, raising skill levels, and positively influencing those who are participating.

BACKGROUND

Gamification, a term coined in the early 2000s, refers to the use of video-game logic and psychology in real-world environments, most prominently in marketing, education, and the corporate world. The theory of gamification holds that people—whether consumers, coworkers, or students—respond naturally and efficiently to competition, reward, and simulated risk of the type that have made video games such a cultural phenomenon since the 1980s. The concept applies especially to the generation of Americans born after 1975, many of whom were raised playing video games, who began assuming positions of prominence in businesses and organizations in the early twenty-first century. These video-game aficionados brought with them many of the assumptions and strategies of gaming—incentivized decision making, rapid problem solving, the self-evident logic of specific tasks and short-term rewards, an adrenaline response to simulated risk, and the perception of achievement as a measure of self-expression—all guided by the assumption that operating in such a matrix is both fun and profitable.

OVERVIEW

Though the paradigm of gamification emerged in a very short time through weekend seminars, online training sessions, and social media, its application soon became widespread. Examples of gamification include fast-food restaurants that offer a free meal after ten visits, community organizations that attract volunteers by creating an artificial system of participation levels, and banks that create automated teller machine (ATM) games that award points for savvy deposit decisions. Further examples of gamification include entrepreneurs who offer a system of points to potential investors as a way to measure their short-term returns and to map otherwise entirely artificial levels of success; fitness clubs that sponsor competitions that pit members against each other to meet their fitness goals; and companies that maximize productivity by engaging coworkers in competition against one another, not to earn traditional benefits such as better office space or salary bumps, but rather in the simple spirit of competition. In all these situations, gaming principles are brought into an otherwise non-gaming context.

Psychologists point out that the essential strategy of gamification is just a newer version of old-school operant conditioning that leverages desired behavior by reducing people to pawns and supervisors to crass manipulators. Some companies and educational institutions that must consider long-term developments fear that gamification narrows people's focus to short-term gains and to necessarily limited, even vague, objectives. A number of educators in particular have expressed concerns that the emphasis on external rewards may decrease students' intrinsic motivation for learning, although the extent to which this is a genuine concern remains in dispute; many argue that proper gamification of learning involves using extrinsic rewards to support, rather than supplant, intrinsic motivation.

Some more traditional individuals, particularly those born before the advent of video games, find the premise of gamification insulting and do not believe that the serious work of real life should mimic the simulated thrills of video games. Critics of gamification believe that simply motivating people to act in a certain way—whether customers, students, or coworkers—is a potentially catastrophic strategy because motion can be mistaken for progress, activity for achievement, and competition for teamwork.

—*Joseph Dewey*

Further Reading

Burke, Brian. *Gamify: How Gamification Motivates People to Do Extraordinary Things*. Bibliomotion, 2014.

Dowling, David. *The Gamification of Digital Journalism: Innovation in Journalistic Storytelling*. Routledge, 2020.

McCormick, Ty. "Gamification: A Short History." *Foreign Policy*, 24 June 2013, foreignpolicy.com/2013/06/24/gamification-a-short-history. Accessed 13 July 2023.

McGonigal, Jane. *Reality Is Broken: Why Games Make Us Better and How They Can Change the World*. Penguin, 2011.

Peláez, Carlos Alberto, and Andrés Solano. "A Practice for the Design of Interactive Multimedia Experiences Based on Gamification: A Case Study in Elementary Education." *Sustainability*, vol. 15, no. 3, 2023, doi.org/10.3390/su15032385. Accessed 13 July 2023.

Werbach, Kevin, and Dan Hunter. *For the Win: How Game Thinking Can Revolutionize Your Business*. Wharton, 2012.

Zichermann, Gabe, and Christopher Cunningham. *Gamification by Design: Implementing Game Mechanics in Web and Mobile Apps*. O'Reilly, 2011.

Zichermann, Gabe, and Joselin Linder. *The Gamification Revolution: How Leaders Leverage Game Mechanics to Crush the Competition*. McGraw, 2013.

Google

ABSTRACT

Google began as an internet search engine and grew rapidly once it developed path-breaking models for advertising as well for the collection and monetization of user data. With that revenue, the company expanded into numerous other fields such as online maps, email services, productivity tools, video and document sharing, and mobile devices.

BACKGROUND

Although the World Wide Web had been around since 1989, the general public did not begin to become aware of all the possibilities it offered for finding information until the mid-1990s. Internet users then looked for something to guide them through the increasingly large maze of sites. Search engines such as Excite, HotBot, Infoseek, Lycos, Magellan, and WebCrawler all took different approaches, but each returned results in a random pattern. These search engines were designed to find as many sites as possible without determining the quality of those sites. Two Stanford University graduate students, Sergey Brin and Larry Page, decided to devise a different approach.

Brin and Page met when Brin conducted an orientation tour for new graduate students at Stanford in 1995. The two young men came from similar academic backgrounds. Page's father was a computer science professor at Michigan State University, and his mother was a database consultant who taught computer programming at Michigan State. Brin's father was a mathematics professor at the University of Maryland, and his mother was a scientist at the National Aeronautics and Space Administration's Goddard Space Flight Center. The Brins had immigrated to the United States from the Soviet Union in 1979.

The graduate students were lucky to find themselves at Stanford, where such technological firms as Hewlett-Packard and Sun (for Stanford University Network) Microsystems had received their inspirations. Such entrepreneurship was encouraged by the university. In 1994, another pair of Stanford graduate students, David Filo and Jerry Yang, created a different method of organizing the web from the search engines that employed programs called crawlers to locate sites. Filo and Yang's Yahoo! used a team of editors to find the best sites on all subjects and place them into categories. This web-portal approach bypassed the unpredictability of other search engines but did not address the need for speedy results. Many internet users did not have the patience to make their way through directories and subdirectories to find the information, products, and services they wanted.

One of the first steps toward the creation of Google was the establishment by Brin and his faculty adviser, Professor Rajeev Motwani, of MIDAS

Photo via iStock/Prykhodov. [Used under license.]

(mining data at Stanford), a research group aimed at finding how the data-mining concept (which includes predicting the behavior of consumers) could be applied to organizing the internet. After discovering that the search engine AltaVista, which claimed to find more sites than other search engines, merely reported on the links it found, Page told one of his professors that he planned to download the entire web onto his computer, thinking that would take a week, only to find that the web was much larger than he had expected. Page and Brin then set about analyzing web links, estimating that they cost Stanford's computer science department $20,000 every time they sent a spider program out to report back on the entire internet. Brin was attracted to the project because of his interest in extracting information from large sets of random data.

OVERVIEW

As the web became increasingly commercialized in the second half of the 1990s, webmasters eager to improve their rankings in the search engine (and thus increase their revenue by getting more traffic) found ways to manipulate the search engines into giving them higher rankings than were really merited. Common tricks included repeating one's description multiple times in the header and setting large blocks of text in the same color as the background at the bottom of the body text, where it would be seen by the search engine crawlers but not the page's human audience.

When these questionable tactics led pages to rank high for commonly searched keywords that were in fact unrelated to the actual content of the page, users who arrived at them would often feel frustrated and betrayed, and they would typically blame the search engine. Computers dealt best with facts. Making value judgments about information was a matter generally left to human minds. However, the World Wide Web was simply too large to hire enough people to evaluate all the pages to be indexed. Therefore, one had to find some easily recognizable characteristic that would effectively indicate the value of a web page.

Page decided that the number of links pointing toward a particular site indicated its popularity, and he and Brin had to come up with a way of measuring the relevance of sites. They arrived at a mathematical formula, first named BackRub and later PageRank, hoping that it might lead to a dissertation topic, and, together with Motwani, they created a search engine for internal use at Stanford in 1996. Needing a name, they decided on Google, a misspelling of "googol," a mathematical term signifying an enormous number. Unlike the websites of many search engines at the time, instead of including numerous links on the Google home page, Brin and Page decided to go with a clean, uncluttered look (later, other search engines also adopted this look). Still academics, they published two papers about how their creation worked.

In 1998, Brin and Page tried to sell PageRank for $1 million to AltaVista, planning to resume their studies afterward, but their offer was rejected because AltaVista's parent company, Digital Equipment Corporation, was not receptive to outside technology. Excite, Yahoo!, and other search engines also turned them down. Yahoo! was designed to keep its users on its site to see advertisements, shop, and play games, not go off immediately to other sites.

INCORPORATION

David Cheriton, head of Stanford's Distributed Systems Group, arranged a meeting for Brin and Page with Andy Bechtolsheim, a founder of Sun Microsystems and a vice president at Cisco Systems. Excited about Google's potential, Bechtolsheim gave Brin and Page $100,000 for new equipment. Encouraged by Bechtolsheim's enthusiasm and generosity, they incorporated the company as Google Inc. in a friend's garage in Menlo Park on September 7, 1998, with Page as chief executive officer (CEO) and Brin as president. They raised additional capital, mostly from Silicon Valley investors, and then, in 1999, two firms, Sequoia Capital and Kleiner Perkins Caufield & Byers, provided $25 million to take Google to the next level.

To justify these investments, Brin and Page had to create a business plan. Strongly opposed to the type of advertising seen on Yahoo! and other search engines, they decided that ads would not clutter their site but would appear in their users' search results. A search for "Ford cars," for example, would lead first to text-only advertisements for the Ford Motor Company and its dealerships. Revenue was based on the number of visits Google users made to these commercial sites. Rapid growth followed, with the search engine going from 500,000 daily queries in 1999 to 200 million a day by 2004; the company grew from three employees in 1998 to more than nine thousand by 2006.

Brin and Page tried to model their company after the academic environment with which they were so familiar. The increasingly large staff was composed of mathematicians, technologists, and engineers, without the usual layers of middle managers found in most companies. Google expected its employees to spend 20 percent of their time exploring whatever interested them, with the assumption that some of this research would eventually enhance the company's operation. Its corporate motto was "Don't be evil." In 2007, *Fortune* magazine ranked Google as the best American company to work for.

On August 19, 2004, Google went public at $85 per share, raising nearly $2 billion. In less than a year, Google's stock price increased to more than $300 per share, and the value of the company reached $80 billion.

The growing popularity of the Google search engine in the early 2000s provided the company with a lucrative market for the launch of its advertising services. AdWords, a platform that works by serving context-sensitive text advertisements based on keywords chosen by the advertiser, debuted in 2002. The cost depended on the popularity of the keyword and the number of people who clicked on the advertisement. The advertising platform was flexible and affordable enough for small businesses, and it became highly successful. Google also expanded its corporate customer base by offering tools such as Site Search, a search engine using Google's efficient technology to locally search large organizational websites for relevant information.

Consumer tools also continued to appear, notably including the free email service Gmail, first launched to the public in beta form in early 2004 and more widely in 2007. Gmail quickly proved popular and eventually attracted over 1 billion users. Another service was Google Books, launched as Google Print in 2004, which made scanned versions of texts freely available online.

CONTINUED GROWTH AND ACQUISITIONS

Google made its initial public offering (IPO) on August 18, 2004. Following its policy of innovation, Google priced the shares by an unusual Dutch auction method. This gave small investors a greater chance to compete with major banks and financial institutions. The IPO managed to secure $1.6 billion for Google, providing the company with much-needed resources for expansion. Shares given to employees and early investors made many people very rich overnight. The IPO also gave Google the leverage needed to acquire smaller companies as its stock valuation reached $29.4 billion. (By mid-2015, when the company was reorganized as a subsidiary of holding company Alphabet Inc., its stock valuation was $440 billion.)

Bolstering its own success, Google also started buying other tech companies. The company heads often sought out start-ups offering new technologies that they wanted to develop further and integrate with Google's own services. One such case was Deja.com's Usenet discussion service, which found a new place in Google's portfolio of online products rebranded as Google Groups. Keyhole, makers of three-dimensional mapping software, and Where 2 Technologies, maker of a desktop mapping software, joined the list of acquisitions in 2004 and later resurfaced as Google Earth and Google Maps, respectively. In 2005, Urchin, a web analytics company, formed the basis of Google Analytics. Writely, a web-based word-processing software company, later became a part of Google Docs. Google acquired Grand Central, a voice-communication service using the internet, in 2007 and relaunched it as Google Voice.

The online video-sharing service YouTube became a notable exception to Google's rebranding strategy when it was acquired in 2006. Instead of being integrated into Google Video, the new asset continued to exist under its original name, and Google Video was itself discontinued in 2012. By November 2016, Google had purchased more than two hundred companies.

Meanwhile, Google also moved toward packaging and marketing multiple services for a particular audience of users. Google for Education, for example, bundled many of Google's productivity tools, along with the Google Classroom portal through which educators and students could interact, collaborate, and view results. Similarly, Google Politics & Elections, which was targeted toward political candidates and their campaign managers, combined various advertisement placement opportunities with data analytics and market research tools. Google for Nonprofits likewise offered productivity tools, grants for advertising, enhanced map capabilities, and fund-raising campaigns through Google One Today. The company's first attempt at a social media platform was Google Buzz in 2010, which made little impact. It was replaced by Google+ in 2011, which had strong early growth but eventually lagged in competition with social media giants Facebook and Twitter (and was largely discontinued in 2019).

While remaining predominantly a software company, Google increasingly became involved with hardware. In 2007, Google entered the mobile-device industry with its smartphone operating system Android, an open platform project backed by a joint effort of many leading hardware manufacturers and telecommunication companies. Google acquired phone manufacturer Motorola in 2012 in order to produce its own Android-loaded smartphones. Within just a few years, Android controlled a significant majority of the smartphone market worldwide, with Apple's iOS as its primary competitor. Google's original hardware smartphones were replaced with a model line known as the Pixel in 2016.

In 2011 the company first introduced its Chromebook laptops, which ran its Linux-based Chrome operating system (itself first offered in 2008). The laptops were notable for doing away with expensive, high-capacity hard drives in favor of a reliance on cloud computing. Tablet computers running the Android operating system also appeared. Google Glass, essentially a computer system attached to eyeglasses, was released to the public in 2014 but was widely considered a failure. More successful was Google Home, a smart speaker with digital personal assistant capabilities that debuted in 2016 and went through different iterations in subsequent years. In an even more ambitious project, Google in 2014 announced efforts to develop a driverless car, though in late 2016 the venture was spun off as its own company, Waymo.

By the time Google underwent the major restructuring in 2015 that established parent company Alphabet, it was among the most influential technology companies in the world. Its search engine remained its core product, so commonly used that "google" became widely known as a verb meaning to search something online. The company's founders also remained involved; Brin became president of Alphabet and Page served as CEO. Eric Schmidt took on the executive chair position of the holding company until early 2018, and remained on the board until announcing he would step down in 2019. During the restructuring Sundar Pichai became the CEO of Google itself.

In 2019, Google announced a new contribution to the gaming world when it unveiled its cloud-based gaming service Google Stadia. This platform was designed to allow users to access and play streamed games across connected devices. Stadia remained in operation until January of 2023, when it was shut down. In addition to introducing new projects, parent company Alphabet remained active in corporate acquisitions as well: in 2021, for instance, the company acquired the wearable technology company Fitbit, while the following year, Alphabet acquired several cybersecurity firms as well as businesses working in fields such as artificial intelligence and robotics.

CORPORATE CULTURE AND SOCIAL RESPONSIBILITY

From Google's inception, Page and Brin prided themselves on the eccentric culture of their company. Playing pranks and breaking conventions were encouraged. Interviews were conducted while roller skating. April Fools' jokes became a part of corporate branding when Google started incorporating them annually in its search engine as early as 2000.

In its early years, Google became known for the positive environment it sought to create for its employees, setting a precedent that many other tech, start-up, and other progressive companies emulated. The company headquarters in Mountain View, California, included a number of amenities, including laundry and dry-cleaning services, gyms, volleyball and basketball courts, medical services, and day care. Indoor climbing walls, bowling alleys, pub-style lounges, and massage rooms ensured that employees had little reason to go far from their work spaces in search of recreation. Food, prepared by professional chefs, was provided free for most employees at the Mountain View campus. Because employees remained onsite for the most part, productivity was high, though workdays were frequently long.

The benefits of working at Google went beyond gourmet meals and extravagant facilities. Employees enjoyed retirement benefits in which personal savings were matched by the company. The value of company stock, which all early employees received, was another significant draw. Yet perhaps the most unique perk at Google was the company's policy of allowing employees the freedom to pursue their own ideas—rather than assigned projects—for up to 20 percent of the workday.

As the company's profits grew, Google also became more involved in charitable initiatives. Google Grants, a program to provide free advertising to nonprofit organizations, was established in 2003 and went on to assist thousands of nonprofit companies all over the world. Google.org debuted in 2004 as the company's centralized effort to address global philanthropic and ecological challenges. The nonprofit organization sought to improve the quality of life and the environment by leveraging Google's core competencies in the technology field. Google.org concentrated on providing computer and technology education, protecting wildlife and the environment, and combating human trafficking. Google itself also took on renewable energy projects, such as wind power development.

Another initiative that Google avidly supported was network neutrality. The goal of the movement is to keep the internet open to all people, free of government or corporate censorship or control. In 2006, Schmidt officially stated the company's commitment to the cause of network neutrality in a letter to all Google users. He urged users to oppose any proposed legislation that would limit the freedom of the internet. However, in 2014, Google teamed with Verizon to create enforceable rules for internet usage, a move seen by some as in opposition to its previous stance on net neutrality.

Indeed, the apparent move away from support for total net neutrality was just one sign of what critics began to suggest was a significant shift in Google's corporate image. Although the company had long sought to distance itself from the stereotypical view of large corporations interested only in profit, even adopting the slogan "Don't be evil," it increasingly found itself mired in controversy over its business practices. For example, observers noted that its spending on lobbying efforts increased drastically over the years, it took advantage of numerous loopholes to avoid taxes, and it adhered to state censorship rules in China. The company also faced criticism for cooperating with a US National Security Agency mass surveillance program.

By the late 2010s concerns over online privacy and data collection were frequent subjects of debate, even crossing into political territory as some activists suggested big tech companies such as Google should be subject to greater regulation. The European Union (EU) led the way in attempting to install and enforce such regulations. Google was subject to multiple investigations by the European Commission regarding antitrust violations and received record-setting $2.7 billion and $5 billion fines in 2017 and 2018. In 2019 another hefty fine was imposed on the company by France's data regulation agency for violation of the EU's General Data Protection Regulation (GDPR) framework. Later that year the European Commission followed with a further $1.7 billion antitrust fine.

IMPACT

Growing from a small company to one of the leaders in the field of technology, Google has had a significant impact on the lives of nearly everyone around the world with access to the internet. It has been credited with helping to usher in an age of unprecedented global connectivity and information access with products such as its core search engine, Google Maps, and Gmail. It also has had a hand in many other hugely popular internet services, as through its ownership of YouTube. Apart from providing many businesses and individuals with valuable online tools, the company has had a measurable economic impact on the United States and, indeed, in many other countries the world over. Yet this success came at a cost, as many people have begun to question the apparent loss of personal privacy and internet-obsessed culture generated by the pervasiveness of Google and similar technology companies.

—*Miroslav Farina*

Further Reading

Brin, Sergey, and Lawrence Page. "The Anatomy of a Large-Scale Hypertextual Web Search Engine." *Stanford University InfoLab*, 1998, snap.stanford.edu/class/cs224w-readings/Brin98Anatomy.pdf. Accessed 13 July 2023.

Davis, Yvette. *Google Secrets: Do What You Never Thought Possible with Google.* Wiley, 2011.

D'Onfro, Jillian. "Here's a Reminder of Just How Huge Google Search Truly Is." *Tech Insider*, 27 Mar. 2016, www.businessinsider.com/google-search-engine-facts-2016-3. Accessed 13 July 2023.

"Google Dominates Search: But the Real Problem Is Its Monopoly on Data." *The Guardian*, 19 Apr. 2015, www.theguardian.com/technology/2015/apr/19/google-dominates-search-real-problem-monopoly-data. Accessed 13 July 2023.

Levy, Steven. *In the Plex: How Google Things, Works, and Shapes Our Lives.* Simon, 2011.

Lohr, Steve. "At Google, Cube Culture Has New Rules." *New York Times*, 5 Dec. 2005, www.nytimes.com/2005/12/05/technology/at-google-cube-culture-has-new-rules.html. Accessed 13 July 2023.

O'Connell, Brian. "History of Google: How It Began and What's Happening Beyond 2019." *Street*, 31 Dec. 2018, www.thestreet.com/technology/history-of-google-14820930. Accessed 13 July 2023.

Peters, Jay, and Alex Cranz. "Google Is Shutting Down Stadia." *Verge*, 29 Sept. 2022, www.theverge.com/2022/9/29/23378713/google-stadia-shutting-down-game-streaming-january-2023. Accessed 13 July 2023.

Porter, Jon, and Nick Statt. "Google Completes Purchase of Fitbit." *Verge*, 14 Jan. 2021, www.theverge.com/2021/1/14/22188428/google-fitbit-acquisition-completed-approved. Accessed 13 July 2023.

Schmidt, Eric, and Jonathan Rosenberg. *How Google Works*. Grand Central, 2017.

Wilson, Mark. "Exclusive: A First Look Inside Google's Top-Secret Design Lab." *Fast Company*, 6 Aug. 2019, www.fastcompany.com/90378288/exclusive-a-first-look-inside-googles-top-secret-design-lab. Accessed 13 July 2023.

Graphic Designer and Illustrator

ABSTRACT

Graphic designers and illustrators create visually appealing products and illustrative materials using a variety of tools, including design and illustration software. They may be employed by specific companies or creative agencies or may work on a freelance basis. Students aspiring to enter the profession should take courses in subjects such as art, graphic design, computer science, and the social sciences.

BACKGROUND

Graphic designers and illustrators create visually appealing products and illustrative materials that range from simple logos or business cards to picture books and entire corporate branding campaigns.

Their work is intended to convey a commercial or educational message or otherwise draw attention to an idea, which they accomplish mostly with sophisticated graphic design and illustration techniques. Traditional artistic mediums, such as printmaking or painting, continue to be used, but sporadically. A designer interested in a long-term career must also learn skills in animation, digital video production, and web design, or collaborate frequently with people who possess these skill sets or perform these job functions.

OVERVIEW

Graphic designers and illustrators work mostly in the publishing, advertising, and marketing industries, but some work for graphic design firms or government agencies. Many are self-employed. They spend much of their time working on computers but may also have access to a full art studio. If self-employed, they interact heavily with clients. If employed in a design firm or design department, they interact with a team of professionals and staff and have less direct contact with clients; however, some customer service is necessary when choosing the final design—a process that can take anywhere from a couple of days to several weeks.

People who are attracted to graphic design and illustration tend to be creative thinkers who are interested in solving problems with images or in making the world more visually interesting. They are artistic and have a good eye for detail but are also capable of seeing the big picture and being flexible in their

Photo via iStock/amenic181. [Used under license.]

presentation of multiple design or pictorial ideas for a single project. They must have excellent interpersonal and communication skills since they almost always work closely with members of a team. They must be able to handle criticism and work under pressure to meet deadlines.

DUTIES AND RESPONSIBILITIES

Graphic designers and illustrators are responsible for planning and carrying out projects that fulfill their clients' needs. A major specialty, for example, is branding, in which the designer works with a team of writers, artists, market researchers, and others to create a company's image, including a recognizable logo, stylized advertisements, catchy slogans, and high-tech trade show displays. The designer will suggest colors, images, fonts, and other artistic elements, then create several sample designs for each element of the branding campaign.

Graphic designers and illustrators create the covers and interior layouts for magazines, books, brochures, newspapers, and other print materials. Typically, they work with editors to acquire the articles and advertisements, and then fit them into the allotted space in the most appealing manner. Graphic designers select fonts, graphics, and other design elements (in collaboration with the client or author if the project is a book) and also design internal advertisements that are presented with the final product.

Study of graphic design and illustration includes the option to learn website design skills, so some designers/illustrators specialize in web design and are able to earn a living without working on any print design projects. Some graphic designers/illustrators collaborate with programmers to create video games, with authors to create graphic novels, and even with interior decorators and architects to design building spaces.

Self-employed graphic designers/illustrators must also spend some of their time marketing their services, billing customers, preparing contracts, and handling other administrative and business-management tasks.

OCCUPATION SPECIALTIES

Cartoonists. Cartoonists draw political cartoons, newspaper comics, or comic books. Their work is used in commercial applications, as well. They may work with animators in creating digital productions.

Drafters. Drafters use software to convert the designs of engineers and architects into technical drawings and plans.

Fashion Artists. Fashion artists draw stylish illustrations of new clothing fashions for newspapers or related advertisements.

Illustrators. Illustrators create pictures for books, magazines, billboards, posters, and packaging. They may work with multimedia artists in creating digital productions.

Medical and Scientific Illustrators. Medical and scientific illustrators draw precise illustrations of machines, plants, animals, or parts of the body for business and educational purposes.

Typographers. Typographers create type fonts for use in print and online publications. One major use is in signage, including "wayfinding" signs for drivers and other travelers.

WORK ENVIRONMENT

Physical Environment. Graphic designers and illustrators usually work in studios or offices surrounded by art samples and design reference materials. If a project has a tight deadline, the designer/illustrator can expect to work some evening hours or work on the design from their home computer until the project is done.

Human Environment. In larger design firms and departments, a graphic designer or illustrator is often one member of a creative team composed of photographers, multimedia artists, web developers, and others who collaborate on projects under the supervision of a creative director. Designers/illustrators may also work with market researchers, architects, interior designers, content editors, clients, authors, and other professionals outside the firm.

Technological Environment. Graphic designers use computer software for cover design, manipulating and creating illustrations, page layout design, editing and placing digital photographs, and other purposes. They also use digital photography and video cameras, scanners, printmaking equipment, printing and publishing equipment, and other tools. Each design project is different and may require different resources, so graphic designers should enjoy learning new skills. Illustrators make use of computer software and traditional art materials in creating their works.

EDUCATION AND TRAINING

High School/Secondary. Students should pursue a comprehensive college-preparatory program that includes courses in art, graphic design, computer science, and the social sciences. Other relevant courses include film, new media, photography, and industrial arts. Awareness of contemporary graphic design, web design, and animation software programs is extremely important. Most college admissions programs require a portfolio of artistic work, which might include digital designs, as well as hand-drawn sketches or paintings, sculpture, and other sample work.

College/Postsecondary. Most entry-level positions require a bachelor's degree from an art school or program. Graphic design programs include courses in studio art, design, computer graphics, printing, and other graphic design specialties. Programs should include the option to work an internship. A general awareness of contemporary design is also helpful. Illustration programs focus more on drawing, painting, and other manual methods while not excluding digital technologies. In either case, the development of a portfolio for use in future job searches is essential, so many graphic design and illustration students try to obtain freelance work.

Adult Job Seekers. Adults with a background in fine art, illustration, photography, typography, or another creative discipline can learn the fundamentals of graphic design and illustration and update their skills by taking graphic arts courses, which some schools offer in the evenings and on weekends to accommodate adult professionals. A portfolio can be assembled independently and/or in conjunction with classes.

Advancement for graphic designers and illustrators comes with experience or taking classes in new software or techniques to supplement current skills. Some designers/illustrators choose to establish their own firms. Advanced degrees can help experienced artists begin to obtain work in a different specialty, such as web animation.

Professional Certification and Licensure. There are no state licenses or nationally recognized certificates required for graphic designers or illustrators; however, the idea of professional certification has gained popularity, so it is advisable to follow the issue as it progresses. Getting certified on the use of a particular software program or type of equipment, for example, may be in order as one's career advances.

Additional Requirements. Graphic artists and illustrators must have a good eye for aesthetics, be extremely creative, enjoy art and design, and find satisfaction in continuing professional development. They should be willing to follow trends in advertising, web media, design, and illustration and should enjoy brainstorming for a single project. Although creativity is a plus, graphic designers/illustrators must be able to distance themselves from their work enough to accept a client's criticism and revisions of their ideas. Excellent people skills are a must in this collaborative field, which can be highly competitive and requires the ability to make and maintain good contacts with clients and colleagues.

—Briana Nadeau

Further Reading

Costello, Vic. *Multimedia Foundations: Core Concepts for Digital Design*. 3rd ed., Focal Press, 2023.

Prust, Z. A., and Peggy B. Deal. *Graphic Communications: Digital Design and Print Essentials*. 7th ed., Goodheart-Willcox, 2022.

Wagner, Karin. *From ASCII Art to Comic Sans: Typography and Popular Culture in the Digital Age*. MIT Press, 2023.

GRAPHICAL USER INTERFACE (GUI)

ABSTRACT

Graphical user interfaces (GUIs) are human-computer interaction systems. In these systems, users interact with the computer by manipulating visual representations of objects or commands. GUIs are part of common operating systems such as Windows and macOS. They are also used in other applications.

BACKGROUND

A user interface is a system for human-computer interaction. The interface determines the way that a user can access and work with data stored on a computer or within a computer network. Interfaces can be either text-based or graphics-based. Text-based systems allow users to input commands. These commands may be text strings or specific words that activate functions. By contrast, graphical user interfaces (GUIs) are designed so that computer functions are tied to graphic icons such as folders, files, and drives. Manipulating an icon causes the computer to perform certain functions.

The earliest computers used a text-based interface. Users entered text instructions into a command line. For instance, typing "run" in the command line would tell the computer to activate a program or process. One of the earliest text-based interfaces for consumer computer technology was known as a "disk operating system" (DOS). Using DOS-based systems required users to learn specific text commands, such as "del" for deleting or erasing files or "dir" for listing the contents of a directory. One of the earliest known GUIs was the interactive interface provided in Sketchpad, an early computer-aided design program created in the 1960s. Further early GUIs were created in the 1970s as visual "shells" built over DOS systems.

GUIs transform the computer screen into a physical map on which graphics represent functions, programs, files, and directories. In GUIs, users control an onscreen pointer, usually an arrow or hand symbol, to navigate the computer screen. Users activate computing functions by directing the pointer over an icon and "clicking" on it. For instance, GUI users can cause the computer to display the contents of a directory (the "dir" command in DOS) by clicking on a folder or directory icon on the screen. Twenty-first-century GUIs combine text-based icons, such as those found in menu bars and movable windows, with linked text icons that can be used to access programs and directories.

OVERVIEW

Computer programs are built using coded instructions that tell the computer how to behave when given inputs from a user. Many different programming languages can be used to create GUIs. These include C++, C#, JavaFX, XAML, and XUL, among others. Each language offers different advantages and disadvantages when used to create and modify GUIs.

User-centered design focuses on understanding and addressing user preferences, needs, capabilities, and tendencies. According to these design principles, interface metaphors help make GUIs user friendly. Interface metaphors are models that represent real-world objects or concepts to enhance user understanding of computer functions. For example, the desktop structure of a GUI is designed using the metaphor of a desk. Computer desktops, like actual desktops, might have stacks of documents (windows) and objects or tools for performing various functions. Computer folders, trash cans, and recycle bins are icons whose functions mirror those of their real-world counterparts.

Object-oriented user interfaces (OOUIs) allow a user to manipulate objects onscreen in intuitive ways based on the function that the user hopes to achieve. Most modern GUIs have some object-oriented functionality. Icons that can be dragged, dropped, slid, toggled, pushed, and clicked are "objects." Objects include folders, program shortcuts, drive icons, and trash or recycle bins. Interfaces that use icons can also be direct manipulation interfaces (DMIs). These interfaces allow the user to adjust onscreen objects as though they were physical objects to get certain results. Resizing a window by dragging its corner is one example of direct manipulation used in many GUIs.

FUTURE OF INTERFACE DESIGN

GUIs have long been based on a model known as WIMP. WIMP stands for "windows, icons, menus, and pointer" objects, which describes the ways that users can interact with the interface. Modern GUIs are a blend of graphics-based and text-based functions, but this system is more difficult to implement on modern handheld computers, which have less space to hold icons and menus. Touch-screen interfaces represent the post-WIMP age of interface design. With touch screens, users more often interact directly with objects on the screen, rather than using menus and text-based instructions. Touch screen design is important in many application-specific GUIs. These interfaces are designed to handle a single process or application, such as self-checkout kiosks in grocery stores and point-of-sale retail software.

Computer interfaces of the early twenty-first century typically require users to navigate through files, folders, and menus to locate functions, data, or programs. However, voice activation of programs or functions is also available on many computing devices. As this technology becomes more common and effective, verbal commands may replace many functions that have been accessed by point-and-click or menu navigation.

—Micah L. Issitt

Further Reading

"Graphical User Interface." *Techopedia*, 28 May 2021, www.techopedia.com/definition/5435/graphical-user-interface-gui. Accessed 13 July 2023.

Jackson, Daniel. *The Essence of Software Design: Why Concepts Matter for Great Design*. Princeton UP, 2023.

Johnson, Jeff. *Designing with the Mind in Mind*. 2nd ed., Morgan Kaufmann, 2014.

Long, Simon. *An Introduction to C & GUI Programming*. Raspberry Pi Press, 2019.

Norman, Jeremy M. "Ivan Sutherland Creates the First Graphical User Interface." *Historyofinformation.com*, historyofinformation.com/detail.php?id=805. Accessed 13 July 2023.

Reimer, Jeremy. "A History of the GUI." *Ars Technica*, 5 May 2005, arstechnica.com/features/2005/05/gui. Accessed 13 July 2023.

Tidwell, Jenifer, Charles Brewer, and Aynne Valencia. *Designing Interfaces: Patterns for Effective Interaction Design*. 3rd ed., O'Reilly Media, 2020.

"User Interface Design Basics." *Usability.gov*, www.usability.gov/what-and-why/user-interface-design.html. Accessed 13 July 2023.

Wood, David. *Interface Design: An Introduction to Visual Communication in UI Design*. Fairchild Books, 2014.

Graphics Formats

ABSTRACT

Graphics formats are standardized forms of computer files used to transfer, display, store, or print reproductions of digital images. Digital image files are divided into two major families, vector and raster files. They can be compressed or uncompressed for storage. Each type of digital file has advantages and disadvantages when used for various applications.

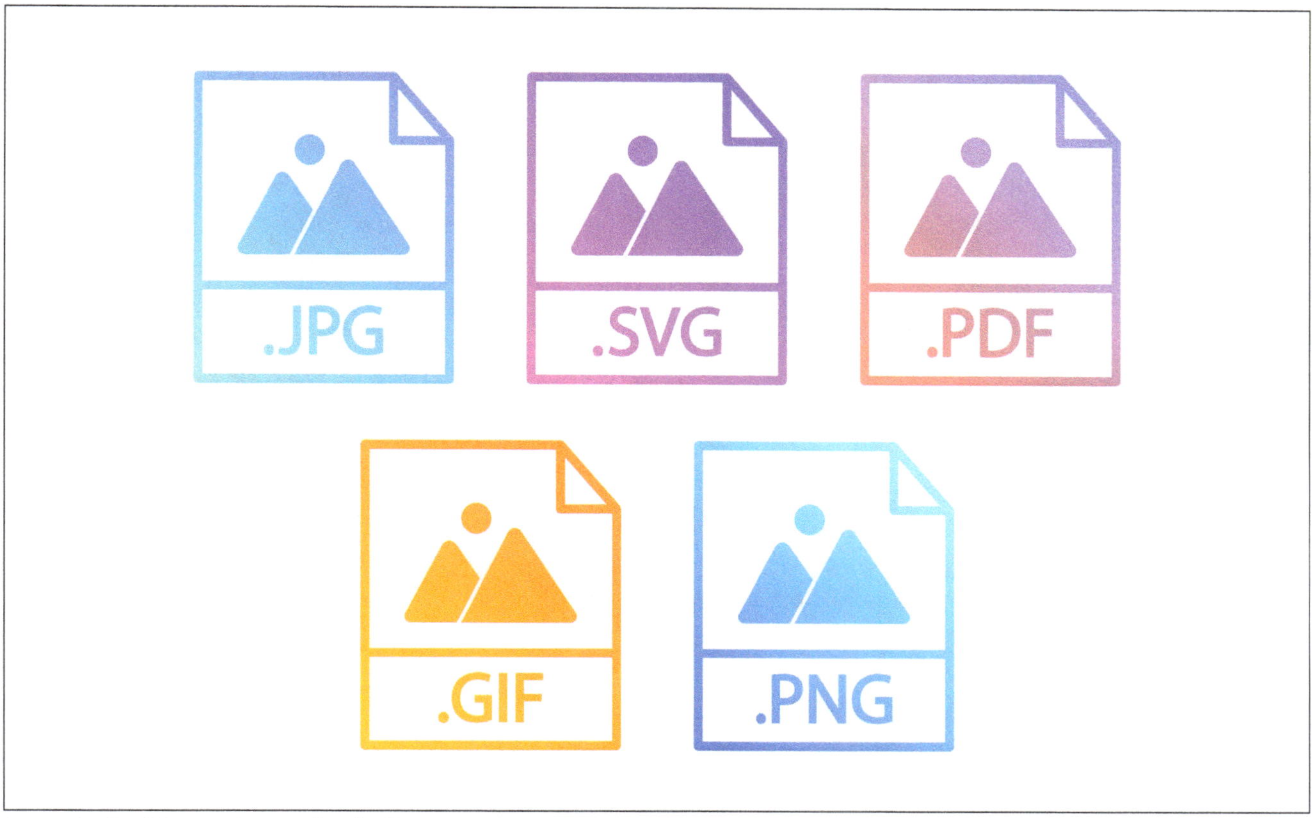

Image via iStock/irem soyler. [Used under license.]

BACKGROUND

A digital image is a mathematical representation of an image that can be displayed, manipulated, and modified with a computer or other digital device. It can also be compressed. Compression uses algorithms to reduce the size of the image file to facilitate sharing, displaying, or storing images. Digital images may be stored and manipulated as raster or vector images. A third type of graphic file family, called "metafiles," uses both raster and vector elements.

The quality and resolution (clarity) of an image depend on the digital file's size and complexity. In raster graphics, images are stored as a set of squares, called "pixels." Each pixel has a color value and a color depth. This is defined by the number of "bits" allocated to each pixel. Pixels can range from 1 bit per pixel, which has a monochrome (two-color) depth, to 32-bit, or "true color." 32-bit color allows for more than 4 billion colors through various combinations. Raster graphics have the highest level of color detail because each pixel in the image can have its own color depth. For this reason, raster formats are used for photographs and in image programs such as Adobe Photoshop. However, the resolution of a raster image depends on size because the image has the same number of pixels at any magnification. For this reason, raster images cannot be magnified past a certain point without losing resolution.

Vector graphics store images as sets of polygons that are not size-dependent and look the same at any magnification. For relatively simple graphics, such as logos, vector files are smaller and more precise than raster images. However, vector files do not support complex colors or advanced effects, such as blurring or drop shadows. Depending on the image format, data may be lost after compression and restoration. Loss of image data reduces the quality of the image.

Two basic color models are used to digitally display various colors. The RGB color model, also called "additive color," combines red, green, and blue to create colors. The CMYK model, also called "subtractive color," combines the subtractive primary colors cyan, magenta, yellow, and black to absorb certain wavelengths of light while reflecting others.

Image compression reduces the size of an image to enable easier storage and processing. Lossless compression uses a modeling algorithm that identifies repeated or redundant information contained within an image. It stores this information as a set of instructions that can be used to reconstruct the image without any loss of data or resolution. One form of lossless compression commonly used is the LZW (Lempel-Ziv-Welch) compression algorithm developed in the 1980s. The LZW algorithm uses a "code table" or "dictionary" for compression. It scans data for repeated sequences and then adds these sequences to a "dictionary" within the compressed file. By replacing repeated data with references to the dictionary file, space is saved but no data is lost. Lossless compression is of benefit when image quality is essential but is less efficient at reducing image size. Lossy compression algorithms reduce file size by removing less "valuable" information. However, images compressed with lossy algorithms continue to lose resolution each time the image is compressed and decompressed. Despite the loss of image quality, lossy compression creates smaller files and is useful when image quality is less important or when computing resources are in high demand.

OVERVIEW

The joint photographic experts group (JPEG) is a type of lossy image compression format developed in the early 1990s. JPEGs support RGB and CMYK color and are most useful for small images, such as those used for display on websites. JPEGs are automatically compressed using a lossy algorithm. Thus, some image quality is lost each time the image is edited and saved as a new JPEG.

Graphics interchange format (GIF) files have a limited color palette and use LZW compression so that they can be compressed without losing quality. Unlike JPEG, GIF supports "transparency" within an image by ignoring certain colors when displaying or printing. GIF files are open source and can be used in a wide variety of programs and applications. However, most GIF formats support only limited color because the embedded LZW compression is most effective when an image contains a limited color palette. Portable network graphics (PNGs) open-source alternatives to GIFs that support transparency and 24-bit color. This makes them better at complex colors than GIFs.

Scalable vector graphics (SVGs) are an open-source format used to store and transfer vector images. SVG files lack built-in compression but can be compressed using external programs. In addition, there are "metafile" formats that can be used to share images combining both vector and raster elements. These include portable document format (PDF) files, which are used to store and display documents, and the Encapsulated PostScript (EPS) format, which is typically used to transfer image files between programs.

—Micah L. Issitt

Further Reading

Brown, Adrian. "Graphics File Formats." *National Archives*, Aug. 2008, www.nationalarchives.gov.uk/documents/information-management/graphic-file-formats.pdf. Accessed 13 July 2023.

Costello, Vic. *Multimedia Foundations: Core Concepts for Digital Design*. 3rd ed., Focal Press, 2023.

Dale, Nell, and John Lewis. *Computer Science Illuminated*. 7th ed., Jones & Bartlett Learning, 2019.

Gambetta, Gabriel. *Computer Graphics from Scratch*. No Starch Press, 2021.

"Introduction to Image Files Tutorial." *Boston University Information Services and Technology*, www.bu.edu/tech/support/research/training-consulting/online-tutorials/imagefiles/image101-1. Accessed 13 July 2023.

Marschner, Steve, et al. *Fundamentals of Computer Graphics*. 4th ed., CRC Press, 2016.

Prust, Z. A., and Peggy B. Deal. *Graphic Communications: Digital Design and Print Essentials*. 7th ed., Goodheart-Willcox, 2022.

Graphics Technologies

ABSTRACT

Graphics technology, which includes computer-generated imagery (CGI), has become an essential technology of the motion picture and video-gaming industries, of television, and of virtual reality (VR). The production of such images, and especially of animated images, is a complex process that demands a sound understanding of not only physics and mathematics but also anatomy and physiology.

BACKGROUND

Imaging is as old as the human race. Static graphics have historically been the norm up to the invention of the devices that could make a series of still pictures appear to move. The invention of celluloid in the late nineteenth century provided the material for photographic film, with the invention of motion picture cameras and projectors to follow. Animated films, commonly known as "cartoons," have been produced since the early twentieth century by repeatedly photographing a series of hand-drawn "cels." With the development of the digital computer and color displays in the second half of the twentieth century, it became possible to generate images without the need for hand-drawn intermediaries.

While graphics technologies include all of the theoretical principles and physical methods used to produce images, the term more specifically refers to the principles and methods associated with digital or computer-generated images. Digital graphics are displayed as a limited array of colored picture elements (pixels). The greater the number of pixels that are used for an image, the greater the resolution of the image and the finer the detail that can be portrayed. The data that specifies the attributes of each individual pixel are stored in an electronic file using one of several specific formats. Each file format has its own characteristics with regard to how the image data can be manipulated and utilized.

Because the content of images is intended to portray real-world objects, the data for each image must be mathematically manipulated to reflect real-world structures and physics. The rendering of images, especially for photorealistic animation, is thus a calculation-intensive process. For images that are not produced photographically, special techniques and applications are continually being developed to produce image content that looks and moves as though it were real.

Computer graphics in the twenty-first century can produce images that are indistinguishable from traditional photographs of real objects. The methodology continues to develop in step with the development of new computer technology and new programming methods that make use of the computing abilities of the technology.

OVERVIEW

Images are produced initially as still or static images. Human perception requires about one-thirtieth of a second to process the visual information obtained through the seeing of a still image. If a sequential series of static images is displayed at a rate that exceeds the frequency of thirty images per second, the images are perceived as continuous motion. This is the basic principle of motion pictures, which are nothing more than displays of a sequential series of still pictures. Computer-generated still images have the same relationship to computer animation.

Images are presented on a computer screen as an array of colored dots called "pixels" (an abbreviation of "picture elements"). The clarity, or resolution, of the image depends on the number of pixels that it contains within a defined area. The more pixels within a defined area, the smaller each pixel must be and the finer the detail that can be displayed. In the 2010s, commercially available digital cameras typically captured image data in an array of between eight and fifty megapixels, though some specialty digital cameras manufactured later in the decade were capable of capturing that data in one hundred megapixels or more. The electronic data file of the image contains the specific color, hue, saturation, and brightness designations for each pixel in the associated image, as well as other information about the image itself.

To obtain photorealistic representation in computer-generated images, effects must be applied that correspond to the mathematical laws of physics. In still images, computational techniques such as ray tracing and reflection must be used to imitate the effect of light sources and reflective surfaces. For the virtual reality (VR) of the image to be effective, all of the actual physical characteristics that the subject would have if it were real must be clearly defined as well so that when the particular graphics application being used renders the image to the screen, all of the various parts of the image are displayed in their proper positions.

To achieve photorealistic effects in animation, the corresponding motion of each pixel must be coordinated with the defined surfaces of the virtual object, and their positions must be calculated for each frame of the animation. Because the motions of the objects would be strictly governed by the mathematics of physics in the real world, so must the motions of the virtual objects. For example, an animated image of a round object bouncing down a street must appear to obey the laws of gravity and Newtonian mechanics. Thus, the same mathematical equations that apply to the motion and properties of the real object must also apply to the virtual object.

Other essential techniques are required to produce realistic animated images. When two virtual objects are designed to interact as though they are real, solid objects, clipping instructions identify where the virtual solid surfaces of the objects are located; the instructions then mandate the clipping of any corresponding portions of an image to prevent the objects from seeming to pass through each other. Surface textures are mapped and associated with underlying data in such a way that movement corresponds to real body movements and surface responses. Image animation to produce realistic skin and hair effects is based on a sound understanding of anatomy and physiology and represents a specialized field of graphics technology.

APPLICATIONS AND PRODUCTS

Software. The vast majority of products and applications related to graphics technology are software applications created specifically to manipulate electronic data so that it produces realistic images and animations. The software ranges from basic paint programs installed on most personal computers (PCs) to full-featured programs that produce wireframe structures, map surface textures, coordinate behaviors, movements of surfaces to underlying structures, and 360-degree, three-dimensional (3D) animated views of the resulting images.

Other types of software applications are used to design objects and processes that are to be produced as real objects. Computer-aided design (CAD) is commonly used to generate construction-specification drawings and to design printed circuit boards, electronic circuits and integrated circuits, complex machines, and many other real-world constructs. The features and capabilities of individual applications vary.

Hardware. The computational requirements of graphics can quickly exceed the capabilities of any particular computer system. This is especially true of PCs. Graphics technology in this area makes use of separate graphics processing units (GPUs) to handle the computational load of graphics display. This allows the PC's central processing unit (CPU) to carry out the other computational requirements of the application without having to switch back and forth between graphic and nongraphic tasks.

Many graphics boards also include dedicated memory for exclusive use in graphics processing. This eliminates the need for large sectors of a PC's random-access memory (RAM) to be used for storing graphics data, a requirement that can render a computer practically unusable.

Another requirement of hardware is an instruction system to operate the various components so that they function together. For graphics applications, with the long periods of time they require to carry out the calculations needed to render a detailed image, it is essential that the computer's operating system (OS) be functionally stable. The main operating systems of PCs are Microsoft Windows, Apple's macOS, the open-source OS Linux, and Google's Chrome OS. Some versions of other operating systems, such as Oracle Solaris, have been made available but do not account for a significant share of the PC market.

The huge amounts of graphics and rendering required for large-scale projects such as motion pictures demand the services of mainframe computers. The operating systems for these units have a longer history than do PC operating systems. Mainframe computers function primarily with the UNIX operating system, although many run under some variant of the Linux operating system. UNIX and Linux are similar operating systems, the main difference being that UNIX is a proprietary system whereas Linux is open source.

Motion Pictures and Television. The motion picture industry of the twenty-first century would not be possible without the digital technology that has been developed since 1980. While live-action images are often recorded on standard photographic film in the traditional way, motion picture special effects and animation have become the exclusive realm of digital graphics technologies. The use of CGI in motion pictures has driven the development of new technology and continually raised the standards of image quality. Amalgamating live action with CGI through digital processing and manipulation enables filmmakers to produce motion pictures in which live characters interact seamlessly with virtual characters, sometimes in entirely fantastic environments. Examples of such motion pictures are numerous in the science-fiction and fantasy film genres, but the technique is finding application in all areas, especially in educational programming. In addition, motion capture technology combines graphics with recorded human movement to create animated characters.

Video Gaming and Virtual Training. The most graphics-intensive application is video gaming. All video games, in all genres, exist only as the graphic representation of complex program code. Video games can range from straightforward computer versions of simple card games to games featuring complex 3D virtual worlds.

Many graphics software applications are developed for the use of game designers, but they have also made their way into many other imaging uses. The same software that is used to create a fictional virtual world can also be used to create virtual copies of the real world. This technology has been adapted for use in pilot- and driver-training programs in all aspects of transportation. Military, police, and security personnel are given extensive practical and scenario training through the use of virtual simulators. A simulator uses video and graphic displays of actual terrain to give the person being trained hands-on experience without endangering either personnel or actual machinery.

SOCIAL CONTEXT AND FUTURE PROSPECTS

Graphics technology is inextricably linked to the computer and digital electronics industries. Accordingly, graphics technology changes at a rate that at minimum equals the rate of change in those industries. Since 1980, graphics technology using computers has developed from the display of just sixteen colors on color television screens, yielding blocky image components and very slow animation effects, to photorealistic full-motion video, with the capacity to display more colors than the human eye can perceive and to display real-time animation in intricate detail. The rate of change in graphics technology exceeds that of computer technology because it also depends on the development of newer algorithms and coding strategies. Each of these changes produces a corresponding new set of applications and upgrades to graphic technology systems, in addition to the changes introduced to the technology itself.

Each successive generation of computer processors has introduced new architectures and capabilities that exceed those of the preceding generation, requiring that applications update both their capabilities and the manner in which those capabilities are performed. At the same time, advances in the technology of display devices require that graphics applications keep pace to display the best renderings possible. All of these factors combine to produce the unparalleled value of graphics technologies in modern society and into the future. With advances in 3D computer graphics and products such as smart glasses and VR headsets, graphics technologies have become even more immersive, allowing for a continually growing range of functions from education to game play.

—Richard M. Renneboog

Further Reading

Brown, Eric. "True Physics." *Game Developer*, vol. 17, no. 5, 2010, pp. 13-18.

Costello, Vic. *Multimedia Foundations: Core Concepts for Digital Design*. 3rd ed., Focal Press, 2023.

Jimenez, Jorge, et al. "Destroy All Jaggies." *Game Developer*, vol. 18, no. 6, 2011, pp. 13-20.

Marschner, Steve, et al. *Fundamentals of Computer Graphics*. 5th ed., CRC Press, 2021.

Prust, Z. A., and Peggy B. Deal. *Graphic Communications: Digital Design and Print Essentials*. 7th ed., Goodheart-Willcox, 2022.

Wang, Yongtian, Xueming Li, and Yuxin Peng, editors. *Image and Graphics Technologies and Applications*. Springer Singapore, 2020.

Zhao, Pengfei, et al., editors. *Advanced Graphic Communications and Media Technologies*. Springer Singapore, 2017.

H

HTML

ABSTRACT

HTML, which stands for hypertext markup language, is a code used to control the functionality of web pages. It was invented in the late twentieth century and has become an indispensable part of online technology. Hypertext consists of a set of instructions for the creation of display pages on browsers, which made the World Wide Web possible. By the mid-1990s, HTML embedded codes that defined fonts, layouts, graphics, and hypertext links provided a standard protocol that allowed web page designers to distribute content to any computer.

BACKGROUND

Timothy Berners-Lee developed hypertext markup language (HTML) over several years during the 1980s while working as a software engineering consultant at the Conseil Européen pour la Recherche Nucléaire (CERN; later known as the Organisation Européen pour la Recherche Nucléaire, or the European Organization for Nuclear Research) in Geneva, Switzerland. He attempted to organize laboratory research documents and statistics from incompatible computer systems submitted by physicists from around the world. In order to pool all of these files for sharing information, Berners-Lee developed a set of formatting codes to work with hypertext protocol by linking text within the files. His invention was called hypertext.

Later in the decade, Berners-Lee invented systems that enabled computers within the same network to communicate with each other. In 1989, he proposed developing a network that would allow computers all over the world to connect. Rather than communicating by email, files could be located on a page on the web, where they could be accessed by anyone with the right technology and credentials—the code that would make this possible was HTML. Although the specific name for this language has changed significantly since its invention, its purpose has remained the same: to facilitate the smooth operation of web pages of all types.

Hypertext enables the computer user to cross-reference information and link formats together through multiple gateways on the World Wide Web. Each page is provided with a unique location address known as a universal document locator (URL). Robert Cailliau, who worked in the Office Computing Systems, Data Handling Division, at CERN, collaborated with Berners-Lee to get the web under way. Cailliau's contributions were essential to the development of the web. He rewrote the original proposal, lobbied administrators for funding, presented papers at conferences, and got programmers to work on the project.

Berners-Lee derived HTML from standard generalized markup language (SGML), an international standard that emphasized document structure and textual relationships. However, SGML proved too complex for the average web page creator, so HTML was developed as a nonproprietary format in order to embed code for text, images, and other files to make them easily accessible through the web. Berners-Lee's prototype for hypertext was the NeXT computer station, but he encouraged others to program, design, and improve software for displaying HTML documents. In 1993, Marc Andreessen, an undergraduate student at the National Center of Supercomputing Applications (NCSA) at the Univer-

Photo via iStock/virusowy. [Used under license.]

sity of Illinois, designed a web browser to display HTML documents, called Mosaic, which was widely adopted and accelerated internet traffic over the following three years.

OVERVIEW

Although the language of HTML is not necessarily second nature to all users, it is extremely systematic, with a few basic, important components. The file in which HTML code is written is called a document, which usually has an .html extension. The term HTML is in brackets that look like this——at the beginning and end of a document. The words inside the brackets are called tags. Tags are a crucial component of an HTML document.

Tags are generally in pairs. The first tag, called the start tag, looks like this: <tag>, and the second tag, called the end tag, looks like this: </tag>. Keywords, the text between tags, depend on the content of a page. Keywords may be of many different types, with many different degrees of significance. Many tags will simply designate the exact words and phrases that will appear on a page. For example, the words <title>My Favorite Job Ever</title> would probably appear at the top of the screen on the web version of an html document. This code might appear in the first paragraph under the title: <p> I have had many jobs. But one stood out as a favorite. </p>

It is important to remember that HTML is a unique language. The text within tags is read and interpreted by a computer's software, so the rules of the language must be followed. For example, the tag <!DOCTYPE html> must be used at the beginning of an HTML document to indicate the document type. When an HTML document is opened within a web browser, the language of the document guides the appearance and function of the page. The tags <title>text</title> indicate the title of a document, <body> and</body> indicate the start and end of the main text of a document, and <p>text</p> indicate the starting and ending points of a paragraph within a document. To place headers within a document, these tags must be used: <h1> and</h1>. To create headers of different sizes, these tags are used: <h2> and </h2>, <h3> and </h3>. Images to be inserted in a document are preceded and followed with the tags <img> and </img>.

The characteristics of an HTML document, such as the size of an image, are called attributes. They are usually placed between the brackets of a start tag and are followed by an equals (=) sign; the exact attribute itself is in quotes. For example, to display an image that is 600 pixels wide, this text would be used: width="600." Every element of a web page has its own designation in HTML. HTML has codes for tables, animations, different tabs within a page, and the size and shape of a page. Once an HTML document is complete, it is opened in a browser, which is when the person writing the HTML code can see whether or not the code has been properly written.

As time has passed, HTML has been modified and used in different ways, such as in mobile devices. A notable development is XHTML. The X stands for Extensible, indicating that the language can be used in a wider variety of situations than web development. As programmers continue to write code and as the internet continues to become more pervasive in daily life, the true significance of Berner-Lee's invention becomes more and more evident.

SIGNIFICANCE

The World Wide Web became a global, economic, and social phenomenon by the end of the 1990s. However, American businesses shied away from the internet in the early part of the decade, believing that it would be used only for research in the academic environment. In 1994, Berners-Lee, who was now working at the Massachusetts Institute of Technology (MIT), founded the World Wide Web Consortium (W3C) to promote guidelines regarding the growth of network infrastructure and strict language syntax with HTML, especially with the emerging browser battles between Netscape and Microsoft Explorer. HTML promoted interoperability, causing internet traffic to explode.

Search engines had difficulty trying to track all of the pages found on the web. Companies soon embraced the World Wide Web for commercial ventures, and in response, HTML continued to improve support. Browsers adopted cascading style sheets (CSS) to improve document appearance. Dynamic HTML added capabilities to respond interactively with JavaScript. Extensible markup language (XML), a relative of HTML, created additional tags to identify structures and relationships within a document. XML had two important features. First, web page creators had more flexibility to create their own tags. Second, XML separated content from formatting through the use of sophisticated style sheets, ensuring that all data structures and relationships were identified within the XML tags in which they were enclosed. XML made search engines more powerful for cataloging contents, enabling computers to become even more interactive and responsive to user actions.

—Max Winter and Gayla Koerting

Further Reading

Bell, Mary Ann, Mary Ann Berry, and James L. Van Roekel. *Internet and Personal Computing Fads*. Haworth Press, 2004.

Berners-Lee, Tim. *Weaving the Web: The Original Design and Ultimate Destiny of the World Wide Web by Its Inventor*. Harper, 1999.

Cailliau, Robert, and Helen Ashman. "Hypertext in the Web: A History." *ACM Computing Surveys*, vol. 31, Dec. 1999, pp. 1-6.

Campbell-Kelly, Martin, William F. Aspray, Jeffrey R. Yost, Honghong Tinn, and Gerardo Con Díaz. *Computer: A History of the Information Machine*. 4th ed., Routledge, 2023.

Costello, Vic. *Multimedia Foundations: Core Concepts for Digital Design*. 3rd ed., Focal Press, 2023.

"Deprecated HTML5 Tags and Attributes." *DESE*, 14 Mar. 2018, www.doe.mass.edu/nmg/html5-deprecated.html. Accessed 13 July 2023.

Frain, Ben. *Responsive Web Design with HTML5 and CSS*. 4th ed., Packt, 2022.

"HTML Tutorial." *W3Schools*, www.w3schools.com/html. Accessed 13 July 2023.

Mowery, David C., and Timothy Simcoe. "Is the Internet a US Invention? An Economic and Technological History of Computer Networking." *Research Policy*, vol. 31, Dec. 2002, pp. 1369-87.

Musciano, Chuck, and Bill Kennedy. *HTML and XHTML: The Definitive Guide*. 6th ed., O'Reilly, 2006.

Nielsen, Jakob. *Multimedia and Hypertext: The Internet and Beyond*. SunSoft Press, 1995.

HUMAN-COMPUTER INTERACTION

ABSTRACT

Human-computer interaction (HCI) is a field concerned with the study, design, implementation, evaluation, and improvement of the ways in which human beings use or interact with computer systems. The importance of HCI within the field of computer science has grown in tandem with technology's potential to help people accomplish an increasing number and variety of personal, professional, and social goals. For example, the development of user-friendly interactive computer interfaces, websites, games, home appliances, office equipment, art installations, and information distribution systems such as advertising and public awareness campaigns are all applications that fall within the realm of HCI.

BACKGROUND

Before the advent of the personal computer, those who interacted with computers were largely technology specialists. In the 1980s, however, more and more individual users began making use of software such as word-processing programs, computer games, and spreadsheets. Human-computer interaction (HCI) as a field emerged from the growing need to redesign such tools to make them practical and useful to ordinary people with no technical training. The first HCI researchers came from a variety of related fields: cognitive science, psychology, computer graphics, human factors (the study of how human capabilities affect the design of mechanical systems), and technology. Among the thinkers and researchers whose ideas have shaped the formation of HCI as a science are John M. Carroll, best known for his theory of minimalism (an approach to instruction that emphasizes real-life applications and the chunking of new material into logical parts), and Adele Goldberg, whose work on early software interfaces at the Palo Alto Research Center (PARC) was instrumental in the development of the modern graphical user interface.

In the early days of HCI, the notion of usability was simply defined as the degree to which a computer system was easy and effective to use. However, usability has come to encompass a number of other qualities, including whether an interface is enjoyable, encourages creativity, relieves tension, anticipates points of confusion, and facilitates the combined efforts of multiple users. In addition, there has been a shift in HCI away from a reliance on theoretical findings from cognitive science and toward a more hands-on approach that prioritizes field studies and usability testing by real participants.

OVERVIEW

Human-computer interaction is an interdisciplinary science with the primary goal of harnessing the full potential of computer and communication systems for the benefit of individuals and groups. HCI researchers design and implement innovative interactive technologies that are not only useful but also easy and pleasurable to use and anticipate and satisfy the specific needs of the user. The study of HCI has applications throughout every realm of modern life, including work, education, communications, health care, and recreation.

The fundamental philosophy that guides HCI is the principle of user-centered design. This philosophy proposes that the development of any product or interface should be driven by the needs of the person or people who will ultimately use it, rather than by any design considerations that center around the object itself. A key element of usability is affordance, the notion that the appearance of any interactive element should suggest the ways in which it can be manipulated. For example, the use of shadowing around a button on a website might help make it look three-dimensional, thus suggesting that it can be pushed or clicked. Visibility is closely related to affordance; it is the notion that the function of all the controls with which a user interacts should be clearly mapped to their effects. For example, a label such as "Volume Up" beneath a button might indicate exactly what it does. Various protocols facilitate the creation of highly usable applications. A cornerstone of HCI is iterative design, a method of

Photo via iStock/master1305. [Used under license.]

Photo via iStock/gremlin. [Used under license.]

development that uses repeated cycles of feedback and analysis to improve each prototype version of a product, instead of simply creating a single design and launching it immediately. To learn more about the people who will eventually use a product and how they will use it, designers also make use of ethnographic field studies and usability tests.

HOW IT WORKS

Input and Output Devices. The essential goal of HCI is to improve the ways in which information is transferred between a user and the machine he or she is using. Input and output devices are the basic tools HCI researchers and professionals use for this purpose. The more sophisticated the interaction between input and output devices—the more complex the feedback loop between the two directions of information flow—the more the human user will be able to accomplish with the machine.

An input device is any tool that delivers data of some kind from a human to a machine. The most familiar input devices are the ones associated with personal computers: keyboards and mice. Other commonly used devices include joysticks, trackballs, pen styluses, and tablets. Still more unconventional or elaborate input devices might take the shape of head gear designed to track the movements of a user's head and neck, video cameras that track the movements of a user's eyes, skin sensors that detect changes in body temperature or heart rate, wearable gloves that precisely track hand gestures, or automatic speech recognition devices that translate spoken commands into instructions that a machine can understand. Some input devices, such as the sensors that open automatic doors at the fronts of banks or supermarkets, are designed to record information passively, without the user having to take any action.

An output device is any tool that delivers information from a machine to a human. Again, the most familiar output devices are those associated with personal computers: monitors, flat-panel displays, and audio speakers. Other output devices include wearable head-mounted displays or goggles that provide visual feedback directly in front of the user's field of vision and full-body suits that provide tactile feedback to the user in the form of pressure.

Perceptual-Motor Interaction. When HCI theorists speak about perceptual-motor interaction, what they are referring to is the notion that users' perceptions—the information they gather from the machine—are inextricably linked to their physical actions, or how they relate to the machine. Computer systems can take advantage of this by using both input and output devices to provide feedback about the user's actions that will help them make the next move. For example, a word on a website may change in color when a user hovers the mouse over it, indicating that it is a functional link. A joystick being used in a racing game may exert what feels like muscular tension or pressure against the user's hand in response to the device being steered to the left or right. Ideally, any feedback a system gives a user should be aligned to the physical direction in which he or she is moving an input device. For example, the direction in which a cursor moves on screen should be the same as the direction in which the user is moving the mouse. This is known as kinesthetic correspondence.

Another technique HCI researchers have devised to facilitate the feedback loop between a user's perceptions and actions is known as augmented reality. With this approach, rather than providing the user with data from a single source, the output device projects digital information, such as labels, descriptions, charts, and outlines, on the physical world. When an engineer is looking at a complex mechanical system, for example, the display might show what each part in the system is called and enable them to call up additional troubleshooting or repair information.

APPLICATIONS AND PRODUCTS

Computers. At one time, interacting with a personal computer required knowing how to use a command-line interface in which the user typed in instructions—often worded in abstract technical language—for a computer to execute. A graphical user interface, based on HCI principles, supplements or replaces text-based commands with visual elements such as icons, labels, windows, widgets, menus, and control buttons. These elements are controlled using a physical pointing device such as a mouse. For instance, a user may use a mouse to open, close, or resize a window or to pull down a list of options in a menu in order to select one. The major advantage graphical user interfaces have over text-based interfaces is that they make completing tasks far simpler and more intuitive. Using graphic images rather than text reduces the amount of time it takes to interpret and use a control, even for a novice user. This enables users to focus on the task at hand rather than to spend time figuring out how to manipulate the technology itself. For instance, rather than having to recall and then correctly type in a complicated command, a user can print a particular file by selecting its name in a window, opening it, and clicking on an icon designed to look like a printer. Similarly, rather than choosing options from a menu in order to open a certain file within an application, a user might drag and drop the icon for the file onto the icon for the application.

Consumer Appliances. Besides computers, a host of consumer appliances use aspects of HCI design to improve usability. Graphic icons are ubiquitous parts of the interfaces commonly found on cameras, stereos, microwave ovens, refrigerators, and televisions. Smartphones such as Apple's iPhone rely on the same graphic displays and direct manipulation techniques as used in full-sized computers. Many also add extra tactile, or haptic, dimensions of usability such as touch-screen keyboards and the ability to rotate windows on the device by physically rotating the device itself in space. Entertainment products such as video game consoles have moved away from keyboard and joystick interfaces, which may not have kinesthetic correspondence, toward far more sophisticated controls. The hand-held device that accompanied the iterations of the Nintendo Wii, for instance, allowed players to control the motions of avatars within a game through the natural movements of their own bodies, and the Joy-Con controllers for the later Nintendo Switch console likewise offered motion-control capabilities. Finally, HCI research influences the physical design of many household devices. For example, a plug for an appliance designed with the user in mind might be deliberately shaped so that it can be inserted into an outlet in any orientation, based on the understanding that a user may have to fit several plugs into a limited amount of space, and many appliances have bulky plugs that take up a lot of room.

Increasingly, HCI research is helping appliance designers move toward multimodal user interfaces. These are systems that engage the whole array of human senses and physical capabilities, match particular tasks to the modalities that are the easiest and most effective for people to use, and respond in tangible ways to the actions and behaviors of users. Multimodal interfaces combine input devices for collecting data from the human user (such as video cameras, sound recording devices, and pressure sensors) with software tools that use statistical analysis or artificial intelligence to interpret these data (such as natural language processing programs and computer vision applications). For example, a multimodal interface for a global positioning system (GPS) installed in an automobile might allow the user to simply speak the name of a destination aloud rather than having to type it in while driving. The system might use auditory processing of the user's voice as well as visual processing of their lip movements to interpret speech more accurately. It might also use a camera to closely follow the movements of the user's eyes, tracking their gaze from one part of the screen to another and using this information to helpfully zoom in on particular parts of the map or automatically select a particular item in a menu.

Similarly, in 2015, Amazon took the technology of the Bluetooth speaker one step further with its release of the Echo device. This speaker has a built-in program that allows the user to give voice commands to instruct the device to play certain music or to sync up with other applications and devices. In the 2020s, Amazon continued to expand upon the functionality of this program and the devices with which it is compatible. With greater interest in the use of digital assistants to create a "smarter" home, Amazon released new iterations of the Echo (including the smaller Echo Dot), and other devices were made to support this digital assistant to allow people to work lighting, temperature, and even door locks and more through them. As further proof of their value, digital assistants and accompanying devices were also developed by companies such as Google and Apple by that time.

Workplace Information Systems. HCI research plays an important role in many products that enable people to perform workplace tasks more effectively. For example, experimental computer systems have been designed for air traffic control to increase safety and efficiency. Such systems work by collecting data about the operator's pupil size, facial expression, heart rate, and the forward momentum and intensity of his or her mouse movements and clicks. This information helps the computer interpret the operator's behavior and state of mind and respond accordingly. When an airplane drifts slightly off its course, the system analyzes the operator's physical modalities. If their gaze travels quickly over the relevant area of the screen, with no change in pupil size or mouse click intensity, the computer might conclude that the operator has missed the anomaly and attempt to draw attention to it by using a flashing light or an alarm.

Other common workplace applications of HCI include products that are designed to facilitate communication and collaboration between team members, such as instant messaging programs, wikis (collaboratively edited websites), and videoconferencing tools. In addition, HCI principles have contributed to many project management tools that enable groups to schedule and track the progress they are making on a shared task or to make changes to common documents without overriding someone else's work.

Education and Training. Schools, museums, and businesses all make use of HCI principles when designing educational and training curricula for students, visitors, and staff. For example, many school districts have been moving away from printed textbooks and toward interactive electronic programs that target a variety of information-processing modalities through multimedia. Unlike paper and pencil worksheets, such programs also provide instant feedback, making it easier for students to learn and understand new concepts. Businesses use similar programs to train employees in such areas as the use of new software and the company's policies on issues of workplace ethics. Many art and science museums have installed electronic kiosks with touch screens that visitors can use to learn more about a particular exhibit. HCI principles underlie the design of such kiosks. For example, rather than using a text-heavy interface, the screen on an interactive kiosk at a science museum might display video of a museum staff member talking to the visitor about each available option.

CAREERS AND COURSEWORK

The paths toward becoming an HCI professional are extraordinarily varied. Bachelor's degrees in cognitive science, neuroscience, computer science, graphic design, psychology, engineering, art, and many other fields could serve as appropriate preparation for a career as someone who uses HCI principles. No matter which concentration an aspiring HCI researcher or student chooses, it is important to acquire basic programming skills, a broad understanding of human psychophysiology, and some practical experience or training with either graphic design or product design. Common areas of work include developing websites; computer operating systems; interfaces for consumer appliances such as cell phones, printers, or cameras; and educational materials such as interactive employee training courses, advertising campaigns, or any other applications that demand accessible, learnable, and usable computer systems. Although a graduate degree is not required for entering the field, many universities offer specialized master's programs in HCI.

SOCIAL CONTEXT AND FUTURE PROSPECTS

As HCI moves forward with research into multimodal interfaces and ubiquitous computing, notions of the computer as an object separate from the user may eventually be relegated to the archives of technological history, to be replaced by wearable machine interfaces that can be worn like clothing on the user's head, arm, or torso. Apple released its eighth version of a "smartwatch" in 2022, which is designed to have all of the features of smartphones, and more, in a wearable, theoretically more convenient format. Much like other wearable gadgets such as the Fitbit, playing into society's increased concern with exercise and overall health, the watch has the ability to track human components such as heart rate and blood oxygen levels and serve as a GPS that can map running, walking, and biking routes. Virtual reality interfaces have been developed that are capable of immersing the user in a 360-degree space that looks, sounds, feels, and perhaps even smells like a real environment—and with which they can interact naturally and intuitively, using their whole bodies. As the capacity to measure the physical properties of human beings becomes ever more sophisticated, input devices may grow more and more sensitive; it is possible to envision a future, for instance, in which a machine might "listen in" to the synaptic firings of the neurons in a user's brain and respond accordingly. Indeed, it is not beyond the realm of possibility that a means could be found of stimulating a user's neurons to produce direct visual or auditory sensations. The future of HCI research may be wide open, but its essential place in the workplace, home, recreational spaces, and the broader human culture is assured. As further evidence of the significance of HCI and its place in modern technology, the 2020, 2021, and 2022 editions of the International Conference on Human-Computer Interaction took place virtually due to the COVID-19 pandemic, while the 2023 conference was held as a hybrid-format event.

—M. Lee

Further Reading

Augstein, Mirjam, Eelco Herder, and Wolfgang Wörndl, editors. *Personalized Human-Computer Interaction*. 2nd ed., De Gruyter Oldenbourg, 2023.

Bainbridge, William Sims, editor. *Berkshire Encyclopedia of Human-Computer Interaction: When Science Fiction Becomes Fact*. Berkshire, 2004. 2 vols.

Bogdon, Razuan, et al. "A Practical Experience on the Amazon Alexa Integration in Smart Offices." *MDPI*, 22 Jan. 2021, www.mdpi.com/1424-8220/21/3/734. Accessed 13 July 2023.

Helander, Martin. *A Guide to Human Factors and Ergonomics*. 2nd ed., CRC Press, 2006.

Hughes, Brian. "Technology Becomes Us: The Age of Human-Computer Interaction." *Huffington Post*, 20 Apr. 2016, www.huffingtonpost.com/brian-hughes/technology-becomes-us-the_b_9732166.html. Accessed 13 July 2023.

Jacko, Julie A., editor. *The Human-Computer Interaction Handbook: Fundamentals, Evolving Technologies, and Emerging Applications*. 3rd ed., CRC Press, 2012.

Jokinen, Jussi P. P. "Emotional User Experience: Traits, Events, and States." *International Journal of Human-Computer Studies*, vol. 76, 2015, pp. 67-77.

Pressman, Aaron. "Alexa Turns 5: What Amazon's Kindergarten-Aged Assistant Can Teach the Rest of Tech." *Fortune*, 6 Nov. 2019, fortune.com/2019/11/06/amazon-alexa-echo-5-anniversary-dave-limp-interview. Accessed 13 July 2023.

Purchase, Helen C. *Experimental Human-Computer Interaction: A Practical Guide with Visual Examples*. Cambridge UP, 2012.

Rogers, Yvonne, Helen Sharp, and Jennifer Preece. *Interaction Design: Beyond Human-Computer Interaction*. 6th ed., Wiley, 2023.

Soegaard, Mads, and Rikke Friis Dam, editors. *The Encyclopedia of Human-Computer Interaction*. 2nd ed., Interaction Design Foundation, 2014.

Thatcher, Jim, et al. *Web Accessibility: Web Standards and Regulatory Compliance*. Springer, 2006.

Tufte, Edward R. *The Visual Display of Quantitative Information*. 2nd ed., Graphics, 2007.

I

Image Editing

ABSTRACT

Image-editing software uses computing technology to change digital images. Image editing can involve altering the appearance of an image, such as showing a hot-air balloon underwater, or improving the quality of a low-resolution image. Images may also be compressed so that they require less computer storage space.

BACKGROUND

There are as many ways of digitally altering images as there are uses for digital art. The first step in image editing is to obtain an image in digital format. The easiest method is to use a digital camera to take a photograph and then transfer the photograph to a computer for editing. Another approach that is occasionally necessary is to scan a print photograph or film negative. This converts the photo to a digital image ready for editing. It is also possible to create an image by hand in native digital format, such as by using a tablet and stylus to draw or paint. Finally, rendering makes it possible for a computer to produce a digital image from a two-dimensional (2D) or three-dimensional (3D) model.

Once the digital image is available, the next step is to determine what will be done to it. Most often, the image will be enhanced (improving the image quality through interpolation or other techniques), compressed (decreasing the file size by sacrificing some image quality or clarity), or altered (made to depict something that was not originally there). These changes may be through destructive or nondestructive editing. In destructive editing, the changes are applied to the original file. By contrast, in nondestructive editing, they are saved in a separate version file.

OVERVIEW

In the early days of the internet, image compression was an especially important type of image editing. Bandwidth was limited then, and it could take several minutes to transmit even a medium-sized image file. Image compression algorithms were invented to help reduce the size of these files, with some loss of quality. Lossless compression can avoid degradation of the image, but in most cases, it does not reduce file size as much as lossy compression does.

Computers store image data as sets of numeric values. Each pixel onscreen is lit in a particular way when an image is displayed, and the colors of each pixel are stored as numbers. For example, if the color black were represented by the number eight, then anywhere in a picture that has three black pixels in a row would be stored as 8, 8, 8. Because an image is composed of thousands of pixels, all of the numbers needed to describe the colors of those pixels, when combined, take up a lot of storage space. One way to store the same information in less space is to create substitutions for recurring groups of numbers. The symbol *q1* could be used to represent three black pixels in a row, for instance. Thus, instead of having to store three copies of the value "8" to represent each of the three black pixels, the computer could simply store the two-letter symbol *q1*, thus saving one-third of the storage space that otherwise would be required. This is the basis for how digital images are compressed.

Most images are compressed using lossy compression algorithms, such as the Joint Photographic Ex-

Before and after editing. Photo via iStock/Hdesert. [Used under license.]

perts Group (JPEG) standard. Compression thus usually requires the sacrifice of some image quality. For most purposes, the reduction in quality is not noticeable and is made up for by the convenience of more easily storing and transmitting the smaller file. It is not uncommon for compression algorithms to reduce the file size of an image by 75 to 90 percent, without noticeably affecting the image's quality.

Image enhancement typically relies on the mathematical adjustment of the numeric values that represent pixel hues. For instance, if an image editor were to desaturate a photograph, the software would first recognize all of the pixel values and compare them to a grayscale value. It would then interpolate new values for the pixels using a linear operation. Similarly, a filtering algorithm would find and apply a weighted average of the pixel values around a given pixel value in order to identify the new color codes for each pixel being adjusted. The median or the mode (most common) value could also be used. The type of filter being applied determines which mathematical operation is performed. Filters are often used to correct for noise, or unwanted signal or interference.

Image editing is possible on multiple platforms, including both personal computers (PCs) and mobile devices such as smartphones and tablets. Certain programs work only on PCs, others strictly on mobile devices, and still others on both. By the early 2020s, popular image-editing programs and applications included Adobe products such as Photoshop, Lightroom, and the online tool Adobe Express; the open-source software Gimp; and the design tool Canva, which was accessible through both web browsers and a dedicated mobile app. Some social media apps, such as Instagram, also offered limited image-editing capabilities.

—*Scott Zimmer*

Further Reading

Freeman, Michael. *Digital Image Editing & Special Effects: Quickly Master the Key Techniques of Photoshop & Lightroom*. Focal, 2013.

Gonzalez, Rafael C., and Richard E. Woods. *Digital Image Processing*. 4th ed., Pearson, 2019.

Hardy, David Leicester. *Introduction to Digital Media Design: Transferable Hacks, Skills, and Tricks*. Bloomsbury Visual Arts, 2022.

Holleley, Douglas. *Photo-Editing and Presentation: A Guide to Image Editing and Presentation for Photographers and Visual Artists*. Clarellen, 2009.

Prust, Z. A., and Peggy B. Deal. *Graphic Communications: Digital Design and Print Essentials*. 7th ed., Goodheart-Willcox, 2022.

Xue, Su. *Data-Driven Image Editing for Perceptual Effectiveness*. Yale UP, 2013.

Imgur

ABSTRACT

Online media sites that enable posting of pictures and videos have become very popular. Such sites as Facebook, TikTok, and Instagram are well-known venues for pictures, videos, and other features of digital communication. Imgur, introduced in 2009, was among the most popular image sharing and hosting services of the early twenty-first century.

BACKGROUND

Alan Schaaf developed Imgur in 2009, while studying computer science at Ohio University. His goal was to produce an online image sharing and hosting service that could avoid the usability problems users were encountering with other such services. Initially, his service held only pictures and images that had gained a lot of traction on other sites such as Reddit. Traction is acquired by garnering high numbers of views and popularity votes, or even going viral on those sites.

According to Schaaf, his Imgur (which may be pronounced as imager) site became very popular very quickly, garnering a million total page views within five months of its release. The service was financed initially by donations toward the cost of hosting the service's website. Revenue from display ads, introduced in 2009, added to the level of support funding, and this was further augmented by the introduction of sponsored images and self-service ads in 2013.

The first years of Imgur's existence were full of change. Three different web-hosting services were used in the first year alone, until Schaaf settled the site with Voxel as the hosting service. That lasted until late 2011, when the hosting service was switched again from Voxel to Amazon Web Services. In 2012, Imgur introduced direct image sharing so that users could share images directly to Imgur without the need to first become popular on Reddit and other sites. This followed the move of the company from Ohio to San Francisco in 2011.

The success of the service earned it the 2012 Crunchies Award as the Best Bootstrapped Startup, and in June 2013, Imgur had a total of 40 employees. In 2014, Imgur raised a $40 million-funding boost from Andriessen Horowitz, which was subsequently represented on Imgur's board of directors by Lars Dalgaard. Imgur's popularity and success was, at the time, generating good profits from advertising revenue and the sale of "Pro" subscriptions. The nature of business competition, however, guarantees that profitability is fleeting. When Reddit introduced native image hosting in 2016, the number of submissions to Imgur decreased by a substantial amount, and Imgur's popularity, as listed in Alexa's Top Sites in the United States, dropped from 16th place in 2016 to 51st over the next five years. In 2021, Imgur was acquired by MediaLab AI Incorporated, its current owner. As of January 2022, Imgur had 300 million users, with 85 percent of those users identifying as male, and 35 percent of all users being younger than thirty-five years old.

In April of 2023, Imgur announced plans to delete large quantities of content from its site, including both explicit images and content that the site's leadership described as "old, unused, and inactive." The latter announcement sparked widespread concerns among online communities that had long used Imgur as an image-hosting source, as the deletion of old content could create broken links and render the images themselves unviewable.

OVERVIEW

A logical way of organizing images is to arrange them in galleries, a feature that became part of the Imgur structure as Imgur Accounts in 2010. This was improved by the addition of Albums later the same year. The user account provides complete image management that includes image editing, deletion, creation and embedding of Albums, and allows users to comment on viral images and submit them to the public gallery. Users could then view their past public activity as well, much like Facebook's personal timelines.

With paid Pro Imgur accounts, added in 2010, users were not subject to image upload restrictions and could store an unlimited number of images. Initially, images posted to Imgur were hosted indefinitely unless they went for a period of three months without being viewed, at which time they would become subject to removal from the site. In 2015, Imgur chose instead to host all images permanently unless the poster specifically requested removal, even if they had not been posted from a Pro account. This policy changed with the announcement of planned deletions in 2023.

Memes are popular images that are often posted to media sites. A meme is typically a regular image or screen grab that has a pithy quote or a humorous comment applied to it. In June 2013, Imgur introduced a built-in meme generator that allows users to create and share their own memes for posting.

The public gallery in Imgur collects the most viral of images from around the web. It uses an algorithm that calculates the number of times an image has been viewed, the number of times it has been shared, and the number of votes it has received from Imgur users relative to time. Users have been allowed to browse the entire public gallery at random since 2012. Since 2014, Imgur has automatically converted animated .gif files to WebM and MP4 video files, and beginning in 2015, users became able to link video URLs to create .gif files directly via Imgur. The Imgur's Topics utility, also implemented in 2015, allowed users to search for and view specific images according to the tags that have been applied to them. Mobile apps for iOS and Android were also introduced in 2015.

In use, the Imgur website opens directly to a simple landing screen displaying public content that users can scroll through indefinitely. This is reminiscent of an individual's Timeline in Facebook but is much broader in scope, and presumably in extent as well. The image stream is interspersed with unobtrusive ads that scroll along with the image stream. Visitors have the choice of opening the website with specific purposes or ordering of the image stream. A visitor can also open the website at the sign-in page to either log into their existing account or to open a new account.

—Richard M. Renneboog and Jake D. Nicosia

Further Reading

"The Magic of the Internet." *Imgur*, imgurinc.com. Accessed 13 July 2023.

Pollard, Barry. *HTTP/2 in Action*. Simon & Schuster, 2019.

Purdy, Kevin. "Hosting Site Imgur Will Remove Explicit and Anonymous Content Next Month." *Ars Technica*, 20 Apr. 2023, arstechnica.com/gadgets/2023/04/hosting-site-imgur-will-remove-explicit-and-anonymous-content-next-month. Accessed 13 July 2023.

Smith, Craig. "Imgur Statistics and User Count (2023)." *DMR*, 8 Mar. 2023, expandedramblings.com/index.php/imgur-statistics. Accessed 13 July 2023.

Summers, Danny. *Imgur: 42 Success Secrets—42 Most Asked Questions—Imgur: What You Need to Know*. Emereo Pty Ltd., 2014.

Swarts, Jason. *Wicked, Incomplete and Uncertain: User Support in the Wild and the Role of Technical Communication*. UP of Colorado, 2018.

INDUSTRIAL DESIGNER

ABSTRACT

Industrial designers develop the concepts for a wide range of manufactured products. They work closely with clients and other professionals, including engineers, production experts, and market research analysts. Students aspiring to enter the profession should pursue studies in areas such as drawing, computer-aided design and drafting (CADD), and three-dimensional (3D) modeling.

BACKGROUND

Industrial designers develop the concepts for manufactured products, such as cars, home appliances, and toys. They combine art, business, and engineering to make products that people use every day. Industrial designers consider the function, aesthetics, production costs, and usability of products when developing new product concepts.

OVERVIEW

Industrial designers primarily work in offices, but may also travel to meet with clients, as well as locations where the product is manufactured. Industrial designers are both creative and technically minded individuals who want to use their skills to create the next generation of products that people use on a daily basis. Their interests may vary widely, from creating toys to state-of-the-art home appliances that involve complex inner workings such as artificial intelligence (AI) components.

DUTIES AND RESPONSIBILITIES

The duties and responsibilities of an industrial designer typically include consulting with clients to determine requirements for designs; researching the various ways a particular product might be used, and who will use it; examining materials and manufacturing requirements to determine production costs; evaluating product safety, appearance, and function to determine if a design is practical; and presenting designs to clients for approval. Industrial designers may also create sketches or renderings of their designs, use computer software to develop virtual models of different designs, and create physical prototypes.

Some industrial designers focus on a particular product category. For example, they may design medical equipment or work on consumer electronics products, such as computers and smart phones. Other designers develop ideas for products such as new bicycles, furniture, housewares, and snowboards.

Other designers, sometimes called user interface designers or interaction designers, focus on the usability of a product, such as an electronic device, and ensure that the product is both simple and enjoyable to use, even if its components are complex.

WORK ENVIRONMENT

Physical Environment. Work spaces for industrial designers often include work tables for sketching designs, meeting rooms with whiteboards for brainstorming with colleagues, and computers and other office equipment for preparing designs and communicating with clients. Although industrial designers work primarily in offices, they may travel to testing facilities, design centers, clients' exhibit sites, users' homes or workplaces, and places where the product is manufactured.

Industrial designers who are self-employed or work for firms that hire them out to other organizations may need to adjust their workdays frequently in order to meet with clients in the evenings or on weekends. In addition, they may spend some of their time looking for new projects or competing with other designers for contracts.

Human Environment. Industrial designers imagine how consumers might use a product and test different designs with consumers to see how each design looks and works. Industrial designers often work with engineers, production experts, and market research analysts to find out if their designs are feasible. They apply the input from their colleagues' professional expertise to further develop their designs. For example, industrial designers may work with market research analysts to develop plans to market new product designs to consumers.

Technological Environment. Computers are a major tool for industrial designers. Industrial designers use two-dimensional (2D) computer-aided design and drafting (CADD) software to sketch ideas, because computers make it easy to make changes and show alternatives. 3D CAD software can be used by industrial designers as a tool to transform their 2D designs into physical models with the help of 3D printers. If they work for manufacturers, they also may use computer-aided industrial design (CAID) software to create specific machine-readable instructions that tell other machines exactly how to build the product. Depending on their area of expertise, they may also need to understand at least the basics of advanced components such as AI.

EDUCATION AND TRAINING

High School/Secondary. Industrial designers must be well-rounded in their grasp of both arts and science, so high school students interested in pursuing this field should take a mix of both, as well as English to build communication and writing skills. Classes in mathematics may include algebra, trigonometry, and calculus, and sciences may include computer science, chemistry, and physics.

College/Postsecondary. A bachelor's degree is usually required for entry-level industrial design jobs. Common fields of degree include fine arts, engineering, or architecture. Most industrial design programs include courses in drawing, CADD, and 3D modeling, as well as courses in business, industrial materials and processes, and manufacturing methods.

The National Association of Schools of Art and Design accredits more than three hundred postsecondary colleges, universities, and independent institutes with programs in art and design. Many schools require successful completion of some basic art and design courses before granting entry into a bachelor's degree program. Applicants also may need to submit sketches and other examples of their artistic ability.

Many programs provide students with the opportunity to build a professional portfolio of their designs from classroom projects, internships, or other experiences. Students can use these examples of their work to demonstrate their design skills when applying for jobs and bidding on contracts for work.

Adult Job Seekers. Adults seeking to enter the field of industrial design will stand the best chance if they come from a related background, such as visual art, engineering, or architecture. Coursework may be required in order to learn the latest industry standards and requirements, including technology such as computerized design programs.

—*Stuart Paterson*

Further Reading

The Design Book. Rev. ed., Phaidon, 2020.

Fiell, Charlotte, and Peter Fiell. *Industrial Design A-Z.* Taschen, 2016.

Ulrich, Karl, Steven Eppinger, and Maria C. Yang. *Product Design and Development.* 7th ed., McGraw-Hill, 2020.

Information Visualization

ABSTRACT

Information visualization, also known as data visualization, helps the human eye perceive huge amounts of data in an understandable way. It enables researchers to represent, draft, and display data visually in a way that allows humans to identify, visualize, and remember it more efficiently. Techniques of information visualization take advantage of the broad bandwidth pathway of the human eye. Information visualization is an indispensable part of applied research and problem solving, helping people graphically display, spatially organize, and locate huge amounts of data and information to facilitate fast and efficient categorization, classification, organization, and systematization of information.

BACKGROUND

A number of fields and disciplines have contributed to the formation of information visualization, including computer science, psychology, visual design, human-computer interaction, graphics, and business methods. Computer graphics gave rise to the emergence of information visualization and its extensive use in scientific problem solving and research. The 1987 special issue of *Computer Graphics* on visualization in scientific computing heralded the emergence of information visualization as a scientific graphical technique and opened up vast possibilities for researchers and explorers.

OVERVIEW

Both quantitative and qualitative methods of research use information visualization techniques extensively. Graphs, charts, scatterplots, box plots, tables, diagrams, maps, three-dimensional figures, pictograms, schemes, tree diagrams, and coordinate plots are just few examples of information visualization techniques that help illustrate data. Information visualization reinforces human cognition by intuitively representing abstract information, including both numerical and nonnumerical data, textual information, and geographical dislocation.

Statistical research uses information visualization most vividly. The results of hypothesis testing, regression analysis, and instrumental variables analysis are all displayed via information visualization techniques to make the numbers and figures meaningful for data interpretation, analysis, and relevant decision making. Data grouping, clustering, classifying, and organizing are other examples of data mining and visual presentation.

Information visualization helps a human eye creatively structure and order seemingly unstructurable information. It yields unexpected insights into a vast array of data and opens up the orders and systems hidden from the human eye while gathering the information. Information visualization also facilitates the formation of a hypothesis-generating scheme, which is often followed by a more complex method of hypothesis testing.

Information visualization is used in scientific research, data representation, digital libraries, data mining, financial data analysis, social data analysis, and many other fields. Cloud-based visualization services, digital media, computer-mediated communication, dynamic interactive representations, cognitive ethnography, multiscale software, and video data analysis are all fast-growing and developing fields that use information visualization extensively. In addition, effective system design is based on dynamic interactive representations, and scientists refer to information visualization in order to understand their cognitive and computational characteristics. Information visualization is also increasingly used in works of art, and its principles are applied to exhibit displays and configurations in modern art museums throughout the world.

—*Mariam Orkodashvili*

Further Reading

Freeman, Eric, and David Gelernter. "Lifestreams: A Storage Model for Personal Data." *SIGMOD Record*, vol. 25, no. 1, 1996, pp. 80-86.

Gong, Yihong. *Intelligent Image Databases: Towards Advanced Image Retrieval*. Kluwer, 1998.

Healy, Kieran. *Data Visualization: A Practical Introduction*. Princeton UP, 2018.

Koblin, Aaron. "Q&A: Aaron Koblin: The Data Visualizer." Interview by Jascha Hoffman. *Nature*, vol. 486, no. 7401, 2012, p. 33.

Lima, Manuel. *Visual Complexity: Mapping Patterns of Information*. Princeton Architectural, 2013.

Ware, Colin. *Visual Thinking for Information Design*. 2nd ed., Morgan Kaufmann, 2021.

Wilke, Claus. *Fundamentals of Data Visualization: A Primer on Making Informative and Compelling Figures*. O'Reilly, 2019.

INSTAGRAM

ABSTRACT

Instagram (sometimes abbreviated as IG) is a photograph-sharing and social media mobile application (app) for smartphones and tablets that launched in October 2010. Instagram also supports videos and in 2021 introduced Instagram Reels as a competitor to the popular video-sharing app TikTok.

BACKGROUND

Software engineer Kevin Systrom developed the basis of Instagram while attending Stanford University in Stanford, California. After working a day job, Systrom learned how to code and built a hypertext markup language (HTML) prototype app called Burbn. This app allowed users to check in at locations, make plans, and earn points for hanging out with other users and posting pictures.

Systrom met backers from venture capital investment firms Baseline Ventures and Andreessen Horowitz. Two weeks after meeting with these investors, Systrom had raised $500,000. Engineer Mike Krieger was brought on as a partner, and a team of other engineers was assembled. Systrom and Krieger decided that if they wished to establish themselves as a legitimate company among the many other app developers, they would have to focus on their app performing one task very well. They chose to focus on the mobile photograph aspect of Burbn and spent eight weeks developing a new photography app, which they called Instagram, a combination of the words, "instant" and "telegram."

After beta testing and fixing some malfunctions of the app, Instagram was launched on October 6, 2010. Within hours, over ten thousand users had downloaded the app, and by the end of the first week, it had over 100,000 downloads. By the end of 2010, over a million users were using the app; within four years, that number reached approximately 200 million active monthly users and 20 billion photos shared.

In February 2011, Instagram was valued at $20 million and was receiving a great amount of media coverage. Investors continued to finance Instagram,

Image via Wikimedia Commons. [Public domain.]

and in April 2012 Facebook (later renamed Meta) acquired the company for $1 billion in cash and stock. A decade later, Instagram's brand value would be estimated in excess of $30 billion.

OVERVIEW

For several years, Instagram went through only slight changes in its features. Users can take a photograph, add a filter to it to give it a desired effect, share it, and "like" and comment on other users' photographs. In June 2014, the app added ten new features so that users could modify their photographs even further. These additional features allow users to adjust the contrast, brightness, and saturation of their photographs. In addition, they can also use a slide bar to adjust the intensity of the preset filters. For these additional features, Instagram engineers examined film cameras to study how light is affected by different lenses. In August 2014, Instagram introduced the Hyperlapse app, which allowed users to post time-lapse videos, and the following year, the company introduced the Boomerang app, which facilitated the creation of looping videos. Both apps remained available until early 2022, when Instagram delisted them from multiple app stores.

In August 2015, through the company blog, Instagram announced its biggest change up until that point. For the first several years of the app's existence, Instagram exclusively permitted the posting of photographs in a square shape, rather than the 4:3 aspect ratio typically found on smartphone cameras. Despite the app's beginnings as a program dedicated to preserving the nostalgia of the historical square photograph format, the company acknowledged that a large number of photos are rectangular, causing frustrated users to turn to outside apps to format for posting. With the change to allow the posting of rectangular photos, users can share both photos and videos in portrait and landscape orientation.

Just as Facebook had changed its "feed" method in 2009, Instagram also followed suit in March 2016 and began testing an algorithm that arranges items in a user's feed according to photos users would most want to see based upon who they follow rather than in reverse chronological order. In further recognition of such transformations, the app instituted another large and controversial alteration when it presented a new logo. The company replaced its iconic retro camera logo with a more modern pink, purple, and orange logo, which a spokesperson explained is meant to capture the "diverse storytelling" for which the app has come to serve as a platform. These changes were met with mixed reviews from users and commentators.

In 2016, Instagram introduced the ability to create Stories, collections of photos and videos that expire after a day. This functionality is similar to that of the app Snapchat, one of Instagram's competitors. The feature quickly became popular; Instagram stated in 2017 that over 250 million people used Stories every day.

Instagram announced in June 2018 that the company would soon be launching a long-form video hub similar to YouTube. The resulting standalone app, IGTV, was launched in June of 2018 and remained in use until it was shut down in 2022. Instagram delved further into video content with the introduction of Reels, developed in 2020 to compete with TikTok. Users post Reels, or short videos, for a variety of informational and entertainment purposes. This allows content creators to reach more users and to transmit their messages to larger and more engaged audiences. While Reels were initially limited to fifteen seconds in length, Instagram increased that limit several times in the subsequent years and by mid-2022 permitted a maximum Reel length of ninety seconds.

CULTURAL INFLUENCE

After its launch, Instagram users quickly adapted the use of hashtags to enhance their photograph commenting. The hashtag is a word or words grouped together without spaces and prefixed with the hash (#) symbol. Also popular on social networking services such as Twitter, hashtags enable users to search for and locate posts concerning a specific topic.

One of the most popular hashtags used on Instagram is #selfie, a term for when people use their smartphone's camera to photograph themselves. By December 2014 more than 200 million photos on Instagram were tagged "#selfie." *The Oxford English Dictionary* even named "selfie" its 2013 word of the year. Numerous studies have been undertaken concerning their effect on ego and self-image. For example, some believe that since Instagram is image-driven, it can have a negative effect on self-esteem when users do not "like" a person's selfie.

Another prevalent hashtag is #TBT, an abbreviation for "Throwback Thursday." Using this hashtag, users post old photographs of themselves. Posting baby photographs is a popular practice with this hashtag. Other frequently used hashtags on Instagram include #WCW ("Woman Crush Wednesday," in which users post photographs of women they admire), #MCM ("Man Crush Monday," in which users post photographs of men they admire), and #blessed, in which users post photographs of people or things for which they feel grateful.

Numerous celebrities use Instagram, and many of them take frequent selfies, which some attribute to the rise in the hashtag's popularity. As of July 2023, the top five most-followed celebrity accounts on Instagram were those of soccer players Cristiano Ronaldo and Lionel Messi, singer and actor Selena Gomez, television personality Kylie Jenner, and actor and retired wrestler Dwayne "The Rock" Johnson. Celebrities commonly use their accounts to offer a glimpse into their private lives, communicate with fans, and promote causes they support.

Many major brands also use Instagram for marketing purposes. Retailers, restaurant chains, and sports teams are examples of some of the types of brands that widely use Instagram to market themselves. Ways of marketing that have been embraced on Instagram include posting behind-the-scenes photographs, holding contests in which users are asked to repost photographs and hashtags, promoting new products, and displaying customers using their products. In 2016, Instagram began adding "shoppable tags," small icons of a shopping bag appearing on advertising posts that users can click to find out how to buy products depicted in that post, to posts made by a small number of corporate partners. Use of shoppable tags expanded over the next several years, and in 2018 they were introduced into Stories.

Instagram has also made an impact on the world of photography. It widely popularized the art of photography and exposed people who did not regularly take photographs to it. While some see this as a positive movement, some professional photographers believe Instagram has undermined the true art of photography and created a homogenized brand of photograph.

Some critics have identified Instagram as harmful to mental health, particularly that of young people. In May 2017, a survey of nearly fifteen hundred people between the ages of fourteen and twenty-four, conducted by the UK Royal Society of Public Health, found that of all social networking apps, Instagram was the most detrimental for young people's mental health. The survey, called "#StatusofMind," reported that Instagram negatively affected body image (especially among young women) and sleep patterns and contributed to "fear of missing out," or FOMO, a term coined by Patrick McGinnis in 2004. The author of the survey report, Matt Keracher, noted that young women who use Instagram are more driven to "compare themselves against unrealistic, largely curated, filtered, and photoshopped versions of reality." In response, the report called for social media platforms to identify images that have been digitally altered. Despite these negative effects, however, the survey also found that Instagram had a positive effect on self-expression and self-identity for many of those surveyed.

—Patrick G. Cooper and Laura Nicosia

Further Reading

Aslam, Salman. "Instagram by the Numbers: Stats, Demographics & Fun Facts." *Omnicore*, 21 June 2017, www.omnicoreagency.com/instagram-statistics. Accessed 13 July 2023.

Bevan, Kate. "Instagram Is Debasing Real Photography." *Guardian*, 19 July 2012, www.theguardian.com/technology/2012/jul/19/instagram-debasing-real-photography. Accessed 13 July 2023.

Bradford, Alina. "Everything You Need to Master Instagram Stories." *CNet*, 24 Apr. 2018, www.cnet.com/how-to/how-to-use-instagram-stories. Accessed 13 July 2023.

"Brand Value of Instagram Worldwide from 2020 to 2022." *Statista*, May 2022, www.statista.com/statistics/1324427/instagram-brand-value. Accessed 13 July 2023.

Constine, Josh. "Instagram Plans June 20th Launch Event for Long-Form Video Hub." *TechCrunch*, 11 June 2018, techcrunch.com/2018/06/11/instagram-long-form-video. Accessed 13 July 2023.

Constine, Josh, and Kim-Mai Cutler. "Facebook Buys Instagram for $1 Billion, Turns Budding Rival into Its Standalone Photo App." *TechCrunch*, 9 Apr. 2012, techcrunch.com/2012/04/09/facebook-to-acquire-instagram-for-1-billion. Accessed 13 July 2023.

Coulthard, Charissa. "Self-Portraits and Social Media: The Rise of the 'Selfie.'" *BBC News*, 7 June 2013, www.bbc.com/news/magazine-22511650. Accessed 13 July 2023.

Fox, Kara. "Instagram World Social Media App for Young People's Mental Health." *CNN*, 19 May 2017, www.cnn.com/2017/05/19/health/instagram-worst-social-network-app-young-people-mental-health/index.html. Accessed 13 July 2023.

Gillett, Rachel. "How the Most Successful Brands Dominate Instagram, and You Can Too." *Fast Company*, 22 Apr. 2014, www.fastcompany.com/3029395/how-the-most-successful-brands-dominate-instagram-and-you-can-too. Accessed 13 July 2023.

Hamburger, Ellis. "Instagram's New Editing Features Could Make It Your Only Photo App." *Verge*, 3 June 2014, www.theverge.com/2014/6/3/5771860/instagram-new-editing-features-could-make-it-your-only-photo-app-vsco-cam-afterlight. Accessed 13 July 2023.

Hatmaker, Taylor. "Instagram Adds Shopping Tags Directly into Stories." *TechCrunch*, 12 June 2018, techcrunch.com/2018/06/12/instagram-adds-shopping-tags-directly-into-stories. Accessed 13 July 2023.

Hutchinson, Andrew. "Instagram Announces Closure of Separate IGTV App, Removal of In-Stream Video Ads." *SocialMediaToday*, 28 Feb. 2022, www.socialmediatoday.com/news/instagram-announces-closure-of-separate-igtv-app-removal-of-in-stream-vide/619532. Accessed 13 July 2023.

_____. "Instagram Has Removed Its Separate Boomerang and Hyperlapse Apps from App Stores." *SocialMediaToday*, 7 Mar. 2022, www.socialmediatoday.com/news/instagram-has-removed-its-separate-boomerang-and-hyperlapse-apps-from-app-s/619945. Accessed 13 July 2023.

"Instagram Accounts with the Most Followers Worldwide as of July 2023." *Statista*, July 2023, www.statista.com/statistics/421169/most-followers-instagram. Accessed 13 July 2023.

Isaac, Mike. "Instagram May Change Your Feed, Personalizing It with an Algorithm." *New York Times*, 15 Mar. 2016, www.nytimes.com/2016/03/16/technology/instagram-feed.html. Accessed 13 July 2023.

Markowitz, Eric. "How Instagram Grew from Foursquare Knock-Off to $1 Billion Photo Empire." *Inc.*, 10 Apr. 2012, www.inc.com/eric-markowitz/life-and-times-of-instagram-the-complete-original-story.html. Accessed 13 July 2023.

Monckton, Paul. "Instagram Makes Biggest Ever Photo Changes." *Forbes*, 27 Aug. 2015, www.forbes.com/sites/paulmonckton/2015/08/27/instagram-makes-biggest-ever-photo-changes. Accessed 13 July 2023.

Titcomb, James. "Instagram Is Changing Its Iconic Logo—Here's Why." *Telegraph*, 11 May 2016, www.telegraph.co.uk/technology/2016/05/11/instagram-is-changing-its-iconic-logo—heres-why. Accessed 13 July 2023.

Internet and Web Engineering

ABSTRACT

The term "internet" is often used to describe the web of computer networks connected to each other digitally that is accessible to the public around the world. The World Wide Web was developed by Oxford University graduate Tim Berners-Lee in 1989 while working at the European Council for Nuclear Research. The World Wide Web makes it possible for users to browse through various documents on different websites by clicking on hyperlinks located on web pages. Numerous hardware and software advances have made the internet an indispensable tool for business transactions and personal communication.

BACKGROUND AND HISTORY

The precursor to the internet was created in 1969 by the Advanced Research Projects Agency (now the Defense Advanced Research Projects Agency) of the United States Department of Defense. This project proposed a method to link together the computers at several universities (University of California, Los Angeles and Santa Barbara, Stanford, and the University of Utah) to share computational data via networks. This network became known as Advanced Research Projects Agency Network (ARPANET). In the 1980s the International Organization for Standardization (ISO) implemented its Open Systems Interconnection—Reference Model (OSI—RM) to facilitate the interoperability of different hardware components. The creation in the 1990s of new software and programming languages, such as browsers and hypertext markup language (HTML), made the explosive growth of internet activity and websites possible.

The internet is a type of wide area network (WAN) because digital devices of many different types, such as personal computers (PCs), cell phones, and tablets, can all connect to the internet from locations all over the world. Although sharing information between computers via the network connections that comprise the internet is free, connecting to the internet itself is not free in North America and requires the use of a private company that functions as an internet service provider (ISP). Some ISPs were originally phone companies, as during the 1990s, most connections to the internet were made via analogue dial-up connections through the phone lines. Those connections were made possible through a device called a modem. A modem functions by translating the digital signal from a sending computer to the analogue signals transmitted by phone lines and then back into the digital signal required by the receiving computer.

New hardware devices and software have continuously been developed to make sharing information via the internet easier and faster. HTML was created

Tim Berners-Lee, inventor of the world wide web. Photo by Paul Clarke, via Wikimedia Commons.

to design and format the user interface to web page content for transmission over the internet, based on hypertext transfer protocol (HTTP). By the third decade of the twenty-first century, there were many web applications available to computer users and users of internet-enabled mobile devices all over the world.

Along with the increasing use of microcomputers, there has also been an increase in the need for trained workers because companies need to be able to share information. This need to share information was one of the driving forces in the development of the internet and the client-server model common to web architecture, which now allows workers to share files and access centralized databases even when working remotely.

OVERVIEW

To transmit data across the internet, there must be a level of communication possible between devices, analogous to two people shaking hands. For computers, this ability to communicate is called interoperability, and it can be classified as connection-oriented or connectionless. The connection-oriented mode of communication is somewhat similar to the process of a phone call because it requires that the sender of a message wait for the intended recipient to answer before any data is actually sent. It is more secure, but slower, for transmission of data over the internet than the connectionless mode of data transmission. The connectionless mode has more of a broadcast nature, with data being transmitted before any secure connection is established. The connectionless mode is analogous to sending a letter or postcard through the regular mail. Different protocols, developed according to the Institute of Electrical and Electronics Engineers (IEEE), are used for communication across the internet, depending on whether the connection or connectionless mode is best for a given situation.

Logical and Physical Topology Classifications. The different protocols function via alternative topologies that can be classified as logical or physical. Logical topologies describe the theoretical, but not visible, pathways for transmission of data signals throughout the physical topologies, which describe the tangible hardware connections of the actual pieces of equipment on a network. Logical topologies include bus and ring, and physical topologies include bus, star, ring, mesh, and cellular.

The Seven Layers of the OSI Web Architecture Model. The OSI model was developed to be the web architectural model for designing components that allow communication across the internet. It continues to be useful as a theoretical construct to facilitate the development of different pieces of software and hardware from different manufacturers according to widely accepted standards. If one single piece of hardware equipment within the network for one home or business fails, then that one piece can be purchased, even from a different manufacturer, and successfully replace the faulty equipment. This network architecture model was developed by the ISO in 1977, primarily to provide standardization among different manufacturers and vendors to allow data communication to flow uninterrupted across the internet through different nations with complete interoperability of the software and hardware components. Therefore, the OSI model also helps with the diagnosis of connection problems by dividing the communication between two computers connected by a network into seven different layers.

The lowest level is called the physical layer because it consists of the most fundamental hardware devices that connect computers and transmit bits, including coaxial, twisted-pair, and fiber-optic cabling and connectors. Next is the data link layer, which manages timing, flow control, error control, and encryption. It also recognizes the physical device addresses, which consist of 48 bits provided by manufacturers. Most bridges and switches are used to

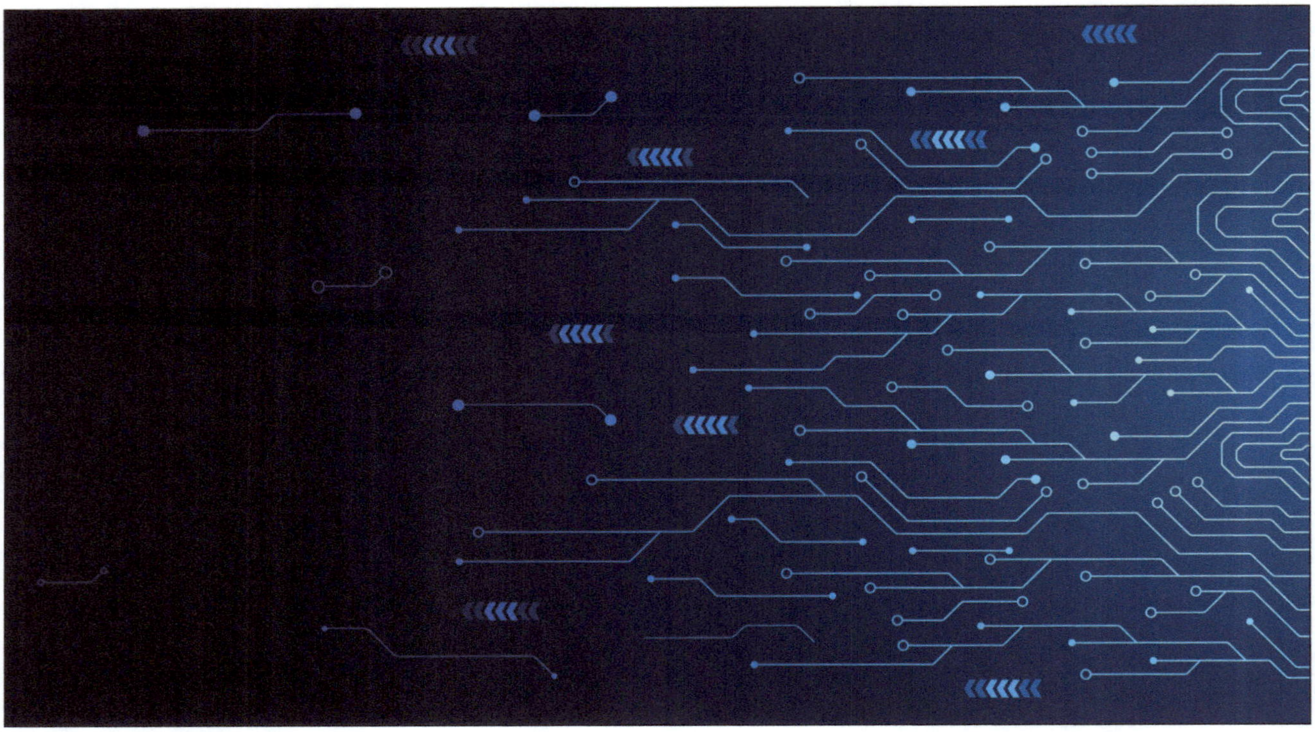

Image via iStock/Foryou13. [Used under license.]

repeat data across different networks according to protocols such as Ethernet through this layer. The network layer uses the internet protocol (IP) for five classes of addresses for devices, based on a 32-bit logical (not physical) address. These addresses are typically assigned by a network administrator for an organization. Routers consist of both hardware and software programs that can make decisions regarding the choice of routes for data transmission across different networks. The transport layer uses the transmission control protocol, which is a reliable, connection-oriented protocol that transmits data between two points at the same time. The session layer coordinates individual connection sessions and uses Structured Query Language (SQL) to retrieve information from databases. The presentation layer uses image, sound, and video formats such as graphics interchange format (GIF), musical instrument data interface (MIDI), and moving picture experts group (MPEG). It also encrypts data and compresses data for secure transmission. The top layer, the application layer, has the most direct interaction with a computer user. It contains the protocols for email, web access, operating systems, and file transfer protocol. These seven layers work together through encapsulation to make the transmission of data from one device to another device seamless.

APPLICATIONS AND PRODUCTS

In 1990, the World Wide Web began as a public web of networks linking digital content described using HTML at different digital locations defined by a uniform resource locator (URL). Initially the digital content was static and slow to display. To display forms that could interact with the user, Brendan Eich of Netscape developed the language initially called LiveScript in 1995, which later became widely known as JavaScript.

Web Applications. HTML was useful for allowing static web pages to be displayed but not dynamic web pages. Technologies were developed to provide dynamic web page access, including Active Service Pages (ASP), Perl/CGI, JavaScript, and ColdFusion. ColdFusion and ASP are examples of server-side technologies, which means that the computer of an individual user (client) is not required to have any special software or hardware to access the information from a remote geographical location other than any type of browser that allows access to the internet. JavaScript, on the other hand is a client-side technology. Server-side technologies are more secure because individual clients do not have any direct access to the code. Java libraries can be imported into web pages by using platforms such as ColdFusion, which can take Java byte code and directly compile it into ColdFusion code and thus speed up the web page interactions.

Linus Torvalds created Linux in 1991 while he was a student at the University of Helsinki. Linux has continued to grow in popularity as an operating system ideal for internet applications because it is free, open source (without any hidden or proprietary interfaces), and compatible with many devices and web applications, including Oracle and IBM databases, and many browsers. Moreover, it is powerful enough to be able to handle many of the applications that were originally handled by the UNIX operating system. Some of the applications that use Linux include digital camera programs, sound cards, printers, modems, smartphones, and tablets. Android, one of the most popular mobile operating systems as of 2023, is based on Linux.

To decrease the time for web pages to be displayed, frames were introduced in the mid-1990s to divide a document into pieces and display them gradually. Navigator 2.0, a browser created by Netscape, was the first to use both frames and JavaScript. Microsoft's Explorer was the next browser to implement frames, and the creation of Dynamic HTML (DHTML) along with Cascading Style Sheets (CSS) and the Document Object Model (DOM) in the late 1990s revolutionized web pages by allowing parts of web pages to be modified and also provided structures to web pages. Hidden frames, extensible markup language (XML), and ActiveX controls such as XMLHttp became widespread after 2001.

These technologies were taken to the next level to make web pages more dynamic in 2005 by Jesse James Garrett, cofounder of Adaptive Path. He called this new technology Ajax, an acronym for asynchronous JavaScript and XML. The XML parser is software that allows one to use XML to create one's own markup that can read and interpret XML data.

Google was one of the first companies to make use of this technology, incorporating it into Google Maps, Google Suggest, and Gmail web applications. Many additional companies adopted Ajax as well, including Amazon, Microsoft, and Yahoo. Ajax increases interactivity with web pages for users over the traditional "click and wait" process. JavaScript embedded within a web page sends a message to a web server that allows data to be retrieved and transmitted to the user without having to wait for the entire web page to be displayed.

Flash is a programming language introduced by Macromedia to create dynamic images on a web page by using vector graphic images instead of the bitmap images (common in digital cameras) that are smaller and slow down the dynamic animations and graphics of a web page. Throughout the first two decades of the twenty-first century, stand-alone applications such as Flash Player were used to display movies and similar content on the internet. Support for applications such as Flash Player waned amid the rise of the HTML5 markup standard, and browsers such as Google Chrome, Mozilla Firefox, and Microsoft Edge ceased their support for Flash by the end of 2020.

Common Gateway Interface. Common Gateway Interface (CGI) is used to write programs that interact with web pages to process information a person types into a form for registration, credit card purchase, or search on a website. CGI is a tool that functions as an interface between code written in Perl (Practical Extraction and Report Language) and a web server. These CGI applications are referred to as scripts instead of programs. Because CGI is able to interact dynamically to web users, it is designed to locate URLs instantly using HTTP and HTML. Embedded with CGI scripts are tags to create cache information, so that a user can quickly return to websites. Because the general use of CGI is to process dynamic data for forms and to assist with searches through forms, it also interacts with Perl. Variants of CGI include FastCGI and the Simple Common Gateway Interface (SCGI).

Active Server Pages, ColdFusion, and PHP (hypertext preprocessor). The primary problem with CGI is that it requires a new and separate execution of the CGI script each time updated input is necessary. Therefore, several alternatives have been developed. One is called the ASP engine, which is included as part of a web server, so the extra step of accessing the web server is not needed. In addition, ASP can be used with several programming languages, including JavaScript and Visual Basic, making ASP more versatile than CGI. ColdFusion allows functions to be programmed and then called from within a web page, allowing more flexibility. Both ColdFusion and PHP are integrated into web servers. Hypertext preprocessor was developed in 1994 as the acronym for personal home page because it was used as a scripting language to facilitate the development of web pages. Hypertext preprocessor can be embedded within the HTML code of a web page to make it more dynamic because the PHP code is interpreted rather than compiled.

CAREERS AND COURSEWORK

The occupation of internet or web engineer falls within a broader career field that also encompasses network administrators, network engineers, systems engineers, database administrators, help desk engineers and technicians, web designers, security analysts, programmers, and project managers. There are ample opportunities for employment as help desk engineers and technicians, web designers, database administrators, and network administrators for individuals who have an associate's degree in an appropriate field or work experience and certifications. In addition to some of the vendor-specific certifications offered by companies such as Cisco, Oracle, and Microsoft, there are several certifications that are vendor neutral, such as CompTIA's Network+, Security+, and Cloud+. Network administrators design networks for organizations and maintain the security of these networks on a daily basis, as well as install shared software, assign logical IP addresses, and back up data. Web designers use web page languages such as HTML, Perl, and Java. Certifications in these languages indicate useful skills as well. A bachelor's degree in computer science or a computer-related field is especially helpful to secure employment as a security analyst or project manager.

SOCIAL CONTEXT AND FUTURE PROSPECTS

In the twenty-first century, the economy can be described as a digital one because the internet is used as a tool to facilitate all kinds of financial transactions, including those involving bank accounts, credit cards, and the stock market. These financial transactions can be completed using both wired and wireless computer internetworking. The rapid growth of internet applications that facilitate communication and financial transactions has also been accompanied by a growth in identity theft and other cybercrimes. Web architecture has developed both in terms of providing convenience for consumer transactions and social networking sites, such as Facebook, and also in terms of providing security through the use of firewalls and new security software to try to prevent cybercrimes.

Despite the security risks, the usage of the internet for personal, financial, educational, and business-related activities is expected to increase, especially in countries in which internet infrastructure remains in development.

—*Jeanne L. Kuhler*

Further Reading

Biffl, Stefan, and Marta Sabou, editors. *Semantic Web Technologies for Intelligent Engineering Applications*. Springer International Publishing, 2016.

Campbell-Kelly, Martin, William F. Aspray, Jeffrey R. Yost, Honghong Tinn, and Gerardo Con Díaz. *Computer: A History of the Information Machine*. 4th ed., Routledge, 2023.

Casabona, Joe. *HTML and CSS: Visual QuickStart Guide*. 9th ed., Peachpit Press, 2020.

Dean, John. *Web Programming with HTML5, CSS, and JavaScript*. Jones & Bartlett Learning, 2018.

Duckett, Jon. *Web Design with HTML, CSS, JavaScript and jQuery*. 2 vols. John Wiley & Sons, 2014.

Love, Chris. *ASP.NET 3.5 Website Programming: Problem-Design-Solution*. John Wiley & Sons, 2010.

O'Leary, Timothy, Linda O'Leary, and Daniel O'Leary. *Computing Essentials 2023*. McGraw-Hill, 2022.

Prayaga, Lakshmi, and Hamsa Suri. *Programming the Web with ColdFusion MX 6.1 Using XHTML*. McGraw-Hill, 2004.

Spaanjaars, Imar. *Beginning ASP.NET 3.5 in C# and VB*. Wiley Publishing, 2008.

Underdahl, Brian. *Macromedia Flash MX: The Complete Reference*. McGraw-Hill/Osborne Media, 2003.

Internet of Things (IoT)

ABSTRACT

The Internet of Things (IoT) is a term used to describe the trend of more and more devices being connected to the internet and to each other. As smartphones, wearable data collection devices, and other "connected" products become more common, the IoT is becoming more useful and more popular. The IoT became a real factor beginning in the early 2010s, as high-speed internet became more readily available and hardware and software became less expensive to produce. It has been said that IoT could potentially affect all devices with an on/off switch. The impact of the IoT will most likely be far-reaching, as it will affect people's personal lives as well as businesses.

BACKGROUND

The Internet of Things (IoT) is a term used to describe a trend in which numerous devices are designed to be connected to the internet and to each other. The IoT has become possible for a number of reasons, perhaps most notably the increasing availability of broadband internet throughout the first decades of the twenty-first century. As more people become connected to the web, the demand for data and connectivity has increased. At the same time, the cost of hardware and software production has decreased. Another reason the IoT has become more prevalent is that more people are using smartphones, wearable data collection devices, and "smart" devices such as smart televisions and smart speakers. Computing technology gives people access to the internet and data at all times and in all places.

People are also becoming more aware of the IoT's importance. The IoT will have far-reaching effects, and many people believe that it will also have great economic value. Connected devices that send information can help reduce the amount of time that humans spend collecting and analyzing data. In other words, the IoT can help businesses and people be more efficient and save them money.

OVERVIEW

Objects that can be connected using the IoT include almost anything that can be turned on and off. These devices include home security systems, vehicles, medical devices, appliances, smartphones, and much more. Some examples of the IoT at work are a thermostat that a homeowner can change using a smartphone or a wearable device that tells a person how far she ran during a workout.

The IoT depends a great deal on data. Many modern devices collect and track data. When devices are connected through the IoT, they can transmit the data they collect. The IoT will help devices connect with people and other devices. In addition, it will help people connect with other people.

EFFECTS OF THE IOT

The IoT could possibly affect every business sector in the world. The fields that will most likely see important effects include manufacturing and production, health and medicine, and transportation.

Manufacturing and Production. The IoT will affect many business sectors, but it has the potential to change manufacturing and production greatly. A manufacturing facility could use the IoT to reorder supplies when they are running low without a human having to check inventory. Company machines connected to the IoT could inform an employee when they need to be repaired, a capability known as "predictive maintenance." Such practices can increase efficiency, production uptime, and product quality while lowering costs and energy use. IoT manufacturing equipment may also enhance workers' safety and productivity.

Health and Maintenance. Wearable devices that track exercise and fitness are already a significant part of the IoT; others serve as medical emergency bracelets. Still other medical devices connected to the IoT have the potential to be even more important for health and medicine. It is possible that someday scanning devices could analyze results and help doctors prescribe treatments for patients. Medical devices connected to the IoT could also connect doctors located in different parts of the world. These types of collaborations could help patients receive improved care.

Transportation. The IoT could help automobiles locate empty parking spaces and drivers locate available vehicles via car-sharing apps. In addition, cars could be fitted with technology that would automatically send a message to friends, family members, or colleagues when drivers are stuck in heavy traffic. They could also use diagnostic sensors to alert drivers and manufacturers to maintenance and repair needs. IoT technology might also help airplanes avoid deadly collisions or enable the planes to report when they require maintenance. Onboard devices have expanded travelers' wireless internet connectivity for cell phones and computers, and popular smartphone navigation and music apps have increasingly been integrated into automotive electronics systems.

Everyday Effects. Most people can see the effects that the IoT will have on business. However, the IoT also has the potential to impact everyday life. Homeowners are already using security systems, smoke detectors, and lighting that can be accessed from smartphones, as well as virtual personal assistants via smart speakers. Both smart speakers and smart televisions facilitate the consumption of digital media, including music, podcasts, web videos, and films. Indoor air quality, energy management, and water use are other areas in which IoT assists property owners. People could also see their lives change in other small ways thanks to the IoT. For example, an IoT refrigerator could tell people when they are running low on milk or eggs. A person could locate a wallet, purse, phone, remote control, or lost pet using an app on a computer or smartphone.

The IoT could also affect towns and cities. Smart cities could use the IoT to control traffic lights to make roads less congested. Devices connected with the IoT could also track water and air pollution in an attempt to identify the causes. Cities could easily provide visitors with information about local attractions, including real-time data. For instance, an amusement park could provide guests with data about the current wait times at particular attractions.

SECURITY CONCERNS AND THE IOT

The IoT relies a great deal on the sharing of data. Because of that, some experts are concerned about data security on the IoT. According to the statistics platform *Statista*, 11.28 billion IoT connected devices were in use worldwide as of 2021, and that number was projected to increase to more than 29 billion by 2030. The widespread adoption of such devices gives hackers access to a large quantity of potentially sensitive personal information, including medical records, city utility details, and more. Experts have pointed out that many devices are unencrypted, which means the data is easy for hackers to read and understand. In addition, many devices and software platforms require, or even provide, only weak passwords. A number of factors leave IoT devices and their users vulnerable to hackers, surveillance, and malware, including the difficulty or infeasibility of upgrading many IoT devices and the inclusion of "backdoors" by manufacturers. To ensure data, software, and hardware are secure in the future, IoT users, businesses, and governments will have to make internet security a priority and develop new ways to protect data on the IoT.

—*Elizabeth Mohn*

Further Reading

Burgess, Matt. "What Is the Internet of Things? WIRED Explains." *Wired*, 16 Feb. 2018, www.wired.co.uk/article/internet-of-things-what-is-explained-iot. Accessed 13 July 2023.

Greengard, Samuel. *The Internet of Things*. Rev. ed., MIT Press, 2021.

Kumar, Raghvendra, Rohit Sharma, and Prasant Kumar Pattnaik, editors. *Multimedia Technologies in the Internet of Things Environment*. Springer, 2021.

Meola, Andrew. "Smart Buses, Trains, Cars & Planes: How IoT Will Create Private and Public Smart Transportation." *Business Insider*, 30 Jan. 2020, www.businessinsider.com/smart-transportation-iot. Accessed 13 July 2023.

Morgan, Jacob. "A Simple Explanation of 'The Internet of Things.'" *Forbes*, 13 May 2014, www.forbes.com/sites/jacobmorgan/2014/05/13/simple-explanation-internet-things-that-anyone-can-understand. Accessed 13 July 2023.

"Number of Internet of Things (IoT) Connected Devices Worldwide from 2019 to 2021, with Forecasts from 2022 to 2030." *Statista*, July 2022, www.statista.com/statistics/1183457/iot-connected-devices-worldwide. Accessed 13 July 2023.

O'Marah, Kevin, and Pierfrancesco Manenti. "The Internet of Things Will Make Manufacturing Smarter." *Industry Week*, 14 Aug. 2015, www.industryweek.com/manufacturing-smarter. Accessed 13 July 2023.

Interpreter and Translator

ABSTRACT

Interpreters and translators facilitate communication between people who speak different languages or between hearing and deaf people. A translator translates written materials, while an interpreter translates oral communication. Students aspiring to enter either field must work to achieve proficiency in the desired language or languages.

BACKGROUND

Interpreters and translators facilitate communication between people who speak different languages or between hearing and deaf people. While the terms are commonly thought to be interchangeable, translators and interpreters work in different media. A translator translates written materials, usually into his or her native language, while an interpreter translates oral communication and may switch between languages. Some do both types of work. Among the most popular languages being translated into English in the twenty-first century are Spanish, Arabic, Chinese, and American Sign Language (ASL).

OVERVIEW

Translators often work by themselves at home, where they receive assignments via the internet or mail. Interpreters work in a variety of settings, such as hospitals, courtrooms, schools, airports, and government offices. They may work alone with just their clients or with partners. In some cases, a translator or interpreter might work the night shift or odd hours, especially when communicating with people who live and work in other time zones. Some interpreters work with reporters in combat zones, risking their lives to do so.

Interpreting and translating attract those who are linguistically gifted and enjoy foreign cultures. Translators tend to be introverts who enjoy reading and writing and prefer solitary work, while interpreters tend to be extroverts who love being around people, have excellent hearing and listening skills, and are quick thinkers. Translators must manage deadlines while interpreters comply with variable schedules. In either case, the work demands strong cognitive skills and a sharp memory. Sign language interpreters also need excellent hand dexterity.

DUTIES AND RESPONSIBILITIES

A translator spends most of his or her day translating documents at a computer. Documents might range from a few paragraphs to hundreds of pages in length. The translator takes time to reflect on what he or she reads. The translator then tries to communicate the message with as much of its natural rhythm and nuances intact as possible. Such work requires full knowledge of each language, including slang, subject-specific jargon, and colloquialisms, as well as a deep understanding of each culture. Usually a translator has several dictionaries and style guides at his or her disposal, as grammar and punctuation are extremely important.

Interpreters work closely with their clients, in person or via phone, videophone, or microphone. Interpretation may be simultaneous or consecutive. Simultaneous interpreting involves listening to a speaker and translating orally, or signing, at the same time. In some cases, the interpreter is given a written speech or paper to consult in advance for general ideas and language. In other cases, there is no time to think. Consecutive interpreting involves listening to a speaker complete a few words or a sentence and then translating it orally. Depending on the speaker's pace, the interpreter might have time to consider various interpretations of a word or phrase.

Interpreters at United Nations conventions or other types of conferences often sit in the audience and whisper their translations into a microphone. Sign language interpreters sometimes use videophones and a computer to communicate with deaf clients.

Self-employed translators and interpreters spend part of the day keeping up with marketing, billing, and other administrative tasks.

OCCUPATION SPECIALTIES

The services of interpreters and translators are needed in a number of different subject areas. Although these professionals often do not specialize in any particular field or industry, many do focus on one area of expertise. The following are examples of occupational specialties.

Conference Interpreters. Conference interpreters work at conferences that have non-English-speaking attendees. The work is often in the field of international business or diplomacy, although conference interpreters can interpret for any organization that works with speakers of foreign languages. Employers prefer high-level interpreters who have the ability to translate from at least two languages into one native language—for example, the ability to interpret from Spanish and French into English. For some positions, such as those with the United Nations, this qualification is required.

Guide or Escort Interpreters. Guide or escort interpreters accompany either US visitors abroad or foreign visitors in the United States to ensure that they are able to communicate during their stay. These specialists interpret informally and on a professional level. Frequent travel is common for these workers.

Health or Medical Interpreters and Translators. Health or medical interpreters and translators typically work in health-care settings and help patients communicate with doctors, nurses, and other medical staff. Both interpreters and translators must have a strong grasp of medical terminology and the common words for those medical terms in both languages.

Legal or Judiciary Interpreters and Translators. Legal or judiciary interpreters and translators typically work in courts and other legal settings. At hearings, arraignments, depositions, and trials, they help people who have limited English proficiency. They must understand legal terminology. Many court interpreters must sometimes read documents aloud in a language other than that in which they were written, a task known as sight translation.

Literary Translators. Literary translators rewrite journal articles, books, poetry, and short stories from one language into another language. They strive to keep the author's tone and style intact as well as his or her meaning. Whenever possible, literary translators work closely with authors to capture their intended meaning and literary characteristics.

Localization Translators. Localization translators adapt text for a product or service from one language into another. Localization specialists strive to make it appear as though the product originated in the country where it will be sold. They must not only know both languages but also understand the technical information they are working with and the culture of the people who will be using the product or service.

Sign Language Interpreters. Sign language interpreters help people who are deaf or hard of hearing and people who can hear communicate with each other. In the United States, sign language interpreters must typically be fluent in English and in American Sign Language (ASL), which combines signing, finger spelling, and specific body language. ASL is a separate language from English, with its own grammar.

WORK ENVIRONMENT

Physical Environment. Interpreters tend to work in diverse interior and exterior environmental conditions, including potentially dangerous or unhealthy job sites. Travel is often required. Translators, on the other hand, work in offices with less variable conditions or at home.

Human Environment. Unless self-employed, translators and interpreters report to supervisors or directors and usually interact with various office staff and professionals. Their clients may change from day to day. Interpreters also interact with the public at conventions or while touring cities with their clients.

Technological Environment. A translator uses translation software on a computer and might use a transcription machine. An interpreter sometimes uses a microphone and might rely on a smart phone, tablet computer, or laptop to access the internet or a digital dictionary. Some sign language interpreters use a videophone along with a video-relay service (VRS) or video-interpreting service.

EDUCATION AND TRAINING

High School/Secondary. Achieving proficiency in a foreign language takes many years. A college-preparatory program with four years of at least one foreign language, along with courses in English, speech, and the social sciences (political science, anthropology, and world cultures), will provide the best foundation for a career in interpretation or translation. Those students interested in translating technical material should consider additional courses in science and technology. Foreign exchange programs and travel, volunteer work with relevant organizations, and other independent educational experiences can prove invaluable.

College/Postsecondary. While a bachelor's degree is the minimum requirement for most jobs, the selection of a major is a personal decision based on the type of work desired. Students might consider double majoring in a foreign language and in another subject, such as English literature, history, or a social science, or in two foreign languages, such as Spanish and French. Some translators and interpreters need an advanced degree to translate subject-specific concepts and vocabulary.

Study abroad programs, foreign travel, and participation in international clubs are some ways to gain important hands-on experience. An internship might be needed for some jobs; some employers offer on-the-job training.

Adult Job Seekers. Bilingual adults should be able to transition well into an interpreting or translating career, especially with relevant experience. For example, a bilingual nurse would have an advantage translating or interpreting in a medical setting. Continuing education courses can refresh or teach new skills. Prospective interpreters and translators should expect to be tested in their language abilities as a prerequisite for employment.

Advancement is highly dependent on experience. Advancement opportunities might include better work hours, higher pay, or more interesting assignments. Those with experience may also consider moving into editorial positions or starting their own companies.

Professional Certification and Licensure. Professional certification is voluntary, although many employers only hire those who are certified in a particular subject or specialty. Professional associations, such as the American Translators Association, offer certification. Translators typically must pass a written examination. Some states require interpreters working with deaf or hard of hearing clients to register with the state government and attain certification from an organization such as the Registry of Interpreters for the Deaf.

Additional Requirements. Interpreters and translators who wish to work for government agencies may be required to pass a civil service exam, while freelancers need good business skills as well as experience in the field. Interpreters and translators should consider membership in professional associations, which often provide opportunities for networking and professional development. Work experience or certification is required for membership in some organizations.

—Sally Driscoll

Further Reading

Milton, John, and Silvia Cobelo. *Translation, Adaptation and Digital Media*. Routledge, 2023.

Munday, Jeremy, Sara Ramos Pinto, and Jacob Blakesley. *Introducing Translation Studies*. 5th ed., Routledge, 2022.

Nolan, James. *Interpretation: Techniques and Exercises*. 2nd ed., Multilingual Matters, 2012.

iOS

ABSTRACT

Apple's iPhone operating system (iOS) is an operating system designed for mobile computing. It is used on the company's iPhone and iPod Touch products and was previously used on the iPad. The system, which debuted in 2007, was based on the company's OS X. iOS was the first mobile OS to incorporate advanced touch-screen controls.

BACKGROUND

Apple's iOS is an OS designed for use on Apple's mobile devices, including the iPhone and iPod Touch. In 2020, iOS was the world's second most popular mobile OS after the Android OS. Between 2010 and 2019, Apple's iPad tablet devices also used iOS. However, Apple renamed the variant of iOS used for the iPad the iPadOS in mid-2019.

Introduced in 2007, iOS (then known as iPhone OS) was one of the first mobile OSs to incorporate a capacitive touch-screen system. The touch screen allows users to activate functions by touching the screen with their fingers. The Apple iOS was also among the first mobile OSs to give users the ability to download applications (apps) to their mobile devices. The iOS is therefore a platform for numerous apps, including the approximately 1.8 million apps available in the App Store by mid-2023. The first iOS system and iPhone were unveiled at the 2007 Macworld Conference.

OVERVIEW

The original iOS had a number of limitations. For example, it was unable to run third-party apps, had no copy and paste functions, and could not send email attachments. It was also not designed for multitasking, forcing users to wait for each process to finish before beginning another. However, iOS introduced a sophisticated capacitive touch screen. The iOS touch features allowed users to activate most functions with their fingers rather than needing a stylus or buttons on the device. The original iPhone had only five physical buttons. All other functions, including the keyboard, were integrated into the device's touch screen. In addition, the iOS system supports multitouch gestures. This allows a user to use two or more fingers (pressure points) to activate additional functions. Examples include "pinching" and "stretching" to shrink or expand an image.

JAILBREAKING

Computer hobbyists soon learned to modify the underlying software restrictions built into iOS, a process called jailbreaking. Modified devices allow users greater freedom to download and install apps. It also allows users to install iOS on devices other than Apple devices. Apple has not typically pursued legal action against those who jailbreak iPhones or other devices. In 2010, the US Copyright Office authorized an exception permitting users to jailbreak their legally owned copies of iOS for select legal purposes. However, jailbreaking iOS voids Apple warranties.

VERSION UPDATES

The second version of iOS was launched in July 2008. With iOS 2, Apple introduced the App Store, where users could download third-party apps and games. In 2009, iOS 3 provided support for copy and paste functions and multimedia messaging. A major advancement came with the release of iOS 4 in 2010. This update introduced the ability to multitask, allowing iOS to begin multiple tasks concurrently without waiting for one task to finish before initiating the next task in the queue. The iOS 4 release was also the first to feature a folder system in which similar apps could be grouped together on the device's home screen (called the "springboard"). FaceTime video calls also became available with iOS 4.

The release of iOS 5 in 2011 integrated the voice-activated virtual assistant Siri as a default app. Other iOS 5 updates include the introduction of iMessage, Reminders, and Newsstand. In 2012, iOS 6 replaced Google Maps with Apple Maps and redesigned the App Store, among other updates. Released in 2013, iOS 7 featured a new aesthetic and introduced the Control Center, AirDrop, and iTunes Radio.

FURTHER INNOVATIONS

With the release of iOS 8, Apple included third-party widget support for the first time in the company's history. Widgets are small programs that do not need to be opened and continuously run on a device. Examples include stock tickers and weather widgets that display current conditions based on data from the web. Widgets had been a feature of Android and Windows mobile OSs for years. However, iOS 8 was the first iOS version to support widgets for Apple. Since their release, Apple has expanded the availability of widgets for users.

The release of iOS 9 in 2015 marked a visual departure for Apple. This update debuted a new typeface for iOS called San Francisco. This specially tailored font replaced the former Helvetica Neue. The release of iOS 9 also improved the battery life of Apple devices. This update introduced a low-power mode that deactivates high-energy programs until the phone is fully charged. Low-power mode can extend battery life by as much as an hour on average.

Coinciding with the release of iOS 9, Apple also debuted the iPhone 6S and iPhone 6S Plus, which introduced 3D Touch. This new feature is built into the hardware of newer Apple devices and can sense how deeply a user is pressing on the touch screen. 3D Touch is incorporated into iOS 9 and enables previews of various functions within apps without needing to fully activate or switch to a new app. For instance, within the camera app, lightly holding a finger over a photo icon will bring up an enlarged preview without needing to open the iPhoto app.

Apple added additional features to iOS over the subsequent years, including implementing the use of apps within iMessage in iOS 10. The OS continued to evolve through the releases of iOS 11 and iOS 12, the latter of which introduced a function known as Screen Time, used to track and limit an individual's time using a device or a specific app. Released in September of 2019, iOS 13 introduced Dark Mode as well as a variety of tweaks to existing apps and features. In September of 2020, iOS 14 introduced features such as home-screen widgets, pinned conversations in Messages, and an App Library.

The iOS 15 update, released in September 2021, added features such as Live Text and new focus modes, while iOS 16, introduced the following year, made changes to the lock screen and notifications, among other elements of the OS. In June of 2023, Apple announced that iOS 17 would be released later in the year and would include a variety of new features, including a Personal Voice feature that would enable users to train their phones to mimic their voices.

—Micah L. Issitt

Further Reading

"App Store." *Apple*, 2023, www.apple.com/app-store. Accessed 13 July 2023.

Chokkattu, Julian. "The Top New Features in Apple's iOS 17 and iPadOS 17." *Wired*, 8 June 2023, www.wired.com/story/apple-iphone-ios-17-ipados-17-new-features. Accessed 13 July 2023.

Costello, Sam. "The History of iOS, from Version 1.0 to 17.0." *Lifewire*, 5 June 2023, www.lifewire.com/ios-versions-4147730. Accessed 13 July 2023.

Heisler, Yoni. "The History and Evolution of iOS, from the Original iPhone to iOS 9." *BGR*, 19 Dec. 2018, bgr.com/2016/02/12/ios-history-iphone-features-evolution. Accessed 13 July 2023.

"iOS: A Visual History." *Verge*, 16 Sept. 2013, www.theverge.com/2011/12/13/2612736/ios-history-iphone-ipad. Accessed 13 July 2023.

Williams, Rhiannon. "Apple iOS: A Brief History." *Telegraph*, 17 Sept. 2015, www.telegraph.co.uk/technology/apple/11068420/Apple-iOS-a-brief-history.html. Accessed 13 July 2023.

Wuerthele, Mike. "Apple Unveils iPadOS, Adding Features Specifically to iPad." *AppleInsider*, 2 June 2019, appleinsider.com/articles/19/06/03/apple-supplements-ios-13-with-new-tablet-specific-ipad-os-branch. Accessed 13 July 2023.

Isometric Drawing

ABSTRACT

Isometric drawing is the process of representing a physical or conceptual object as a three-dimensional (3D) view. Isometric drawings are traditionally prepared using graphing paper in which the grid lines are angled at 30 degrees, and height, width, and depth scales use identical dimension scales. Hence, they are isometric. Manual conversion of a two-dimensional (2D) drawing into an isometric drawing requires point-to-point transfer of features. Computer graphics programs are capable of rendering 2D views into 3D views, an essential feature of computer-generated imagery (CGI) in many applications.

BACKGROUND

Isometric drawing is the process of representing a physical or conceptual object as a three-dimensional (3D) view. While dynamic manipulation of isometry is a relatively recent development in historical terms, the principle of isometric drawing is not. Isometric drawings appeared in Chinese art as early as the twelfth century CE, and many examples of isometric drawings of machinery construction emerged by the early 1800s.

Isometric drawings are traditionally prepared using graphing paper in which the grid lines are angled at 30 degrees, and height, width, and depth scales use identical dimension scales. Hence, they are isometric. Using different scales on the depth axis results in a perspective drawing that is not isometric. Manual conversion of a two-dimensional (2D) drawing into an isometric drawing requires point-to-point transfer of features. This allows the 2D, or orthographic, drawing to be transferred in a one-to-one manner from regular square-grid graphing paper to the 3D, or isometric, graphing paper. The isometric paper itself does not have to be on the same scale as the orthographic drawing, which allows the isometric drawing to be larger or smaller than the orthographic drawing as desired. Computer graphics programs can render 2D views into 3D views, an essential feature of computer-generated imagery (CGI) used in animations, films, video games, motion-based simulators, and many other applications.

The simplest manual drafting techniques that are taught to students use a set of basic drafting tools: a precision-made drafting table, very sharp pencils, a T square, a 30-60 triangle template, a 45-90 triangle template, and a gum eraser. Typically, the drafting instructor will present an object with a specific shape and specific dimensions and task the students to produce an accurate orthographic drawing consisting of the front, side, and top views of the object. Beginners often face some confusion at first regarding the three different views and may even attempt to represent the object's three dimensions in a single view before catching on to the correct method. Part of the problem in this is that from birth, people see the world in perspective view due to the phenomenon of visual parallax. One's left and right eyes see slightly different images of the same thing due to their separation from each other, and these different views are combined into a single perspective image by the visual processing center of the brain. It requires some training to overcome this tendency and understand orthographic imaging. For example, a person who looks at one side of a cube understands the cube's overall shape, and the brain expects to see that shape. However, an orthographic view of one side of a cube is a simple 2D square and nothing more. It takes a degree of focus to remove the brain's expectation of the cube's shape. The same principle applies to every 3D object seen in an orthographic view, no matter how complex the structure of the object may be.

The basic principle of orthographic drawing is that the draftsperson draws only the features that are visible on the object when looking directly at it along one of its length, width, or depth axes. If a cube had a hole drilled through it from one side to the other, that hole would not be visible in the front and top views, only in the side view. The side view of this perforated block would therefore appear simply as a square with a circle in it at the relative location of the hole. In keeping with drafting conventions such as the use of dashed lines to show such hidden features of an object, the front and top views could have dashed lines extending from one side to the other to show the dimensions of the hole.

With the front view of the object determined, the remainder of the orthographic drawing can then be completed. The drafters will draw barely visible "construction lines" vertically and horizontally from pertinent features of the front view to transfer those dimensions to the top and side views, then draw

45-degree construction lines to complete the transfer of structural details between the top and side views.

OVERVIEW

In basic drafting, an isometric drawing is generated from the orthographic drawing using isometric graphing paper that is scaled to represent the 3D axes (x, y, and z, corresponding to length, width, and depth) at mutual angles of 120 degrees. This results in a view that declines at a 15-degree angle for the length and inclines at a 15-degree angle for width and depth. It is important to note that an isometric drawing therefore has no horizontal lines, except in the rare case that an angular feature of the object being represented coincidentally falls horizontally in the isometric drawing. In essence, an isometric drawing represents the object as it would appear orthographically if it had been tilted 15 degrees clockwise and then rotated 15 degrees toward the viewer.

To transfer an orthographic drawing to isometric graphing paper, the draftsperson chooses an appropriate starting point, often the lower left corner of the front view. A corresponding starting point is then chosen on the isometric graphing paper. Each scale point on the orthographic drawing is then transferred to the corresponding point on the isometric graphing paper and connected by drawing lines accordingly. This produces a 3D view of the object much as it would appear in reality. Naturally, certain features and shapes will have a different appearance in an isometric view than they have in an orthographic view. For example, a circle that is transformed from its perfectly round orthographic shape will become an ellipse in an orthographic view. Similarly, an equilateral triangle will transform into an isosceles triangle, all the while retaining its relative location in the overall proportions of the drawing.

ISOMETRIC DRAWING IN COMPUTER GRAPHICS

Conversion of an orthographic drawing to an isometric drawing is equivalent to rotating the object in the drawing slightly about its depth and length axes. Mathematically, coordinates are transformed rotationally using transformation matrix operations. The exact form of the transformation matrix uses the location of the point being rotated and the angle through which it is to be transformed. For static isometric renderings of an object, this transformation is carried out just once for each point to generate the 3D view of the object. Repeated, or iterative, rotational transformations allow for the 3D view to be rotated dynamically and manipulated in various ways to view potential changes in shape and function without first having to create the actual physical object. Many examples of the capability of this process can be seen in the supplementary material included with many home releases of films that contain significant CGI footage.

Similarly, dynamic isometric rendering is a staple of video games and motion simulators, especially those that employ a 360-degree world view. In such applications, however, the rendering is termed axonometric (of which isometric is a subset) to maintain the perspective view by using different scales on the 3D axes. In such games, as well as in advanced simulators (flight simulators, driving simulators, and others), the player's point of view is able to turn in any direction and see the surroundings in the appropriate perspective. It requires a considerable amount of dedicated graphics processing power to achieve this, but the capability to do so increases almost daily with advances in computer technology.

—Richard M. Renneboog

Further Reading

Congdon-Fuller, Ashleigh, Antonio Ramirez, and Douglas Smith. *Technical Drawing 101 with AutoCAD 2022: A*

Multidisciplinary Guide to Drafting Theory and Practice. SDC Publications, 2021.
Costello, Vic. *Multimedia Foundations: Core Concepts for Digital Design*. 3rd ed., Focal Press, 2023.
Marschner, Steve, and Peter Shirley. *Fundamentals of Computer Graphics*. 4th ed., CRC Press, 2016.
Rathnam, K. *A First Course in Engineering Drawing*. Springer, 2018.
Sexton, Timothy J. *A Concise Introduction to Engineering Graphics Including Worksheet Series B*. 6th ed., SDC Publications, 2019.
Singh, Lakhwinder Pal, and Harwinder Singh. *Engineering Drawing: Principles and Applications*. Cambridge UP, 2021.

iTunes

ABSTRACT

iTunes is a type of digital media software developed by Apple Inc. It allows users to add, organize, and play digital media files such as songs, videos, television shows, movies, audiobooks, or podcasts. These can be accessed on a computer or portable device such as the iPod portable media player or iPhone. Apple launched iTunes in 2001 and introduced the iTunes Store in 2003. In 2019, the company replaced iTunes for macOS with several new media apps, including Apple Music.

BACKGROUND

To commercialize on the digital music industry, computer technology company Apple Inc. debuted iTunes in January of 2001. iTunes was software that allowed Macintosh users to add, organize, and play digital media files on their computers. Less than a year later, the company introduced the iPod, a portable digital media player that worked exclusively with iTunes to permit customers to download and play media files. Apple also released the second version of iTunes to coincide with the introduction of the iPod.

The third upgrade came soon after and included audiobooks. In 2003, Apple debuted its iTunes Store, allowing customers to purchase digital files. In just one week, the iTunes Store sold more than 1 million songs. A few months later, more than 1 million people were seen sporting the iconic white iPod earbuds, a testament to how popular the iPod had become in such a short period. At this time, however, only Mac users could utilize iTunes. To widen its market, Apple released a Windows version of iTunes.

OVERVIEW

The iTunes 4 update came in 2004 and included a shuffle feature as well as support for digital photos and videos. The following year, Apple released an upgrade that permitted consumers to access podcasts. iTunes 6, which debuted in 2006, added minor enhancements, fixed bugs, and introduced

Photo via iStock/SKrow. [Used under license.]

users to video streaming. iTunes 7 launched in September of 2006 and featured iPod games and gapless (or uninterrupted) playback of music files.

In 2007, the company released its first iPhone smartphone, which linked to iTunes and the iTunes Store—and later to the App Store. Upgrades iTunes 8 and iTunes 9 followed, both of which helped to streamline all of Apple's services. People were able to rent and purchase television shows and movies, download courses from iTunes U, and buy digital books.

iTunes 10 came in 2010 and included a new social media network called Ping; however, Ping failed to attract followers and was quickly discarded. In 2011, Apple debuted iTunes Match, which worked with iCloud and iTunes in the Cloud to help match songs with subscribers. The iTunes 11 upgrade launched in 2012 with the addition of iTunes Radio and minor enhancements and fixes to the software. In 2014, Apple debuted iTunes 12, which changed the interface to make the software easier to navigate.

The following year, Apple released an updated version of iTunes 12 to embrace digital music streaming with the debut of Apple Music. Throughout 2016, the company continued to release other versions of iTunes 12 in an effort to fix several weaknesses in the system. It also worked on streamlining its iTunes and other services to make them more user friendly.

Far iTunes as titled, was, for all intents and purposes, phased out for macOS in 2019, in favor of the more generally named Apple Music app. The app encompassed the Apple Music streaming service, the iTunes store, and the contents of the user's existing iTunes music library, while television shows and podcasts previously downloaded from the iTunes store were relocated to the apps Apple TV and Apple Podcasts, respectively. However, Apple continued to offer iTunes for Windows until 2023, when the company announced the introduction of an Apple Music app for that operating system.

—Angela Harmon

Further Reading

Cunningham, Andrew. "Apple Previews a Trio of Apps That Will Finally Replace iTunes for Windows." *Ars Technica*, 12 Jan. 2023, arstechnica.com/gadgets/2023/01/apple-previews-a-trio-of-apps-that-will-finally-replace-itunes-for-windows. Accessed 13 July 2023.

Jarnow, Jesse. "Apple's iTunes Is Alienating Its Most Music-Obsessed Users." *Wired*, 17 Nov. 2015, www.wired.com/2015/11/itunes-alternatives. Accessed 13 July 2023.

Layton, Julia, and Jonathan Strickland. "How iTunes Works." *HowStuffWorks*, electronics.howstuffworks.com/itunes.htm. Accessed 13 July 2023.

McElhearn, Kirk. "iTunes: A Look at Apple's Media App and Its Influence on an Industry." *Macworld*, 8 Jan. 2016, www.macworld.com/article/3019878/software/15-years-of-itunes-a-look-at-apples-media-app-and-its-influence-on-an-industry.html. Accessed 13 July 2023.

McGarry, Caitlin. "How Apple Plans to Make iCloud, Siri, and iTunes Better." *Macworld*, 6 Oct. 2016, www.macworld.com/article/3128754/software/how-apple-plans-to-make-icloud-siri-and-itunes-better.html. Accessed 13 July 2023.

Pepitone, Julianne, and David Goldman. "The Evolution of iTunes." *CNN Money*, 26 Apr. 2013, money.cnn.com/gallery/technology/2013/04/25/itunes-history.

Perez, Sarah. "Redesigned Version of iTunes Launches with iCloud Music Glitches." *Tech Crunch*, 13 Sept. 2016, techcrunch.com/2016/09/13/redesigned-version-of-itunes-launches-with-icloud-music-glitches. Accessed 13 July 2023.

Roettgers, Janko. "Apple Is Officially Killing iTunes for Mac, Replacing It with Three Dedicated Media Apps." *Variety*, 3 June 2019, variety.com/2019/digital/news/itunes-shutting-down-1203231598. Accessed 13 July 2023.

"What Happened to iTunes?" *Apple*, support.apple.com/en-us/HT210200. Accessed 13 July 2023.

M

Machine Learning

ABSTRACT

Machine learning is a branch of computer-science algorithms that allow the computer to display behavior learned from past experience, rather than human instruction. Machine learning is essential to the development of artificial intelligence, but it is also applicable to many everyday computing tasks. Common examples of programs that employ machine learning include email spam filters, optical character recognition, and news feeds on social networking sites that alter their displays based on previous user activity and preferences.

BACKGROUND

One of the earliest attempts to enable machine learning was the perceptron algorithm, developed by Frank Rosenblatt in 1957. The algorithm was intended to teach pattern recognition and was based on the structure of a neural network, which is a computing model designed to imitate an animal's central nervous system. While the perceptron model showed early promise, Marvin Minsky and Seymour Papert demonstrated in their 1969 book *Perceptrons: An Introduction to Computational Geometry* that it had significant limitations, as there were certain classes of problems the model was unable to learn. Consequently, researchers did not pursue this area of study for some time.

In the 1980s, after other avenues for developing artificial intelligence had resulted in only limited success, scientists began to revisit the perceptron model. Multilayer perceptrons, or neural networks composed of multiple layered computational units, proved to have the processing power to express problems that Rosenblatt's single-layer, or linear, perceptrons could not. Around the same time, John Ross Quinlan introduced decision-tree algorithms, which use predictive models to determine a variable's value based on available data.

OVERVIEW

Since then, numerous machine-learning algorithms have been developed. Among those most commonly used are support vector machines (SVMs) and naive Bayes classifiers. SVMs, introduced by Vladimir N. Vapnik and Corinna Cortes in 1995 and based on an algorithm previously created by Vapnik, are used to recognize patterns in data and classify the various data points. Naive Bayes classifiers are applications of Bayes's theorem, named for the eighteenth-century mathematician and reverend Thomas Bayes, which deals with conditional probabilities. This algorithm was used in one of the earliest email spam filters, iFile, released by Jason Rennie in 1996. Many email clients still employ Bayesian spam filtering, which works by determining the probability that an email containing certain keywords is spam.

Machine-learning algorithms can be divided into categories based on how they train the machine. These categories include supervised learning, in which the machine learns from inputs that are mapped to desired outputs; unsupervised learning, in which the machine analyzes input without knowledge of the desired output; semisupervised learning, in which some of the input is paired with a desired output and some is not; transduction, in which the machine tries to predict new outputs based on training with previous inputs and outputs; reinforcement learning, in which the machine must form a policy

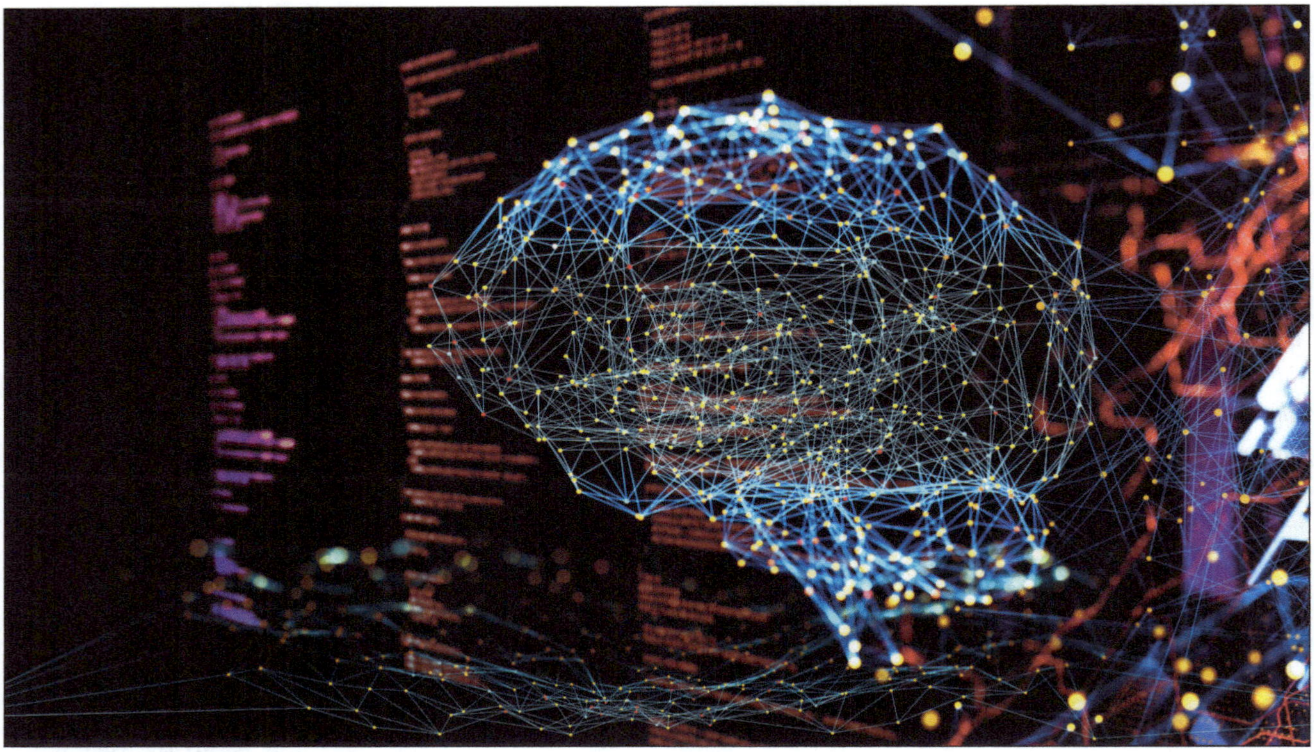

Image via iStock/MF3d. [Used under license.]

on how to act based on observing how certain actions affect its environment; and learning to learn, which teaches inductive bias based on previous experience. SVMs, multilayer perceptrons, decision trees, and naive Bayes classifiers all fall into the category of supervised learning.

It is important to distinguish between machine learning and data mining. Although the two concepts are related and use similar, often overlapping, methods, data mining is focused on discovering information about given data, while machine learning focuses more on learning from the given data in order to make predictions about other data in the future. Many consider data mining to be a subset of machine learning.

—*Randa Tantawi*

Further Reading

Abu-Mostafa, Yaser S. "Machines That Think for Themselves." *Scientific American*, July 2012, pp. 78-81.

Alpaydin, Ethem. *Machine Learning*. Rev. ed., MIT Press, 2021.

Audry, Sofian. *Art in the Age of Machine Learning*. MIT Press, 2021.

Brodley, Carla E. "Challenges and Opportunities in Applied Machine Learning." *AI Magazine*, vol. 33, no. 1, 2012, pp. 11-24.

Domingos, Pedro. "A Few Useful Things to Know about Machine Learning." *Communications of the ACM*, vol. 55, no. 10, 2012, pp. 78-87.

Heaven, Douglas. "Not Like Us: Artificial Minds We Can't Understand." *New Scientist*, 7 Aug. 2013, www.newscientist.com/article/mg21929290-700-not-like-us-artificial-minds-we-cant-understand. Accessed 13 July 2023.

James, Mike. "The Triumph of Deep Learning." *I Programmer*, 14 Dec. 2012, www.i-programmer.info/programming/article-intelligence/5206-the-triumph-of-deep-learning.html. Accessed 13 July 2023.

Marsland, Stephen. *Machine Learning: An Algorithmic Perspective*. Taylor, 2009.

Miller, Arthur I. *The Artist in the Machine: The World of AI-Powered Creativity*. MIT Press, 2019.

Mitchell, Tom M. *Machine Learning*. McGraw, 1997.

Piore, Adam. "The Quest to Build a Silicon Brain." *Discover*, 24 May 2013, www.discovermagazine.com/mind/the-quest-to-build-a-silicon-brain. Accessed 13 July 2023.

MacOS

ABSTRACT

MacOS is a family of operating systems—programs used to run computers—pioneered by the Apple computer company and utilized on a wide variety of Apple systems. Noted for being the first successful graphical interface operating system and for making its debut on the original 1984 Macintosh, MacOS played a key role in the emergence of the personal computer as a popular consumer product. Though later overshadowed by Microsoft Windows, MacOS continued to be one of the most recognized and widely used operating systems on the market. Over the years, it evolved through a variety of iterations, each of which has offered new features and an improved user experience. From 2016 on, MacOS was referred to as macOS.

BACKGROUND

Although MacOS often receives credit for being the first graphical interface operating system, it actually had two predecessors. The first of these was the operating system featured on the Xerox Alto. The Alto was a computer manufactured in the 1970s with a graphical operating system that was the first of its kind. This software replaced the traditional command-line operating system that required users to issue text commands. However, its staggering cost of $32,000 ensured that it would never be a commercial success. Regardless, after visiting Xerox's facilities in the early 1980s, Apple engineers recognized the potential of the graphical interface to the future of computing. Eventually, the Apple team members incorporated lessons from Xerox when they produced the Lisa, the first Apple computer to feature a graphical interface operating system. Like the Xerox Alto, however, the Lisa was a commercial failure.

Still determined to make an affordable personal computer that would appeal to the average consumer, Apple introduced the Macintosh in 1984. Powering the Macintosh was a graphical interface operating system that, although unnamed at the time, would eventually become known as System 1, the first version of MacOS. With a virtual desktop, windows, icons, menus, scrollbars, a file manager called Finder, and mouse control, System 1 represented a revolutionary shift in the user experience that would soon become the industry standard.

OVERVIEW

When the Macintosh proved to be a success, Apple engineers began working to improve its operating system. The result of their efforts was System 2, released in 1985. Though very similar to System 1, System 2 featured several key upgrades, such as the inclusion of key-commands for creating new folders and shutting down the computer, as well as an enhanced version of Finder. Thanks to these and other tweaks, System 2 ran 20 percent faster than System 1.

After Microsoft released the inaugural version of Windows in late 1985, Apple once again revisited the design of its own operating system. When System 3 debuted the following year, it included a new version of Finder that came equipped with file-nesting capabilities that allowed users to create folders within folders for the first time. It also included Disk Cache, a program that improved performance by storing commonly used commands in memory. While these improvements were helpful for users, System 3 was plagued by bugs and, as a result, did not work as well as hoped.

Apple released System 4 in 1987. Among other updates, System 4 offered support for disk drives and multiple monitors. It also included MultiFinder, a new version of Finder that allowed systems with enough memory to run more than one program at a time. The following year, Apple skipped System 5 and jumped ahead to System 6. The most notable improvement in System 6 was the inclusion of color support. Although the desktop and Finder were still rendered in black and white as they had been since System 1, System 6 made it possible for third-party applications to run in full color.

With the release of System 7 in 1991, MacOS took its first major step forward. In addition to including a new version of Finder that combined the previously separate Finder and MultiFinder into a single file manager, System 7 introduced virtual memory, which turned unused disc drive space into random-access memory (RAM), a type of memory designed to improve system performance. Until the release of Mac OS X in 2001, System 7 was Apple's longest running operating system.

In 1997, Apple made the decision to officially rename its operating system series MacOS. Thus, MacOS 8 was the first title in the series not to bear the "System" name. Along with a number of other upgrades, MacOS 8 featured the debut of Sherlock, an integrated web search function that worked in tandem with Finder.

Although not substantially different from macOS 8, 1999's MacOS 9 offered several key improvements. Most notably, it allowed multiple users to create individual accounts on a single machine. MacOS 9 also gave users the ability to download system updates from the internet.

The launch of MacOS X in 2001 marked the debut of the modern MacOS. Touting a completely redesigned user interface and a brand-new code base, MacOS X radically changed the Apple experience.

Photo via iStock/Ake Ngiamsanguan. [Used under license.]

Along with several new features, MacOS X also introduced the concept of protected memory, which prevented applications from corrupting one another's data. The debut of MacOS X also enabled the creation of an array of new programs, including the Safari web browser. OS X, the tenth version of the Macintosh OS, was released on March 24, and it combined a stable Unix-based platform with a visually appealing user interface called Aqua. For software developers, OS X introduced the Carbon application programming interface (API) and the X-code integrated development environment (IDE) for writing code. OS X was a key step in making Macs a central focus of consumer computing culture.
Cat-themed names for OS X releases were popular with enthusiasts and eventually became marketing devices. OS X versions progressed from Cheetah (version 10.0) and Puma (10.1) in 2001 to Jaguar (10.2) in 2002, Panther (10.3) in 2003, Tiger (10.4) in 2005, Leopard (10.5) in 2007, Snow Leopard (10.6) in 2009, Lion (10.7) in 2011, and Mountain Lion (10.8) in 2012.

Beginning with Mavericks (10.9) in 2013, the company moved away from the cat-themed names to names associated with California, subsequently introducing Yosemite (10.10) in 2014 and El Capitan (10.11) in 2015. In 2016, macOS Sierra (10.12) was released. Using the name "macOS" rather than "Mac OS," the release represented Apple's attempt to bring some consistency to the naming of operating systems for all of its products, which also included iOS and watchOS. MacOS High Sierra (10.13) followed in 2017, and macOS Mojave (10.14) followed in 2018.

In 2019, iTunes was dismantled, and macOS Catalina (10.15) allowed users to connect an iPad screen to their Mac. In 2020, macOS Big Sur (11) allowed users to run iOS apps natively on their Mac computers. These were called universal apps. Shortly afterward, in fall 2021, macOS Monterey (12) enabled Shortcuts and a wide range of FaceTime augmentations. The next version of macOS, macOS Ventura (13), was released in late 2022 and featured several new or renamed applications. In June of 2023, Apple announced that version 14 of macOS, known as macOS Sonoma, would be released later that year.

—*Jack Lasky*

Further Reading

Edwards, Benj. "The Little-Known Apple Lisa: Five Quirks and Oddities." *Macworld*, 30 Jan. 2013, www.macworld.com/article/2026544/the-little-known-apple-lisa-five-quirks-and-oddities.html. Accessed 13 July 2023.

Haslam, Karen. "macOS Versions: Every Version Including the Latest." *Macworld*, 13 June 2023, www.macworld.com/article/672681/list-of-all-macos-versions-including-the-latest-macos.html. Accessed 13 July 2023.

"An Illustrated History of Mac OS X." *Tower*, 12 Jan. 2016, www.git-tower.com/blog/history-of-osx. Accessed 13 July 2023.

Moretti, Marcus. "Before Mac OS X, There Was OS 1 Through 9: A History of Apple's Operating System." *Business Insider*, 10 July 2012, www.businessinsider.com/mac-os-i-through-x-2012-7?op=1. Accessed 13 July 2023.

Porter, Jon. "Apple Announces macOS Sonoma with Game Mode and Support for Desktop Widgets." *Verge*, 5 June 2023, www.theverge.com/2023/6/5/23745460/apple-macos-14-sonoma-features-updates-wwdc-2023. Accessed 13 July 2023.

Warren, Christina. "The Evolution of Mac OS, From 1984 to Mountain Lion." *Mashable*, 17 Feb. 2012, mashable.com/2012/02/17/mac-os-timeline/#KKsyA8f26qqR. Accessed 13 July 2023.

MAPS AND MAPPING

ABSTRACT

Mapping encompasses methods for representing geospatial information in paper and digital forms as well as newer digital technologies such as remote sensing and geographic information systems. Maps are needed in areas ranging from civil engineering to regional planning. Computer technologies and the internet have brought profound changes to the ways maps are created and used, and a growing number of mapping applications have been developed for portable devices such as car navigation systems.

BACKGROUND

For more than four thousand years, maps have been used to record and communicate information about Earth. The ancient Greeks created maps to characterize Earth's spherical shape, while legions of Roman soldiers used them as tools in the conquest of new territories. Drawn by hand, early maps showed landscape features using pictorial symbols to represent mountains, rivers, and other physical features. Later technologies such as printing eliminated the need for maps to be painstakingly copied and increased the speed at which maps could be reproduced and disseminated. Exploration, especially during the fifteenth and sixteenth centuries, created a demand for maps to assist in ocean navigation and to document newly discovered locations.

In the twenty-first century, maps assume many forms, including spherical globes, folded paper charts, wall-sized murals, floor plans in architectural drawings, genealogical "family trees," images displayed on tiny electronic screens, "heads up" displays in vehicles, and even holographic projections. As symbolic representations at varying scales, maps show a selection of information including some component of space. The most well-known form of map is a geographic two-dimensional (2D) depiction of a three-dimensional (3D) area for means of orientation, as in the standard road map. However, maps may also be 3D themselves, be interactive, or present much more abstract spaces or concepts. In addition to describing features on the surface of Earth, maps can represent underwater areas, the interiors of caves, celestial bodies such as planets, the structure of deoxyribonucleic acid (DNA), the relationships between the characters in a work of fiction, and the layout of a virtual world, among many other examples. The term "cartography" refers to the process of creating maps and related geographic products. According to the International Cartographic Association, cartography is the art, science, and technology of making maps.

OVERVIEW

Maps can be divided into two general categories: general-purpose and thematic maps. General-purpose maps show a variety of information, commonly including lakes, rivers, roads, cities, and administrative or political boundaries. Most general-purpose maps, such as road and topographic maps, are designed to serve as reference tools. In contrast, the purpose of thematic maps is to show patterns and distributions corresponding to a single topic or theme, such as rainfall or corn production.

Modern maps play many roles in society, from guiding aircraft to providing reference information within school textbooks. Under most circumstances, maps are superior to charts and graphs in their ability to represent distances, directions, and the relative sizes of objects over space. In addition to showing locations of features, maps are able to efficiently represent geographic patterns and the spatial extent of physical entities such as rivers and mountains. They are also useful for showing cultural features such as property boundaries and political or administrative areas. As archives of spatial information, maps are used extensively for recording the locations of historical events as well as spatially referenced scientific data. In addition to their use in storing information, maps serve as important research tools for visualizing geographic data needed for evaluating hypotheses about spatial distributions. The development of computer technology allowed geographic information systems (GISs) and other

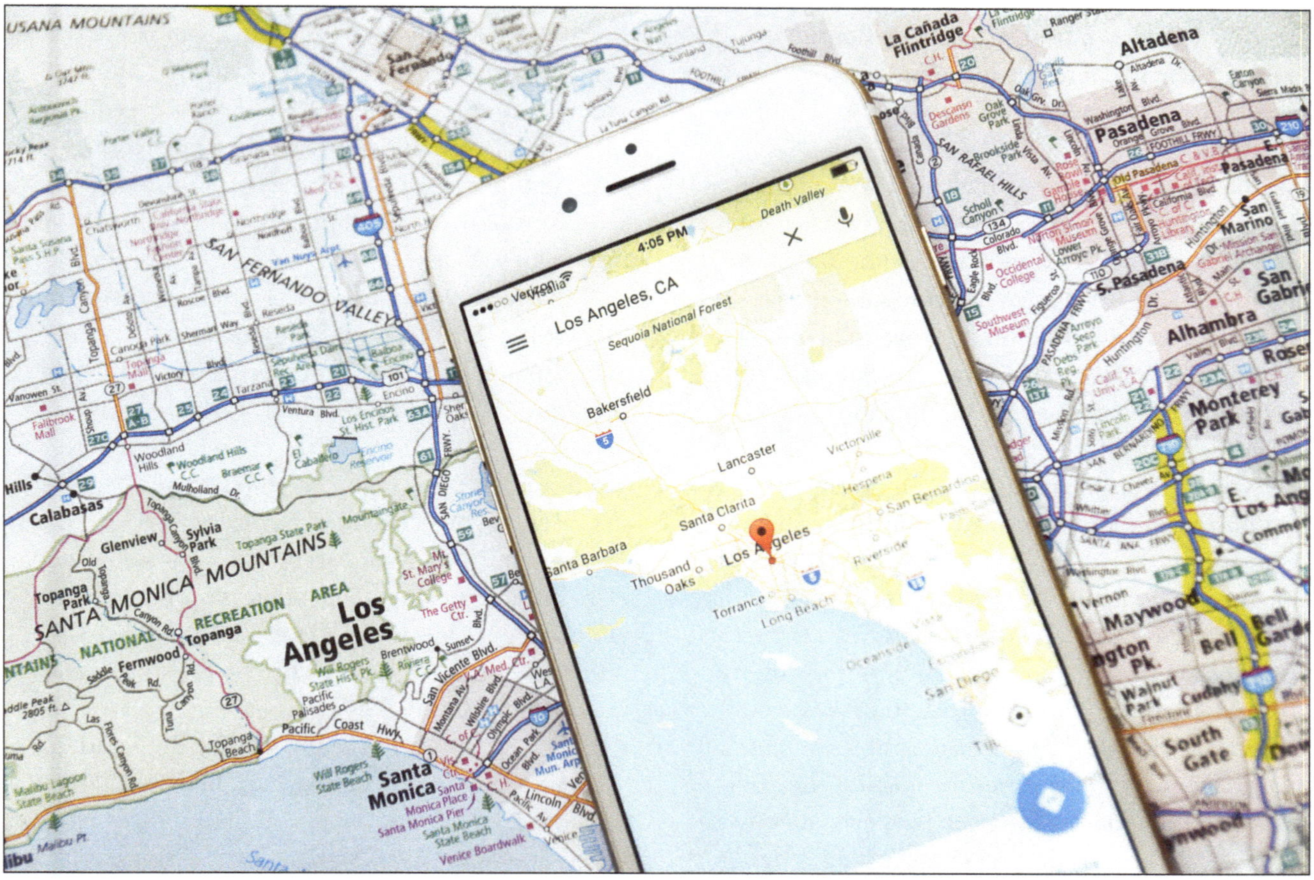

Photo via iStock/bgwalker. [Used under license.]

mapmaking tools to drastically increase the power and potential applications of maps and mapping.

HOW IT WORKS

Scale and Projection. All maps of Earth's surface require some level of reduction. Map scale is an important concept to maps and mapping because it determines the level of map reduction necessary. Scale is the ratio of a distance measured on a map to the distance it represents in the real world, using the same units of measurement. For example, one inch (2.5 centimeters) measured on a map with a scale of 1:24,000 represents 24,000 in (60,960 cm) (or 2,000 feet [610 meters]) in the real world. The scale used on any given map determines the size of the area that can be represented as well as the level of detail that can be depicted. In addition to the representative fraction (e.g., 1:100,000), scale can be shown using a verbal statement such as "one inch equals one mile" or a graphic or bar scale printed directly on the map.

Cartographers draw a distinction between small- and large-scale maps. Small-scale maps represent very large areas of the world and are useful for representing entire, countries, continents, or the globe. In contrast, large-scale maps are capable of showing very small areas at higher levels of detail. City road maps and topographic maps are examples of large-scale maps. Given that geographic mapmaking involves reductions, the cartographer must make choices about the features to be shown and the symbols used to represent those features. This process, called "selection," must be closely tied to the purpose of the map. In addition, complex objects such as coastlines must be simplified to reduce unnecessary detail.

While most maps use a single scale, some may vary in scale throughout in order to achieve a desired effect. Others intentionally distort scale, whether to simplify use (as in many subway maps that stylize the train lines to improve readability) or to emphasize a point (such as a cartogram that displays countries' sizes based on population).

Another important consideration is the choice of projection for displaying mapped information. Projection refers to a mathematical transformation of geographic information from the spherical Earth to a flat or developable surface. Because spherical surfaces cannot be flattened without distortion, the cartographer must select the most important properties to maintain during this transformation. For example, on a world map, it may be desirable to maintain the relative sizes of countries for making comparisons. Although maintaining the property of equivalence, this type of projection distorts the shapes of countries near the poles. The well-known Mercator projection, first published in 1569, is useful for ocean navigation because it preserves shapes as well as lines of constant geographic direction as straight lines. However, a disadvantage of the Mercator projection is that it severely distorts the relative sizes of polar areas, making the island of Greenland in the North Atlantic look larger than the continent of South America, which is actually fifteen times larger than Greenland.

Geographic Grids. Cartographers use geographic grids, or coordinate systems, for the placement of points, lines, and areas to systematically organize geographic information. Most maps are created using either the latitude and longitude coordinate system or a rectangular coordinate system. Latitude-longitude and rectangular grids use Cartesian coordinates (x, y). The point where the prime meridian (0 degrees longitude) and the equator (0 degrees latitude) cross serves as an origin against which all other locations are referenced. Smaller geographic areas are often mapped with rectangular coordinates that use an origin located outside and to the southwest of the area of interest to make all coordinate values positive.

Map Design. Map design is a systematic process involving the selection and arrangement of map elements in a way that facilitates a user's correct interpretation of ideas and concepts. Cartographic design has benefited from developments in both graphic design and cognitive psychology. Because the principal objective of mapmaking is communication, it is important for cartographers to select visual elements and arrangements that present concepts and spatial relationships in ways that can be easily interpreted by the viewer. The design process can require several stages as alternatives are tested and modified. Basic graphic elements manipulated in the design process include points, lines, and area symbols. In addition, the selection and placement of text is an important design element. As suggested by French cartographer Jacques Bertin, graphic elements used in cartography can be made more or less prominent through changes in hue (color), value (darkness or lightness), texture, orientation, size, and shape. Design is affected not only by the choice and design of symbols but also by the arrangement of map elements. In the design process, cartographers must be careful to use space efficiently and to avoid symbols that create unwanted attention or noise. Other important concepts within map design include achieving balance in the arrangement of map elements and creating a visual hierarchy that draws the map reader's attention to features most important to the map's purpose.

Map Output. Developments in map reproduction have followed innovations in printing, beginning with the invention of the printing press by Johannes Gutenberg (1400-1468) in 1440. Early printing technology employed a process called "letterpress," in which ink was applied to raised portions of wooden blocks that were then pressed against paper. Introduced in the early 1700s, engraving involved the application of ink to depressions in a metal printing plate. Engraving improved map reproduction because it enabled the use of finer line widths and improved gray tones and patterns.

Lithography, a printing technique based on the incompatibility of grease and water, was introduced in the 1800s. Most modern paper maps are reproduced using lithographic plates, which are applied to a rotating drum to transfer an image to a moving sheet of paper. Four-color maps use lithographic plates corresponding to the subtractive primary colors of cyan, magenta, and yellow plus black. A wide variety of colors can be created by combining percentages of each primary color. Modern printing presses use computer files to create plates needed to produce color maps, with sheets of paper passing through the press once for each plate.

Maps can also be reproduced in smaller quantities using low-cost output devices such as laser printers and ink-jet plotters. Electrostatic (or xerographic) copying is a common method for reproduction. Laser and ink-jet printers have also become a popular method for reproducing small numbers of maps. The readability of map lettering and the sharpness of lines and gray tones are a function of the printer's resolution as represented in dots per inch (dpi). Low-end printers typically offer 300-dpi output, meaning that each square inch has 90,000 dots. Higher-end printers and plotters offer resolutions exceeding 2,000 dpi.

The introduction of personal communications devices with high-resolution screens has led to an increasing number of maps that are never printed. These virtual maps may be used for tasks such as checking weather conditions or finding a street address. Internet-based Google Maps has been an innovator in making virtual maps, aerial photographs, and satellite images accessible to the public. Google Maps and Google Earth enable the simultaneous viewing of aerial photographs and map information in the form of roads, contour lines, and political boundaries. Mapping tools enable the user to zoom in or out, pan in any direction, and select specific map features and points of interest to be displayed.

Computers and Mapmaking. The single most significant change to cartography during the twentieth century was the introduction of computers that enabled tasks previously requiring hours to be carried out in seconds. Computer cartography, or the use of computers in map production, has been particularly beneficial in enabling cartographers to experiment. Using mapping software, the cartographer can resize and reposition map elements such as text labels, scale bars, and legend boxes. Computers also facilitate mathematical and statistical transformations of tabular data needed for making thematic maps. Some programs enable map images to be combined with other spatial data such as aerial photographs or satellite images. Increasingly, digital databases are replacing paper maps as the principal method for storing and retrieving geographic information.

Before geographic information can be manipulated by a computer, it must be converted to a digital form. The process of capturing points, lines, and areas shown on a paper map in a digital format is called "digitizing." Analogue information can be digitized by tracing features of interest individually using a tablet or table-sized digitizer. Digitizing can also be accomplished by scanning an entire map sheet. Before scanned data are useful, they must be transformed into a system of coordinates such as latitude and longitude. This process is called "geocoding" or "georeferencing."

APPLICATIONS AND PRODUCTS

Since 1985, a significant amount of data used in computer mapping has been converted into digital format. For example, the US GeoData digital line graphs are computer files containing transportation and other line information that has been captured from topographic maps maintained by the US Geological Survey (USGS). The set of digital files called "digital raster graphics" includes information scanned from USGS topographic maps. USGS digital elevation model data files contain grids of elevation values captured every 98 ft (30 m) and are useful for constructing contour maps, 3D terrain diagrams, and computer simulations such as aerial flybys.

Computer technology also makes it possible to create new types of cartographic visualizations such as animated maps. Maps that use a sequence of digital images are called "temporal animations." This type of map is useful for representing changes over time, such as the movement of military fronts during a battle.

Computer technologies have influenced maps and mapping through developments in closely related fields such as aerial photography, GIS, and remote-sensing satellites. Remote sensing involves the capture of information about Earth using sensors and is commonly divided into photographic methods such as aerial photography and nonphotographic imaging that include sensors looking outside visible portions of the electromagnetic spectrum.

Digital orthophoto quadrangles (DOQs) have emerged as an important source of information for constructing maps. DOQs are aerial photographs in digital form that have been rectified (corrected) for camera distortion. In addition to serving as a source from which to extract the location of physical and cultural features, they can be used as background images for maps. Multispectral scanners used by Landsat and other satellites capture large amounts of data using scanners that are sensitive to narrow wavelengths of energy emitted from Earth's surface. Such data can be useful for revealing information that cannot be seen in photographic images, including stress or disease in vegetation.

Radio detection and ranging (radar) is another nonphotographic imaging system that can be used for mapping surface features, underground areas, and atmospheric conditions. Radar uses transmitted energy that is reflected by objects and then captured and converted into an image. Imaging radar used on space shuttle missions has become an important source of elevation data used in a variety of mapping applications. Light detection and ranging (lidar) is an innovation for mapping Earth by measuring the properties of scattered light as it comes into contact with distant objects. Lidar images are created by measuring the time needed for laser pulses to reach Earth's surface. Some applications for lidar include 3D mapping of urban areas and assessments following natural disasters.

CAREERS AND COURSEWORK

College and university courses in cartography are most often taught within departments of geography. Students seeking careers in the mapping sciences should complete college coursework in geography, mathematics, statistics, and technical writing. Depending on the type of maps they will be involved in producing, they may also take coursework in graphic arts, computer programming, physics, remote sensing, computer-aided drafting, database management, planning, or geology. A wide variety of agencies and organizations employ cartographers, GIS analysts, and remote-sensing specialists. The USGS employs cartographers to maintain both paper and digital map products. Its National Imagery and Mapping Agency supports mapping and imagery needs for the military and policy making. The US Department of Agriculture's Natural Resources Conservation Service maintains the Soil Survey Geographic Database, an important spatial data set defining soil boundaries within the United States. The Central Intelligence Agency (CIA) provides mapping products used internally and by other intelligence agencies. State and local government agencies and private organizations also employ geospatial experts. State agencies produce road maps used for promoting tourism, and local government departments maintain property and planning maps. Maps and proprietary data used in mapping are also created by private companies.

SOCIAL CONTEXT AND FUTURE PROSPECTS

Maps and other cartographic products are important within areas ranging from civil engineering and planning to the humanities. For example, maps are used in weekly magazines, in newspapers, and as part of websites. Journalistic maps typically show the countries, cities, or accident sites that are the focus of news events.

Transportation maps such as nautical and aviation charts are of critical importance to the safe movement of ships and aircraft. Within the United States, the National Ocean Service of the National Oceanic and Atmospheric Administration (NOAA) publishes four types of nautical charts: general, sailing, coast, and harbor charts. The Federal Aviation Administration publishes several aviation charts, including the world aeronautical charts at a scale of 1:1,000,000 and sectional charts that represent smaller areas in greater detail at a scale of 1:500,000. Nautical and aviation charts must be updated frequently to reflect changes that can affect safety. Other types of transportation maps include street and highway maps, road atlases, maps showing airline or train routes, and bicycle or walking trail maps.

Topographic maps show rivers, lakes, mountains, and other surface features, as well as cultural features such as roads, cities, levees, and towers. Elevations on topographic maps are shown using contours defined as lines that join points of equal elevation. Bathymetric or hydrographic charts are a related type of map that uses lines to represent underwater depths. The USGS maintains topographic maps as well as maps showing geological cross sections, groundwater areas, earthquake faults, and seismically active areas. Geologic maps have a wide range of uses, from mineral exploration to civil engineering. Most geologic maps use colors, gray tones, patterns, and sometimes letter codes to represent different rock types. In addition to government agencies, energy and mining companies also maintain maps showing fossil fuels, mineral reserves, and areas with the potential to generate power using solar or wind energy.

Maps are important for recording the location of property boundaries and for planning applications. Cadastral maps are maintained by local government agencies to show ownership boundaries as a basis for property taxes. This type of map is also important to insurance companies. Many local government planning agencies maintain maps showing the city or county infrastructure, such as water resources, sewers, electric lines, and gas lines. City and regional planners use maps for zoning changes and to maintain 911 emergency response systems.

The NOAA's National Weather Service is the principal federal agency responsible for disseminating information about weather conditions. Weather maps are used by forecasters and others to monitor or predict atmospheric conditions including temperature, precipitation, wind direction and speed, humidity, cloud cover, and anomalies such as thunderstorms, hurricanes, and tornadoes. Weather maps may show a single parameter such as surface pressure or multiple types of information. Climate maps include some of the same information shown on weather maps but for longer timeframes. Examples include maps depicting monthly temperature highs and lows or average yearly rainfall.

Widespread availability of the internet has created possibilities for the publication of new types of mapping products such as electronic atlases that allow users to generate customized maps by selecting the data to be shown. Electronic atlases enable users to search on place-names and easily locate cities and find other information. An added benefit of the internet is its ability to incorporate updates as soon as new information becomes available. In the twenty-first century, maps also feature prominently in numerous forms of media, including films, television programs, video games, and virtual and augmented reality projects. In the field of the visual arts, some artists use cartographic practices and the visual form of the map to explore themes such as social, political, and economic inequity.

—*Thomas A. Wikle*

Further Reading

Amadeo, Ron. "Google Maps Gets Augmented Reality Search Results." *Ars Technica*, 28 Sept. 2022, arstechnica.com/gadgets/2022/09/google-maps-gets-augmented-reality-search-results. Accessed 13 July 2023.

Cartwright, William, Michael P. Peterson, and Georg F. Garner, editors. *Multimedia Cartography*. 2nd ed., Springer, 2007.

Dodge, Martin, Rob Kitchin, and Chris Perkins, editors. *Rethinking Maps: New Frontiers in Cartographic Theory*. Routledge, 2009.

Field, Kenneth. *Cartography: A Compendium of Design Thinking for Mapmakers*. ESRU, Press, 2018.

Hancock, James G., Hennie Haworth, Stuart Hill, and Sarah King. *The Art of Map Illustration*. Quarto Publishing, 2018.

Kent, Alexander, and Peter Vujakovic, editors. *The Routledge Handbook of Mapping and Cartography*. Routledge, 2017.

Kraak, Menno-Jan, and Ferjan Ormeling. *Cartography: Visualization of Spatial Data*. 3rd ed., Prentice-Hall, 2010.

Okada, Alexandra, Simon J. Buckingham Shum, and Tony Sherborne, editors. *Knowledge Cartography: Software Tools and Mapping Techniques*. Springer, 2008.

Pearson, Luke Caspar, and Sandra Youkhana. *Videogame Atlas: Mapping Interactive Worlds*. Thames & Hudson, 2022.

Relf, Emily. "Maps Providing a Sense of Direction." *Art21*, 29 June 2021, art21.org/read/maps-providing-a-sense-of-direction. Accessed 13 July 2023.

Robinson, Arthur H., et al. *Elements of Cartography*. 6th ed., Wiley, 1995.

Wilford, John N. *The Mapmakers: The Story of the Great Pioneers in Cartography from Antiquity to the Space Age*. Pimlico, 2002.

Measurement and Units

ABSTRACT

Measurement is the act of quantifying a physical property, an effect, or some aspect of them. Seven fundamental properties are recognized in measurements: length, mass, time, electric current, thermodynamic temperature, amount of a substance, and luminous intensity. In addition, two supplementary or abstract fundamental properties are defined: plane and solid angles. The base units for the seven fundamental properties can be manipulated to produce derived units for other quantities that are the effect of combinations of these fundamental properties. For instance, a Newton is a derived unit measuring force and weight, and a square meter is a derived unit used to measure area.

Historically, a number of units representing different amounts of the same properties have been used in various cultures around the world. For example, the United States traditionally uses the US customary system (miles, cups, pints, ounces, and so on), while most other industrialized nations use the metric system (kilometer, milliliter, liters, grams, and so on). The metric system, developed in France in the late eighteenth century, represents the first true standard measurement system. The theory and physical practice of measurement is constant no matter what system of units is being used.

BACKGROUND

Measurement has the purpose of associating a dimension or quantity proportionately with some fixed reference standard. Such an association is intended to facilitate the communication of information about a physical property in a manner that allows it to be reproduced in concept, or in actuality if needed. The function of an assigned unit that is associated with a definite dimension is to provide the necessary point of reference for someone to comprehend the exact dimensions that have been communicated. For example, a container may be described as having a volume of eight cubic feet. Such a description is incomplete, however, because it does not state the shape or relative proportions of the container. The description applies equally well to containers of any shape, whether they be cubic, rectangular, cylindrical, conical, or some other shape. At a more basic level, however, there is the assumption that the person who receives the description has the same understanding of what is meant by "cubic feet" as does the person who provided the description. This is the fundamental principle of any measurement system: to provide a frame of reference that is commonly understood by a large number of people and that indicates exactly the same thing to each of them.

A measurement system, no matter what its basic units, must address a limited group of fundamental properties. These are length, mass, temperature, time, electric current, amount of a substance, and luminous intensity. In addition, it must also be able to describe angles. All other properties and quantities can be described or quantified by a combination of these fundamental properties.

Historically, measurement systems were generally based on various parts of the human body, and some of these units have remained in use in the modern world. The height of horses, for example, is generally given as being so many hands high. In many languages, the word for "thumb" and the word for "inch" are closely related, if not identical: In French, both words are *pouce*; in Norwegian, *tomme* and *tommelfinger*; and in Swedish, *tum* and *tumme*. Although using the thumb as a basis for measurement is convenient in that almost everyone has one, in practice, the generally accepted thumb size tended to vary from city to city, making it impossible to interchange parts made in different locations by separate artisans.

Units of measurement traditionally have been defined by a decree issued by a political leader. In ancient times, units of length often corresponded to certain parts of the human body: The foot was based on the human foot and the cubit on the length of the forearm. Other units used to weigh various goods were based on the weight of commonly available items. Examples include the stone (still used widely but not officially in Britain) and the grain. Invariably, the problem with such units lay in their variability. A stone weight may be defined as equivalent to the size of a stone that a grown man could enclose by both arms, but grown men come in different sizes and strengths, and stones come in different densities. A grain of gold may be defined as equivalent to the weight of a single grain of wheat; however, in drier years, when grains of wheat are smaller and lighter, the worth of gold is significantly different from its worth in years of good rainfall, when grains of wheat are larger and heavier.

In practice, any of these fundamental properties can be defined relative to any randomly selected relevant object or effect. Logically, though, for a measurement system to be as effective as possible, the objects and effects that are selected as the defined units of fundamental properties must be readily available to as many people as possible and readily reproducible. If, for example, a certain king were to decree that the "foot" to be used for measurement was the human foot, a great deal of confusion would result because of the variability in the size of the human foot from person to person and from a person's left foot to right foot. Should he instead decree that the "foot" would correspond to his own right foot and no other, the unit of measurement becomes significantly more precise, but at the same time, the decree raises the problem of how to verify that measurements being taken are in fact based on the decreed length of the foot. A physical model of the length of the king's right foot must then be made available for comparison. The same logic holds true for any and all defined properties and units. All measurements made within the definitions of a specific system of measurement are therefore made by comparing the proportional size of an effect or property to the defined standard units.

OVERVIEW

Measurements and units are most useful when they are standardized so that measurements mean the same thing to everyone and are comparable. One method of solving the problem of variability in measurement is to establish a standard value for each unit and to regularly compare all measuring devices to this standard.

Historically, many societies required that measuring devices be physically compared with and calibrated against official standards. Physical representations of measuring units were made as precisely as possible and carefully stored and maintained to serve as standards for comparison. In ancient Egypt, the standard royal cubit was prepared as a black granite rod and most likely kept as one of the royal treasures by the pharaoh's chief steward. That simple stone object would have been accorded such a status because of its economic value to trade and construction and because it helped maintain the pharaoh's reputation as the keeper of his kingdom.

Later nations and empires also kept physical representations of most standard units appropriate to the economics of trade. The British made and kept definitive representations of the yard and the pound, just as the French made and kept definitive representations of the meter and the kilogram after developing the metric system in the late 1700s. Early on, countries recognized that the representative of the unit must not be subject to change or alteration. The Egyptian standard royal cubit was made of black granite; the standard meter and the kilogram, as well as the foot and the pound, were made from a platinum alloy so that they could not be altered by corrosion or oxidation.

France adopted the metric system as its official measuring system in 1795, and that system was standardized in 1960 as the International System of Units (*Le Système International d'Unités*, known as SI). The units of the metric system were defined based on unchanging, readily reproducible physical properties and effects rather than any physical object. For example, the standard SI unit of length, the meter, was originally defined to be one ten-millionth of the distance from the equator to the north polar axis along the meridian of longitude that passed through Paris, France. In 1960, for greater accuracy, the definition of the meter was based on wavelengths of light emitted by the krypton-86 isotope. In 1983, it was changed to the length traveled by light in a vacuum during 1/299,792,458 of a second. Similarly, the length of a standard second of time had been defined as a fraction of one rotation of the earth on its axis, until it was realized that the rate of rotation was not constant but rather was slowly decreasing, so that the length of a day is increasing by 0.0013 second per hundred years. In 1967, it was formally redefined as a duration of 9,192,631,770 periods of the radiation corresponding to the transition between the two hyperfine levels of the ground state of the cesium-133 atom. As technology develops, it becomes possible to measure smaller quantities with finer precision. This capability has been the principal that permits ever more precise definitions of the basic units of measurement.

In 2017, it was announced that the kilogram, an SI base unit, would be redefined by the General Conference on Weights and Measures in November 2018, as a constant rather than based on a physical object. Three other SI base units were also under consideration for redefinition. In November 2018, the sixty countries participating in the conference unanimously voted to redefine the kilogram, kelvin, mole, and ampere in terms of constants. The new definitions came into effect in May of the following year.

The application of measuring procedures is of fundamental importance in the economics of trade, especially in the modern global economy. In manufacturing, engineering, the sciences, and other fields, the accuracy and precision of measurement are essential to statistical process control and other quality-control techniques. All such measurement is a process of comparing the actual dimensions or properties of an item with its ideal or design dimensions or properties. The definition of a standard set of measuring units greatly facilitates that process.

APPLICATIONS AND PRODUCTS

It is quite impossible to calculate the economic effects that various systems of measurement, both good and bad, have had throughout history. Certainly, commonly understood and accepted units of weight, distance, and time have played a major role in facilitating trade between peoples for thousands of years. In many ways, all human activity can be thought of as dependent on measurement.

The study of measurement and measurement processes is known as "metrology." In essence, metrology is the determination and application of more precise and effective means of measuring quantities, properties, and effects. The value in metrology derives from how the information obtained is used. This is historically and traditionally tied to the concepts of fair trade and of well-made products. Measurement serves to ensure that trade is equitable, that people get exactly what they are supposed to get in exchange for their money, services, or other trade goods, and that as little goes to waste as possible.

By far, the largest segment of metrology deals with the design, production, and calibration of the various products and devices used to perform measurements. These devices range from the simplest spring scale or pan balance to some of the most sophisticated and specialized scientific instruments ever developed. In early times, measurements were restricted to those of mass, distance, and time because these were the foremost quantities used in trade. The remaining fundamental properties of electric current, amount of a substance, temperature, and luminous intensity either remained unknown or were not of consequence.

Weight Determination. Originally, weights were determined with relative ease by the use of the pan balance. In the simplest variation of this device, two pans are suspended from opposite ends of a bar in such a way that they are at equal heights. The object to be weighed is placed in one of the pans and objects of known weights are placed in the other pan until the two pans are again at equal heights. The precision of the method depends on the ability of the bar to pivot as freely as possible about its balance point; any resistance will skew the measurement by preventing the pans from coming into proper balance with each other.

Essentially, all balances operate on the principle of comparing the weight of an unknown object against the weight of an accurately known counterweight, or some property such as electric current that can in turn be measured very accurately and precisely. The counterweight may not actually be a weight but rather an electronic pressure sensor or something that can be used to indicate the weight of an object such as the tension of a calibrated spring or a change in electrical resistance. Scales used in commercial applications, such as those in grocery stores, grain depots, and other trade locations, are inspected and calibrated on a regular basis according to law and regulations that govern their use.

Length Determination. For many practical purposes, a linear device such as a scale ruler or tape measure is all that is needed to measure an unknown length. The device used should reflect the size of the object being measured and have scale markings that reflect the precision with which the measurement must be known. For example, a relatively small dimension being measured in a machine shop would be measured by a trained machinist against a precision steel scale ruler with dimensional markings of high precision. Training in the use of graduated scale markings typically enables the user to read them accurately to within one-tenth of the smallest division on the scale. High-precision micrometers are generally used to make more precise measurements of smaller dimensions, and electronic versions of such devices provide the ability to measure dimensions to extremely precise tolerances. Smaller dimensions, for which the accuracy of the human eye is neither sufficient nor sufficiently reproducible, are measured using microscope techniques and devices. Accurate measurements of larger dimensions have always been problematic, especially when the allowable tolerance of the measurement is very small. This has been overcome in many cases by the measuring machine, a semirobotic device that uses electronic control and logic programming to determine the distance between precise points on a specific object.

The 1960 definition of the meter was achieved through the use of a precision interferometer, a device that uses the interference pattern of light waves such that the number of wavelengths of a specific frequency of light can be counted. Using this device, the meter was precisely defined as the distance equal to 1,650,763.73 wavelengths of the $2p_{10}$-$5d_5$ emission of krypton-86 atoms. In 1983, however, the meter was defined as the distance traveled by light in vacuum in 1/299,792,548 of a second.

Time Measurement. Essentially, the basic unit of time measurement in all systems is the second. This is a natural consequence of the fact that the length of a day is the same everywhere on the planet. The natural divisions of that period of time according to the observed patterns of stars and their motions almost inevitably results in twenty-four equal divisions. A natural result of the metric system is that a pendulum that is one meter in length swings with a period of one second. Pendulum clockworks have been used to measure the passage of time, coordinated to the natural divisions of the day, for thousands of years. More precise time measurements have become possible as better technology has become available. With the development of the metric system, the unit duration of one second was defined to correspond to the appropriate fraction of one rotation of the planet. Until the development of electronic methods and devices, this was a sufficient definition. However, with the realization that the rotation of the planet is not constant but is slowly decreasing, the need to redefine the second in terms that remain constant in time led to its redefinition in 1967. According to the new definition, a second is the time needed for a cesium-133 atom to perform 9,192,631,770 complete oscillations.

Temperature Determination. Of all properties, measurement of thermodynamic temperature is perhaps the most relative and arbitrary of all. The thermometer was developed before any of the commonly used temperature scales, including the Fahrenheit, Celsius, Kelvin, and Rankine scales. All are based on the freezing and boiling points of water, the most readily available and ubiquitous substance on the planet. The Fahrenheit scale arbitrarily set the freezing point of saturated salt water as 0 degrees. The physical dimensions of the scale used on Fahrenheit's thermometer resulted in the establishment of the boiling point of pure water as 212 degrees. The Celsius scale designated the freezing point of pure water to be 0 degrees and the boiling point of pure water to be 100 degrees. The relative sizes of Fahrenheit and Celsius degrees are thus different by a factor of 5 to 9. Conversion of Fahrenheit temperature to Celsius temperature is achieved by subtracting 32 degrees and multiplying the result by 5/9; to convert from Celsius to Fahrenheit, first multiply by 9/5, then add 32 degrees to the result. The temperature of -40 degrees is the same in both scales.

Each temperature scale recognizes a physical state called "absolute zero," at which matter contains no thermal energy whatsoever. This must physically be the same state regardless of whether Fahrenheit or Celsius degrees are being used. Because of this, two other scales of temperature were developed. The Rankine scale, established as part of the school of British engineering, uses the Fahrenheit degree scale, beginning at 0 degrees at absolute zero. The Kelvin scale uses the Celsius degree scale, beginning at 0 degrees at absolute zero.

Temperature is accurately measured electronically by its effect on light in the infrared region of the electromagnetic spectrum, although less accurate physical thermometers remain in wide use.

Amount of a Substance. Of all the fundamental properties, amount of a substance is the least precisely known. The concept is intimately linked to the modern atomic theory and atomic weights, although it predates them by almost one hundred years. Through studies of the properties of gases in the early 1800s, Italian Amedeo Avogadro concluded that a quantity of any pure material equal to its molecular weight in grams contained exactly the same number of particles. This number of particles came to be referred to as the Avogadro constant. Thus, 2 grams of hydrogen gas (molecular weight = 2) contains exactly the same number of molecules as does 342 grams of sucrose (common white table sugar, molecular weight = 342) or 18 milliliters of water (molecular weight 18, density = 1 gram per milliliter). Calculations have determined the value of the Avogadro constant to be about 6.02214×10^{23}. Because all twenty-three decimal places are not known, the absolute value has not yet been determined, and therefore amount of a substance is the least precisely known unit of measurement.

The amount of material represented by the Avogadro constant (of atoms or molecules) is termed the "mole," a contraction from "gram molecular weight." The number is constant regardless of the system of measurement that is used, as a result of the indivisibility of the atom in modern chemical theory. Thus, a gram molecular weight of a substance and a pound molecular weight of a substance both contain a constant number of molecules.

Electric Current. Of course, atoms are not indivisible in fact. They consist of a nucleus of protons and neutrons, surrounded by a cloud of electrons. The electrons can move from atom to atom through matter, and such movement constitutes an electric current. More generally, any movement of electronic charge between two different points in space defines an electric current. Although some controversial evidence suggests that electricity may have been known in ancient times, any serious study of electricity did not begin until the eighteenth century, after the discovery of the electrochemical cell. At that time, electricity was thought to be a mysterious fluid that permeated matter. With the discovery of subatomic particles (electrons and protons) in 1898 and the subsequent development of the modern atomic theory, the nature of electric currents came to be better understood. The ampere, named after one of the foremost investigators of electrical phenomena, is the basic unit of electric current and corresponds to the movement of one mole of electrons for a period of one second.

The development of the transistor—and the electronic revolution that followed—made it possible to measure extremely small electric currents as well as corresponding values of voltage, resistance, induction, and other electronic functions. This made it possible to precisely measure fundamental properties by electronic means rather than physical methods.

Measurement of Luminous Intensity. Until it became important and necessary to know precisely the intensity of light being emitted from a light source, particularly in the fields of astronomy and physics, there was no need for a fundamental unit of luminous intensity. Light intensities were generally compared, at least in postindustrial Revolution Europe, to the intensity of light emitted from a candle. This sufficed for general uses such as lightbulbs, but the innate variability of candle flames made them inadequate for precision measurements. The candela was set as the standard unit of luminous intensity and corresponds to the output of energy of 0.00146 watts. Modern lightbulbs typically are rated at a certain number of watts, but this is a measure of the electric power that they consume and not of their luminous intensity.

CAREERS AND COURSEWORK

Measurement and units, as they apply to metrology, are part of every technical and scientific field. The student planning on a career in any of these areas can expect to learn how to use the metric system and the many ways in which measurement is applied in a specific field of study. Because measurement is related to the fundamental properties of matter, a good understanding of the relationships between measurements and properties will be essential to success in any chosen field. In addition, the continuing use of multiple measuring systems, such as the metric, US customary, and British Imperial systems, in the production of goods means that an understanding of those systems and how to convert between them will be necessary in many careers.

Specific courses that involve measurement include geometry and mathematics, essentially all of the physical science and technology courses, and design and technical drawing courses. Geometry, which literally means "earth measurement," is the quintessential mathematics of measurement and is essential for careers in land surveying, architecture, agriculture, civil engineering, and construction. Gaining an understanding of trigonometry and angular relationships is particularly important. The physical sciences, such as physics, chemistry, and geology, employ analytical measurement at all levels. Specializations in which measurement plays a prominent role are analytic chemistry and forensic research. Technology programs such as mechanical engineering, electronic design, and biomedical technology all rely heavily on measurement and the application of metrological techniques in the completion of projects that the engineer or technologist undertakes.

SOCIAL CONTEXT AND FUTURE PROSPECTS

Measurement and an understanding of the units of measurement are an entrenched aspect of modern society, taught informally to children from early childhood and formally throughout the course of their schooling. They are fundamental to the continued progress of technology and essential to the determination of solutions to problems as they arise, as well as to the development of new ideas and concepts. The need for individuals who are trained in metrology and who understand the relationship and use of measurements and units will be of increasing importance in ensuring the viability of both new and established industries. Particular areas of growth and continuing development include the fields of medical research and analysis, transportation, mechanical design, and aerospace. The accuracy and precision of measurement in these fields is critical to the successful outcome of projects.

—Richard M. J. Renneboog

Further Reading

Butcher, Kenneth S., Linda D. Crown, and Elizabeth J. Gentry, editors. *The International System of Units (SI): Conversion Factors for General Use*. National Institute of Standards and Technology, 2006.

Cardarelli, Francois. *Encyclopedia of Scientific Units, Weights and Measures: Their SI Equivalences and Origins*. Springer, 2003.

De Courtenay, Nadine, Olivier Darrigol, and Oliver Schlaudt, editors. *The Reform of the International System of Units (SI): Philosophical, Historical and Sociological Issues*. Routledge, 2019.

Kuhn, Karl F. *In Quest of the Universe*. 6th ed., Jones and Bartlett, 2010.

"The Redefinition of the SI Units." *NPL*, www.npl.co.uk/si-units/the-redefinition-of-the-si-units. Accessed 13 July 2023.

"SI Redefinition." *NIST*, www.nist.gov/si-redefinition. Accessed 13 July 2023.

Tavernor, Robert. *Smoot's Ear: The Measure of Humanity*. Yale UP, 2007.

MEDIA LITERACY

ABSTRACT

Media literacy is defined as the ability to access, communicate, interpret, and evaluate messages or texts across a range of digitally mediated forms. Individuals who are media literate can identify how, why, and for what purposes messages or texts are constructed. Scholars and educators are calling for the development of new media literacy: a form of critical literacy that attends to the ongoing development of media and information and communication technologies (ICTs), multimodality, and the convergence of media toward the privatization and commercialization of knowledge.

BACKGROUND

The first wave of media literacy research in the 1960s and 1970s was primarily concerned with investigating the ideological messages embedded in print and digital texts. The second wave of research, in the 1980s, shifted the focus to the reader or the audience of texts. In 1982, the United Nations Educational, Scientific and Cultural Organization (UNESCO) Grünwald Declaration established media literacy as a global concern. It called for educational and political systems to promote the development of a critical understanding of communication and information technologies for all citizens, specifically in relation to images, words, and sounds. The third wave, in the 1990s, highlighted the ways in which audiences negotiate the meaning of various texts in relation to their social locations—for example, their social class, gender, ethnicity, and sexuality—and social and cultural contexts.

The rise of the internet in the 1990s catalyzed debates regarding how to use digital technology for learning and education. Since 2000, governmental, educational, and institutional support for media literacy has continued to grow, as has the everyday use of ICTs for communication, corporate and domestic labor, education, and entertainment. New media literacy highlights, among other things, the convergence of media ownership and the privatization and commercialization of media over time.

OVERVIEW

Media literacy builds on a foundation of active inquiry and critical thinking and includes the ability to access, communicate, interpret, evaluate, create, reflect on, and interact with digitally mediated texts. Interdisciplinary research has contributed literature that examines the relationships between media literacy and cultural practices, public health and well-being, societal norms and values, and education and work opportunities.

The following themes were identified across research: (1) mass media influences both individuals and society; (2) this influence is perceived to be subtle, and individuals who are passive consumers are more susceptible to media influence; (3) a goal of media literacy is to enable individuals to become aware of the role and influence that media has on their lives and to empower them to use media as a tool through which they can achieve their own goals; (4) media literacy is learned and develops over time, and; (5) to increase media literacy individuals must be engaged in active knowledge construction and skill development.

Media shapes individually lived experiences, cultures, and societies. Cultural and corporate norms, values, ideologies, and practices are mediated by modern day modes of storytelling from movies, videogames, television, music, and the internet.

Drawing on educational reformer John Dewey's argument in the early twentieth century that a strong democracy relies on an educated and literate citizenry, scholars of new media position media literacy as a vital ability for a democratic citizenry and socially just world.

—Kristen P. Goessling and Jennifer A. Vadeboncoeur

Further Reading

Alvermann, Donna E., and Margaret C. Hagood. "Critical Media Literacy: Research, Theory, and Practice in 'New Times.'" *Journal of Educational Research*, vol. 93, no. 3, 2000, pp. 193-205.

Buckingham, David. *The Media Literacy of Children and Young People: A Review of the Research Literature on Behalf of Ofcom.* Ofcom, 2005.

Christian, Sue Ellen. *Everyday Media Literacy: An Analog Guide for Your Digital Life.* 2nd ed., Routledge, 2023.

"Core Principles of Media Literacy Education in the United States." *NAMLE*, namle.net/resources/core-principles. Accessed 13 July 2023.

Luke, Carmen. "As Seen on TV or Was That My Phone? New Media Literacy." *Policy Futures in Education*, vol. 5, no. 1, 2007, pp. 50-58.

"Media Literacy: A Definition and More." *Center for Media Literacy*, www.medialit.org/media-literacy-definition-and-more. Accessed 13 July 2023.

Potter, W. James. *Media Literacy.* 10th ed., SAGE Publications, 2021.

_____. "Review of Literature on Media Literacy." *Sociology Compass*, vol. 7, no. 6, 2013, pp. 417-35.

Share, Jeff. *Media Literacy Is Elementary: Teaching Youth to Critically Read and Create Media.* Lang, 2009.

Metadata

ABSTRACT

Metadata is descriptive, structural, or administrative information included as part of a digital file, electronic record, or other resource. It often includes information such as the file or resource's title, creator, and structure. Metadata is used to catalog and preserve information. By including keywords and descriptions within the metadata, creators or archivists enable users to search for specific records as well as discover similar ones. Metadata is related to the fields of both computer science and library science, as it combines traditional cataloging methods with new electronic file types and distribution methods.

BACKGROUND

Often described as data about data, metadata consists of information either embedded in a digital file or stored as a separate record in a database or similar directory. The concept is closely related to the cataloging systems used for centuries in libraries and archives, which typically consist of collections of listings—either physical or, following the advent of computers, digital—that include the title, the creator, and a description of each work or artifact, among other key facts. Metadata is intended to assist people in searching for and retrieving the files or information they seek. Similarly, this data enables archivists and catalogers to ensure that information will be well organized and preserved for future reference.

OVERVIEW

Metadata is an evolving concept, and the term has consequently been used in various ways by different organizations and in different fields. Considered broadly, metadata can be divided into three areas based on the type of information being preserved. Information such as the title and author of an e-book falls under the classification of descriptive metadata, while data regarding the structure of the e-book—the order of the pages and chapters—is considered structural metadata. Administrative metadata is somewhat more loosely defined but generally includes information regarding when the file was created. This category may also include the subcategories of technical metadata, which concerns file formats and sizes; preservation metadata, which can include information about the file's relationships to other files; and intellectual property rights metadata, which is used to capture copyright information. In other cases, technical, preservation, and intellectual property rights metadata are considered distinct categories.

When an item exists in only physical form, as in the case of physical artifacts and texts that have not been digitized, the metadata for that item is typically stored in a searchable database. When dealing with electronic files and other resources, however, creators and catalogers often embed metadata within the files. In hypertext markup language (HTML) files such websites, metadata is typically embedded in the files through the use of specialized HTML tags. Image, audio, and video files can likewise include embedded metadata, usually added by the creator or, in the case of commercially distributed films or music, the studio or distributor. Some cameras, including many smartphone cameras, automatically embed metadata such as the date, time, and place a photograph was taken in each file they produce. This capability has led to concerns that sensitive personal information could be distributed in the form of metadata without an individual's knowledge or permission, and the use of metadata thus remains a topic of debate among individuals and institutions concerned about privacy.

In 2013, National Security Agency (NSA) contractor Edward Snowden revealed that the agency was collecting, analyzing, and storing bulk telephone and online metadata from US citizens and others under the auspices of the USA PATRIOT Act (2001). Though the agency argued that the practice

Photo via iStock/atakan. [Used under license.]

was necessary to detect and fight terrorism, the US Court of Appeals ruled that such bulk data collection and retention was illegal in May 2015. In June 2015 President Barack Obama signed the USA Freedom Act, which put new restrictions on the NSA surveillance of phone metadata. The issue of the US government's access to metadata remained a contentious one in the years that followed, particularly in light of the fact that government agencies did not need to obtain a warrant in order to access some metadata stored in remote cloud storage.

—Joy Crelin

Further Reading

Baca, Murtha, editor. *Introduction to Metadata*. 3rd ed., Getty Research Institute, 2016.

Cox, Mike, Ellen Mulder, and Linda Tadic. *Descriptive Metadata for Television*. Focal, 2006.

Garciìa-Barriocanal, Elena, et al., editors. *Metadata and Semantic Research*. Springer, 2011.

Hider, Philip. *Information Resource Description: Creating and Managing Metadata*. American Library Association, 2012.

Horodyski, John. *Metadata Matters*. CRC Press, 2022.

Lima, Cristiano. "Want Our Metadata? Get a Warrant, Rep. Ted Lieu Says." *Washington Post*, 20 Apr. 2022, www.washingtonpost.com/politics/2022/04/20/want-our-metadata-get-warrant-rep-ted-lieu-says. Accessed 13 July 2023.

Miller, Steven Jack. *Metadata for Digital Collections*. 2nd ed., ALA Neal-Schuman, 2022.

National Information Standards Organization. *Understanding Metadata*. NISO Press, 2004.

Park, Jung-ran. *Metadata Best Practices and Guidelines*. Routledge, 2011.

Roberts, Dan, and Spencer Ackerman. "NSA Mass Phone Surveillance Revealed by Edward Snowden Ruled Illegal." *The Guardian*, 7 May 2015, www.theguardian.com/us-news/2015/may/07/nsa-phone-records-program-illegal-court. Accessed 13 July 2023.

Smiraglia, Richard P., editor. *Metadata: A Cataloger's Primer*. Routledge, 2012.

Steinhauer, Jennifer, and Jonathan Weisman. "U.S. Surveillance in Place since 9/11 Is Sharply Limited." *New York Times*, 2 June 2015, www.nytimes.com/2015/06/03/us/politics/senate-surveillance-bill-passes-hurdle-but-showdown-looms.html. Accessed 13 July 2023.

Metaverse

ABSTRACT

A metaverse is a virtual world in which users can interact with one another. Metaverses can allow individuals who are far apart to interact; for example, coworkers could have a virtual meeting. In the third decade of the twenty-first century, popular video games such as Fortnite and Minecraft operated in metaverses, and notable companies such as Meta sought to bring metaverses to life through virtual reality (VR) technology. As technology advances, metaverse connections are expected to become increasingly common.

BACKGROUND

The word metaverse comes from the prefix *meta*, which means "beyond," and *verse*, short for "universe." Author Neal Stephenson coined the term and used it in 1992 in *Snow Crash*, a science fiction novel. In his book, people wear headsets that allow them to interact in the Metaverse, a virtual world that operates like a video game.

A near-future, cyberpunk science fiction novel, *Snow Crash* is also a dystopian novel in which the US government and its agencies, including intelligence agencies, are privately owned. The protagonist, Hiro Protagonist—he uses a pseudonym—is a skilled hacker who sells information to the Central Intelligence Agency (CIA). He was one of the earliest citizens of the Metaverse and wrote some of the code during the development stage, so he knows some of its secrets. In real life, he ekes out an existence, but in the Metaverse, he is a VIP living the good life. Hiro's mentor in hacking, Da5id, encounters a drug called Snow Crash that both destroys Da5id's mind and crashes his computers. Hiro realizes Snow Crash is more than a drug; it is also a computer virus and an ancient physical-world virus that affects individuals' mental functions and brainwashes them.

The novel is known mostly for Stephenson's vision at a time when technological advances were developing rapidly. Cell phones existed but were much larger than those of the twenty-first century, and without smartphone capabilities. Use of the internet was limited to the few who could plug into it for research. Finding information could be tedious; for example, libraries relied on card catalogs, which individuals had to sift through to find books, periodicals, and other sources. Stephenson imagined a host of future developments, including avatars that represent users in the Metaverse; global internet access; wireless access; smartphones; VR goggles; digitization of books, videos, and other information; digital currency; Google Earth; digital rights management (DRM) to protect data; and rogue hackers. Stephenson went on to serve as chief futurist for the VR start-up Magic Leap for several years, and in 2022, he announced the creation of Lamina1, a form of blockchain technology intended for use with metaverse projects.

OVERVIEW

Technology available in modern times offers users two means of access to the metaverse. These are virtual reality (VR) and augmented reality (AR). Virtual reality involves a completely created digital world, such as a Sims community or a battleground in an adventure video game. The user is completely immersed in the metaverse through the use of a VR headset. VR devices commercially available by 2023 included the Meta Quest 2, the Sony PlayStation VR2, and the HTC Vive Pro 2. AR is the real world overlaid with digital elements. The *Pokemon Go* game, for example, has players use their smartphone cameras to "find" digital creatures. In addition to smartphones, devices used to access AR experiences include smart glasses and certain VR headsets, which may offer what is sometimes referred to as mixed reality (MR) capabilities. All the technologies involved may be described as extended reality (XR) technology.

Many Silicon Valley innovators, including Facebook creator Mark Zuckerberg, have been inspired to develop metaverses. His company, renamed Meta in October of 2021, has invested heavily in headsets for VR and AR and in 2021 introduced Horizons Workrooms, where people can work together virtually. Microsoft is another big player in the push to build metaverses. However, a great deal of attention is being focused on finding ways to loop people into metaverses without headsets, which only a small percentage of the population owns. Instead, companies hope people can access metaverses using phones, tablets, and other common devices.

Possibilities for metaverses are virtually limitless. Beyond gaming, individuals might visit a travel destination or attend a concert. For example, rapper Lil Nas X performed virtual concerts in *Roblox*, a children's game, in November 2020, garnering 33 million views in a few days. The metaverse around the performer changed to match each song. In 2021, Ariana Grande recorded a concert presented in *Fortnite* in which attendees were snatched mid-show by a roller coaster-type ride, which whisked them away to shoot at a monster in the game. *Fortnite* was a veteran at staging virtual concerts by that time, having first presented a show featuring Marshmello in 2019.

The metaverse field has attracted many investors and involves a wide range of companies. Republic Realm, for example, is a company that manages digital real estate nonfungible tokens (NFTs) in multiple metaverses. This includes digital fantasy islands, destinations such as art galleries and pools, and a shopping district called Metajuku, with stores leased and operated by digital wearables brands.

Some critics have cautioned against embracing the metaverse unreservedly. The companies developing these worlds are investigating means to profit in them, such as marketing digital and real-world goods as product placements in metaverses or signing deals with companies to include digital merchandise for sale to consumers. Prominent indications of this trend have emerged as luxury fashion houses team up with video game companies. For example, fashion company Balenciaga commissioned a game, *Afterworld: The Age of Tomorrow*, and gave gamers the chance to dress their avatars in Balenciaga attire. The fashion house also sent VR headsets to select individuals in 2020 so journalists and VIPs could enter and experience its digital runway show during the global COVID-19 pandemic. Although it was not the first fashion brand to use VR and gaming, the game and digital runway show were very well-received. Other companies create digital versions of designers' physical fashions and, for a fee that might be several hundred dollars, "dress" a customer's avatar in them. Some limited-edition digital items are sold with blockchains that ensure the customer always owns the items, even if the merchant ceases to exist, and can sell the goods if desired. Virtual fashions are also being developed for use on Zoom, in much the way users can change their backgrounds.

Critics have also expressed particular concern about discrimination in the metaverse, including discriminatory practices deriving from in-metaverse factors. In *Snow Crash*, for instance, Stephenson calls attention to the potential for in-metaverse discrimination: people in the novel's Metaverse judge individuals based on their avatars, and those with more privilege are able to access better Metaverse experiences. Critics of the metaverse have likewise noted that a lack of real-world diversity among metaverse users could lead to disparities within the metaverse itself. While the community has become more diverse, leaders in metaverse development have asserted that the digital realm must be as diverse as the real world, including gender, ethnicity, and age, if it is to be embraced by the masses.

—Josephine Campbell

Further Reading

"11 Ways VR and AR Stand to Impact Advertising, Marketing and PR." *Forbes*, 31 Aug. 2021, www.forbes.com/sites/forbesagencycouncil/2021/08/31/11-ways-vr-and-ar-stand-to-impact-advertising-marketing-and-pr/?sh=25e7879c5201. Accessed 13 July 2023.

Greenwald, Will. "Augmented Reality (AR) vs. Virtual Reality (VR): What's the Difference?" *PCMag*, 31 Mar. 2021, www.pcmag.com/news/augmented-reality-ar-vs-virtual-reality-vr-whats-the-difference. Accessed 13 July 2023.

Gupton, Nancy, and Patrick J. Kiger. "What's the Difference between AR, VR, and MR?" *Franklin Institute*, 21 Sept. 2017, www.fi.edu/difference-between-ar-vr-and-mr. Accessed 13 July 2023.

Heath, Alex. "This Is Meta's AR/VR Hardware Roadmap for the Next Four Years." *Verge*, 28 Feb. 2023, www.theverge.com/2023/2/28/23619730/meta-vr-oculus-ar-glasses-smartwatch-plans. Accessed 13 July 2023.

Lee, Alexander. "With the Metaverse on the Horizon, Its Creators Reflect on the Importance of Diversity in the Virtual World to Come." *Digiday*, 17 Aug. 2021, digiday.com/marketing/with-the-metaverse-on-the-horizon-its-creators-reflect-on-the-importance-of-diversity-in-the-virtual-world-to-come. Accessed 13 July 2023.

McDowell, Maghan. "These Platforms Want to Be the Farfetch of Digital Fashion." *Vogue Business*, 25 Jan. 2021, www.voguebusiness.com/technology/these-platforms-want-to-be-the-farfetch-of-digital-fashion. Accessed 13 July 2023.

Robinson, Joanna. "The Sci-Fi Guru Who Predicted Google Earth Explains Silicon Valley's Latest Obsession." *Vanity Fair*, 23 June 2017, www.vanityfair.com/news/2017/06/neal-stephenson-metaverse-snow-crash-silicon-valley-virtual-reality. Accessed 13 July 2023.

Stein, Scott. "Microsoft, Facebook and Everyone Else are Building the Metaverse. Will You Want to Live There?" *CNet*, 4 Aug. 2021, www.cnet.com/tech/computing/microsoft-facebook-and-everyone-else-are-building-the-metaverse-but-will-you-want-to-live-there. Accessed 13 July 2023.

"Visionary Futurist Neal Stephenson and Crypto Pioneer Peter Vessenes Announce Lamina1, the Layer-1 Blockchain for the Open Metaverse." *BusinessWire*, 8 June 2022, www.businesswire.com/news/home/20220608005423/en. Accessed 13 July 2023.

Microprocessor

ABSTRACT

Microprocessors are one of the critical components of the personal computer. They process all of the information entered into the computer through input devices, including mice, keyboards, cameras, and microphones. A more powerful microprocessor allows a computer to process more information at once. Microprocessors have become increasingly powerful over the years. Smaller, more powerful microprocessors allow devices such as smartphones to exist.

BACKGROUND

Computers were not always a big business. In the era before personal computers, building and maintaining computers was done by rare hobbyists. At this time, computers required specialized knowledge to build. They were usually ordered in kits, which had to be assembled by a skilled technician. Computer owners would have to solder their own computer components into place and write their own code. Because of these limitations, computers appealed to a very limited market. For this reason, they were not mass produced.

Employees at major technology firms such as Intel and Hewlett-Packard urged their employers to enter the personal computer market. Both companies began manufacturing various computer chips, including different types of memory and processors. At the time, computer processors were large, inefficient, and composed of several separate devices.

In 1969, a small technology firm approached Intel with a simplified computer design. Their computer would have only four chips: a chip for input and output devices, a read-only memory (ROM) chip, a random-access memory (RAM) chip, and a general processor chip. The general processor chip would handle sorting information between the other parts. Intel bought the rights to one of these chips, the 4004, in 1970 and began commercial production the following year. While the 4004 was designed to serve as a tiny central processing unit for digital calculators, engineers at Intel believed it could be applied to many of their devices. Under Intel, the 4004 became the first commercially marketed microprocessor.

Intel continued to improve their chips, dominating the fledgling microprocessor market. Their 8008 processor chip helped grow the computer-building hobbyist movement and was used in the famous Altair 8800 computer build. The Altair 8800 was the first machine to use the Beginner's All-purpose Symbolic Instruction Code (BASIC) programming language. BASIC was the first programming language written specifically for use on personal computers. This led to the development of the Intel 8080 chip, which was used on some of the first popular personal computers. Eventually, chip manufacturers developed even more powerful chips. This began in 1978 with Intel's 16-bit microprocessors. These were followed by 32-bit microprocessors in the 1980s, which slowly gained market share until the 1990s. That decade, 32-bit microprocessors were replaced by 64-bit microprocessors, which first became available in 1992. Over the following decades, increasingly powerful 64-bit microprocessors became the industry standard.

OVERVIEW

Modern microprocessors are used in a variety of devices. They are found in personal computers, smartphones, microwaves, dishwashers, refrigerators, digital versatile disc (DVD) and Blu-ray players, televisions, alarm systems, electric toothbrushes, and washing machines. They also may be found in any other electronic device that uses a small computer for basic operations.

The most recent generations of microprocessors have a variety of advantages over older models. Despite being exponentially more powerful, modern microprocessors are small, low-profile, and easy for computer engineers to utilize in a variety of hardware setups. They are cheap to manufacture, even though the process of doing so is extremely complex and delicate. Finished, tested products are very reliable and should not fail for many years. Lastly, microprocessors require very little electrical power to operate, making them energy efficient.

The microprocessor manages all data that enters a device. It then sorts the data through the appropriate parts. For example, any information typed on a keyboard first would enter the microprocessor. The microprocessor then would know to send that information to the video card, showing it on the computer's screen, and to one of the computer's memory units. The microprocessor usually has to communicate with every chip in the device, and it is one of the core parts of the modern personal computer. Other central parts include RAM, hard drives or flash drives, the motherboard, and the power supply.

PARTS OF A MICROPROCESSOR

All microprocessors can be divided into six parts: the arithmetic and logic unit (ALU), the registers, the control unit, the instruction register, the program counter, and the bus. The ALU is the part of the microprocessor that performs all of the mathematical calculations necessary for the device to function. Higher-bit processors can perform more calculations at once. Most computer functions are broken down into mathematical functions and are processed here. The registers are the internal storage device of the microprocessor. Any data that the microprocessor needs to temporarily store is stored in this part. The control unit sorts all data entering and leaving the microprocessor, directing it to the appropriate parts. The instruction register is a separate temporary memory storage device used only for storing task instructions. It makes sure the microprocessor remembers how to perform all of its necessary operations. The program counter temporarily stores the address of the next instruction to be executed. Lastly, the bus is a set of connective cables that allows all of the microprocessor parts to communicate with one another.

—Tyler Biscontini

Further Reading

Campbell-Kelly, Martin, William F. Aspray, Jeffrey R. Yost, Honghong Tinn, and Gerardo Con Díaz. *Computer: A History of the Information Machine*. 4th ed., Routledge, 2023.

"A Computer in Your Pocket: The Rise of Smartphones." *Science Museum*, 13 Nov. 2018, www.sciencemuseum.org.uk/objects-and-stories/computer-your-pocket-rise-smartphones. Accessed 13 July 2023.

Gordon, Whitson. "How to Build a Computer, Lesson 1: Hardware Basics." *Lifehacker*, 1 Aug. 2011, lifehacker.com/5826509/how-to-build-a-computer-from-scratch-lesson-1-hardware-basics. Accessed 13 July 2023.

"Intel's First Microprocessor: Its Invention, Introduction, and Lasting Influence." *Intel*, www.intel.com/content/www/us/en/history/museum-story-of-intel-4004.html. Accessed 13 July 2023.

O'Leary, Timothy, Linda O'Leary, and Daniel O'Leary. *Computing Essentials 2023*. McGraw-Hill, 2022.

Singer, Graham. "The History of the Microprocessor and the Personal Computer." *TechSpot*, 1 Oct. 2020, www.techspot.com/article/874-history-of-the-personal-computer. Accessed 13 July 2023.

MOBILE APPS

ABSTRACT

Mobile apps are programs designed to run on smartphones, tablets, and other mobile devices. These apps usually perform a specific task, such as reporting the weather forecast or displaying maps for navigation. Mobile devices have special requirements because of their small screens and limited input options. Furthermore, a touchscreen is the primary means of entering information into a mobile device. Programmers need special knowledge to understand the mobile platform for which they wish to create apps.

BACKGROUND

Mobile applications, or apps, are computer programs designed specifically to run on smartphones, tablets, and other mobile devices. Apps must be designed for a specific platform. A mobile platform is the hardware and system software on which a mobile device operates. Some of the most widely used mobile platforms include Google's Android and Apple's iOS. Mobile devices support a variety of software. At the most basic level, a platform's system software includes its operating system (OS). The OS supports all other programs that run on that device, including apps. In addition to the OS, smartphones and tablets come with a variety of preinstalled utility programs, or utilities, that manage basic functions. Examples of utilities include a clock and a calendar, photo storage, security programs, and clipboard managers. While utilities are not essential to the functionality of the OS, they perform key tasks that support other programs.

However, the real power of mobile devices has come from the huge assortment of mobile apps they can run. The app stores for various mobile devices contain hundreds of thousands of different apps to download. Each app has been designed with different user needs in mind. Many are games of one sort or another. There are also vast numbers of apps for every pursuit imaginable, including video chat, navigation, social networking, file sharing and storage, and banking.

Developers of mobile apps use various approaches to design their software. In some cases, an app is little more than a mobile website that has been optimized for use with small touchscreens. Other mobile apps are developed specifically for the mobile devices they run on. Programmers must program their apps for a specific mobile platform. App developers usually use a special software development kit and an emulator for testing the app on a virtual version of a mobile device. Emulators provide a way of easily testing mobile apps. Emulators generate detailed output as the app runs, so the developer can use this data to diagnose problems with the app.

OVERVIEW

Mobile apps have evolved into a multibillion-dollar business. Before the advent of mobile devices, software was developed for use on personal computers, and a software package often sold for hundreds of dollars. The mobile app marketplace has adopted a very different model. Instead of creating apps that cost a large amount of money and try to provide a wide range of functions, the goal is to create an app that does one thing well and to charge a small amount of money. Many apps are free to download, and paid apps are typically priced anywhere from ninety-nine cents to a few dollars. Despite such low price points, developers of successful apps can still earn large sums of money. This is in part due to the large numbers of smartphone and tablet users.

Mobile apps also have a low cost of distribution. In the past, software was sold on physical media such as floppy disks or compact discs read-only memory (CD-ROMs). These had to be packaged and shipped to retailers at the software developer's expense. With mobile apps, there are no such overhead costs because apps are stored online by the platform's app store and then downloaded by users. Aside from the time and effort of developing an app, the only other financial cost to the developer is an annual registration fee to the platform's app store and a percentage of revenue claimed by the platform. App stores typically vet apps to make sure they do not contain malware, violate intellectual property, or make false advertising claims. App developers can earn revenue from the download fee, in-app advertisements, and in-app purchases, the latter of which are particularly common in mobile games.

MOBILE APPS AND SOCIAL CHANGE

Mobile apps can do much more than entertain and distract. For example, Twitter is a microblogging app that allows users to post short messages and read updates from others. Twitter played a significant role in social movements such as the Arab Spring of 2011 and the Black Lives Matter protests of 2020 and likewise served as a means of disseminating information—and misinformation—during the COVID-19 pandemic. Because it relies on telecommunications technology that continues to function even when there are disruptions in other media, the Twitter app has allowed people to communicate even during major disasters and political upheavals. Other apps allow users to report and pinpoint environmental damage or potholes for government agencies to fix.

PLATFORM LOCK

Mobile apps must be designed for a particular mobile platform. Sometimes, a developer will make a version of an app for each major platform. In other cases, however, developers only create an app to run on a single platform. This leaves users of other mobile platforms unable to use the app. In the case of popular apps, this can cause frustration among those who want to be able to use whatever apps they want on their platform of choice.

—*Scott Zimmer*

Further Reading

Banga, Cameron, and Josh Weinhold. *Essential Mobile Interaction Design: Perfecting Interface Design in Mobile Apps*. Addison-Wesley, 2014.

Glaser, J. D. *Secure Development for Mobile Apps: How to Design and Code Secure Mobile Applications with PHP and JavaScript*. CRC Press, 2015.

Goggin, Gerard. *Apps: From Mobile Phones to Digital Lives*. Wiley, 2021.

Heckman, Rocky. *Designing Platform Independent Mobile Apps and Services*. Wiley-IEEE Computer Society Press, 2016.

Knott, Daniel. *Hands-On Mobile App Testing*. Pearson Education, 2015.

Miller, Charles, and Aaron Doering. *The New Landscape of Mobile Learning: Redesigning Education in an App-Based World*. Routledge, 2014.

Platz, Cheryl. *Design Beyond Devices: Creating Multimodal, Cross-Device Experiences*. Rosenfeld Media, 2020.

Whitaker, Rob. *Developing Inclusive Mobile Apps*. Apress, 2020.

MOBILE WEB TECHNOLOGY

ABSTRACT

The mobile web refers to the use of the internet through handheld mobile devices such as smartphones and tablets. Thanks to increasingly sophisticated technology, these small devices now boast more computing power than National Aeronautics and Space Administration's (NASA's) original mainframes. Although challenges remain for both users and developers of mobile web technology, it allows for much the same online access as desktop or laptop computers, and its use is expected only to grow.

BACKGROUND

Mobile technology has come a long way since 1973, when Motorola engineer Martin Cooper became the first person to make a call on a portable device—one that weighed more than two pounds, cost $4,000, and took ten hours to charge for just thirty-five minutes of power. Mobile communications devices are, by definition, small enough to be mobile. Screen size can be an issue, particularly for older people with declining vision. Their small size also has ramifications for those with less dexterity, who may find it difficult to manipulate keyboards and pointing devices. Technologists must therefore take those factors into consideration when designing mobile interfaces, keeping in mind that mobile users expect the same quality experience as that available on desktop or laptop devices but operate in a much different context.

As mobile devices became increasingly ubiquitous, the World Wide Web Consortium (W3C) created a set of universal standards for mobile web access and developed web technologies that consider the challenges associated with mobile devices, thus allowing designers to create pages that look good on both handheld and larger devices. These tools include CSS Mobile and XHTML Mobile, versions of the Cascading Style Sheets (CSS) language and Extensible Hypertext Markup Language (XHTML) optimized for mobile use.

Besides screen size, speed poses a major issue for mobile users, many of whom have experienced the frustratingly slow download of a movie, game, or other such content. Following the commercial introduction of fifth-generation (5G) mobile service in 2019, however, latency (the time it takes for information to travel from the browser to the server) was vastly reduced.

OVERVIEW

Developers must decide whether a mobile app or mobile-optimized website best meets their needs. Although any website can be viewed with a mobile browser, unoptimized sites can be uncomfortable or unreadable for the user. Websites optimized for mobile devices, on the other hand, have readable fonts, images that scale well on a small screen, and a vertical layout.

Just like traditional websites, mobile sites consist of browser-based hypertext markup language (HTML) pages linked together and display text, data, images, and video. They can also access mobile-specific features, including one-button click-to-call and location-based mapping. So-called "responsive sites," which are designed to be compatible with different platforms and adjust to different screen sizes and layouts, are also common.

Mobile apps, by contrast, are installed on a user's mobile device and can download content so that it can be accessed without an Internet connection. A mobile app is often an attractive choice for a company seeking to build brand loyalty, as it serves as an ever-present extension of a company's website and allows a consumer to set up preferences and customize their experience. Mobile apps can also seamlessly interface with devices' cameras, global positioning system (GPS) functions, and other features.

While they have limited access to a device's features and require an internet connection to run, optimized mobile websites have their own benefits, including broader reach (since unlike an app, a website does not have to be located in and downloaded from an app store) and cost efficiency. Mobile websites may also offer greater visibility in search engine results.

USABILITY IS KEY

Whether a designer chooses to build a mobile app or mobile website, overall usability is central, along with availability and accessibility. Mobile sites must be hosted on reliable servers with good uptime to ensure that users do not receive an error message while trying to load them, and designers should regularly check for and eliminate dead links. Both mobile sites and apps should contain clear, easy-to-explore, relevant content, formatted within a sensible information architecture. Intuitive interfaces that do not require instructions or repeated trial and error to figure out are also imperative.

—Mari Rich

Further Reading

Abraham, Nikhil. "Building Mobile Web Apps." *Dummies*, 26 Mar. 2016, www.dummies.com/programming/building-mobile-web-apps. Accessed 13 July 2023.

Goggin, Gerard. *Apps: From Mobile Phones to Digital Lives*. Wiley, 2021.

Moscaritolo, Angela. "5G Will Save You Almost 24 Hours of Download Time Per Month." *PCMag*, 16 Oct. 2018, www.pcmag.com/news/5g-will-save-you-almost-24-hours-of-download-time-per-month. Accessed 13 July 2023.

O'Leary, Timothy, Linda O'Leary, and Daniel O'Leary. *Computing Essentials 2023*. McGraw-Hill, 2022.
"Principles for Useable Design." *Usability Body of Knowledge*, www.usabilitybok.org/principles-for-usable-design. Accessed 13 July 2023.
Rouse, Margaret. "Mobile Web." *Techopedia*, 4 Apr. 2017, www.techopedia.com/definition/23588/mobile-web. Accessed 13 July 2023.
"Web Standards." *W3C*, 2023, www.w3.org/standards. Accessed 13 July 2023.

MULTIMEDIA ARTIST AND ANIMATOR

ABSTRACT
Multimedia artists and animators create two- and three-dimensional models, animations, and effects for television, film, video games, and other media. They may perform their work as employees of production companies or as self-employed artists. Students aspiring to enter the profession should pursue studies in the visual arts as well as in computer science.

BACKGROUND

Multimedia artists and animators create visual effects and animations for television, movies, video games, and other media. They create two- and three-dimensional models and animations. They may be involved in the creation of media advertisements and other marketing campaigns or create illustrations for websites, online magazines, and other forms of media. Some are self-employed, working from home offices, while others work for film, television, and video production companies of varying sizes.

OVERVIEW

Multimedia artists and animators often work in studio environments that are generally clean and comfortable, or they work in a home office/studio. Their hours vary based on the size and scope of the project on which they are working, as well as the time constraints established in a contract. Smaller companies and independent, self-employed artists tend to work longer hours to manage not only their projects but also the issues associated with running a small business.

Multimedia artists and animators must combine a talent for art and creative thinking with good research and communication skills, close attention to detail, and the ability to meet deadlines and work in a competitive atmosphere, all while remaining true to the needs of the client. They should be aware of general public attitudes and keep up with current trends. Many independent commercial artists set their own hours and act as small business entrepreneurs as well as creative artists.

DUTIES AND RESPONSIBILITIES

The duties and responsibilities of multimedia artists and animators vary based on the area and the size of the business in which they work. These artists may specialize in video games or animated movies, or they may create visual effects for television shows or online venues. They can further specialize in particular elements, such as characters, scenery, or pieces that contribute to the overall look and feel of a digital production.

In general, multimedia artists and animators first confer with clients or supervisors to establish the preferred design approach, budget, and anything else that needs to be considered. They then develop the requested design, often showing samples to the client at different points in the process. The artist may work independently or as part of a team overseen by an art director. In a team setting, the art director's job is to assign tasks, give the artists advice and feedback, and approve and present the final product.

WORK ENVIRONMENT

Physical Environment. Multimedia artists and animators work primarily in video production firms, studios, or office spaces in film and television compa-

Photo via iStock/Avalon_Studio. [Used under license.]

nies. These environments are well lit and well ventilated, with computers and other production and display technologies. Many artists/animators are independent consultants who work from studios and office spaces in their own private residences.

Human Environment. Depending on their areas of expertise, multimedia artists and animators meet and interact with a wide range of individuals. These parties include creative professionals, business executives, editors, designers, and other specialized commercial artists.

Technological Environment. Multimedia artists and animators not only might use computer programs in their work but also may be required to write programming code in order to create or mount their art. For this reason, a solid grounding in digital graphics technology is a must. Some animation studios have their own software and computer applications that they use to create films. They give workers on-the-job training to use this software.

EDUCATION AND TRAINING

High School/Secondary. High school students should study art, including drawing, photography, and design; math, including geometry; and computer science, including graphic design and drafting. They should also take advantage of any subject areas of interest to them as artists; for example, future animators are advised to take cartooning and media classes.

College/Postsecondary. Aspiring multimedia artists and animators may pursue a bachelor's degree in graphic art, design, computer graphics, or a similar field. Alternatively, they may enroll in art or design institutes for programs with more studio time and a greater focus on graphic design, digital imaging, or illustration. Further education may be warranted depending on how an artist chooses to specialize. For example, a prospective art director may also study management or art administration, while someone interested in animation would be well served by specialist programs in that field.

Adult Job Seekers. An internship or apprenticeship is a good way to gain necessary experience. Individuals looking for work can apply directly to the art or advertising director of a particular company and may also find opportunities through professional organizations such as the American Institute of Graphic Arts (now known as AIGA). Any potential commercial artist must have a portfolio showing his or her best work.

Professional Certification and Licensure. Some organizations provide certification programs to help multimedia artists and animators become specialists in their particular fields. Such certification can provide a competitive edge for job candidates.

Additional Requirements. Multimedia artists and animators should be both creative and extremely knowledgeable of the wide range of media options available to them to meet a client's needs. They should be willing to listen to and communicate with clients who may or may not agree with their ideas. Such artists must have self-discipline and a strong work ethic, especially in light of the fact that many are self-employed.

—*Michael Auerbach*

Further Reading

Costello, Vic. *Multimedia Foundations: Core Concepts for Digital Design*. 3rd ed., Focal Press, 2023.

Hardy, David Leicester. *Introduction to Digital Media Design: Transferable Hacks, Skills, and Tricks*. Bloomsbury Visual Arts, 2022.

Heidmets, Rao. *The Animation Textbook*. CRC Press, 2022.

Multitouch Displays

ABSTRACT

Devices with multitouch displays allow a user to touch a screen to enter commands and data. Traditional touch screens could respond to only a single touch, as when a user touches a button on the screen. Multitouch displays can interpret touches from more than one conductive surface, such as two or more fingers, allowing for a greater range of inputs and responses.

BACKGROUND

Multitouch displays are an innovation in computer input technology. Input technology takes several forms. Users can enter information into a computer using a keyboard. They type commands that the computer executes. Users can also move a mouse to control the movement of a cursor on the screen. Users can likewise make use of touch screens, which display functions that users can activate by touching graphics on the screen.

Multitouch displays take this capability a step further. They allow users to touch the screen with more than one finger at once. Users can combine touch and gestures to issue commands to the computer. For example, if a user wished to move a photo on a multitouch display screen, the user could touch the screen and slide the photo across it, just as if they were sliding a piece of paper across a table. A multitouch display senses the user's touches and movement and determines how the image should appear to move.

OVERVIEW

From an engineering standpoint, there are several ways to design multitouch displays. Each has its own method for the computer to sense multiple touches and movements and then to use software to interpret what these touches and movements mean. Some methods rely primarily on electrical conductivity between the user (a person's fingers, most often) and the display. Surface capacitive technology uses a conductive film to coat the display. When the user touches the film, a circuit is completed, and the display can detect where it was touched. Projected capacitive touch uses the same basic idea of detecting completed circuits, but instead of a conducting film coating the display, a grid of conductive material is etched into the display's glass.

OPTICS AND PRESSURE-SENSING

Some multitouch displays rely on optics instead of capacitive touch. Optical touch screens use visual sensors, which are like tiny cameras. They detect when an object is about to touch the display and at what location. These systems can be preferable when touch-screen input from nonconductive objects may be necessary. For instance, a major drawback of capacitive touch screens is that one cannot use them while wearing gloves. An optical touch screen, on the other hand, will still respond to a touch from a gloved hand because it detects the touch without relying on conductivity. This means that multitouch displays located outdoors in cold climates will be much more user-friendly if they have optical touch screens.

Another category of multitouch displays uses neither optics nor capacitive touch. Instead, it can sense the amount of force the user employs when touching the display. This category is described as force-sensing touch technology. Within this category, resistive touch screens are the most common form. Resistive touch screens can tell how hard a user is pressing because they contain several layers of material separated by tiny spaces. When a user touches the display lightly, only a few layers are compressed enough to come into contact with one another. A touch with greater force behind it will cause more layers to compress. As layers touch one another, they complete a circuit and send a signal to the device about the location and amount of force used.

ADVANTAGES

One main advantage to using multitouch displays is that they allow users with little or no training to input information that would otherwise be time-consuming to enter into a computer or that would require more technical knowledge than the average user has. For example, a user wishing to rotate an image to a more pleasing orientation would probably not know that they want to turn the image precisely 37 degrees to the right. Even if the user did know how many degrees to rotate the image, typing this into a computer or using a mouse to make the adjustment would take much more care than simply touching the corner of the image and sliding a finger in an arc to the right.

PINCH TO ZOOM

Not surprisingly, much of the growth in multitouch displays has been driven by the mobile computing industry. As ever more computing power is packed into ever smaller devices, consumers find themselves navigating websites on ever smaller screens. One of the more frequent multitouch gestures used is the pinch-to-zoom option. This feature enables one to magnify a particular part of the information being displayed on the screen, as if one were zooming in with a microscope.

—Scott Zimmer

Further Reading

Cipolla-Ficarra, Francisco. *Optimizing Human-Computer Interaction with Emerging Technologies*. IGI Global, 2018.

Gray, Tony. *Projected Capacitive Touch: A Practical Guide for Engineers*. Springer International Publishing, 2019.

Hasegawa, Shoichi, et al., editors. *Haptic Interaction: Science, Engineering and Design*. Springer Nature Singapore, 2018.

Horspool, Nigel, and Nikolai Tillmann. *Touchdevelop: Programming on the Go*. Apress, 2013.

Hughes, John F. *Computer Graphics: Principles and Practice*. Addison-Wesley, 2014.

Moss, Frank. *The Sorcerers and Their Apprentices: How the Digital Magicians of the MIT Media Lab Are Creating the Innovative Technologies That Will Transform Our Lives*. Crown Business, 2011.

O'Leary, Timothy, Linda O'Leary, and Daniel O'Leary. *Computing Essentials 2023*. McGraw-Hill, 2022.

Rogers, Scott. *Swipe This!: The Guide to Great Touchscreen Game Design*. John Wiley & Sons, 2012.

Saffer, Dan. *Designing Gestural Interfaces*. O'Reilly Media, 2008.

Soares, Marcelo M., and Francisco Rebelo. *Advances in Usability Evaluation*. CRC Press, 2013.

MUSIC EDITING

ABSTRACT

Music editing involves the use of computer technology to alter files of recorded sound. Music files can be compressed, enhanced, combined, or separated through editing software. Both the recording and film industries rely on music editing software to create their products. Compression has enabled the music industry to move away from physical media and has driven the growth of online music transfer and streaming. The sharing of audio files has also led to a more collaborative music scene.

BACKGROUND

Music editing software is used by sound engineers working with musicians to record and produce music for commercial purposes. Performers record their music in studios, often in separate sessions. Once all of the recording has been completed, sound engineers begin mixing the tracks together. Mixing combines different sounds into a single audio recording. One track might include the vocals, another the backup singers, and a third the instruments. Recording these separately can help preserve the full depth of sound produced by each source (or input). Sound engineers mix these recordings together such that they support and enhance one another instead of competing. Mixers allow them to adjust the volume of each input (channel), equalize frequencies, reduce noise, add effects, and control channel subgroups (buses). Music editing software usually includes a virtual mixer and a timeline with a scroll bar, allowing the editor to move back and forth along an audio track in a process called scrubbing. Scrubbing is particularly helpful for aligning vocal tracks with other audio tracks or with film visuals.

When the mixing stage has been completed for an album, the next step is mastering. This is the creation of a master recording that can be used to produce copies for sale thereafter. Mastering was much more complex when music was mainly sold on physical media, such as record albums or cassette tapes. The recording master had to be protected from theft, damage, and the ravages of time. Today, most music editing is done entirely digitally. In digital mastering, the mixed audio recording is checked for errors, reformatted for distribution, and then subtly adjusted for the best sound quality possible.

OVERVIEW

Music editing can be destructive, meaning that changing a recording destroys the original version of the recording, or it can be nondestructive, in which editing preserves the original recording. The technology that made it possible for music to move away from physical media was compression, specifically the MP3 compression algorithm and file type. MP3 was developed in the early 1990s as a type of lossy compression for audio files. It can greatly reduce the size of an ordinary audio file while losing only a negligible amount of sound quality.

MPEG-1 Audio Layer-3, or MP3 is an audio codec that makes music portable by compressing it into files small enough for transfer over the internet and for digital storage. A codec can reduce the size of audio files by using shorthand to describe repeated data within the file more concisely. When a compressed recording is played back, it is decoded by the device playing it, provided that the device is equipped with the proper codec. The recording's fidelity (similarity to the original) depends on how much information is kept and how much of the original frequency and dynamic range are captured. For instance, 16 bits of data are encoded for a high-fidelity compact disc (CD) as compared to 24 bits in a studio recording.

MOTION PICTURE MUSIC EDITING

Motion pictures rely on the skills of sound engineers to mix the film's dialogue, musical score and soundtrack, and sound effects into a single audio accompaniment to the visuals. The process is like that used in a recording studio. However, there is the added complication that the audio tracks must all not only be in harmony but also follow the timing of the movie's scenes. The dialogue usually takes priority, with the music and sound effects being mixed in later.

CULTURE OF REMIXING

The development and adoption of the MP3 audio codec made it easier and more common than ever for music to be shared illegally. People found it preferable to make illegal copies of music than to buy it, since those copies essentially retain the same quality as the original. Despite this, some praise the sharing of music online as giving rise to a culture of remixing. Artists release their creative output into the world so that audiences can enjoy it and other artists can build off it and even transform it. In remixing, other artists add their own perspectives to produce a collaborative work that is greater than the sum of its parts. Consumer audio editing software, such as Audacity or Adobe Audition, enables such artists to record and mix music using their personal computers rather than studio equipment. Some mobile applications also offer music-editing capabilities, enabling users to create or edit music using a tablet or smartphone.

—Scott Zimmer

Further Reading

Ashcroft, Fran. *The Analogue Approach to Digital Recording and Mixing*. Crowood Press, 2023.

Collins, Mike. *Pro Tools for Music Production: Recording, Editing, and Mixing*. 2nd ed., Routledge, 2017.

Cross, Mark. *Audio Post Production for Film and Television*. Berklee Press, 2013.

Kefauver, Alan P., and David Patschke. *Fundamentals of Digital Audio*. A-R Editions, 2007.

Lakatos, Anna. *The Music Producer's Creative Guide to Ableton Live 11*. Packt, 2023.

Pinch, T. J., and Karin Bijsterveld. *The Oxford Handbook of Sound Studies*. Oxford UP, 2013.

Saltzman, Steven. *Music Editing for Film and Television: The Art and the Process*. Focal, 2015.

Shen, Jialie, John Shepherd, Bin Cui, and Ling Liu. *Intelligent Music Information Systems: Tools and Methodologies*. IGI Global, 2008.

MUSIC TECHNOLOGY

ABSTRACT

Music technology is the application of computer software, electronic instruments, and other sound-manipulation equipment in the composition, performance, and recording of music. The boundaries of personal expression through music are expanding with computer-instrument hybrids, such as those based on the piano and the guitar. Software facilitates the composition of scores with automatic notation and immediate playback features. Technology also allows the precise reproduction and efficient storage of music as well as advanced editing, mixing, and mastering capabilities.

BACKGROUND

Music technology is the integration of personal expression through instrumental and vocal music and cutting-edge applied science that enhances rather than replaces traditional music theory, composition techniques, and philosophy. It also facilitates the invention of novel instrumental and choral sounds and styles of music. Music technology often incorporates mathematic underpinnings to create synergy in composition and performance among electronic instruments, such as synthesizers, sequencers, and samplers, and classic instruments. Musicians of the early twenty-first century have access to technical equipment that enables them to make stronger harmonic decisions and more fully manifest their artistic visions. The social culture of music has changed with the advent of personal MPEG Audio Layer-3 (MP3) players, downloadable open-sharing files, and music-streaming services; more people have immediate access to new music from around the world than ever before.

In the broadest sense, any technological development used in music can be seen as music technology. The development of instruments, from prehistoric drums to the classical forms of the violin and piano, has traditionally been one of the driving forces in musical innovation. Beginning in the late nineteenth and early twentieth centuries, however, with most acoustic instruments well established in standard designs, the leading edge of technology and music began to focus mainly on electronics. A major breakthrough was the invention of the triode vacuum tube in 1906, which allowed the development of new electronic musical instruments. Further twentieth-century developments in the field of music technology included the invention of the theremin, created by Russian engineer Léon Theremin in 1920; the development of magnetic-tape recording, amplification technologies, and electric guitars in the 1930s; and the completion of the RCA Mark II synthesizer in 1959. While music technology as a whole continued to evolve throughout the latter half of the twentieth century, synthesizer technology proved particularly influential during that era. Computer technology evolved alongside digital synthesizers, and by the 2000s many software applications were developed for music synthesis.

Also of note was the development of hyperinstruments, which began in 1986 in the Massachusetts Institute of Technology (MIT) Media Lab under the direction of renowned composer and pro-

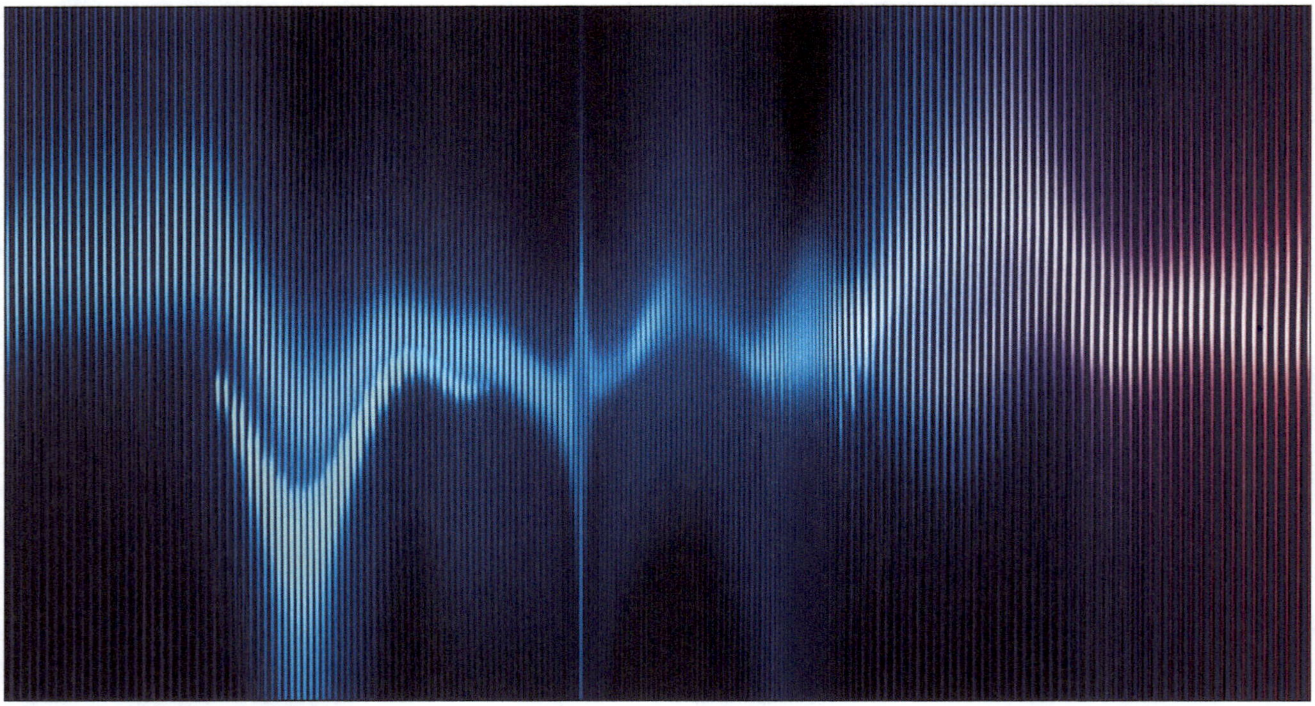

Photo via iStock/piranka. [Used under license.]

fessor of media arts and sciences Tod Machover. The goal of the project was to find ways to use technology to increase the power and finesse of musical instruments for superior performers. In 1992, the focus of this project shifted toward creating innovative interactive musical instruments for use by nonprofessional musicians and students, thus making music technology more accessible to the general public.

OVERVIEW

Music technology is often confused with sound technology. Sound technology involves the reproduction of sounds with adjustments for volume and clarity and is applied in the development of hearing aids and acoustic halls. It also differs from audio engineering, or recording engineering, which is concerned with the capturing, editing, and storing of music files. Music technology considers aesthetic composition and innovative performance, as well as reproduction with fidelity.

Composition. Aesthetic music compositions are typically layers of patterns that can be translated into mathematical expressions. Computer programs are able to perform this translation quickly, and the results can then be applied as objective parameters for developing new musical creations. Software can assist with rapid, clear musical notation and real-time playback during the composing period as well as with subsequent complex harmonious orchestration. Songwriting is similarly facilitated, and appropriate vocal keys, harmonies, and arrangements can be easily derived.

Performance. Traditional instruments may be enhanced with computer parts that create novel yet related voices. Most electronic instruments are also available for live performance. A music synthesizer is an electronic keyboard instrument that produces electrical signals of various frequencies to generate new voices, some of which imitate other traditional instruments. Synthesizers operate on programmed algorithms, and these synthesis techniques may be additive, subtractive, modulate frequency, or cause phase distortion. A music sequencer is a software program or an electronic device for recording, editing, and transmitting music, often but not exclusively in musical instrument digital interface (MIDI) format. A music sampler is similar to a synthesizer, except that instead of generating music, it plays back preloaded music samples, such as those from a sequencer, to compose as well as perform. Like synthesizers, samplers can modify the original sound signals to generate new voices, and many can generate multiple notes simultaneously.

Recording. In the recording process, music may be modified in its analogue form or converted to digital data before alteration. Sound mixing is the process of blending sounds from multiple sources into a desired end product. It often starts with finding a balance between vocal and instrumental music or dissimilar instruments so that one does not overshadow the other. It involves the creation of stereo or surround sound from the placement of sound in the sound field to simulate directionality (left, center, or right). Equalizing adjusts the bass and treble frequency ranges. Effects such as reverberation may be added to create dimension. Auditory electrical signals may then be sent to speakers, where they are converted into acoustical signals to be heard by an audience. Digital signals may be broadcast in real time over the internet as streaming audio. Otherwise, the processed signals may be stored for future reproduction and distribution. Analogue signals may be stored on magnetic tape or phonograph disc. Digital signals may be stored on a compact disc or subjected to compression and encoding for storage on a computer or personal music player or for transmission via a music-streaming service.

APPLICATIONS AND PRODUCTS

Hyperinstruments. Music technology products include hyperinstruments, a term coined in 1986 for electroacoustic instruments. They were originally intended for virtuosos; customized instruments were fabricated for musicians such as Yo-Yo Ma and Prince. However, their accessibility is being expanded for use by nonprofessional musicians and children. The theremin is an archetypal electronic musical instrument that is played without actual physical contact. The performer uses fine motor control to wave his or her hands in proximity to a pair of antennae to control the pitch and volume produced. The instrument responds to every movement, whether deliberate or unintentional. A chameleon guitar has interchangeable soundboards in the central cavity that allow the same body, neck, and frets to have a familiar feel while producing various voices that imitate other instruments. Hyperpianos yield MIDI data that are augmented as solo music or keyboard accompaniment carefully matched to other instruments and voices. With the hypercello, the bow pressure, string contact, wrist measurements, and neck fingering are measured, the data are processed mathematically, and an enhanced signal is generated.

A hyperviolin makes no sound of its own but creates electronic output when it is played with a hyperbow. The speed, force, and position of the hyperbow are measured wirelessly, the resulting data are adjusted, and the sound is appropriately enhanced for consistently intentional expression. The hyperviolin and hyperbow may be seen in performances by virtuoso Joshua Bell.

Innovative Musical Productions. The technology that has emerged from the MIT Media Lab, an influential institution in the twentieth- and twenty-first-century field of music technology, has been channeled by MIT professor Tod Machover into a new form of opera in the broadest sense, telling a story through music. One such project, *The Brain Opera*, debuted in 1996. Each night's event occurred in two parts. In the first half, the audience was invited to experiment with electronic instruments such as the Rhythm Tree (by punching a node on the tree, a thump comes from a nearby speaker), Gesture Walls (advanced theremins), and Harmonic Driving devices (video games that people drive to make musical choices). In the second half, the music created in the first half was incorporated into a multimedia presentation for a unique show every night. The project toured worldwide, finding a permanent home in Vienna in July 2000. Another major production, *Death and the Powers*, debuted in September 2010. This one-act opera tells the story of an inventor, Simon Powers, who builds an electronic system into which he downloads his memories and humanity, which is then expressed by robots, animated bookcases, and a musical chandelier. Typically inanimate objects become personified with electronic music and voices in what Machover calls "disembodied performance." Music technology is responsible for connecting audiences with the emotion of the story.

CAREERS AND COURSEWORK

In the United States, music technology degrees are offered at many colleges and universities. Students interested in seriously pursuing a musical technology career may choose to emphasize the musical aspects at a music-focused institution such as the Juilliard School in New York City, or focus on the technical side in an engineering program such as at the MIT in Cambridge, Massachusetts.

Students of music technology may pursue various levels of education. Although some associate's and doctoral degree programs are available, most are bachelor's and master's degree programs. Admission into these programs often requires the ability to read music, the ability to play a traditional instrument, some experience with computer hardware and software, and some experience with music recording. Programs offer courses in topics such as music history, traditional and contemporary music theory, music appreciation and critical evaluation, analogue and digital signal processing, recording, mixing, mastering, synchronization, songwriting, orchestration and ensemble performance, and music business. Many colleges and universities facilitate internships with local recording studios and radio stations to provide practical application.

Graduates may pursue many career paths in music technology. One is studio production, engineering, and recording for film, television, radio, video, internet, and record labels. Another is composing, arranging, and orchestrating for film, television, theater, church, advertising, and video games. Some choose to perform in live music venues. Some work in education and research, studying music design based on how the mind interprets pitch, rhythm, melody, and harmony. Similarly, some become music critics who must stay abreast of the current trends in both popular music and advancing technology. Another option is multimedia collaboration, integrating audio and video. This collaboration extends to web design, digital video editing, and interactive media. A similar option is equipment design, creating the next generations of synthesizers, sequencers, and samplers. Another popular career path is music business and administration involving sales, marketing, and management for music retailers, production companies, and record labels.

SOCIAL CONTEXT AND FUTURE PROSPECTS

Music technology is going in several exciting and novel directions beyond those of audio engineering. While audio engineering is developing music information retrieval systems, advanced digital signal processing, and mobile recording studios, music technology is pursuing more intimate avenues that will strengthen the bond between the artists and the audience.

One such area is music cognition. Technology is developing the capability to interpret and map expressive gestures made by professional musicians to personify electronic music. Similarly, researchers are mapping the brain's responses to different pieces of music to predict the effect of new compositions on an audience. Conversely, bioengineers are working to make brain waves audible so that happy, sad, fearful, or angry thoughts can be heard as musical patterns and subsequently evaluated by mental health professionals.

Another emerging area is interactive music systems, which take the audience from listening passively to thinking actively about how they react to music as they create it by making choices. This ranges from live musical performances, in which the audience is deliberately invited to participate, to blind research in which musical stairs are placed in proximity to escalators, and observers measure the rate at which pedestrians progressively choose to take the stairs, thus getting more exercise for the reward of music. Interactive music systems overall make music an inclusive experience.

—Bethany Thivierge

Further Reading

Ballora, Mark. *Essentials of Music Technology*. Oxford UP, 2015.

Brown, Andrew R. *Music Technology and Education: Amplifying Musicality*. 2nd ed., Routledge, 2015.

Diduck, Ryan. *Mad Skills: MIDI and Music Technology in the Twentieth Century*. Repeater, 2018.

Hosken, Dan. *An Introduction to Music Technology*. 2nd ed., Routledge, 2014.

Katz, Mark. *Capturing Sound: How Technology Has Changed Music*. Rev. ed., U of California P, 2010.

King, Andrew, Evangelos Himonides, and S. Alex Ruthman, editors. *The Routledge Companion to Music, Technology, and Education*. Routledge, 2019.

Manzo, V. J. *Foundations of Music Technology*. Oxford UP, 2015.

Masters, Marc. *High Bias: The Distorted History of the Cassette Tape*. U of North Carolina P, 2023.

Middleton, Paul, and Steven Gurevitz. *Music Technology Workbook: Key Concepts and Practical Projects*. Focal Press, 2008.

Payling, Dave. *Electronic Visual Music: The Elements of Audiovisual Creativity*. Focal, 2023.

Musician and Composer

ABSTRACT

Musicians and composers express themselves through the use of instruments and/or voice. They may compose or perform music for a wide range of multimedia projects, including films, television series, web videos, and video games. Students aspiring to enter either profession must typically attain proficiency in at least one musical instrument.

BACKGROUND

Musicians and composers express themselves through the use of instruments and/or voice. While many musicians write their own music, recognition as a composer is usually reserved for those who compose original works meant to be performed by other musicians. They each tend to specialize in a particular musical genre, although crossovers are common. In addition to contemporary compositions, musicians and composers are able to draw on a vast collection of music in the public domain for new arrangements and interpretations.

OVERVIEW

Musicians compose and practice in studios, often located in their homes, and perform in diverse environments, from cruise ships, nightclubs, and churches to large concert halls and stadiums. Some perform outdoors when weather permits. Performance makes up only a fraction of the musician's workweek, with the bulk of their time taken up by practice, rehearsals, composing, travel, and other responsibilities, including recording, if they are a recording artist. With the exception of orchestra members, church organists, and a few others, most musicians and composers are self-employed. Musicians have the highest chance of succeeding professionally if they live near major cities, where the greatest number of employment opportunities can be found. Composers who create compositions for use in specific multimedia projects, such as video games, usually work in both an office and a studio, whether one belonging to a development company, an independent studio hired on a freelance basis, or a home studio.

Most people who choose to become a professional musician or composer do so after becoming proficient at one or more instruments. They enjoy expressing themselves creatively in this "nonverbal language," and, in the case of musicians, enjoy performing for an audience. While they have certainly learned that practice is the key to success, they demonstrate other necessary qualities as well, including creativity, stamina, confidence, and the ability to cooperate with other musicians.

DUTIES AND RESPONSIBILITIES

Musicians divide much of their time between practice and rehearsals, performing, and any related travel, and, in many cases, recording and promotion. A practice session normally involves an individual musician learning a brand new piece, polishing a technique, or gaining proficiency on a new instrument, whereas rehearsal involves the whole band or ensemble, if there is one, practicing a complete piece of music for performance. Established musicians with a degree of professional recognition may have a manager to handle the details of promotion, scheduling, recording, and production, while those who are just starting out may have to do everything themselves.

A successful professional composer may be offered commissions to write advertising jingles; film, television, and video game scores; orchestral works; or other types of music. Some composers are also hired to arrange music—that is, take an existing work and have it performed in a novel way. For example, a composer might take a pop song and turn it into an orchestral work. When not busy with their creative work, composers meet with clients and spend time promoting their talents.

Musicians may also set aside time for composing music or writing songs. They also listen to record-

Photo via iStock/Pekic. [Used under license.]

ings by other musicians and keep abreast of new equipment and technologies. Because the field is so competitive, making a living as a full-time musician or composer is difficult. Therefore, many musicians and composers also hold day jobs that divide their attentions even further.

WORK ENVIRONMENT

Physical Environment. Musicians play in a variety of settings, ranging from elegant concert halls to bars to weddings and other private events. Depending on the type and size of venue as well as the type of music they play, musicians need to pay attention to the effect of the performance environment on their instruments and especially on their bodies, as they can be at risk for ailments such as hearing loss and repetitive motion injuries. Composers typically work in an office or studio.

Human Environment. Musicians interact with many different people, most notably their audiences and other musicians or singers. A session musician will also work closely with studio technicians, a touring artist may travel with a staff and stage crew, and a church organist will collaborate with the music director and minister. Composers often work closely with film directors, advertising teams, video game developers, choreographers, and other professionals.

Technological Environment. Musicians playing instruments that are not electric or electronic, including string, wind, and other traditional instruments, are concerned mainly with the operation and maintenance of those instruments, although any performing artist typically needs to also work with microphones, amplifiers, and other electrical equipment. Additional musical technologies are involved in working with electric guitars, keyboards, and the like, and for recording artists there is a world of sound recording technology with which to become familiar. Composers often compose electronically and must often be familiar with composition and music production software.

EDUCATION AND TRAINING

High School/Secondary. Musicians generally need at least a high school diploma. Most will want to pursue a strong college preparatory program, with electives in music and courses specific to their interests, such as creative writing, electronics, or film studies. Private lessons, summer music camps, and extracurricular performance opportunities are extremely important. Proficiency in one or more instruments will be required for admission to a college music program or conservatory.

College/Postsecondary. In general, musicians must find the balance of formal and informal education that best meets their needs. An unusually talented musician may choose to forego college and move directly into their performing career. On the other hand, some musicians benefit from at least a bachelor's degree in music. Music programs include courses in theory, composition, arranging, and performance. Instead of music, one might consider majoring in music education, audio engineering, or even double majoring in music and business. Continuing education courses in piano tuning, instrument repair and construction, or other related courses might be good choices as well.

Adult Job Seekers. Adults who take up an instrument later in life face stiff competition; however, it is common for a musician to become a composer later in adulthood, and for musicians to switch instruments later in life. Although musicians can work part time, the hours, and often the pay, are not always conducive to parenting and other adult responsibilities.

Advancement opportunities are limited and, in most cases, highly dependent on one's level of success. They include chaired positions or opportunities for solos; more lucrative recording contracts, performance fees, or commissions; and/or the ability to hire a manager to take care of business responsibilities, thus opening up more time for creative pursuits. Only a very small percentage of musicians and composers reach celebrity status.

Job seekers might consider joining professional organizations such as the American Federation of Musicians, considered to be the largest such organization dedicated to furthering the careers of professional musicians, to seek career advancement opportunities.

Professional Certification and Licensure. There are no certificates or licenses needed for most performers or composers, although a few organizations offer certification that may be required for jobs in that specialty, such as the American Guild of Organists.

Additional Requirements. Musicians and composers are usually artists with a passion for their craft, but forging a professional career in this field is no easy task. Aspiring professional musicians and composers need to be prepared for all the financial hazards of freelance work, including an inconsistent income (or backup income from another source) that needs to be spread over periods of no musical work. As with any fine arts career, passion and persistence are often the most important qualifications.

—*Sally Driscoll*

Further Reading

Fairchild, Charles. *Sounds, Screens, Speakers: An Introduction to Music and Media*. Bloomsbury Academic, 2019.

Hallgren, Scott W., editor. *Composing for the Screen*. Routledge, 2022.

Stone, Stephen C. *Music Theory and Composition: A Practical Approach*. Rowman & Littlefield, 2018.

N

Net Neutrality

ABSTRACT

Net neutrality is the idea that internet service providers and governments should treat all online data the same, regardless of the website, platform, user, or application it comes from. Proponents of net neutrality believe it should be assured by law so that service providers cannot fragment access to users in order to make them pay more for different services, such as using File Transfer Protocol (FTP), playing online games, or establishing high-quality peer-to-peer (P2P) communications. To do so would mean establishing "tiered service," which many advocates of net neutrality say curtails the ideals of freedom of information that the internet was founded on. Tiered service would require a service provider's customers to purchase different packages in order to obtain the level of access they want to information and services usually freely available to everyone on the internet. Service providers would be able to block certain content and slow down or prohibit certain types of services to those who do not pay for such package upgrades.

BACKGROUND

The most marked characteristic of the internet is its decentralized structure. Theoretically, all nodes in the network are potentially equal in power and influence, although in practice this rarely happens because of preexisting social and economic structures. There are three dimensions to the internet: hardware, software, and infrastructure. Hardware consists of the devices people use to access the internet, whether it is a smartphone, a desktop computer, or the massive server farms where multitudes of data are stored. Software consists of the programs and applications people use to access or interact online. Infrastructure refers to the systems that connect hardware to other hardware. There are two main types of infrastructure: wired systems and wireless systems. Wired systems use phone lines or fiber optic cable networks, while wireless systems use allotted frequencies on the radio spectrum to transmit internet access.

The explosion of the internet as a tool of mass communication was accompanied by socioeconomic trends such as vertical integration. This means that the companies that provide access to the internet for a fee have also taken over other companies that supply content to the internet. This has led to concerns over the concentration of power and the control of information flowing through the internet. Some argue that concentration of power limits the democratic exchange of data and the quality of civic participation on the internet. Net neutrality refers to the struggle to keep the internet a public good, rather than a resource to be used for the personal gain of a relatively small group of people.

OVERVIEW

One of the emergent issues in the net neutrality debate is the new challenges created by data-rich media forms transmitted via the World Wide Web. For instance, video transmission can congest internet infrastructures at peak use periods, and vertically integrated internet service providers (ISPs) have the ability to favor their own content during these times. This gives them an unfair advantage in the competition for consumers' attention. ISPs that have developed infrastructure can limit the performance of competing ISPs that use their infrastructure, as well as restricting bandwidth of peer-to-peer (2P2) file sharing programs.

Opponents of net neutrality cite various reasons for their position. One of the most common arguments is that regulation of the internet will lead to decreased competition, which in turn will reduce the type of rapid innovation that made the internet successful and revolutionary in the first place. A related notion is that regulation will cause a drop in investments that support not only internet service providers but many peripheral industries as well. Many claim that the fear of privileged internet access for the wealthy and poor service for others is a false scenario. Other argue that regulation may cause an increase in taxation or that neutrality will cause an overuse of internet bandwidth.

In the debate about net neutrality for wireless communications, one of the crucial issues is the equitable and safe distribution of the frequencies on the radio spectrum. If the radio spectrum is considered a natural resource, how does one make sure one company does not monopolize it and divert internet traffic to sites the company controls? Antitrust laws that ensure fairness in competition among businesses are an integral part of the net neutrality debate at this point.

The debate over net neutrality is affected by the intense and swift technological innovation in the internet and data services industries. For example, in 2013, the company that developed a standardized system for data exchange between computers worked on technology to allow data exchange between radio frequencies, phone lines, and cell phones. This emerging technology is called the Internet Protocol Interoperability and Collaboration System (IPICS). This changes information and communications technology dramatically and will likely impact policy in order to ensure a fair and robust online public sphere.

Because of this rapid pace of technological development, relevant laws and policy often lag behind. For example, though issues regarding net neutrality have been debated since 2000, it was not until 2010 that the US Federal Communications Commission (FCC) proposed the Open Internet Order to preserve net neutrality. In November 2014, President Barack Obama called on the FCC to regulate internet service providers like public utilities, imploring the agency to create strict rules to enforce net neutrality. In response the FCC investigated the possibility of implementing Title II of the Communications Act of 1934, regarding common carriers, and the Telecommunications Act of 1996 to the internet in order to support neutrality. While a vote held on February 26, 2015, went in favor of net neutrality, in April 2018, those rules were rolled back, resulting in what some referred to as the "death of net neutrality."

Attempts to restore net neutrality continued throughout the next several years, though some, such as the federal Save the Internet Act, proved unsuccessful. In July of 2021, President Joe Biden signed an executive order dealing with a variety of issues related to economic consolidation. Among other stipulations, the order included a section encouraging the FCC to "adopt . . . through appropriate rulemaking 'Net Neutrality' rules similar to those previously adopted." In July of 2022, the Net Neutrality and Broadband Justice Act was introduced to Congress in an effort to reclassify broadband internet as a telecommunications service and thus facilitate the codification of net neutrality.

—*Trevor Cunnington*

Further Reading

Ammori, Marvin. "The Case for Net Neutrality." *Foreign Affairs*, vol. 93, no. 4, 2014, pp. 62-73.

Anderson, Steve. "Net Neutrality: The View from Canada." *Media Development*, vol. 56, no. 1, 2009, pp. 8-11.

Biden, Joe. "Promoting Competition in the American Economy." *Federal Register*, 14 July 2021, www.federalregister.gov/documents/2021/07/14/2021-15

069/promoting-competition-in-the-american-economy. Accessed 13 July 2023.

Cook, Vickie S. "Net Neutrality: What Is It and Why Should Educators Care?" *Delta Kappa Gamma Bulletin*, vol. 80, no. 4, 2014, pp. 46-49.

Downes, Larry. "Unscrambling the FCC's Net Neutrality Order: Preserving the Open Internet—But Which One?" *CommLaw Conspectus*, vol. 20, no. 1, 2011, pp. 83-128.

Feder, Samuel L., and Luke C. Platzer. "FCC Open Internet Order: Is Net Neutrality Itself Problematic for Free Speech?" *Communications Lawyer*, vol. 28, no. 1, 2011, pp. 1-26.

Jackson, Charles L. "Wireless Efficiency Versus Net Neutrality." *Federal Communications Law Journal*, vol. 63, no. 2, 2011, pp. 445-80.

Kelly, Makena. "House Approves Save the Internet Act That Would Reinstate Net Neutrality." *Verge*, 10 Apr. 2019, www.theverge.com/2019/4/10/18304522/net-neutrality-save-the-internet-act-house-of-representatives-approval. Accessed 13 July 2023.

Lessig, Lawrence. *Remix: Making Art and Commerce Thrive in the Hybrid Economy*. Penguin, 2008.

"Matsui, Markey, and Wyden Introduce Legislation to Reinstate Net Neutrality, Reverse Trump-Era Deregulation." *Doris Matsui*, 28 July 2022, matsui.house.gov/media/press-releases/matsui-markey-and-wyden-introduce-legislation-reinstate-net-neutrality-reverse. Accessed 13 July 2023.

Renda, Andrea. "Competition-Regulation Interface in Telecommunications: What's Left of the Essential Facility Doctrine." *Telecommunications Policy*, vol. 34, no. 1, 2010, pp. 23-35.

Ruiz, Rebecca R. "F.C.C. Sets Net Neutrality Rules." *New York Times*, 12 Mar. 2015, www.nytimes.com/2015/03/13/technology/fcc-releases-net-neutrality-rules.html. Accessed 13 July 2023.

Shin, Dong-Hee. "A Comparative Analysis of Net Neutrality: Insights Gained by Juxtaposing the US and Korea." *Telecommunications Policy*, vol. 38, no. 11, 2014, pp. 1117-133.

Sommer, Jeff. "What the Net Neutrality Rules Say." *New York Times*, 12 Mar. 2015, www.nytimes.com/interactive/2015/03/12/technology/net-neutrality-rules-explained.html. Accessed 13 July 2023.

United States Federal Communications Commission. *Preserving the Open Internet: Report and Order*. GPO, 2010.

NONFUNGIBLE TOKEN (NFT)

ABSTRACT

A nonfungible token (NFT) is a unique digital asset that serves as a certificate of ownership for a real-world investment, such as artwork, music, videos, or digital images. NFTs are recorded as data on a digital ledger known as a blockchain. They serve as proof that an object is an original or a one-of-a-kind work. The concept of nonfungible tokens was first developed in the mid-2010s, but they increased in popularity in the early 2020s as more people began investing in digital artwork.

BACKGROUND

In economics, the word "fungible" refers to a commodity or an asset that can be exchanged for another commodity or asset of equal value. The term has been used since the seventeenth century and has its origins in the Latin word *fungi*, meaning "to perform." A prime example of a fungible asset is currency. A one-dollar bill can be exchanged for another one-dollar bill or broken down into a combination of coins, such as four quarters or ten dimes. A nonfungible asset is unique in that it cannot be exchanged for a different asset. For example, one house cannot be exchanged for another, because each house would have distinct qualities that separate it, even if the two houses have a relatively equal monetary value.

A blockchain is a digital database that stores electronic information on a computer or a network of computers. The information is stored in digital chunks known as "blocks." When a block is filled with data, it is attached or "chained" to another block. As more blocks are filled, they are in turn chained to the previous block of data. This creates a chronological chain of digital information that is easily identified and cannot be broken. Each block is marked by a timestamp and a mathematical algorithm that defines it as unique.

Blockchains are typically stored on a decentralized network of many computers. Once information has been authenticated and stored on a chain, each computer can access that information, but changing it is impossible without the cooperation of a majority of users. An attempt to alter a specific block would be immediately detected and corrected by cross-referencing the previous data stored on other computers in the network.

OVERVIEW

The concept of the nonfungible token (NFT) was invented in 2014 by digital artist Kevin McCoy and tech entrepreneur Anil Dash. McCoy's artwork was often copied and shared online without attribution or compensation, and Dash was doing consulting for media companies looking to find new ways to copyright artwork. Together, they used then-new blockchain technology to develop digital ownership of a video clip made by McCoy's wife. They called their idea monetized graphics. The first widespread use of NFTs occurred in 2015, when the Ethereum blockchain developed standards to record and store digital tokens. Two years later, Ethereum hosted a video game called *Cryptokitties*, in which users could buy and collect their own virtual cats. The ownership of the cats was authenticated by unique NFTs stored on the blockchain. The effort proved wildly successful, generating $20 million just weeks after launch. Some users spent up to $100,000 for digital ownership of their own cat. From 2017 to 2020, users spent about $175 million on NFTs; in 2020 alone, that figure had skyrocketed to $250 million.

At that time most NFTs were stored on Ethereum, but several other blockchain technologies had been making inroads in the market. NFTs act as a unique "signature," designating that an object is authentic. The NFT does not assign a specific value to an ob-

Illustration of a non-fungible token generated by a smart contract (a program designed to automatically execute contract terms). Image via Wikimedia Commons. [Public domain.]

ject; it only designates authenticity. The object's value, especially in the case of artwork, is still determined by scarcity, market value, or interest by collectors. Buying an NFT is similar to an art collector purchasing a painting and hanging it in their home. It assures the buyer of the status of owning the original artwork or object but does not entitle them to royalties. The creator of an artwork can still receive royalties for copies of his or her work.

In the case of digital art, people can still find a copy of a work online and download that image as often as they want. However, they are only in possession of a copy; the holder of the NFT has the "bragging rights" of owning the real thing. NFTs have been sold for objects as diverse as real estate, sports memorabilia, video clips of famous NBA highlights, and even sneakers. In March 2021, Canadian singer Grimes sold a collection of artwork and music videos for a total of $6 million. Later that month, Jack Dorsey, then the chief executive officer (CEO) of Twitter, sold an NFT of the first-ever tweet on the social media site for $2.9 million. The tweet, which read "just setting up my twttr," was first posted in March 2006.

Digital artwork remained the most popular objects to be bought and sold via NFTs throughout the early 2020s. In 2017 a pair of software engineers had created a program that generated ten thousand pixelated images of unique characters they called CryptoPunks. The characters were inspired by England's punk rock scene and numerous futuristic cyberpunk films. In 2021, nine of the CryptoPunk images were sold as NFTs at the famed auction house Christie's in New York City. Two of the images sold for more than $7 million, while two others went for more than $1 million. The year 2021 also saw the release of other collections of digital artwork, including the Bored Ape Yacht Club series.

By far the largest NFT sale at that time was a collection of thirteen years of work by digital artist Mike Winkelmann, known in the art world as Beeple. Winkelmann began designing digital art in

Some digital art NFTs, like these pixel art characters, are examples of generative art. Image via Wikimedia Commons. [Public domain.]

2007 and, since that time, created an image a day as a way to improve his skill. In March 2021, he sold that collection, titled *Everydays: The First 5000 Days*, as an NFT at Christie's for $69.3 million. The sale was not only the largest sale of digital art in history up to that point but the fourth highest amount ever for the work of a living artist. In December 2021 this record was broken with a transaction that netted a digital artist $91.8 million, the highest amount paid for an NFT up to that point.

In 2020 the NFT market was estimated to be worth about $250 million, and by the end of 2021 the value of the NFT market had surged to an estimated $15.7 billion. Subsequently, however, the market experienced a significant decline as a result of several factors, including a substantial decrease in trading volume over the course of 2022. Between January of 2022 and January of 2023, monthly NFT sales revenue decreased by more than 80 percent overall, with the NFT art market experiencing a particularly notable decline.

—*Richard Sheposh*

Further Reading

Conti, Robyn, and John Schmidt. "What You Need to Know about Non-Fungible Tokens (NFTs)." *Forbes Advisor*, 14 May 2021, www.forbes.com/advisor/investing/nft-non-fungible-token. Accessed 13 July 2023.

Conway, Luke. "Blockchain Explained." *Investopedia*, 1 June 2021, www.investopedia.com/terms/b/blockchain.asp. Accessed 13 July 2023.

Dash, Anil. "NFTs Weren't Supposed to End Like This." *The Atlantic*, 2 Apr. 2021, www.theatlantic.com/ideas/archive/2021/04/nfts-werent-supposed-end-like/618488. Accessed 13 July 2023.

Davis, Riccardo A. "NFTs: What Are They, And How Do They Work?" *Kiplinger*, 6 May 2021, www.kiplinger.com/investing/602743/nfts-what-are-they-and-how-do-they-work. Accessed 13 July 2023.

Fang, Block. "PAK's NFT Artwork 'The Merge' Sells for $91.8 Million." *Penta*, 7 Dec. 2021, www.barrons.com/articles/paks-nft-artwork-the-merge-sells-for-91-8-million-01638918205. Accessed 13 July 2023.

Holland, Oscar. "How NFTs Are Fueling a Digital Art Boom." *CNN*, 10 Mar. 2021, www.cnn.com/style/article/nft-digital-art-boom/index.html. Accessed 13 July 2023.

Iredale, Gwyneth. "List of 10 Most Expensive NFTs Ever Sold." *101 Blockchains*, 27 Apr. 2021, 101blockchains.com/most-expensive-nfts. Accessed 13 July 2023.

"NFT Sales Drop 83% in a Year ($2.37bn Drop in Revenue)." *Yahoo!Finance*, 16 Jan. 2023, finance.yahoo.com/news/nft-sales-drop-83-2-114048086.html. Accessed 13 July 2023.

"Non-Fungible Tokens (NFT)." *Ethereum*, 2 June 2021, ethereum.org/en/nft. Accessed 13 July 2023.

Nover, Scott. "The NFT Market Is Down by Almost Every Metric." *Quartz*, 14 Oct. 2022, qz.com/the-nft-market-is-down-by-almost-every-metric-1849655660. Accessed 13 July 2023.

Thaddeus-Johns, Josie. "What Are NFTs, Anyway? One Just Sold for $69 Million." *New York Times*, 12 May 2021, www.nytimes.com/2021/03/11/arts/design/what-is-an-nft.html. Accessed 13 July 2023.

O

Object-Oriented Design

ABSTRACT

Object-oriented design (OOD) is an approach to software design that uses a process of defining objects and their interactions when planning code to develop a computer program. Programmers use conceptual tools to transform a model into the specifications required to create the system. Programs created through OOD are typically more flexible and easier to write.

BACKGROUND

Object-oriented design (OOD) was developed to improve the accuracy of code while reducing software development time. Object-oriented (OO) systems are made up of objects that work together by sending messages to each other to tell a program how to behave. Objects in software design represent real-life objects and concerns. For example, in the code for a company's human resources website, an object might represent an individual employee. An object is independent of all other objects and has its own status. However, all objects share attributes or features with other objects in the same class. A class describes the shared attributes of a group of related objects.

OOD typically follows object-oriented analysis (OOA), which is the first step in the software development process when building an OO system. OOA involves planning out the features of a new program. OOD involves defining the specific objects and classes that will make up that program. The first OO language, Simula, was developed in the 1960s, followed by Smalltalk in 1972. Examples of well-known OO programming languages include Ruby, C++, Java, PHP, and Smalltalk.

OVERVIEW

OOD consists of two main processes, system design and object design. First, the desired system's architecture is mapped out. This involves defining the system's classes. Using the example of a human resources website, a class might consist of employees, while an object in that class would be a specific employee. Programmers plan out the essential attributes of a class that are shared across all objects within the class. For example, all employees would have a name, a position, and a salary. The class defines these attributes but does not specify their values. Thus, the class would have a field for the employee's name, and the individual object for a specific employee might have the name Joe Smith. All the objects within a class also share methods, or specific behaviors. For example, employee objects could be terminated, given a raise, or promoted to a different position.

Object design is based on the classes that are mapped out in the system design phase. Once software engineers have identified the required classes, they are able to write code to create the attributes and methods that will define the necessary objects. As development progresses, they may decide that new classes or subclasses need to be created. New classes can inherit features from existing classes through class-based inheritance. Inheritance is a form of code reuse. Class-based inheritance draws attributes from an existing class to create a new class. New attributes can then be added to the new class. This is distinct from prototypal inheritance, which involves cloning existing objects, called prototypes, instead of drawing from existing classes.

OOD PRINCIPLES AND APPLICATIONS

Software engineers often follow a set of OOD principles identified by the acronym SOLID. These principles were first compiled by software engineer and author Robert Cecil Martin. The SOLID principles provide guidance for better software development, maintenance, and expansion. These principles are:

- Single responsibility principle: Each class should have only one job.
- Open-closed principle: A class should be open for extension but closed to modification (i.e., it can be easily extended with new code without modifying the class itself).
- Liskov substitution principle: Each subclass can be substituted for its parent class.
- Interface segregation principle: Do not force clients to implement an interface (a list of methods) they do not use. Interfaces should be as small as possible. Having multiple smaller interfaces is better than having one large one.
- Dependency inversion principle: High-level modules must not be dependent on low-level modules. Attempt to minimize dependencies between objects.

OOD is used to create software that solves real-world problems, such as addressing a user's needs when interfacing with an automatic teller machine (ATM). In this example, the software engineer first defines the classes, such as the transaction, screen, keypad, cash dispenser, and bank database. Classes or objects can be thought of as nouns. Some classes, such as transactions, might be further categorized into more specific subclasses, such as inquiry, withdrawal, deposit, or transfer.

Relationships between classes are then defined. For example, a withdrawal is related to the card reader, where the user inserts an ATM card, and the keypad, where the user enters their personal identification number (PIN). Next, classes are assigned attributes. For a withdrawal, these might be the account number, the balance, and the amount of cash to withdraw.

Finally, methods or operations are assigned to each class, based on user needs identified in the OOA phase. Methods can be thought of as verbs, such as execute, display, or withdraw. Detailed diagrams describe changes that happen within the objects, the activities they carry out, and how they interact with each other within the system.

PRECAUTIONS AND DRAWBACKS

Successful development with OOD requires realistic expectations. Possible problems with this approach include insufficient training, which can lead to the belief that OOD will clear up every development delay, resulting in missed deadlines. Further issues arise when timelines are shortened in expectation of OOD's promise of speed. Thorough training is a programmer's best course to understanding the limitations and subtleties of OOD.

OOD has been a mainstay in software programming and will be used for years to come, but advances in technology will require programming languages that can deliver faster search results, more accurate calculations, and better use of limited bandwidth for the applications users increasingly rely upon. In addition, OOD is not the best choice for every requirement. It is appropriate for systems with a large amount of graphics, interfaces, and databases, but for simpler systems, task-oriented design (TOD) might be a better choice.

—*Teresa E. Schmidt*

Further Reading

Adamson, Austin. "Being a S.O.L.I.D. Engineer." *OrdinaryEngineer*, 27 July 2020, medium.com/ordinaryengineer/solid-principles-4a3525ae0ac8. Accessed 13 July 2023.

Booch, Grady, et al. *Object-Oriented Analysis and Design with Applications*. 3rd ed., Addison-Wesley, 2007.

Dennis, Alan, Barbara Haley Wixom, and David Tegarden. *Systems Analysis and Design: An Object-Oriented Approach with UML*. 5th ed., Wiley, 2015.

Gold, Julian. *Object-Oriented Game Development.* Addison-Wesley, 2004.

Jackson, Daniel. *The Essence of Software Design: Why Concepts Matter for Great Design.* Princeton UP, 2023.

Weisfeld, Matt. *The Object-Oriented Thought Process.* 5th ed., Addison-Wesley, 2019.

Online Entertainment and Music Databases

ABSTRACT

Online entertainment and music databases are storehouses of information found in one location that purport to house a complete record of all available records for their respective industries. Popular examples include the Internet Movie Database (IMDb) and AllMusic. In each case, users can access a plethora of information about productions from the most popular to the most obscure.

BACKGROUND

Databases are structured using the tags of eXtensible Markup Language (XML) and other markup languages to provide specific field identification markers. Tags are user-defined for specific field entries so that by selecting or specifying the desired master file, such as a movie title, the database program will return all of the data fields associated with that master file. The user can then select a specific tag, such as one of the actors, to see the information about that actor. This is the basic operating principle of all databases.

One of the most prominent online entertainment databases, the Internet Movie Database (IMDb), was started in 1990 as the rec.arts.movies database on Usenet by Col Needham. Needham initiated the movie database with its first post being his piece, titled "Those Eyes," in a category he defined as "actresses with beautiful eyes." He also became the database's lead programmer, while allowing for other movie fans to participate in the open-source project (similar to more recognizable encyclopedic databases such as Wikipedia). As the database grew, it became more difficult for users to search the more than 10,000 entries, which required knowing how to use a number of UNIX shell scripts to accomplish.

In 1993, the database was moved from Usenet to become part of the World Wide Web, and in 1996 the IMDb website was incorporated as a business entity in the United Kingdom as Internet Movie Database Limited, with Needham as its principal owner. Advertising, licensing, and partnerships provided the income needed to maintain the company and the operation of the website. Two years later, in 1998, the IMDb company was purchased by Jeff Bezos and Amazon. It continued to operate as an independent subsidiary of Amazon and functioned as an advertising resource for sales of movies and other entertainment products on Amazon. It also served to track new projects still in development for a number of fans' favorite actors, writers, directors, and so forth.

OVERVIEW

The IMDb website was originally programmed using the Perl programming language, but for reasons of security, IMDb no longer reveals what programming language is currently used for its programming. All IMDb websites are presented primarily in English, though there are translated sites available in other languages. IMDb uses a mathematical algorithm to assign weighted ratings to movies in its database, and registered visitors to the IMDb website offer a rating for entries on a 1-to-10 scale.

The average of these ratings, multiplied by the number of votes the entry has received, is added to the required vote minimum, multiplied by the overall average score for the entry, and the resulting number is divided by the number of votes the movie received plus the minimum number of votes required. The value obtained by this calculation is the "weighted average rating" for the movie. There is also a Bottom 100 rating for movies, and a Top 25 rating for television shows. Only votes provided by regular voters—registered IMDb users who have voted a certain number of times on IMDb—are used for the ratings calculations. IMDb does not disclose the minimum number of times a registered user must have voted in order to be considered a regular voter.

Particular pages in the IMDb website, as well as corporate IMDb, are readily accessible through any web browser. If, for example, one searched the title of a movie or the name of an actor or director, the corresponding IMDb page becomes available for viewing. The IMDb page provides a wealth of information about the particular entry, including complete cast biographies, movie trailers, story and plot synopses, episode descriptions, all supporting production personnel, suggestions for similar entries, user reviews, and more. Given the extent of the database and the depth of the information it provides, IMDb is the go-to source for anyone researching cinema, television, music, and video games. In June 2023, the IMDb site encompassed more than 15 million titles across all categories as well as more than 12 million actor and personnel records.

OTHER DATABASES

There are several other online databases that are primarily related to movies and television, each with its own area of focus. Of the online databases, IMDb holds the top spot. The Open Movie Database (OMDb), provides a service similar to that of

Photo via iStock/Robert Way. [Used under license.]

IMDb, but all OMDb content is provided by its users rather than through a corporate entity such as Amazon. The Television Database, TVDb, focuses exclusively on television shows. The Rotten Tomatoes database provides brief overviews and cast lists of movies but is primarily concerned with the critique of movies. The Complete Index to World Film, CITWF, is, as the name suggests, the world's largest film database. It references every movie made since 1895 and also provides print versions of its content. The Big Cartoon Database (BCD), contains information about more than 100,000 cartoon productions and animated features. The Box Office Mojo website focuses on tracking the financial aspects of movies rather than their actors and production values. The site Metacritic collects reviews of its entries, much like Rotten Tomatoes does. However, Metacritic covers television, music, and video games in addition to its movie coverage.

Online databases likewise exist for music. Among the most popular was AllMusic, founded in the 1990s by entrepreneur Michael Erlewine. As of 2023, AllMusic featured listings for millions of albums, each of which typically included information such as a track listing; information about writers, composers, and personnel; and an official review, user ratings, or both.

—Richard M. Renneboog

Further Reading

Hutchison, Tom, Paul Allen, and Amy Macy. *Record Label Marketing*. Taylor & Francis, 2012.

"Press Room." *IMDb*, June 2023, www.imdb.com/pressroom/stats. Accessed 13 July 2023.

Sickels, Robert C. *The Business of Entertainment.* 3 vols. ABC-CLIO, 2009.

Smith, Ernie. "The Story of AllMusic, the Internet's Largest, Most Influential Music Database." *Vice*, 24 Sept. 2016, www.vice.com/en/article/53djj8/the-story-of-allmusic-the-internets-largest-most-influential-music-database. Accessed 13 July 2023.

Wacholtz, Larry. *Monetizing Entertainment: An Insider's Handbook for Careers in the Entertainment and Music Industry*. CRC Press, 2016.

Williams, Joe. *Entertainment on the Net*. Que, 1995.

Online Piracy

ABSTRACT

Online piracy refers to the unauthorized downloading, viewing, copying, or distribution of copyrighted digital content. Targeted content includes films, television, music, games, and books. Piracy of this content has become so commonplace that some people consider it a right instead of a crime. The significant financial impact of online piracy on copyright holders has become the concern of individuals, businesses, and the US Congress, while advocates for internet freedom of speech are just as concerned about remedies that might unduly restrict online expression and legitimate use of content.

BACKGROUND

While the use of the term "piracy" in reference to intellectual property has a long history, it only entered widespread use with the advent of the internet, especially when compact disc (CD) burners made copying and distributing music, software, and other material from the internet as easy as a few clicks. The most commonly pirated content includes music, films, television shows, e-books, and software. Digital music files, typically in MPEG-1 Audio Layer 3 (MP3) format, were one of the first types of files to be extensively copied, due to the development of several file-sharing programs in the 1990s that made it easy to transfer such files both legally and illegally.

Internet pirates have turned illegal copying and sharing of intellectual property into thriving businesses, with numerous websites specializing in hosting pirated files for download. One popular site was megaupload.com, hosted by the Hong Kong-based company Megaupload Limited. In 2012, the US Department of Justice shut down the company's sites and indicted its owners for allegedly operating an organization based on copyright infringement, and Hong Kong's Customs and Excise Department froze the company's $42 million worth of assets.

As quickly as authorities shut down pirating services, others just as quickly spring up to take their place. According to a report by Digimarc, a company that provides antipiracy technology for publishers and authors, in the first month after Megaupload shut down, two other sites, Putlocker and Rapidshare, raised the total of pirated books to 13 percent. In another piracy report, brand-protection firm NetNames reported that in Europe and North America, the peer-to-peer file-sharing protocol BitTorrent is the most popular method of downloading pirated material, while in the Asia-Pacific region, pirates prefer direct-download websites.

Software piracy has also continued to advance as technology has improved. The main types of software piracy include "softlifting," in which users share their software with other, unauthorized users; uploading and downloading; software counterfeiting; original equipment manufacturer (OEM) unbundling, in which OEM software is separated from the hardware it was originally sold with and distributed independently; and hard disk loading, in which unauthorized software is preinstalled on hardware before it is sold.

OVERVIEW

According to a 2011 study by Envisional, a subsidiary of NetNames, digital piracy of music, movies, and other copyrighted material accounted for 23.76 percent of internet bandwidth worldwide and 17.53 percent of bandwidth in the United States. Commenting on the study, Republican senator Orrin Hatch of Utah, cochair of the International Anti-Piracy Caucus, argued that online piracy hijacks the earnings of artists and creators of movies, television, and music, which discourages reinvestment in new job-creating projects and thus weakens the American economy.

Internet book piracy affects millions of writers and publishers whose books have been scanned or their digital files copied and uploaded to websites that feature pirated works, sometimes alongside legitimate uploads. A study published in January 2010 by Attributor, which was later acquired by Digimarc, found that writers and publishers had lost an estimated $2.8 billion to date because of internet piracy, assuming that each illegal download represented a lost sale.

Internet theft of music is a constant challenge. The volume of pirated music and the drop in music industry revenues is significant, with rising digital sales not closing the gap. The Recording Industry Association of America (RIAA) reported that a decade after Napster introduced its file-sharing site in 1999, music sales in the United States dropped from $14.6 billion to $7.7 billion. In 2003, with the launch of the Apple iTunes Store, steady options for legal digital music downloads became available.

The Motion Picture Association of America (MPAA) reported that in 2005, the major US motion-picture studios lost $6.1 billion worldwide to internet piracy, and the worldwide motion-picture industry, including foreign and domestic producers, distributors, theaters, video stores, and pay-per-view operations, lost $18.2 billion. With these statistics in mind, the entertainment industry has countered internet piracy of music, television, and movies with more legal options. As broadband speeds have increased, so too have the opportunities to legally watch television shows and movies or listen to music online. Popular legal streaming services include Netflix, Hulu, Amazon Prime Video, Disney+, AppleTV+, Apple Music, Amazon Music, Spotify, and Pandora, among others.

In 2013, major internet service providers and the entertainment industry announced a partnership to attempt to curb piracy of copyrighted material on-

Image via iStock/A Mokhtari. [Used under license.]

line. Under the Copyright Alert System (CAS) or "six strikes" system, internet subscribers accused of online piracy received a series of alerts each time they were found pirating material. The sixth copyright violation potentially resulted in the user being punished by the internet provider. Subscribers who illegally shared movies or songs could be punished by losing their internet access or having the speed of their broadband downloads reduced to a crawl. After several years of operation, however, the system was deemed unsuccessful, and the CAS initiative came to an end in early 2017.

Responding to an outcry from internet users and technology companies including Google, Congress defeated the Stop Online Piracy Act (SOPA) and the PROTECT IP Act (PIPA) in 2012. SOPA and PIPA would have imposed strict penalties on websites carrying copyright-violating material, which Google and other critics claimed would curtail free speech.

Reputable companies who advertise on the internet have also been pulled into the online piracy problem. Because most internet ad placement is generated automatically through algorithms, companies cannot always control where their ads are placed. Often these ads appear on sites that offer illegal downloads of music and films. A 2015 report issued by Digital Citizens alliance and MediaLink, a consulting firm, found that "theft" websites earned more than $200 million from advertising placed on their pages in 2014. This phenomenon has been termed "ad fraud" and is an effect of online piracy, illustrating the ubiquity of the problem.

As the internet has continued to grow more technologically sophisticated, online piracy has kept pace. A study funded by the UK government and released by the Intellectual Property Office found that during January 2021, the overall level of infringement for all content categories (excluding digital visual images) was at 23 percent. While the unauthorized downloading of files remained popular, the use of illegal streaming sites made up a substantial portion of all online piracy during the early 2020s.

—*Kathy Warnes*

Further Reading

Fisk, Nathan W. *Understanding Online Piracy: The Truth about Illegal File Sharing*. ABC-CLIO, 2009.

Gordon, Sherri Mabry. *Downloading Copyrighted Stuff from the Internet: Stealing or Fair Use?* Enslow, 2005.

Hamedy, Saba. "Report: Online Piracy Remains Multi-Hundred-Million-Dollar Business." *Los Angeles Times*, 19 May 2015, www.latimes.com/entertainment/envelope/cotown/la-et-ct-report-online-piracy-digital-citizens-alliance-medialink-20150518-story.html. Accessed 13 July 2023.

Hinduja, Sameer. *Music Piracy and Crime Theory*. LFB, 2005.

Husak, Douglas. *Overcriminalization: The Limits of the Criminal Law*. Oxford UP, 2008.

Johns, Adrian. *Piracy: The Intellectual Property Wars from Gutenberg to Gates*. U of Chicago P, 2009.

Kastrenakes, Jacob. "'Six Strikes' Anti-Piracy Initiative Ends after Failing to Scare Off 'Hardcore' Pirates." *Verge*, 30 Jan. 2017, www.theverge.com/2017/1/30/14445596/six-strikes-piracy-system-failed-ending. Accessed 13 July 2023.

Murray, Brian H. *Defending the Brand: Aggressive Strategies for Protecting Your Brand in the Online Arena*. AMACOM, 2004.

Nhan, Johnny. *Policing Cyberspace: A Structural and Cultural Analysis*. LFB, 2010.

"Research and Analysis: Online Copyright Infringement Tracker Survey (10th Wave) Executive Summary." *Intellectual Property Office*, gov.uk/government/publications/online-copyright-infringement-tracker-survey-10th-wave/online-copyright-infringement-tracker-survey-10th-wave-executive-summary. Accessed 13 July 2023.

Riley, Gail Blasser. *Internet Piracy. Cavendish*, 2010.

Torr, James D. *Internet Piracy*. Greenhaven, 2004.

P

Photographer

ABSTRACT

Photographers use film and digital cameras to capture images of people, objects, and events. They often specialize in a specific area, such as news, portrait, commercial and industrial, scientific, or fine arts photography. Students aspiring to enter the profession should take courses in communications, computers, art, photography, and media.

BACKGROUND

Photographers capture images of various objects, people, and events using a film or digital camera. They must exhibit a solid understanding of technical camera operation and the fundamental processes behind photography, lighting, and the composition of an image. Most photographers focus on one area of photographic specialty. Photographic specialties include news, portrait, commercial and industrial, scientific, and fine arts photography. Because their profession is based on choosing image composition and creating unique images, creativity is a trait common among all types of photographers regardless of their area of specialization.

OVERVIEW

A photographer's work environment depends primarily on his or her area of photographic specialty. Some photographers, such as those who take studio portraits of children and families, work primarily out of comfortable, well-lit, indoor studios. Other photographers work outside in a multitude of environments and are subject to various weather conditions. Photographers who work for the government, advertising agencies, or private companies frequently maintain a forty-hour week. Freelance and newspaper photographers, or photojournalists, generally work irregular hours, travel often, and are expected to be on-call for last-minute projects or emergency events.

Potential photographers should demonstrate a passion for artistic creation. They should be compelled to tell stories through photographs and possess a deep desire to analyze, present, and offer a unique perspective on their photographic subjects. They should express a definitive opinion through their photographs, and that opinion should be easily discernible to an audience examining their photography. They should be able to lead and work with different types of people and personalities.

DUTIES AND RESPONSIBILITIES

Most photographers purchase and maintain their own camera equipment, lenses, and accessories, which can be costly at the outset. Photographers usually work independently or with an assistant. They are responsible for the physical positioning of subjects as well as the arrangement of lighting and camera angles. If the photographs are taken with film (which is increasingly rare), the photographer develops the film and prints in either a darkroom or printing facility. Digital photographs may be edited and retouched prior to publishing or printing.

Photographers' specialties determine what they photograph and how those images are used. Portrait photographers specialize in photographing people or groups of people. They are generally self-employed and often travel to various locations for special events like weddings, school functions, and other special ceremonies. Commercial and indus-

Photo via iStock/Alexander Shelegov. [Used under license.]

trial photographers travel to various locations to photograph landscapes, buildings, and merchandise. Their photographs are usually published in books, advertisements, catalogs, or other media. Scientific photographers make a photographic record of objects or events related to science and medicine. These photographers usually have technical training in the sciences as well as the arts. News photographers, or photojournalists, take pictures of relevant people or events for publication in regular newspapers or periodicals. Fine arts photographers are usually highly technically proficient, and may display their photographs in museums, art galleries, or private art shows.

Self-employed and freelance photographers must perform business and administrative tasks in addition to their creative work. Such tasks might include managing employees, handling billing and payments, setting appointments, and obtaining licenses, copyrights, contracts, and other legal documents as needed. They must also arrange their own advertising, marketing campaigns, and self-promotion.

OCCUPATION SPECIALTIES

Aerial Photographers. Aerial photographers photograph segments of earth and other subject material from aircraft, including drones.

Scientific Photographers. Scientific photographers use specialized equipment to illustrate and record scientific phenomena.

Studio Photographers. Studio photographers photograph subjects in formal studios or similar settings and use a variety of accessories. They normally specialize in a particular area of photography, such as illustrative, fashion, or portrait.

Photojournalists. Photojournalists photograph newsworthy events, locations, people, or other illustrative and educational material for use in publications or telecasts, using a still camera.

Fine Arts Photographers. Fine arts photographers create photographs for sale as art.

Other Photographic Specialties. Other notable photographic specialties include architectural photography, forensic photography, landscape/nature/wildlife photography, sports photography, and wedding photography.

WORK ENVIRONMENT

Physical Environment. A photographer's working conditions vary greatly depending on his or her specialty. Some photographers can work in clean, comfortable, well-ventilated studios. Others work in unpleasant or dangerous outdoor environments. Photographers regularly travel to and from photographic sites. Those who process film and prints, especially in a darkroom, are exposed to potentially harmful chemicals.

Human Environment. Photographers work with numerous clients, customers, and subjects. They must interact easily with others, and they should be comfortable directing, evaluating, and occasionally comforting their photographic subjects. Photographers sometimes collaborate with graphic designers, journalists, reporters, and editors. Some may report to a supervisor or direct an assistant.

Technological Environment. Photographers must learn how to operate camera equipment in order to be successful. To create a photograph, photographers use film and digital cameras, film, digital memory and storage devices, tripods, lenses and filters, floodlights, reflectors, light meters, and electronic flash units. Image processing may require computers, imaging and editing software, printers and scanners, photographic paper, darkroom equipment, and chemicals for developing film and prints from film.

EDUCATION AND TRAINING

High School/Secondary. High school students interested in becoming photographers should devote time to the study of communications, computers, art, photography, and media. Aspiring photographers should also engage in extracurricular activities (like the school newspaper or yearbook) that allow them to practice taking pictures, editing their work, and posting or printing their best photographs. Interested students should pursue part-time work with a photographer or store and consider applying to postsecondary photography programs.

College/Postsecondary. Most photographers find it helpful to have an undergraduate degree or certificate in photography from a university, community college, private art school, or vocational institute. Many vocational education programs offer courses in visual imaging technology as well as in the fundamentals of photography. Other postsecondary programs teach students the practical and technical aspects of photography. Coursework may include the history of photography and cinema, camera maintenance, photojournalism, composition, color printing and print finishing, lighting, retouching, and other related subjects. Prospective freelance photographers may benefit from courses in business, including marketing, public relations, and business management.

Adult Job Seekers. Many prospective photographers find positions as assistants to local, professional photographers after receiving their formal education. Assistants gain valuable technical experience, on-the-job training, and the practical skills needed to start their own businesses. Other job seekers apply for full- or part-time positions at camera shops, local newspapers, or photography studios. Candidates can also participate in apprenticeships, mentorships, or internships through their schools or photography training programs.

Many photographers subscribe to photography newsletters and magazines in order to make contacts in the industry. Networking, mentoring, and professional development opportunities are also frequently available through professional photographer associations.

Professional Certification and Licensure. Photographers are usually not required to obtain professional certification or licensure in their field; to an extent, this is because the work is highly visual, so photographers can easily provide samples of their work to others. Some professional photography organizations offer voluntary certifications, which may enhance a photographer's marketing and job-seeking efforts. To become a Certified Professional Photographer (CPP) through the Professional Photographers of America, candidates must pass a written exam and provide images for critique. Continuing education is typically required for certification renewal.

Additional Requirements. Photography is a well-respected form of artistic expression. Therefore, aspiring and professional photographers should be naturally artistic and able to understand the fundamentals of photographic composition. Because the field is intrinsically subjective, photographers should maintain the integrity and conviction necessary to present effective artwork and subject analysis, without reacting negatively to criticism. Photographers should be patient, have strong eyesight, possess boundless imagination, and demonstrate impeccable communication skills when dealing with clients and subjects.

—*Briana Nadeau*

Further Reading

Forbes, Ted. *Visually Speaking: Mastering Photography as a Visual Language*. Cider Mill Press, 2023.

Long, Ben. *Complete Digital Photography*. CDP Press, 2018.

Prust, Z. A., and Peggy B. Deal. *Graphic Communications: Digital Design and Print Essentials*. 7th ed., Goodheart-Willcox, 2022.

Photojournalism

ABSTRACT

As a profession, photojournalism is dedicated to telling a story with pictures. The term was coined by Cliff Edom, a journalism professor at the University of Missouri, who taught the first class in photojournalism. Photojournalism was born in the mid-nineteenth century as the technologies of cameras and photo-processing advanced and photography could be used to visually record the realities of war and important events as they happened. Certain iconic photographs have become cultural symbols, such as Dorothea Lange's "Migrant Mother" (1936), which depicts a destitute mother and her children during the Great Depression; "Raising the Flag on Iwo Jima" (1945) by Joe Rosenthal; or the many recognizable pictures of the destruction of the Twin Towers in New York City on September 11, 2001. Photojournalists have frequently become activists, launching campaigns that have led to historic social changes. Traditionally, photojournalists have been trained in both journalism and the use of camera techniques, but the digital revolution has produced what has come to be known as the citizen photojournalist, as new technology allows anyone with a smart phone or camera-enabled mobile device to take pictures as events occur.

BACKGROUND

In 1851 Scott Archer of Great Britain developed the process of wet-plate photography, which allowed for the production of multiple copies of a single photograph. Subsequently, British photographers Roger Fenton and William Simpson broke new ground by photographing the Crimean War (1853-56). In the United States, photojournalism was born during the Civil War when General George McClellan named Captain Alexander

Gardner as his official photographer. Gardner's partner, Matthew Brady, with whom he owned a portrait studio, received permission from President Abraham Lincoln to take battlefield photographs. Brady dispatched Alexander Gardner and James Gibson to Maryland to photograph the war. The two captured the Battle of Antietam on September 17, 1862. That battle was the bloodiest day Americans had ever experienced. When it was over, thousands of men lay dead on the battlefield, and the actual devastation was depicted in photographs rather than in paintings or drawings.

Despite the use of photography in warfare, most newspapers continued to use either illustrations or woodgrains to accompany stories. Between 1880 and 1897, however, photography continued to evolve, and the first half-tone photograph was printed in the *New York Daily Graphic* on March 4, 1880. By the beginning of the twentieth century, the process of half-tone reproduction was widely used, and newspaper publishers were able to accompany many news stories with photographs. The introduction of flash powder in 1887 improved the quality of photographs used by photojournalists.

In the late nineteenth and early twentieth centuries, the work of journalists who became known as "muckrakers" was instrumental in initiating many of the reforms of the Progressive Era. Jacob Riis used photojournalism to portray the realities of urban poverty in 1890 in *How the Other Half Lives*. Lewis Hine employed photojournalism to illustrate the need for reforms in areas such as child labor

Photo via iStock/chameleonseye. [Used under license.]

and immigration policy. In 1906, author Upton Sinclair included photographs of Chicago's meatpacking plants in his novel *The Jungle*, which led to the creation of the Food and Drug Administration as well as new laws on food and drug production and worker safety.

Photography became increasingly important in defining the significance of major events. For example, in 1876 the Brooklyn Theater in New York caught fire, causing the deaths of 297 people. News stories about the event were accompanied by a single illustration, and little was done to make theaters safer. In 1911, the Triangle Shirtwaist Factory caught fire. Supervisors had locked doors to keep workers from leaving the building without permission, and as a result, 146 women died. By that time, new technologies allowed new stories to be illustrated with actual photographs. The horrors of the Triangle Factory Fire, starkly depicted in photographs, led to changes in state fire and safety laws.

The 1920s and 1930s are considered the golden age of photojournalism due in large part to the introduction of the Leica camera, 35 millimeter (mm) film, and flash bulbs. As the field continued to evolve, the names of some photojournalists became internationally known. War correspondent Robert Capa (1913-54) used his camera to capture the realities of five wars: the Spanish Civil War, the Second Sino-Japanese War, World War II, the Arab-Israeli War, and the First Indochina War. Alfred Eisenstaedt (1898-1995) photographed everyone from statesmen such as Winston Churchill and John F. Kennedy to celebrities such as Marilyn Monroe and Sophia Loren. Margaret Bourke-White (1904-42) started the first photography lab at *Life* magazine. She became known for her ability to be in the right place at the right time, and her famous photographs included one taken of Mohandas Gandhi only hours before his assassination on January 30, 1948. Bourke-White had also been on hand to photograph the Battle of Moscow in 1941. W. Eugene Smith (1918-78) established his reputation by photographing World War II by sea, land, and air. His experiences led him to become a fierce proponent of social justice. Photography has continued to be an important component of journalism ever since.

OVERVIEW

In the 1960s, the field of photojournalism expanded to include citizen photojournalists who just happened to be present with a camera when major events took place. The first of the citizen photojournalists was Abraham Zapruder, an amateur photographer who was filming President John F. Kennedy's visit to Dallas, Texas, on November 22, 1963. Zapruder captured events of the day that were later used in the investigation of the president's assassination. A similar event occurred in 1991 when a bystander filmed the savage beating of Rodney King by police officers in Los Angeles, turning a public spotlight on police brutality and the racism that continues to be a major problem in many police departments.

In the twenty-first century, anyone with a smartphone or other mobile device with a camera is able to take spontaneous photographs of ongoing events, which can in turn be shared on social media and even republished by traditional news outlets. Citizen photojournalists were instrumental in capturing real-life events during the first decades of the twenty-first century, including numerous instances of police brutality, social and political unrest, and natural disasters. A downside of the ubiquity of amateur photography during that period was that the likelihood of faked photographs also increased significantly, as there were no professional standards governing the field.

While citizen photojournalists played a key role in informing the public about current events during the early twenty-first century, their emergence had a devastating impact on the field of professional photojournalism, as did the decline of newspapers and print magazines. The Great Recession of 2007-2009 also played havoc with the profession of photojournalism, and many news-reporting venues abolished entire departments, expecting reporters with smart phones to fill the gap. In other cases, photojournalists experienced pay cuts and reduced hours. Such actions led to the frequently expressed fear that professional photojournalism may be on its way to becoming obsolete.

—Elizabeth Rholetter Purdy

Further Reading

Allen, Stuart. *Photojournalism and Citizen Journalism: Co-operation, Collaboration and Connectivity.* Routledge, 2019.

Berger, Martin A. *Freedom Now!: Forgotten Photographs of the Civil Rights Struggle.* U of California P, 2013.

Guerrero Garcia, Virginia and Bella Palomo. "The Crisis of Photojournalism: Rethinking the Profession in a Participatory Media Ecosystem." *Communication and Society*, vol. 28, no. 4, 2015, pp. 33-48.

Garza, Melita M., Michael Fuhlhage, and Tracy Lucht, editors. *The Routledge Companion to American Journalism History.* Routledge, 2023.

Hariman, Robert. *No Caption Needed: Iconic Photographs, Public Culture, and Liberal Democracy.* U of Chicago P, 2007.

Harris, John M. "Truthful as the Record of Heaven." *Southern Cultures*, vol. 19, no. 3, 2013, pp. 79-94.

Jensen, Robin E., Erin F. Doss, and Rebecca Ivic. "Metaphorical Invention in Early Photojournalism: New York Times Coverage of the 1876 Brooklyn Theater Fire and the 1911 Shirtwaist Factory Fire." *Critical Studies in Media Communication*, vol. 28, no. 4, 2011, pp. 334-52.

Kamber, Michael. *Photojournalists on War: The Untold Stories from Iraq.* U of Texas P, 2013.

Langton, Loup. *Photojournalism and Today's News: Creating Visual Reality.* Wiley, 2009.

Lucaites, John Louis, and Robert Hariman. "Visual Rhetoric, Photojournalism, and Democratic Public Culture." *Rhetoric Review*, vol. 20, no. 1/2, 2001, pp. 37-43.

PRINTING

ABSTRACT

Printing includes a wide variety of technologies, increasingly computerized, engaged in the transference and manipulation of images—both words and illustrations—from one surface to another. Surfaces range from paper to glass to metal and even apparel. With the emerging demands of the information age, in which literally millions of agents are engaged in the process of moving information and maintaining accurate records, printing technology has evolved into a cornerstone industry.

BACKGROUND

Printing refers to the applied technologies of reproducing texts, specifically the storing and distribution of data. In many ways, printing is an invisible technology despite being surrounded by evidence of its application. It might be assumed that given the rapid rise in digital technology, specifically the unprecedented storage capability of digital technologies and the ubiquitous presence of personal computers, that printing has become obsolete. Although the computer has revolutionized printing, the demand for the organized and controlled transference of information still relies on printing. From publishing (books, magazines, newspapers) to commercial applications (labels, manuals, banking checks, forms, letterhead, as well as glassware and even clothing), printing is central to a culture that relies on maintaining records and disseminating information. Unlike more prominent applied technologies—such as steel manufacturing or auto making—printing maintains its low profile because, unlike those industries, printing concerns are more likely to be smaller, privately owned, independent industrial facilities that seldom employ more than fifty workers.

Given the impractical and unreliable methods of the ancient oral cultures (memory and verbal communication), the desire to print information is a longstanding one. The earliest developments in printing, woodblocks used to transfer images and symbols to cloth, date to first century CE China. In addition, advances in paper production in the Far East were brought back to Europe in the twelfth century as were advances in movable type, efficient systems of reproducing text using revolving stone tablets and numbered boxes of handset type, most often made of wood. In the early fifteenth century, German metalworker Johannes Gutenberg perfected an efficient system of movable type technology using durable type pieces made of a lead and tin alloy.

It is impossible to overestimate the impact of movable type. It enabled information to be more rapidly shared, and science, philosophy, and literature enjoyed unprecedented prosperity. The printing press has been credited with making the Renaissance possible. Knowledge could accumulate across generations: Scientists could build on each other's work; books could circulate widely; governments could maintain public records; education was readily accessible; writers had authority over their work; reading itself became a private act; and languages became part of a national culture. Over three centuries, the basic technology of Gutenberg's printing press remained unchanged, save for being driven by progressively more energy-efficient means: first steam then electricity. In the closing decades of the twentieth century, however, printing began to redefine image reproduction and text manufacturing to accommodate the rapid strides in information gathering and display offered by digital technologies.

Nearly all commercial nondigital printing is involved in one of five basic processes: lithography, screen, flexography, letterpress, or gravure. Over the last generation, however, digital technologies have begun to reshape printing, and microtechnology has allowed printing to diversify into data-management and web-based systems. In fact, with the advent of desktop publishing and laser printing, text manufacture and information display have become part of personal computing with texts that, unlike those produced a generation earlier by the typewriter, are professional in appearance.

OVERVIEW

Offset. By far the most common process for commercial printing is offset lithography, which includes the mass-quantity, high-speed production of mail-order catalogs, newspapers, directories, weekly magazines, and books. The process is designed for competent and speedy production. It involves spools of paper that weigh upwards of a ton and machines capable of printing more than 1,000 impressions every minute.

First, the text to be printed must be converted to plates, a process that involves photo emulsion. This is the same process as photography in that negatives of the text are exposed to controlled light that passes through the negative and in turn triggers a chemical reaction that transfers the transparent image to a thin flat plate, most often made of zinc or aluminum. The process moves to the actual press run. Because oil and water do not mix, the plate surface is broken down into two areas, the actual image area, which repels water (and thus remains dry and can absorb ink), and the nonimage area, which accepts water and is not part of the reproduced image.

Ink is distributed to the prepared plates through high-efficiency rollers. The plates are rolled first with water and then with ink. The ink adheres to the image, and the water rollers keep the ink off the nonimage area. The plate is then pressed to a rubber blanket, transferring the image quickly and cleanly. High-speed presses then roll the paper across the rubber blanket; the paper never touches the actual plate, thus the term offset. Because the paper, once through the press, is wet, the paper then passes through a gas oven at a temperature near 400 degrees Fahrenheit to dry it, and then it is quickly run through a refrigeration unit to cool the paper to avoid smudging.

Rotogravure Printing. The other most widely used industrial printing process (accounting for roughly 20 percent) is rotogravure, or intaglio, printing, in which the image or text is cut into the surface of a rounded (hence "roto") printing plate. The image is actually formed from the sunken surface on the plate. Each cylinder can hold the image of up to sixteen pages of text. As the cylinder is turned at high speeds, the roll of paper (or fabric or foil or plastic) is fed continuously across its surface, in which computerized stressors maintain the registration controls necessary to transfer the images. The paper is passed across the image cylinder. After that, the paper is passed through high-temperature driers to help set the ink quickly.

Although the process is more technologically sophisticated and more expensive than offset, gravure is actually more cost efficient (state-of-the-art presses can run as fast as 3,000 feet per second), and the reproduction is more consistent, better detailed, and of finer resolution. The tempered plates are more durable than the rubber blankets used in offset. The only drawback to gravure from an industry standpoint is the time it takes to prepare the cylinder plates, a process called diffusion etching, which involves a thick copper mask and an acid bath to achieve the appropriate cut in the cylinder.

APPLICATIONS AND PRODUCTS

Printing applications figure in virtually every industry. These industries are routinely held together through the reproduction and dissemination of printed materials. In the 1950s, Marshall McLuhan, one of the pioneer theorists in communication and language, predicted the inevitable end of the age of print and the evolution of the paperless office, and despite the innovations of the last generation of printers driven by digital technology, reproducing and storing data through printing processes has only increased in application.

Lithography. Offset lithography, introduced more than a century ago, is the predominant technology for commercial printing. Sheet-fed offset printing is cost effective and produces reliably clean images and text because the rubber blanket that actually transfers the image adheres tightly and precisely to the texture of the printing surface. Color reproduction is entirely directed by computerized settings that blends the inks together at tremendous speed. Although the color reproduction is not fine point, the types of printing that use lithography offset—daily and weekly newspapers, weekly magazines, advertising and informational brochures, and most paperback books—are not designed for long-term usage and do not require precise image reproduction.

Gravure Printing. Given its history (the process of cutting designs into soft metallic plates was first used by Renaissance artisans interested in fine and expensive engravings), it is not surprising that gravure printing is used for printing projects that, while requiring a significant number of rungs, have more of an aesthetic element than those created with offset printing applications. These include commercial bags and boxes that bear the trademark logo of a company or store; decorative wrapping paper; labels; mass-mailing glossy inserts; Sunday-newspaper inserts, high-end glossy commercial and department-store catalogs, and monthly magazines; specialty limited-run printing, such as ornamental wallpaper rolls, art prints, publicity posters, or commemorative postage stamps; and, finally, printing runs that involve delicate fabric. Because of the durability of the cut plates, gravure printing is also used in the production of long-term forms and applications.

Flexography. Developed in the early 1900s but not widely used until the invention of cellophane during the Depression created a new industry for food packaging, flexography is a variation of rotary printing using an image carrier made of flexible rubber or a soft plastic plate with raised image designs. It accounts for just about 15 percent of commercial printing applications. Flexography is particularly attractive for high-demand printing work that is less ornamental and more functional. It is based on a simple three-stage technology: an in-feeding unit feeds materials such as cardboard or film; a printing unit delivers high-speed drying ink, which is spread by passing the medium through a series of distribution rollers; the out-feed unit rewinds the printed material to be sent to the appropriate plants. Candy wrappers, for instance, would be rewound and sent to the factory. Flexographic printing is versatile in its applications. It is used widely in the production of cardboard products, product wrapping, corrugated board, medical, sanitary, and food packaging, and plastic film. Because printing on cellophane using the gravure process proved prohibitively expensive, flexography made cellophane printing cost effective.

Screen Printing. As with other printing technologies, screen printing is based on a simple principle that dates back to ancient China. The idea is to transfer an image by passing ink through openings in a precut stencil. A stencil is attached to a piece of fabric stretched on a metal or wooden frame. A squeegee-like device is used to force ink through the stencil opening, and then the material, most often paper but also metal, plastic, or wood, is fed and pressed under the stencil and then removed for drying. Although originally used for producing elaborate wallpaper designs and shop signs in the 1800s, screen printing is now used for short-run projects that require high-quality resolution: photograph reproduction, clothing lines and silk-screen work, cardboard cartons, and even printing microcircuits for computer systems.

Inkjet Printing. With the rise of personal computers and the emergence of digital printing in offices, inkjet printing is becoming the most familiar and most versatile application of printing technology. Given the resources of computer programs and the speed at which such software is marketed (any description of existing inkjet printing technology will quickly become obsolete), the applications of inkjet printing are bound only by the resources of the technological imagination. The basic premise—the transfer of images and text using digital data memory and on-screen formatting to create individual and personalized printing—puts printing skills in the hands of anyone with a computer. The effects are clear and readable but relatively low resolution. High-quality resolution is seldom the aim.

In inkjet printing, computers create the desired effect by controlling the spraying of individual drops of ink by using electronic impulses. Although used most frequently with paper for high-speed reproduction of information within a large or small office network, inkjet printing can also be used with corrugated board, plastic, and fabric.

IMPACT ON INDUSTRY

In the information age, workers skilled in the printing process, particularly the burgeoning field of digital printing, have virtually reshaped every industry. Printing is still at the core of the twenty-first-century-corporate model. Significant development in the field of computerized printing processes, particularly developing secure identification cards, secure banking and medical data, more efficient and better resolution mass printings, and developing environmentally friendly, cost-effective food packaging, has been spearheaded in the United States as well as in Japan and Great Britain.

Two concerns—government and business—direct research in the printing field. Commercial printing is spearheading efforts to become more environmentally friendly. Many printing processes rely on petroleum-based inks, organic emissions, and lubricants that create disposal problems and long-term environmental hazards. Given that printing businesses are seldom major industrial plants but rather are small neighborhood ventures, regulating that environmental hazard is a growing concern. Paper itself represents a special challenge to the printing industry. Worry about the availability of raw materials for paper production have encouraged recycling campaigns, but researchers in printing processes are investigating technologies that would free data—images and words—from paper altogether. That, however, creates a second area of concern within the printing field. Printing processes, relying increasingly on paperless technology, need to be brought into compliance with post-September 11 security realities. Given the volume of data recorded by computer imaging and transmitting through, among others, government offices, law firms, banks, and hospitals, information relayed through the printing medium needs to be secured in ways that would have seemed extreme prior to the twenty-first century.

CAREERS AND COURSEWORK

Courses in mechanical engineering, chemistry, physics, computer science, mathematics, and graphic communication account for the foundational requirements for students interested in careers in printing. Training for most hourly wage printing jobs (technicians and operators) is on-site and job specific. Production workers, printing machine operators, and manual laborers who load and maintain the presses typically have associate's degrees, specific vocational training, or bachelor's degrees (most often in engineering). However, significant career advancement is best secured through knowledge of computer skills, desktop publishing, journalism, and digital graphic design. Upper-level jobs require familiarity with business administration and marketing.

SOCIAL CONTEXT AND FUTURE PROSPECTS

A casual glance at a contemporary department store or shopping mall, an office complex or a bank, a hospital or a local government building confirms the place of print technology. Although seldom acknowledged, the printing industry defines twenty-first-century consumer culture. The advent of computerized printing techniques has reshaped the employment profile away from manual work into prepress design and computerized formatting, and digitalized data banks have drastically cut into the volume of printed material produced. Printing and printed material are still central to virtually every service industry and business enterprise.

Newspapers have suffered dramatically because of competition from internet news sources and television. Despite the advent of electronic books and electronic readers, the print book and magazine industries remain culturally relevant, and print-on-demand technologies represent new possibilities for publishers. Researchers have estimated that before the end of the twenty-first century, the enormous volume of data required by an information-hungry global community will be entirely digitalized, virtually eliminating the need for paper and printing for data retrieval. That same technology, however, will increase the efficiency, reliability, and safety of printing presses able to provide the other, wide-ranging printing applications.

—Joseph Dewey

Further Reading

Adams, J. Michael, and Penny Ann Dolin. *Printing Technology*. 5th ed., Delmar, 2002.

Howard, Nicole. *The Book: The Life Story of a Technology*. Greenwood Press, 2005.

Needham, Paul, and Michael Joseph. *Adventure and Art: The First One Hundred Years of Printing*. Rutgers UP, 1999.

Shaoqiang, Wang, editor. *Get Impressed!: The Revival of Letterpress and Handmade Type*. Hoaki, 2021.

Steinberg, S. H. *Five Hundred Years of Printing*. 1955. Dover, 2017.

Whitington, John. *How Computers Make Books: From Graphics Rendering, Search Algorithms, and Functional Programming to Indexing and Typesetting*. Manning, 2023.

Young, Sherman. *The Book Is Dead. Long Live the Book*. U of New South Wales P, 2007.

Programming Languages for Artificial Intelligence

ABSTRACT

Artificial intelligence (AI) is a branch of computer science that creates computers and software with the thinking capacity of humans. The key to the capabilities of these computers is the programming language used in the process. Two major initial programming languages for AI were LISt Processor (LISP) and PROgrammable LOGic (PROLOG). LISP was more widely used than PROLOG, but both became respected and recognized as invaluable parts of this scientific discipline. Subsequent AI languages include Python, R, and Java, among others. In 2022, ChatGPT, a chatbot designed by AI research group OpenAI, utilized the group's proprietary Generative Pretrained Transformer (GPT) family of language models.

BACKGROUND

Computers are programmed by humans to "think." The ability of a computer to perform its function is only as strong as the language used to instruct it. One such language was LISP, which is short for "List Processor." LISP was developed in 1958 by an American computer scientist named John McCarthy, who also coined the term "artificial intelligence," or AI. The foundation for LISP was an older programming language known as Information Processing Language (IPL), developed by Allen Newell, Herbert Simon, and Cliff Shaw in 1956. This language was useful in the early days of computing, but it was highly complicated. A simpler language was needed to enable programmers to understand it better.

McCarthy altered IPL by adding basic principles from what is known as lambda calculus, a branch of mathematical logic that focuses on variables and substitution. An advantage of LISP is that it enables a data management mechanism known as garbage collection in the AI "thinking" process. Items considered "garbage" in the world of AI are units of memory created during the running of an AI operation that serve as temporary holders of information. While they are useful for a time, they eventually outlive that usefulness. LISP is able to identify this used memory and discard it. Unlike IPL, LISP was able to assign storage space for information effectively and conservatively. In time, LISP became the most commonly used programming language for AI in the United States.

The second major early programming language for artificial intelligence was called PROLOG. The name comes from the phrase "programming in logic." PROLOG was first developed in 1973 by French computer scientist Alain Colmerauer at the University of Aix-Marseilles. Robert Kowalski, an American computer scientist at the University of Edinburgh in Scotland, developed it further.

OVERVIEW

The most crucial part of LISP's functionality is an element called the interpreter. The interpreter takes statements presented to it by users and decides the value of the expression through a process known as evaluation. The evaluation process depends on three factors: the identity of the symbols used in a statement, or what the words or figures presented mean at their most basic level; the associations (e.g., other terms, or lists of terms) with each symbol in a statement; and the list of possible functions that might be associated with a statement. Lists are central to LISP, and yet they are not necessarily the same items we would typically call lists; they are more collections of possible functions or equations centering on a term or symbol. It is significant that the programmer in LISP gives the computer commands by constantly defining new functions and new equations. This process could potentially result in the growth of ability in the machine—as it might with a human, engaged in the act of intellectual exploration.

The chief strength of PROLOG is its ability to implement logical processes. If two statements are en-

Photo via iStock/Olemedia. [Used under license.]

tered, the program can determine if there is a logical connection between them. One reason PROLOG is able to do this is that programmers give the computer a set of facts to work with before beginning to enter statements. These facts might be descriptive statements about people or objects, such as "the car is green," or about relationships between people or objects, "Sam is the father of Carla." The facts can also be rules concerning the people or objects, such as "Sam is the grandfather of Emma is true if Sam is the father of Carla is true and Carla is the mother of Emma is true." This gives the program a basic framework to answer questions. People who use PROLOG do not necessarily have to know its code. Embedded in the language is a program called ProtoThinker (PT-Thinker), which translates normal statements into language a computer driven by PROLOG understands.

Another characteristic that distinguishes PROLOG from other programming languages is that it is declarative. To use this language, the programmer must first decide what the ultimate goal is they want to achieve. The goal is then expressed in PROLOG, and the software determines how that goal can best be met. This allows a considerable programming burden to be taken off the back of the programmer. Rather than telling the computer what to do, the PROLOG user lets the computer determine what to do. The computer does this by sorting through the information it has been given previously to find a match for the request made of it by the user.

The structures native to PROLOG have aspects in common with many searchable databases. PROLOG recognizes words in lowercase and recognizes basic numbers. Among other uses, PROLOG is used in the programming of more intelligent machines and certain trainable robots.

SUBSEQUENT LANGUAGES

Additional programming languages emerged as artificial intelligence technology was further iterated upon throughout the twentieth and twenty-first centuries. Popular programming languages in the field included Python and R, among many others. These languages offered new improvements compared to their precursors and as such facilitated new developments in the field of AI research. Then, in 2022, OpenAI released the AI chatbot ChatGPT. The chatbot used OpenAI's proprietary Generative Pretrained Transformer (GPT) family of language models, which used a large volume of textual data to inform the chatbot's responses. ChatGPT proved wildly popular following its release and reached more than 100 million users in just two months.

—Max Winter

Further Reading

Artasanchez, Alberto, and Prateek Joshi. *Artificial Intelligence with Python*. 2nd ed., Packt, 2020.

Campbell-Kelly, Martin, William F. Aspray, Jeffrey R. Yost, Honghong Tinn, and Gerardo Con Díaz. *Computer: A History of the Information Machine*. 4th ed., Routledge, 2023.

Copeland, Jack. "What Is Artificial Intelligence?" *AlanTuring.net*, May 2000, www.alanturing.net/turing_archive/pages/reference%20articles/what_is_AI/What%20is%20AI05.html. Accessed 13 July 2023.

Milmo, Dan. "ChatGPT Reaches 100 Million Users Two Months after Launch." *Guardian*, 2 Feb. 2023, www.theguardian.com/technology/2023/feb/02/chatgpt-100-million-users-open-ai-fastest-growing-app. Accessed 13 July 2023.

Philips, Winfred. "Introduction to Logic for ProtoThinker." *MindProject*, Apr. 1999, www.mind.ilstu.edu/curriculum/protothinker/logic_intro.php. Accessed 13 July 2023.

Woolridge, Michael. *A Brief History of Artificial Intelligence: What It Is, Where We Are, and Where We Are Going*. Flatiron Books, 2021.

Q

QR Codes

ABSTRACT

QR, short for "quick response," codes are two-dimensional (2D) barcodes invented in Japan in 1994. They serve as identifiers but most popularly are scanned by users' internet-capable devices. Once scanned, a QR code automatically links users to a particular website. Unlike traditional one-dimensional (1D) barcodes, which are read only in one direction by a scanner, QR codes are read in two directions, both horizontally and vertically, which allows them to contain much more information than standard barcodes.

BACKGROUND

The quick response (QR) code was invented by Masahiro Hara, for the Japanese automaker Denso Wave, in 1994. The Japanese manufacturing industry is the birthplace of the "lean" manufacturing process of quality control, in which parts are rigorously tracked as they are moved through manufacturing facilities. Time is saved and productivity is enhanced in this process by eliminating the need to create and read paper certificates that must be filled out by hand at every stage of the process. Barcodes could be read and processed so much more quickly than paper certificates. Unfortunately, a barcode cannot contain the amount of information complete tracking required. However, the QR code, once it was put into use, carried far more information and made the process more efficient. Since that time, the use of QR codes has become ubiquitous for tracking goods and encapsulating information that requires more capacity than a simple barcode can provide.

QR codes are commonly scanned with a camera device, such as a smartphone's built-in camera, and interpreted by a QR code reader application. The app then connects to a network or a website location specified within the QR code, where the user is able to find detailed information about the particular product or service.

OVERVIEW

A typical QR code is a square or rectangular design that vaguely resembles a chess board. The square, black dots look to be positioned in an entirely random manner. They are not, however. Each tiny square dot is precisely located according to a QR code-generating algorithm. A greatly oversimplified way of understanding QR codes is to think of them as two standard barcodes overlaid at right angles to each other. Where the barcode lines and spaces cross, a value is assigned to that location according to Boolean logic, which will determine whether the location will appear as a black square or a white square in the corresponding QR code. Each square will be read as a bit of binary code (a 0 or a 1) when it is scanned.

This cross-hatched placement effectively creates a square with an equal number of horizontal and vertical lines, each of which can be thought of as a separate barcode in its own right. In this way, a pair of ten-line barcodes would generate a QR code with ten readable lines instead of just one line. While this is certainly not the way a QR code is actually generated, it does give an idea of how a QR code works relative to a standard barcode. This is what allows a QR code to contain many times more information than a standard barcode is able to contain.

QR codes can be generated in several ways and in different sizes. The vast majority are produced in black and white to take advantage of the high optical contrast of the two colors. However, QR codes can also be produced using any two highly contrasting colors. Part of the reason the colors must have high contrast to each other is due to the nature of pixel coloration for displaying and scanning purposes. A common feature in computer graphics is the occurrence of a feathering effect at the edges of adjacent colored areas in an image. In effect, the color of one area bleeds slightly into the other area where they are in contact. In this way there is a narrow outline in which the color is indistinct, neither one color nor the other. In graphics, colors have an assigned value. If even one bit of the value is not correct, a scanner will identify it only as not the color it expects and therefore will not recognize it. Thus, if a scanner encounters feathered edges in a QR code made with poorly contrasting colors, the program that interprets the color value cannot assign the correct value, resulting in misinterpretation of the QR code's information content. When highly contrasting colors are used, especially black and white, feathering of the edges does not occur and the transitions between colors cannot be misinterpreted.

QR codes contain square structures of different sizes in specific locations. The largest squares, found in three of the corners of a QR code, define the correct positioning of the QR code for the scanning program. Three corners of a square can only be correctly positioned one way, while one, two, or four corners allow for multiple incorrect positions. A number of smaller squares set in a regular array within the QR code define the correct alignment of the code for the scanning program. This array also uses the "missing corner" method for proper alignment. The smallest squares, which make up the remainder of the QR code, are timing markers that the scanning program interprets as data content. A small number of these squares in a specific location within the QR code define error checking and correction data for the scanning program.

Image via iStock/Delpixart. [Used under license.]

How much data can be contained within a QR code depends on the type of data character being used, such as common text, Japanese kanji, Arabic script, or another data character type. The maximum storage capacity of 7,089 characters requiring 31/2 bits per character is obtained in the Version 40 QR code with the lowest level of error correction. The array size for QR codes ranges from 21 X 21 in Version 1 QR codes to 177 X 177 in Version 40 QR codes.

QR code structure and use is very complex, with many different functional components. It is easy for anyone to generate a QR code, as the process has been automated with QR code-generating applications, some of which are free to use. Alternatively, one can install a QR code generator application on a suitable device.

In the years following their invention, QR codes not only gained widespread popularity within the field of manufacturing but also came into use in a vast array of other fields. At times, they were used as a means of providing easy access to important documents: during the COVID-19 pandemic, for instance, the government of Ontario, Canada, used QR codes to aid in the verification of residents' vaccination status. In the visual arts, QR codes were sometimes used to provide those viewing a physical work of art with an easy-to-use link to a website containing further information about the piece, its artist, or its historical and cultural context. QR codes likewise served as a bridge between the physical world and virtual or multimedia content, as in some augmented reality experiences.

—*Richard M. Renneboog*

Further Reading

Akta°, Calalettin. *The Evolution and Emergence of QR Codes*. Cambridge Scholars, 2017.

Burns, Monica. *Deeper Learning with QR Codes and Augmented Reality: A Scannable Solution for Your Classroom*. Corwin Press, 2016.

Dutson, Phil. *Creating QR and Tag Codes*. Sams Publishing, 2013.

Frith, Jordan. *Barcode*. Bloomsbury Academic, 2023.

Kato, Hiroki, Keng T. Tan, and Douglas Chai. *Barcodes for Mobile Devices*. Cambridge UP, 2010.

Miranda, Carolina A. "Newsletter: QR Codes in Museums Can Be a Blessing. They Are Also a Curse." *Los Angeles Times*, 22 Jan. 2022, www.latimes.com/entertainment-arts/newsletter/2022-01-22/essential-arts-qr-codes-in-museums-blessing-and-curse-essential-arts. Accessed 13 July 2023.

Pless, Vera. *Introduction to the Theory of Error-Correcting Codes*. John Wiley & Sons, 2011.

Winter, Mick. *Scan Me—Everybody's Guide to the Magical World of QR Codes*. Westsong, 2011.

R

RADIO

ABSTRACT

Radio is a technology that involves the use of electromagnetic waves to transmit and receive electric impulses. Since its inception as a method of wirelessly transmitting Morse code, radio communications technology has had a tremendous impact on society. Although television has supplanted radio to a significant extent for public broadcasting, radio continues to play a significant role in this arena. Radio broadcasts may be delivered via technologies such as satellites, cable networks, and the internet. Although the term "radio" commonly brings to mind listening to news broadcasts, music, and other audio signals, electromagnetic-wave transmission encompasses other fields, such as radar, radio astronomy, radio control, and microwave technology.

BACKGROUND

Electromagnetic waves were discovered in 1877 by the German physicist Heinrich Hertz, whose name is used to describe radio frequencies in cycles per second. Eight years later, American inventor Thomas Alva Edison obtained a patent for wireless telegraphy by discontinuous (intermittent) wave.

A far superior system was developed in 1894 by the Italian inventor Guglielmo Marconi. Initially, he transmitted telegraph signals over a short distance on land. Subsequently, an improved system was capable of transmitting signals across the Atlantic Ocean. At the start of the twentieth century, Canadian inventor Reginald Aubrey Fessenden began experimenting with voice transmission via discontinuous waves while employed by the United States Weather Bureau, and he eventually pioneered radio broadcasting. In 1902, he switched to continuous waves and successfully transmitted voice and music. In 1906, history was made when Fessenden transmitted voice and music from Massachusetts that was heard as far away as the West Indies. The same year, Lee de Forest developed the Audion tube (later known as the triode vacuum tube), which became a key component of radios.

In 1920, the first radio news program was broadcast by station 8MK in Detroit. Also in 1920, the station WRUC in New York began broadcasting a series of Thursday night concerts, the initial range of which was 100 miles. However, it was soon expanded to 1,000 mi (1,609 km). WRUC also began sports broadcasts in the same year. In the early 1930s, frequency modulation (FM) and single sideband shortwave radio were invented by amateur radio operators. In 1954, the companies Texas Instruments and Regency embarked on a joint venture to launch the Regency TR-1, the first portable transistor radio, which was powered by a 22.5-volt battery.

Radio also played a role in other fields. Radio astronomy began in the 1930s and expanded greatly after World War II. The broadcast of television over radio waves, first demonstrated in the 1920s, became increasingly popular in the 1950s. The cellular radio telephone was invented in 1947 by Bell Laboratories, although it was not until the late 1990s and early 2000s that cell phones became a major cultural influence. Radio technology played a significant role in the space race, and satellites themselves increased the broadcasting of signals. In the later twentieth and early twenty-first centuries, digital technology shaped the radio industry.

OVERVIEW

In contrast to sound waves, which require a medium such as air or water for propagation, electromagnetic waves can travel through a vacuum. In a vacuum such as outer space they travel at the speed of light 186,282 miles (299,800 kilometers) per second. In space, electromagnetic waves conform to the inverse-square law: the power density of an electromagnetic wave is proportional to the inverse of the square of the distance from a point source. Thus, all radio waves weaken as they travel a distance. When traveling through air the intensity of the waves is weakened. At some point, depending on the strength of the signal, the electromagnetic wave will no longer be discernible. Interference can also weaken or destroy a radio signal. Other radio transmitters and accidental radiators (such as automobile ignition systems) produce interference and static. Frequency modulation (FM) radio signals are much more resistant to interference and static than amplitude modulation (AM) radio signal. Radio waves travel in a straight line; therefore, the curvature of the earth limits their range. However, radio waves can be reflected by the ionosphere, which extends their range.

The following steps occur in radio transmission: A transmitter modulates (converts) sound to a specific radio frequency; an antenna broadcasts the electromagnetic wave; the wave is received by an antenna; and a receiver tuned to the radio frequency demodulates the electromagnetic energy back into sound. Radio waves range from 3 kilohertz (kHz) to 300 gigahertz (GHz) and are categorized as very low frequency (VLF; 3 to 30 kHz); low frequency (LF; 30 to 300 kHz); medium frequency (MF; 300 to 3,000 kHz); high frequency (HF; 3 to 30 megahertz [MHz]); very high frequency (VHF; 30 to 300 MHz); ultra-high frequency (UHF; 300 to 3,000 MHz); super high frequency (SHF; 3 to 30 GHz); and extremely high frequency (EHF; 30 to 300 GHz). Each frequency range has unique characteristics and unique applications for which they are best suited.

HOW IT WORKS

Basic Example of a Radio Receiver. The humble crystal set, which first appeared at the close of the nineteenth century, is a radio receiver in its simplest form. It consists of an antenna, a tuned circuit, a crystal detector, and earphones. The tuned circuit consists of a tuning coil (a sequentially wound coil

Image via iStock/Nadya Ustyuzhantseva. [Used under license.]

Photo via iStock/microgen. [Used under license.]

that can be tapped at any point) connected to a capacitor. This pair of components allows tuning of the receiver to a specific frequency, known as the resonant frequency. Only signals at the resonant frequency pass through the tuned circuit; other frequencies are blocked. The crystal is a semiconductor, which extracts the audio signal from the radio frequency carrier wave. This is accomplished by allowing current to pass in just one direction, blocking half the oscillations of the radio wave. This rectifies, or changes, the wave into a pulsing direct current, which varies with the audio signal. The earphones then convert the direct current into sound. The sound power is solely derived from the radio station that originated it. The electrical power and circuitry of more complex receivers serve to amplify this extremely weak signal to one that can power loudspeakers.

Amplitude Modulation (AM). With AM radio, the amplitude (height) of the transmitted signal is made proportional to the sound amplitude captured (transduced) by the microphone. The transmitted frequency remains constant. AM transmission is degraded by static and interference because sources of electromagnetic transmission such as lightning and automobile ignitions, which are at the same frequency, add their amplitudes to that of the transmitted signal. AM radio stations in the United States and Canada are limited to 50 kilowatts (kW). In the early twentieth century, US stations had powers up to 500 kW, some of which could be heard worldwide. Conventional AM transmission involves the use of a carrier signal, which is an inefficient use of power. The carrier can be removed (suppressed) from the AM signal. This produces a reduced-carrier transmission, which is termed a double-sideband suppressed-carrier (DSB-SC) signal. A sideband refers to one side of the mirror-image radio signal. DSB-SC has three times more power efficiency than an unsuppressed AM signal. Radio receivers for DSB-SC signals must reinsert the carrier. Another AM refinement is single-sideband modulation in which both one sideband and the carrier are stripped out. This modification doubles the effective power of the signal.

Frequency Modulation (FM). With FM, the variation in amplitude from the microphone causes fluctuations in the transmitter frequency. FM broadcasts have a higher fidelity than AM broadcasts and are resistant to static and interference. FM requires a wider bandwidth to operate and is transmitted in the VHF (30 to 300 MHz) range. These high frequencies travel in a straight line, and the reception range is generally limited to about 50 to 100 miles. FM signals broadcast from a satellite back to Earth do not have this distance limitation. An FM transmission can contain a subcarrier in which secondary signals are transmitted in a piggyback together with the main program. The subcarrier allows stereo broadcasts to be transmitted. The subcarrier can also transmit other information such as station identification and the title of the current song being played.

Digital Radio. Digital audio transmission consists of converting the analogue audio signal into a digital code of zeros and ones in a process known as digitizing. This technology allows for an increase in the number of radio programs in a given frequency range, improved fidelity, and reduction in fading (signal loss). Satellite radio is a digitized signal and can cover a distance in excess of 22,000 miles.

APPLICATIONS AND PRODUCTS

Radio applications are widespread and include commercial radio, amateur radio, marine radio, aviation radio, radar, navigation, and microwave.

Commercial Radio. Radio broadcasts are available throughout the globe. They offer a variety of products, such as news, music, and political opinion. Many broadcasts are free of charge and available to receivers in range of the station. The station typically derives its revenue from advertising.

Cable television and internet services rebroadcast these stations either at no charge or for a fee. Cable television services and satellite radio also broadcast a variety of commercial-free programs. In these cases, the subscription fee covers the cost of the broadcast.

Amateur Radio. Amateur radio (also known as ham radio) is the licensed use of designated radio bands for noncommercial exchange of messages, private recreation, emergency communication, and experimentation. Ham operators maintain and operate their own equipment. Through the years, they have made numerous contributions to radio technology, including FM and single sideband. They also perform a number of public services at no cost. For example, during the Vietnam War, the Military Auxiliary Radio System (MARS) allowed military personnel to call friends and relatives in the United States. A broadcast emanating from Vietnam was received by a US-based amateur operator who patched the communication into the phone lines. MARS remained active through the second decade of the twenty-first century; however, many of its services had been supplanted by the internet and email.

Shortwave Radio. Shortwave radio, which was first theorized in 1919, has many applications for long-range communication. The term "shortwave" refers to the wavelength of the frequency spectrum in which it operates: high frequency (3,000 to 30,000 kHz). The high-frequency wavelength is shorter than the ones first used for radio communications: medium frequency and low frequency.

Shortwave broadcasts are readily transmitted over distances of several thousand miles (or kilometers), allowing intercontinental communication and ship-to-shore communication. Low-cost shortwave radios are available worldwide and facilitate the transfer of information to individuals where other forms of media are controlled for political reasons. A major disadvantage of shortwave radio is that it is subject to significant interference problems, such as atmospheric disturbances, electrical interference, and overcrowding of wave bands. The internet and satellite radio have impacted shortwave radio, but it is still useful in areas where those services are unavailable or too expensive.

Marine Radio. All large ships and most small oceangoing vessels are equipped with a marine radio. Its purposes included calling for aid, communicating with other vessels, and communicating with shore-based facilities, such as harbors, marinas, bridges, and locks. Marine radio operates in VHF, from 156 to 174 MHz. Channels from 0 to 88 are assigned to specific frequencies. Channel 16 (156.8 MHz) is designated as the international calling and distress channel.

Aircraft Band. The aircraft band (or air band) operates in the VHF range. Different sections of the band are used for commercial and general aviation aircraft, air traffic control, radio-navigational aids, and telemetry (remote transmission of flight information). VHF omnidirectional range (VOR) or an instrument landing system operates in the aircraft band. A VOR is a ground-based station that transmits a magnetic bearing of the aircraft from the station. Unmanned aircraft (drones) are navigated by a radio system in which a pilot sits at a console on the ground and directs the flight.

Radio Telescope. A radio telescope is a large, directional, parabolic (dish) antenna that collects data from space probes and Earth-orbiting satellites. The first purpose-built example was created in 1937 by Grote Reber, an engineer and amateur astronomer. Many astronomical objects emit electromagnetic radiation in the radio frequency range; therefore, radio telescopes can image astronomical objects such as galaxies. Some researchers are using radio telescopes to search for intelligent life forms in the universe.

Navigation. Radio navigation encompasses a number of applications, including radio direction finding (RDF), radar, long-range navigation (LORAN), and global positioning system (GPS).

An RDF system homes in on a radio transmission, which can be a specialized antenna that directs aircraft or commercial transmitters that have a known location.

Radar is an acronym for radio detecting and ranging. The device consists of a transmitter and receiver. The transmitter emits radio waves, which are deflected from a fixed or moving object. The receiver, which can be a dish or an antenna, receives the wave. Radar circuitry then displays an image of the object in real time. The screen displays the distance of the object from the radar. If the object is moving, consecutive readings can calculate the speed and direction of the object. If the object is airborne, and the radio is so equipped, the altitude is displayed. Radar is invaluable in foggy weather when visibility can be severely reduced.

The LORAN system relies on land-based low-frequency radio transmitters. The device calculates a ship's position by the time difference between the receipt of signals from two radio transmitters. The device can display a line of position, which can be plotted on a nautical chart. Most LORAN convert the data into longitude and latitude. Since GPS became available, the use of LORAN has markedly declined.

GPS is a space-based global navigation satellite system that provides accurate location and time information at any location on Earth where there is an unobstructed line of sight to four or more GPS satellites. GPS can function under any weather condition and anywhere on the planet. The technology depends on triangulation, just as a land-based system such as LORAN employs. GPS is composed of three segments: the space segment; the control segment; and the user segment. The US Space Force operates and maintains both the space and control segments. The space segment is made up of satellites, which are in medium-space orbit. The satellites broadcast signals from space, and a GPS receiver (user segment) uses these signals to calculate a three-dimensional (3D) location (latitude, longitude, and altitude). The signal also transmits the current time, accurate within nanoseconds.

CAREERS AND COURSEWORK

Many technical and nontechnical careers are available in radio. The technical fields require a minimum of a bachelor's degree in engineering or another scientific field. Many also require a postgraduate degree such as a master's or doctorate. Coursework should include mathematics, engineering, computer science, and robotics. Positions are available in both the government and private sector. The ability to be a team player is often of value for these positions because ongoing research is often a collaborative effort. Less technical positions such as operation or maintenance of broadcasting equipment can be achieved with less training, which can be obtained at a technical college or trade school. A bachelor's degree in communications or related fields can qualify one for employment in radio or television broadcasting.

SOCIAL CONTEXT AND FUTURE PROSPECTS

In view of the continuous advances in radio technology, further advances are extremely likely. Radio and related technologies are an integral component of everyday life in most societies. Radio has political implications, particularly in nations governed by tyrants and dictators, where radio may be used as a propaganda tool and use of shortwave radio and the internet is strongly discouraged. The Voice of America (VOA), which began broadcasting in 1942, is an international multimedia broadcasting service funded by the US government. The VOA broadcasts news, information, educational, and cultural programming to an estimated worldwide audience of 280 million people. In some nations, the broadcast is jammed, which involves broadcasting noise at the same frequency to prevent reception.

Many associate the term microwave with a useful appliance to heat a frozen meal, but microwave is in the radio wavelength (0.3 GHz to 300 GHz) and is also used in communication. The microwave oven illustrates the fact that electromagnetic waves are a form of energy capable of heating—and damaging—tissue. For example, standing near or touching a powerful radio transmitter can result in severe burns. Microwave ovens are shielded to prevent exposure; other devices are not. The heating effect of an electromagnetic wave varies with its power and the frequency. The heating effect is measured by its specific absorption rate (SAR) in watts per kilogram. Many national governments as well as the Institute of Electrical and Electronics Engineers (IEEE) have established safety limits for exposure to various frequencies of electromagnetic energy based on their SAR.

—Robin L. Wulffson

Further Reading

Anderson, John Nathan. *Radio's Digital Dilemma: Broadcasting in the Twenty-First Century*. 2014. Routledge, 2016.

ARRL Ham Radio License Manual. 5th ed., ARRL, 2022.

Hallas, Joel. *Basic Radio: Understanding the Key Building Blocks*. American Radio Relay League, 2005.

Hendricks, John Allen, and Bruce Mims. *The Radio Station: Broadcasting, Podcasting, and Streaming*. 10th ed., Routledge, 2018.

Rudel, Anthony. *Hello, Everybody: The Dawn of American Radio*. Houghton Mifflin Harcourt, 2008.

Wooley, Scott. *The Network: The Battle for the Airwaves and the Birth of the Communications Age.* HarperCollins Publishers, 2016.

Wulff, Alex. *Beginning Radio Communications: Radio Projects and Theory.* Apress, 2019.

Radio/TV Announcer and Newscaster

ABSTRACT

Announcers and newscasters deliver news and commentary on radio and television. They typically work in studio environments but may also travel to locations where important news events are unfolding. Students aspiring to enter the profession should study subjects such as public speaking, composition, the dramatic arts, and computer science.

BACKGROUND

Announcers and newscasters deliver news and commentary on radio and television. Radio announcers and television newscasters are both also traditionally known as broadcasters. In addition to delivering news information to listeners and viewers, broadcasters conduct interviews, moderate discussions, and provide commentary for live sports. Some broadcasters also veer into journalism, researching and writing about topics for discussion on their particular programs. As such, broadcasting has broadened into a multidisciplinary profession encompassing mass communication, journalism, and reportage.

OVERVIEW

Broadcasters operate primarily out of radio and television studios, where they work in concert with technical and production staff to prepare radio and television programs. It is not uncommon for broadcasters to travel to areas where important news events are unfolding, presenting their programs from a diverse array of locales from show to show. Broadcasters are also often called upon to visit interview subjects and develop stories from a variety of locations in and around their region, the country, or even the globe. In the past, most broadcasters worked nontraditional hours, including early mornings, late nights, weekends, and holidays, but newer technology has enabled more broadcasts to be prerecorded.

While radio announcers and television newscasters traditionally came from media communications backgrounds, the field is now populated by individuals who come from numerous academic and professional backgrounds, including journalism, politics, science, literature, music, and the arts.

Colleges and universities nationwide offer specific academic programs dedicated to both audio and visual broadcasting, which students often reinforce with coursework dedicated to their other academic interests, notably English, politics, or sports management. Excellent time management, judgment, and organization are just as imperative as personality and conversational skills.

DUTIES AND RESPONSIBILITIES

Radio and television broadcasters spend their days planning future shows, filming or recording new broadcasts, and editing new recordings for public broadcast. On air, broadcasters generally introduce and close programs, present information, and lead discussions. Many of the specific occupational duties and responsibilities of radio and television broadcasters depend on the nature and frequency of the program for which they work.

Developing programs that air live on a daily basis predominantly involves preproduction tasks such as fact gathering, organizing specific questions, and preparing for guest interviews. Live television and radio production is often completed in a fast-paced environment under strict deadlines. As a result, radio and television broadcasters who work in live programming often must effectively adapt to evolving situations and on-air conversations.

Documentary-style radio and television programs conduct a large amount of investigative research and information gathering. Documentary programs tend to air on a less frequent basis, usually weekly or monthly; thus, much of the focus for developing such programs is placed on gathering video and audio copy, narrative construction, fact checking, and follow-up interviews with subjects. The protracted nature of documentary radio and television broadcasting necessitates a lot of editing work prior to presentation.

OCCUPATION SPECIALTIES

Anchors/Hosts. News anchors and program hosts work in television or radio and specialize in a certain area of interest, such as politics, personal finance, sports, or health. They contribute to the preparation of the program's content, interview guests, and discuss issues with viewers, listeners, or the studio or radio audience.

Commentators. Commentators analyze and write commentaries, based upon personal knowledge and experience with the subject matter, for broadcast. They interpret information on a topic and record their commentary or present it live during the broadcast.

Critics. Critics write and deliver critical reviews of literary, musical, or artistic works and performances for broadcast.

Disc Jockeys (DJs). Disc jockeys, or DJs, announce radio programs of musical selections and choose the selections to be made based upon knowledge of audience preference or requests. They also comment upon the music and other matters of interest to the audience, such as the weather, time, and traffic conditions.

Newsreaders. Newsreaders read prepared news copy over the air. They may or may not be involved in the writing and editing of that copy.

Photo via iStock/simonkr. [Used under license.]

Public Address Announcers. Public address announcers provide information to the audience at sporting, performing arts, and other events.

WORK ENVIRONMENT

Physical Environment. Television and radio broadcasting studios are the broadcaster's primary work environment. These spaces are generally bright, soundproof, and temperature controlled. A considerable amount of fieldwork may also be required. Broadcasters may work in a variety of locations, including government buildings, sports arenas, and hospitals. They may also serve as station representatives at public events.

Human Environment. Radio and television broadcasters are often the public face of a larger team of technical and production staff with whom they are required to work closely on a daily basis.

Technological Environment. Radio and television broadcasters use a wide range of communication and broadcasting technology, from microphones and teleprompters to sophisticated editing equipment.

EDUCATION AND TRAINING

High School/Secondary. High school students can best prepare for a career in broadcasting with courses in public speaking, composition, the dramatic arts, and computer science. Many high schools have scholastic television and radio stations that instruct students on broadcasting basics. Exposure to local radio and television broadcasting stations through internships or volunteer work may also be highly beneficial. Writing and reporting on local events for a school or community newspaper will provide high school students with reportage and interviewing experience that can benefit a future career in broadcasting.

College/Postsecondary. Hundreds of colleges and universities in the United States offer undergraduate- and graduate-level programs in broadcasting. The majority of entry-level radio and television broadcasting positions require a bachelor's degree in communication, broadcasting, or journalism.

Undergraduate programs in journalism outline the techniques and strategies that apply to television and radio reporters while honing students' reporting and storytelling skills. Journalism majors also learn the basic ethical standards that dictate news production across all types and levels of media. Undergraduate work in broadcasting exposes students to the vast array of media technologies and software used in the field and helps them learn the acoustics of speech, vocal delivery, and camera presence.

Graduate-level programs in broadcasting are usually completed in conjunction with an internship at a radio or television news studio. In addition to studying advanced topics such as media law, news production, and directing, graduate students also conduct research for a master's thesis dedicated to an area of their particular interest. Individuals with master's degrees often go on to professional careers as radio and television broadcasters, media researchers, or college-level academic instructors.

Adult Job Seekers. The educational and professional experience requirements of broadcasting can make it a difficult field for adult job seekers to enter. Due to the highly competitive job market and low turnover rate of established broadcasters, landing a professional role as a broadcaster can require several years of lower-level experience, during which one is expected to master the production, reporting, and editing aspects of the role. Advancement to higher-level, higher-paying positions often depends on proven ratings, contributions to the station's marketing efforts, and the station's size. Relocation is common.

Professional Certification and Licensure. Professional certification and licensure is not required of broadcast professionals, although membership and affiliation with national organizations and associations can boost credentials and improve networking opportunities.

Additional Requirements. Radio announcers and television newscasters are also often entertainers. Therefore, those interested in the field should be comfortable speaking and engaging with interviewees and audiences, have a sense of humor, work well under pressure, and adapt quickly to changing situations and circumstances.

—*John Pritchard*

Further Reading

Hausman, Carl, Philip G. Benoit, and Fritz Messere. *Announcing for Broadcasting and the Internet: The Modern Guide to Performing in the Electronic Media.* Routledge, 2019.

Sidlow, Faith, and Kim Stephens. *Broadcast News in the Digital Age: A Guide to Reporting, Producing and Anchoring Online and on TV.* Routledge, 2022.

Stewart, Peter, and Ray Alexander. *Broadcast Journalism: Techniques of Radio and Television News.* 8th ed., Routledge, 2021.

RANDOM-ACCESS MEMORY

ABSTRACT

Random-access memory (RAM) is a form of memory that allows the computer to retain and quickly access program and operating system data. RAM hardware consists of an integrated circuit chip containing numerous transistors. Most RAM is dynamic, meaning it needs to be refreshed regularly, and volatile, meaning that data is not retained if the RAM loses power. However, some RAM is static or nonvolatile.

BACKGROUND

The speed and efficiency of computer processes are among the most areas of greatest concern for computer users. Computers that run slowly (lag) or stop working altogether (hang or freeze) when one or more programs are initiated are frustrating to use. Lagging or freezing is often due to insufficient computer memory, typically random-access memory (RAM). RAM is an essential computer component that takes the form of small chips. It enables computers to work faster by providing a temporary space in which to store and process data. Without RAM, this data would need to be retrieved from direct-access storage or read-only memory (ROM), which would take much longer.

Computer memory has taken different forms over the decades. Early memory technology was based on vacuum tubes and magnetic drums. Between the 1950s and the mid-1970s, a form of memory called "magnetic-core memory" was most common. Although RAM chips were first developed during the same period, they were initially unable to replace core memory because they did not yet have enough memory capacity.

A major step forward in RAM technology came in 1968, when IBM engineer Robert Dennard patented the first dynamic random-access memory (DRAM) chip. Dennard's original chip featured a memory cell consisting of a paired transistor and capacitor. The capacitor stored a single bit of binary data as an electrical charge, and the transistor read and refreshed the charge thousands of times per second. Over the following years, semiconductor companies such as Fairchild and Intel produced DRAM chips of varying capacities, with increasing numbers of memory cells per chip. Intel also introduced DRAM with three transistors per cell, but over time the need for smaller and smaller computer components made this design less practical. By the 2010s, commonly used RAM chips incorporated billions of memory cells. There are two major categories of random-access memory: static RAM (SRAM) and dynamic RAM (DRAM).

OVERVIEW

Although all RAM serves the same basic purpose, there are a number of different varieties. Each type has its own unique characteristics. The RAM most often used in personal computers is a direct descen-

Photo via iStock/nazarethman. [Used under license.]

dant of the DRAM invented by Dennard and popularized by companies such as Intel. DRAM is dynamic, meaning that the electrical charge in the memory cells, and thus the stored data, will fade if it is not refreshed often. A common variant of DRAM is speed-focused double data rate synchronous DRAM (DDR SDRAM), the fifth generation of which entered the market in 2020.

RAM that is not dynamic is known as static random-access memory (SRAM). SRAM chips contain many more transistors than their DRAM counterparts. They typically use six transistors per cell: two to control access to the cell and four to store a single bit of data. As such, they are much more costly to produce. A small amount of SRAM is often used in a computer's central processing unit (CPU), while DRAM performs the typical RAM functions.

Just as the majority of RAM is dynamic, most RAM is also volatile. Thus, the data stored in the RAM will disappear if it is no longer being supplied with electricity—for instance, if the computer in which it is installed has been turned off. Some RAM, however, can retain data even after losing power. Such RAM is known as nonvolatile random-access memory (NVRAM).

USING RAM

RAM works with a computer's other memory and storage components to enable the computer to run more quickly and efficiently, without lagging or freezing. Computer memory should not be confused with storage. Memory is where application data is processed and stored. Storage houses files and programs. It takes a computer longer to access program data stored in ROM or in long-term storage than to access data stored in RAM. Thus, using RAM enables a computer to retrieve data and perform requested functions faster. To improve a computer's performance, particularly when running resource-intensive programs such as those used for art and multimedia creation, a user may replace its RAM with a higher-capacity chip so the computer can store more data in its temporary memory.

SHADOW RAM

While RAM typically is used to manage data related to the applications in use, at times it can be used to assist in performing functions that do not usually involve RAM. Certain code, such as a computer's basic input/output system (BIOS), is typically stored within the computer's ROM. However, accessing data saved in ROM can be time consuming. Some computers can address this issue by copying data from the ROM and storing the copy in the RAM for ease of access. RAM that contains code copied from the ROM is known as shadow RAM.

—Joy Crelin

Further Reading

Dieny, Bernard, Ronald B. Goldfarb, and Kyung-Jin Lee. *Introduction to Magnetic Random-Access Memory*. IEEE Press/John Wiley & Sons, 2017.

Hey, Tony, and Gyuri Pápay. *The Computing Universe: A Journey through a Revolution*. Cambridge UP, 2015.

McLoughlin, Ian. *Computer Systems: An Embedded Approach*. McGraw-Hill, 2018.

"Media Arts, Design and Technology Computer Requirements." *NC State College of Design*, May 2023, academics.design.ncsu.edu/it/docs/artdesign-computer-requirements. Accessed 13 July 2023.

O'Leary, Timothy, Linda O'Leary, and Daniel O'Leary. *Computing Essentials 2023*. McGraw-Hill, 2022.

Siddiqi, Muzaffer A. *Dynamic RAM: Technology Advancements*. CRC Press, 2013.

"SK Hynix Launches World's First DDR5 DRAM." *HPC Wire*, 7 Oct. 2020, www.hpcwire.com/off-the-wire/sk-hynix-launches-worlds-first-ddr5-dram. Accessed 13 July 2023.

S

The Semantic Web

ABSTRACT

The Semantic Web, which is sometimes called Web 3.0 or the web of data, is a framework that allows different applications and programs to share data such as dates, numbers, formulas, and more. The Semantic Web is an extension of the World Wide Web (WWW), but it does not serve the same purpose as the WWW. While both the Semantic Web and the WWW aim to make knowledge and information available to people, the WWW achieves this goal by presenting users with countless documents that they can read and learn from, while the Semantic Web classifies individual pieces of data in a manner that allows computers and people to use the data in a variety of ways.

BACKGROUND

The Semantic Web is different from the conventional web, but it is not totally separate from it; rather, the Semantic Web is an extension of the conventional web and is often used in conjunction with it. The history of the Semantic Web also is closely tied to that of the World Wide Web, and the founder of the WWW, Tim Berners-Lee, was also influential in the development of the Semantic Web.

In 2001, Lee and others published an article in *Scientific American* that discussed the Semantic Web and its possibilities. That same year, the World Wide Web Consortium (W3C), a group of volunteers and paid workers who develop web standards, launched the Semantic Web Activity, which helped create the groundwork for the Semantic Web. The group saw the Semantic Web as being similar to the WWW, in that people would publish pages of data in the same way they published websites. However, the Semantic Web would ultimately become more important in the background of the web, and few people published large stores of data.

OVERVIEW

Much of the data people use each day is not actually part of the WWW. The goal of the Semantic Web is to make as much data as possible accessible using architecture that is similar to the WWW architecture (e.g., uniform resource identifiers [URIs] and uniform resource locators [URLs]). Most of the data on the WWW is there for humans to read, but the data on the Semantic Web is there for computers and humans to link to, use, and categorize. For example, one day people could use the Semantic Web to find a doctor appointment (data provided by the doctor's office) and their own schedule (data provided by the calendars on their computers or smart phones). The Semantic Web is often compared to a web of data, with data being pulled from many different places.

The building blocks that make up the WWW and the Semantic Web are program languages that help users to write software. Two of the most important building blocks for the Semantic Web are eXtensible Markup Language (XML) and the Resource Description Framework (RDF). Data on the Semantic Web often are categorized and organized so that humans and devices can use them more efficiently. XML and RDF are two important ways people organize data on the Semantic Web.

The WC3 recommends XML as one possible language to use on the conventional web and the Semantic Web. XML is a markup language, like the popular hypertext markup language (HTML). Unlike HTML, XML allows people to develop their own tags to categorize data. For example, a person using HTML must use official HTML tags, but a person using XML can devise any tags (e.g., <firstname>, <zipcode>, <birthdate>, <age>, etc.). XML is an important building block of the Semantic Web because it allows users to code data in more useful ways than some other languages.

Although XML helps people structure data, it does not give the data meaning. Another important building block, RDF, helps give the data on the Semantic Web meaning. RDF encodes pieces of data with information that is similar to the subject, the verb, and the object of a sentence. This three-part structure can be used to describe much of the data computers hold and process. For example, RDF could be used to indicate that a name on a page is the author of a text. The meaning given to the data by the RDF coding could help people find this information on the Semantic Web. People can choose to use either XML or RDF depending on the goals they want to achieve.

Another important language in the development of the Semantic Web is structured query language (SQL). SQL is a type of query language that allows people to manage and gather information from databases. The Semantic Web can, in one way, be seen as a huge database of information. Therefore, people can use SQL to access, organize, and manage the data on the Semantic Web.

GOALS OF THE SEMANTIC WEB

The Sematic Web has a number of goals. The most important goal is to make data easy to access. Another goal of the Sematic Web is to increase data sharing to make it possible for people and technology to find relationships among data. In addition, the Semantic Web tries to make data available so it can be used to create models of problems in everyday life and try to help solve them. Another way the data might be used is to build machines that are able to "think." The Semantic Web has the potential to change the way people use the internet. However, the current Semantic Web runs mostly in the background of the WWW and has not changed the way everyday internet users experience the web.

—Elizabeth Mohn

Further Reading

Allemang, Dean, James Hendler, and Fabien Gandon. *Semantic Web for the Working Ontologist: Effective Modeling for Linked Data, RDFS, and OWL*. 3rd ed., ACM, 2020.

Berners-Lee, Tim, James Hendler, and Ora Lassila. "The Semantic Web." *Scientific American*, 1 May 2021, www.scientificamerican.com/article/the-semantic-web. Accessed 13 July 2023.

Hitzler, Pascal, Markus Krotzsch, and Sebastian Rudolph. *Foundations of Semantic Web Technologies*. Taylor & Francis, 2010.

"Towards the Semantic Web." *Cambridge Semantics*, cambridgesemantics.com/blog/semantic-university/intro-semantic-web/towards-the-semantic-web. Accessed 13 July 2023.

"Web Standards." *W3C*, 2023, www.w3.org/standards. Accessed 13 July 2023.

"Whatever Happened to the Semantic Web?" *Two-Bit History*, 27 May 2018, twobithistory.org/2018/05/27/semantic-web.html. Accessed 13 July 2023.

"W3C Data Activity: Building the Web of Data." *W3C*, www.w3.org/2013/data. Accessed 13 July 2023.

"W3C Semantic Web Frequently Asked Questions." *W3C*, 12 Nov. 2009, www.w3.org/RDF/FAQ. Accessed 13 July 2023.

Smartphones, Tablets, and Handheld Devices

ABSTRACT

Handheld devices are very small computers. They are very common today and easily fit in one or both hands. Many handheld devices have touch screens. Devices with touch screens do not need keyboards to enter information. Some handheld devices make phone calls or send and receive messages. Some search the internet and receive emails. Some play music. Others allow people to buy and read elec-

tronic books, or e-books. Advanced handheld devices, such as the smartphones available in the third decade of the twenty-first century, do all of these tasks and more.

BACKGROUND

Trends in handheld devices change quickly. New devices keep being made. Some devices are special for being the first or the most popular of their kind, and other companies often make similar devices to compete with them.

The IBM Simon was the first smartphone. It was created in 1993 before the word "smartphone" existed. It was a personal digital assistant (PDA) and a cell phone. It even had a touch screen. However, it was large and expensive. It never became popular.

The PalmPilot was one of the most popular PDAs. It first appeared in the mid-1990s. Later PDAs used the PalmPilot as a model.

The BlackBerry was first sold in the early 2000s. It was another early smartphone. The BlackBerry was a PDA and a cell phone, and it could send and receive emails. It became very popular among businesspeople.

The iPod was introduced in 2001. It quickly became one of the most successful portable media players of the twenty-first century.

The Kindle is an e-reader. The Kindle began to be sold in 2007. It was not the first e-reader, but it became popular very quickly. The Kindle is small and light. It can save hundreds of books. It produces

Photo via iStock/hudiemm. [Used under license.]

a very sharp screen image. The original Kindle and many of its successors make use of electronic ink (e-ink) technology, which allows for a reading experience similar to that of reading text on paper.

The iPhone is a smartphone. It was first sold in 2007. The iPhone was considered the best smartphone when it was first sold. The iPhone has been improved many times since then. It remains one of the most popular smartphones.

The iPad was first sold in 2010. The iPad is a tablet. It quickly became one of the most popular tablets.

OVERVIEW

Cell Phones and Smartphones. Cellular phones, or cell phones, are among the most well-known handheld devices. Cell phones are phones that can make phone calls from almost anywhere. In the past, people could make phone calls only from landlines. Landlines are phones inside a person's home. Landlines connect to wires that stretch from a person's house to the road. These wires connect to phone cables. The cables carry the sound of people's voices across great distances.

Cell phones do not need wires to make calls. Cell phone companies build towers in many places. Each tower serves a cell, or small area. The towers send and receive signals to and from nearby cell phones. People often travel from one cell to another while talking on their cell phones. Cell phones simply switch to the closest tower so people can keep talking as they move.

Cell phones may seem like modern inventions. However, the first cell phone call was made in 1973. Cell phones became more popular in the 1980s. Early cell phones were large and weighed more than two pounds. Smaller cell phones arrived in the 1990s. By 2010 many people used cell phones more than they used landlines.

Smartphones are a type of cell phone. They perform the tasks of many handheld devices at once. People can use smartphones to make phone calls, send text messages, take pictures, play games, and listen to music, and use a variety of apps. Smartphones can also connect to the internet. During the second and third decades of the twenty-first century, many smartphones included a feature known as Wi-Fi calling. That feature enabled a user to connect a smartphone to a location's wireless internet and make phone calls over the internet, rather than rely on traditional cellular service.

Tablets. Tablets are thin, handheld computers. Tablets are larger than smartphones but are typically smaller than laptop computers. They perform many of the same tasks as computers and smartphones. However, most tablets cannot make phone calls. Tablets usually have touch screens rather than keyboards, although users may choose to connect external keyboards to their tablets through Bluetooth or other means.

Other Handheld Devices. PDAs are handheld organizers. PDAs first appeared in the early 1990s. They are like electronic day planners and address books. People use them to schedule appointments and to save contact information. People connect PDAs to computers to move information back and forth.

Portable media players are devices that store and play sound, pictures, and video files. They are small enough to carry easily.

E-book readers, or e-readers, are devices that let people save and read books whenever they want. The first e-book readers arrived in the late 1990s. However, people had to connect them to a computer to save books on them. They were large and could save only a few books. E-readers have improved since then. Today they are thin and light. They can save hundreds of books. E-readers may use e-ink technology or may feature screens that resemble those of mainstream tablets.

—Lindsay Rock Rohland

Further Reading

Campbell-Kelly, Martin, William F. Aspray, Jeffrey R. Yost, Honghong Tinn, and Gerardo Con Díaz. *Computer: A History of the Information Machine*. 4th ed., Routledge, 2023.

Goggin, Gerard. *Apps: From Mobile Phones to Digital Lives*. Wiley, 2021.

Lebert, Marie. "eBooks: 1998—The First eBook Readers." *Project Gutenberg News*, 16 July 2011, www.gutenbergnews.org/20110716/ebooks-1998-the-first-ebook-readers. Accessed 13 July 2023.

Merchant, Brian. *The One Device: The Secret History of the iPhone*. Little, 2017.

Vasquez, Emily, and Adam Hardy. "What Is WiFi Calling? Pros, Cons & More." *Forbes Advisor*, 8 June 2023, www.forbes.com/advisor/business/software/what-is-wi-fi-calling. Accessed 13 July 2023.

Woyke, Elizabeth. *The Smartphone: Anatomy of an Industry*. New Press, 2014.

Social Media

ABSTRACT

Internet-based applications and websites that promote the sharing of user-generated content, communication, and participation on a large scale, are grouped into the general term of "social media." Since the early 2000s social media has taken the internet by storm, and by 2021, according to the Pew Research Center, approximately 70 percent of adults in the United States used some type of social media. A large variety of user-generated applications make up what is considered social media. These applications include blogs, social networks such as Facebook, and audio podcasts. Over time, social media has become used for marketing and as an alternative news source. Although the aim of social media is to make it easier for individuals to communicate and engage in conversations, many argue that it has led to a reduction in human interaction.

BACKGROUND

The idea of social media began in the mid-to-late 1990s, when internet users were first given the ability to make their own websites through servers such as GeoCities. Blogging and social networks also began around that time. Sixdegrees.com, launched in 1997, was one of the first websites that allowed users to create a profile and add lists of friends. It was not until 2002 and the launch of Friendster that the concept of a social network become highly popular. Within three months of its launch, Friendster had gained 3 million users. Other popular social media applications launched around this time include the social networking platforms MySpace (later stylized as Myspace) and LinkedIn, the music service iTunes, and the image-hosting website Flickr.

Myspace was launched in August 2003 and quickly became one of the most visited websites in the United States. It was surpassed by Facebook in 2008 but, as of 2015, it still had 50 million users. The website shifted its focus to cater heavily to musicians and was revamped to make it easy for users to upload and share music from established and upcoming artists. Competition and a 2019 report that a server migration initiative had resulted in the loss of a significant amount of early user data meant that the site continued to fail to attract new users.

OVERVIEW

Social media is classified in several ways. Any application that allows users to create a profile and build a friend list is considered a social network. The most popular example of this is Facebook, which was launched in 2004 and had grown to more than 2 billion daily active users by mid-2023. Blogs allow users to generate a variety of content for publication on the internet. Several websites are devoted to hosting users' blogs; one of the most popular of these servers is WordPress. Forums are another classification that allows users to voice their opinion on a range of topics. Video and audio podcasts allow users to record themselves discussing different topics. Internet users can then download or stream these recordings. Collaborative websites known as wikis, which allow

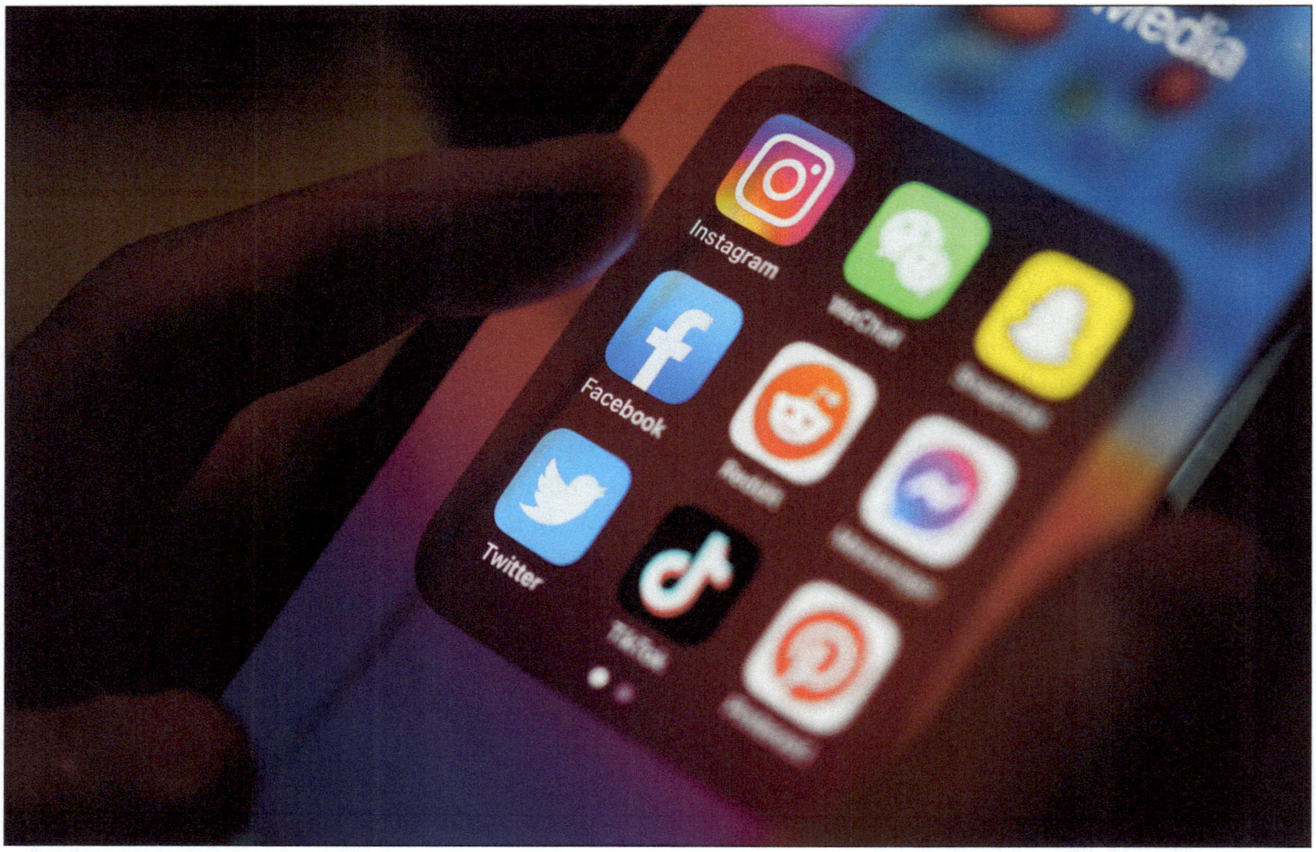

Photo via iStock/P. Kijsanayothin. [Used under license.]

users to generate informational content on a variety of topics, also became very popular; the most famous of these is Wikipedia, a crowdsourced encyclopedia.

Multiplayer online games are also a prevalent type of social media. These games, such as the popular *World of Warcraft*, allow users to communicate with other players while participating in a virtual world. Other popular examples of these games are *The Sims* and *Second Life*, which allowed users to create avatars and interact with other users. Other more general categories that can be considered social media include emails, instant messaging, and video sharing. Many of these varieties have been aggregated with social networks.

The rise in smartphone and tablet computer technology allowed individuals to access an array of social media applications while mobile. Geographic tagging applications allowed users to "tag" themselves at specific locations such as restaurants and stores and post these tags on social media websites such as Facebook and Foursquare.

By the late 2010s, many social media platforms had banned hate speech. Several platforms like Telegram, Parler, and Gab, however, did not, and so became popular with right-wing extremist groups who used the sites to communicate and organize events. In January 2021, such sites were used to coordinate the storming of the US Capitol building in Washington, DC.

Following the storming of the Capitol, many mainstream social media platforms, including Facebook and Twitter, banned President Donald Trump for allegedly inciting the violent event. The sites also banned many more users who violated their rules against hate speech and inciting violence.

FACEBOOK

The most popular social media and networking tool in the world is Facebook. The website was launched on February 4, 2004, by computer programmer and Harvard sophomore Mark Zuckerberg. When the website was first created, it was exclusively for students of Harvard University and others with college-based email addresses. It was opened to everyone on September 26, 2006, and quickly gained hundreds of millions of users. Although the website requires users to be at least thirteen years old, it has been reported that several million users are under age thirteen.

Facebook allows users to build a personal profile that includes pictures and cultural interests, to exchange messages, and to share thoughts, pictures, and videos. The website initially drew comparisons with Myspace, an earlier social networking website. However, Facebook allows for more customization and requires users to give their true identity, which Myspace did not.

TWITTER

The social networking and microblogging website Twitter has risen to immense popularity since it was launched in 2006. It allows users to create a small profile, follow other users, and post brief messages, originally restricted to 140 characters long. (Beginning in November 2017, Twitter users could post messages of up to 280 characters.) These messages are known as tweets. Twitter further encourages interaction and online conversations through the use of hashtags. A hashtag, represented by the pound symbol before a word or group of words, allows users to search for and view every tweet with that specific hashtag attached. The most popular current hashtags are known as "trending topics." The application was launched by web developer Jack Dorsey, and by December 2022, it had about 368 million active monthly users, with hundreds of millions of tweets being sent every day from computers and various mobile devices.

The application is significant for the way it allows people to organize quickly. For example, Twitter was used to rally individuals for political protests around the world, as in the Egyptian Revolution in 2011. Politicians utilized it as a way to garner support and interact with voters. US president Barack Obama used it heavily during his 2008 campaign. He encouraged voters to ask him questions via Twitter and Facebook throughout the campaign. This innovative use was expanded to an unprecedented level by television celebrity Donald Trump, who used his personal Twitter account as a main form of communication during his successful presidential campaign in 2016. During his presidency, he continued to use the platform heavily to express his views to the public, and it became particularly prominent again throughout the 2020 presidential campaign.

Twitter has also been used to report breaking news, although many critics argue that Twitter users rarely perform fact checking before sending out news tweets. For example, during the 2016 US presidential election Twitter served as a highly influential news source for millions of Americans, but it and other social media sites were plagued by accounts spreading misleading or outright false news and analysis. Twitter has also been frequently used in police investigations, education, and public relations.

Technology entrepreneur Elon Musk purchased Twitter in October 2022, and in July of the following year, he changed the name of both the service and its parent company to X.

SOCIAL MEDIA MARKETING

Since social media is accessible through a broad range of applications, it has become a heavily used tool in marketing. Companies can manufacture advertisements that social media users can easily share at no cost to the company, making it an inexpensive alternative to traditional marketing. Additionally, by 2020 many businesses and brands had begun taking advantage of influencer marketing, which typically entailed the involvement of individuals with a large number of social media followers.

Social networking sites such as Facebook allow companies to join them and create a profile for marketing purposes. On Facebook, users can "like" the page of companies to get updates and offers from them, as well as communicate with them. This approach was especially successful for small businesses that used Facebook to promote their brand through special events and offers.

Brands utilize Twitter to market and interact on a more individual level. When users "follow" brands online, short messages posted by the brand appear on the users' main Twitter feed page. Many times, these messages include links that a user could click on to learn more about offers and specific products. This too is a very inexpensive method of online marketing.

Foursquare was another influential social media application frequently used by businesses in online marketing. This application originally allowed users to "check-in" at the location of a store, restaurant, or other business. Their check-in was then posted on the Foursquare website as well as on other applications such as Facebook. Businesses encouraged return customers by offering incentives for checking in many times. While Foursquare itself eventually abandoned this model in favor of focusing on personalized, localized search services, the "check-in" concept was taken up by other social media applications.

The video-sharing website YouTube has become another popular place for advertisers. Marketing on YouTube is personalized through various language-detection programs that analyze users' individual interests and market to them accordingly. Advertisers are able to attach specific advertisements to relevant videos being searched, making it easy for them to reach a target audience. Another popular method of no- or low-cost advertising on YouTube is through "viral videos." These videos, which are oftentimes humorous and culturally relevant, are created with the goal of having users spread them across the web, to gain monetary compensation or temporary fame.

Other popular social media applications that are heavily used in marketing are blogs where advertisements can be posted, business-profile websites such as Yelp, and the business-networking website LinkedIn.

PRIVACY CONCERNS

One of the biggest debates concerning social media remains over online privacy and protection of user information. Those concerned with these issues argue that social networking sites such as Facebook and Twitter do not take the proper steps to protect the information users share with each other. Debates have looked at ownership of the content on social networking websites. Users generate it onto these websites, but corporations own the websites.

In November 2007, Facebook came under heavy negative criticism for implementing Beacon, a system that allowed partner websites to send information concerning the actions of Facebook users on their own websites. This information included items purchased at online retailers and games users played online. Beacon aroused concerns over privacy of information and users' lack of control over how their information was used. Facebook spokespersons argued that it allowed users to further share their interests with friends and help refer them to online retailers. Beacon was discontinued in September 2009, but the company came under similar scrutiny in 2014 for major privacy policy changes as well as for an emotion study that drew the attention and ire of privacy watchdog organizations and individuals.

There also have been concerns over federal and local authorities using Facebook to acquire personal information in the investigation of crimes. Facebook's privacy policy states that they may turn over any information they believe may be related to an illegal or criminal activity, but many argued that personal information has been harvested even when authorities lacked reasonable suspicion. This has raised further questions about what online information falls under the US Electronic Communications Privacy Act (ECPA) of 1986, which regulates what electronic messages and information can be legally seized or intercepted by authorities. Critics argue that the bill is very loose in its language and outdated with respect to contemporary information sharing.

The concept of data mining to extract user information has been another serious concern for social network users. Many worry that companies and individuals have been allowed to freely harvest information and use it for various purposes without necessarily obtaining the consent of users. In 2005, as part of a project on Facebook privacy, two students from the Massachusetts Institute of Technology (MIT) demonstrated the possibility of simple data mining on Facebook. They used an automated script to download the Facebook profiles of over seventy thousand users.

Facebook has since developed higher security methods for their users, including customizable security, but privacy continued to be a concern. Challenges of Facebook's operations and policies were heightened once more in 2018 when it was revealed that the political data firm Cambridge Analytica had been able to harvest a large amount of user data from the site without their knowledge. In early 2020, the platform made concerted efforts to emphasize the importance of information privacy.

IMPACT

Since its inception, social media has changed everything from politics to public relations, from telecommunications to theater. It has fundamentally shifted the way many people experience everyday life, including friendship, shopping, and job searching. While it helps connect people and share their ideas, it continues to come under heavy negative criticism and skepticism. Critics argue that it only creates the illusion of connection and in reality, decreases the need for actual human interaction. Others state that it leads to an increase in cyberbullying, in which people are harassed on social networks.

Many employers have banned social media at work due to concerns that it decreases productivity. The ultimate concern is now one of privacy and the misuse of personal information. Yet supporters of social media claim that these downsides are far outweighed by the social and economic benefits of increased interaction over the internet, a field with ever-growing potential as technology and access improves around the world. For better or worse, social media continues to increase in popularity, particularly with the rise and influx of mobile applications. Following the emergence of the global COVID-19 pandemic in 2020, many people turned increasingly to social media platforms to stay connected while in-person contact was limited in an effort to slow the spread of the virus.

—Patrick G. Cooper

Further Reading

Auxier, Brooke, and Monica Anderson. "Social Media Use in 2021." *Pew Research Center*, 7 Apr. 2021, www.pewresearch.org/internet/2021/04/07/social-media-use-in-2021. Accessed 26 July 2023.

Bryfonski, Dedria. *The Global Impact of Social Media*. Greenhaven, 2012.

Denham, Hannah. "These Are the Platforms That Have Banned Trump and His Allies." *Washington Post*, 14 Jan. 2021, www.washingtonpost.com/technology/2021/01/11/trump-banned-social-media. Accessed 26 July 2023.

Dixon, S. "Number of Twitter Users Worldwide from 2019 to 2024." *Statista*, 14 Dec. 2022, www.statista.com/statistics/303681/twitter-users-worldwide. Accessed 26 July 2023.

Highfield, Tim. *Social Media and Everyday Politics*. Polity, 2016.

Humphreys, Ashlee. *Social Media: Enduring Principles*. Oxford UP, 2016.

Kirkpatrick, David. *The Facebook Effect: The Inside Story of the Company That Is Connecting the World*. Simon, 2011.

Krietzberg, Ian. "Twitter to X: The Complete Timeline of Elon Musk's Twitter Purchase." *TheStreet*, 26 July 2023, www.thestreet.com/technology/twitter-to-x-the-complete-timeline-of-elon-musks-twitter-purchase. Accessed 26 July 2023.

Lovink, Geert. *Networks without a Cause: A Critique of Social Media*. Polity, 2011.

Moe, Wendy, and David A. Schweidel. *Social Media Intelligence*. Cambridge UP, 2014.

Morozov, Evgeny. *The Net Delusion: The Dark Side of Internet Freedom*. Public Affairs, 2011.

Nicosia, Laura. *Educators Online: Preparing Today's Teachers for Tomorrow's Digital Literacies*. Peter Lang, 2013.

Qualman, Erik. *Socialnomics: How Social Media Transforms the Way We Live and Do Business*. Wiley, 2010.

Shirky, Clay. *Here Comes Everybody: The Power of Organizing without Organizations*. Penguin, 2009.

Tuten, Tracy L., and Michael R, Solomon. *Social Media Marketing*. Pearson, 2013.

Sound Engineering Technician

ABSTRACT

Sound engineering technicians record, manipulate, and edit music, speech, and other sounds. They work in studio environments as well as at live events. Students aspiring to enter the profession should pursue coursework in the arts, music, and computers.

BACKGROUND

Sound engineering technicians record, manipulate, and edit music, speech, and other sounds. They work across a wide spectrum of media-related industries, from motion picture and video production to performing arts, recording arts, and television and radio broadcasting. Sound engineering technicians set up microphones to capture live broadcasts, musical recordings, and speaking and news events. They also synchronize and adjust sound levels for optimum audio quality and remove outside interference. Sound engineering technicians in the music industry adjust instrument and vocal volume levels for optimum output, a task also known as mixing and mastering.

OVERVIEW

Sound engineering technicians work primarily in studio environments, on both television and film production sets and in recording studios. Technicians who record live events use mobile audio recording technology capable of being set up wherever an event is taking place. Sound engineering technicians also work in concert venues and theaters. Full-time work is common, but technicians typically work sporadic schedules and hours, depending on the type of media production with which they or their company are involved.

Sound engineering technicians enter the profession from a variety of educational and professional backgrounds. Many are technologically savvy people who transform their love of music and recorded sound into a career, while others are musicians who enter the field by recording their own work or students and professionals who enjoy the process of media production. Because of the duration and intensity of the focus required in sound engineering, strong concentration skills and deft attention to detail are hallmarks of all sound engineering professionals. Professional sound engineering also requires strong collaboration skills and the ability to take instruction from others.

DUTIES AND RESPONSIBILITIES

The responsibilities of sound engineering technicians pertain to anticipating, capturing, and editing audio. A technician works closely with others to capture and produce a desired type of sound and to ensure high quality recording. The tasks involved can change on a day-to-day basis depending on a sound engineering technician's particular realm of work.

Sound engineering technicians who work in television or film usually develop audio recording strategies in collaboration with visual production staff. This traditionally involves setting up, modifying, and testing microphone equipment to ensure proper audio levels. Technicians responsible for recording live events may take several days or weeks of preparation to ensure that quality audio recordings can be captured. These sound engineering technicians record everything from roundtable discussions to lectures, large conferences, important speeches, and presentations at colleges and businesses. They are also often responsible for disassembling the audio or audiovisual (AV) equipment after the event has ended.

Sound engineering technicians who work in the music industry are specially trained to capture the sounds produced by musical instruments and voices. Working primarily in studios, these technicians are skilled at capturing the highest quality sound possible and at arranging their equalization to create the best track possible. They may also add artificial sounds and prerecorded sounds to musical recordings to alter and shape their artistic breadth. Adding sound effects is also an important element of sound engineering for television and film production.

Editing recorded sound for quality and desired effect is a crucial aspect of the job. This can require lengthy and laborious review and computer editing of recorded audio content. Regardless of specialty, sound engineering technicians are responsible for tracking recording events, ensuring the operability

Photo via iStock/yanyong. [Used under license.]

of the audio equipment, and arranging for repairs when necessary.

OCCUPATION SPECIALTIES

Audio and Video (A/V) Equipment Technicians. A/V equipment technicians set up and operate audio and video equipment. They also connect wires and cables and set up and operate sound and mixing boards and related electronic equipment. Audio and video equipment technicians work with microphones, speakers, video screens, projectors, video monitors, and recording equipment. The equipment they operate is used for meetings, concerts, sports events, conventions, and news conferences as well as lectures, conferences, and presentations in businesses and universities.

Broadcast Technicians. Broadcast technicians set up, operate, and maintain equipment that regulates the signal strength, the clarity, and the ranges of sounds and colors of radio or television broadcasts. They operate transmitters to broadcast radio or television programs and use computers to program the equipment and to edit audio and video recordings.

Sound Engineering Technicians. Sound engineering technicians operate machines and equipment that record, synchronize, mix, or reproduce music, voices, or sound effects in recording studios, sporting arenas, theater productions, or movie and video productions. They record audio performances or events and may combine tracks that were recorded separately to create a multilayered final product. Sound engineering technicians operate transmitters to broadcast radio or television programs and use computers both to program the equipment and to edit audio recordings.

Recording Engineers. Recording engineers are sound engineering technicians who operate and maintain video and sound recording equipment. They may operate equipment designed to produce special effects for radio, television, or movies.

Sound Mixers. Sound mixers are sound engineering technicians who produce soundtracks for movies or television programs. After filming or recording is complete, these workers may use a process called dubbing to insert sounds.

WORK ENVIRONMENT

Physical Environment. Studio settings such as television production and musical recording studios predominate. However, many sound engineering technicians travel to record sound at live events or in unusual locations. As such, they must be able to strategize how to maneuver and operate audio recording equipment in harsh conditions and hard-to-reach places.

Human Environment. Sound engineering technicians work with a number of other media production professionals, from writers and camera operations staff to producers, directors, and recording artists.

Technological Environment. Sound engineering encompasses a complex array of technologies. Professionals in the field must be well versed in choosing project-specific recording devices, such as microphones and radio transmitters, and in using numerous audio or audiovisual editing, effects manipulation, and preproduction software programs.

EDUCATION AND TRAINING

High School/Secondary. High school students can best prepare to work in sound engineering with coursework in the arts, music, and computers as well as English, history, mathematics, and physics. Participation in scholastic media arts projects or local media production can also be tremendously beneficial. Summer internships or volunteer work at local music and television studios can bolster the credentials of those interested in exploring audio engineering at the postsecondary level.

College/Postsecondary. Sound engineering and audio production are featured in hundreds of postsecondary programs at technical schools, colleges, and universities throughout the United States. Associate-level programs in sound engineering and audio production familiarize students with the basic tools and technologies of the craft. In addition to teaching critical listening and acoustics, certificate programs familiarize students with popular mixing and mastering software.

Undergraduate audio production courses combine the technical aspects of sound engineering with a survey of music business, audio production, theory and history, and compositional arrangement. Students who major in sound engineering master topics such as advanced mixing techniques, sound design, and film scoring.

Adult Job Seekers. Transitioning to sound engineering is relatively easy for adult job seekers. Those dedicated to learning the field can do so through both formal education and self-instruction. Employment in the field is highly selective, however, so volunteer work, internships, and previous experience in audio production are extremely beneficial.

Skilled sound engineering technicians may advance to higher-level, higher-paying positions in larger studios after gaining experience. Specialized skills are particularly advantageous.

Professional Certification and Licensure. No specific professional certification or licensure is required, although several private organizations, including the Society of Broadcast Engineers, offer voluntary certifications. Broadcast engineering certifications typically require successful completion of an exam.

Additional Requirements. Patience, organization, troubleshooting ability, and the ability to master new technical concepts quickly are key traits of professional sound engineering technicians. The collaborative nature of the position and frequent interactions with other production staff members require excellent teamwork and communication skills.

—*John Pritchard*

Further Reading

Dittmar, Tim. *Audio Engineering 101: A Beginner's Guide to Music Production.* Focal Press, 2017.

Huber, David Miles, and Emiliano Caballero. *Modern Recording Techniques: A Practical Guide to Modern Music Production.* 10th ed., Focal Press, 2023.

Rumsey, Francis, and Tim McCormick. *Sound and Recording: Applications and Theory.* 8th ed., Routledge, 2021.

Speech Recognition

ABSTRACT

The term "speech recognition" refers to the technology behind the ability of some computers, programs, and electronic devices to recognize, interpret, and respond to human speech. It is also referred to as speech recognition technology (SRT), automatic speech recognition (ARS), and speech processing. The term speech recognition also refers to the branch of computer linguistics that deals with the study and development of speech recognition equipment and programs. Speech recognition is different from voice recognition, which refers to technology that specifically determines the unique patterns of individual voices.

BACKGROUND

The earliest forms of speech recognition technology (SRT) could only understand numbers. Bell Laboratories created the Audrey system in 1952; it could identify numbers spoken by a single voice. Researchers continued to refine the technology, and by 1962 a system called Shoebox was developed that could understand sixteen spoken words.

These initial forays into speech recognition were built around words limited to four vowels and nine consonants, but they were enough to attract interest and funding from the US Department of Defense. By 1976, researchers at Carnegie Mellon University had developed Harpy, which had the ability to understand about 1,100 words. That technological leap was made possible in part by improved search tech-

Photo via iStock/Erikona. [Used under license.]

nology that enabled speech recognition programs to find the spoken words in its programming more quickly.

In the 1980s, the capabilities of SRT continued to grow because of the application of a hidden Markov model to search protocols. Technology applying this statistical method was able to determine whether certain sounds were words, rather than merely matching up sounds to known patterns. This enabled the development of some applications for businesses and dictation programs for home use that were able to work with thousands of words. Since the technology dealt with words one at a time, the programs were time-consuming to use; however, it worked well enough for businesses to begin adopting voice responsive automated answering services by the middle of the 1990s.

The ability to integrate speech recognition programs with powerful computer search engines such as Google enabled the technology to take a giant step forward in the early 2000s. Designers were able to use the search engine database, including the millions of searches done by users, instead of a predetermined database. This allowed the speech recognition programming to have access to many more possibilities for predicting and determining spoken words. It also allowed for the programming to incorporate the physical location of a speaker in the speech interpretation process based on information from searches the user had conducted. For example, the SRT may more readily recognize the name of a specific local restaurant if the user has already searched for the same place using the search engine. At the same time, the increased use of mobile devices created a market for applications and programs using speech recognition.

OVERVIEW

SRT has two phases. In the first, the speech sounds are processed and turned into numeric values that represent different vocal sounds. In many cases, the technology also identifies and ignores sounds that are not determined to be part of the voice, such as background noise. These numeric values are then compared to databases to help determine what words have been spoken. These databases include an acoustic model, a lexicon, and a language model.

The acoustic model includes all the sounds used in a language. Some applications of speech recognition include voice recognition and can be "trained" to recognize an individual's voice and the words used most frequently. This allows the technology to comprehend the unique way the person speaks, increasing accuracy of its interpretations and responses.

The lexicon is a list of words and the way they correspond to the numeric values determined in the first step of the process. The lexicon also helps the speech recognition program determine how to pronounce the words.

The language model helps the speech recognition program with correctly combining words to form grammatically correct sentences or phrases. Different forms of SRT have varying levels of language support. For instance, a phone system designed to direct a caller to one of a handful of departments in a small car dealership can have a restricted database that focuses on words callers are most likely to use, such as "service" or "sales." Limiting a system to a set linguistic content in this way can help it respond quicker and more accurately to a caller's input.

FUTURE APPLICATIONS

There are many uses for speech recognition, including automated phone systems used by businesses; talk-to-text software that enables people to speak instructions to a computer or to dictate text for conversion to a written format; and verbal control of devices such as television remotes, handheld electronics, and automobile infotainment systems. SRT is also essential to the operation of virtual assistants such as Apple's Siri as well as to devices such as the Google Nest and Amazon Echo smart speakers.

Both the technology behind speech recognition and the demand for it continued to grow in the third decade of the twenty-first century. Experts theorized that as more people became comfortable with giving verbal directions to cell phones, tablets, and other electronic devices, they would come to expect to have this ability in other places as well. Some experts anticipated that consumers would soon be able to buy appliances that respond to voice commands, and they further suggested that typing could become a thing of the past as voice-to-text speech recognition improves in speed and accuracy.

—Janine Ungvarsky

Further Reading

Juang, B. H., and Lawrence R. Rabiner. "Automatic Speech Recognition—A Brief History of the Technology Development." *University of California, Santa Barbara*, 2004, web.ece.ucsb.edu/Faculty/Rabiner/ece259/Reprints/354_LALI-ASRHistory-final-10-8.pdf. Accessed 13 July 2023.

Oremus, Will. "I Didn't Type This Article." *Slate*, 23 Apr. 2014, slate.com/technology/2014/04/the-end-of-typing-speech-recognition-technology-is-getting-better-and-better.html.

Pinola, Melanie. "Speech Recognition through the Decades: How We Ended Up with Siri." *PCWorld*, 2 Nov. 2011, www.pcworld.com/article/477914/speech_recognition_through_the_decades_how_we_ended_up_with_siri.html. Accessed 13 July 2023.

Posch, Maya. "On Getting a Computer's Attention and Striking Up a Conversation." *Hackaday*, 14 Nov. 2022, hackaday.com/2022/11/14/on-getting-a-computers-attention-and-striking-up-a-conversation. Accessed 13 July 2023.

"What Are the Benefits of Speech Recognition Technology?" *IEEE Signal Processing Society*, 2 Nov. 2018, signalprocessingsociety.org/publications-resources/blog/what-are-benefits-speech-recognition-technology. Accessed 13 July 2023.

Zewe, Adam. "Toward Speech Recognition for Uncommon Spoken Languages." *MIT News*, 4 Nov. 2021, news.mit.edu/2021/speech-recognition-uncommon-languages-1104. Accessed 13 July 2023.

SPOTIFY

ABSTRACT

Spotify is a free and subscription streaming service that makes a vast library of digital music and other audio content available to play on smartphones, computers, tablets, video gaming consoles, and other electronic devices. The name Spotify is a portmanteau of the words "spot" and "identify." In the first decade of its existence, the streaming service transformed the way people listened to music and reshaped the way the music industry makes its money. By 2022, Spotify had set up operations in over 180 countries around the world. Spotify went public in April 2018.

BACKGROUND

Spotify was established in Stockholm, Sweden, in 2006 by Daniel Ek and Martin Lorentzon. Ek, who is considered the driving force behind Spotify, was already a millionaire when he came up with the idea of Spotify. At twenty-two, he had just sold his online marketing firm Advertigo for $1.2 million. Lorentzon was the cofounder of Tradedoubler, the digital marketing company that had acquired Advertigo. Daniel Ek envisioned Spotify as a legal version of Napster, the popular online service from the late 1990s that had offered music selections for free from users' personal databases. The service introduced online music piracy and helped drive a marked decline in music industry revenue. Ek convinced Napster's founder Sean Parker to serve on Spotify's board. Ek obtained an engineer for his new project by acquiring uTorrent, a torrent client service, and he gave cash advances to Swedish artists to convince them to sign deals with the new service.

Spotify operates on what has been termed the "freemium" model, offering both free and paid subscriptions. The free subscriptions offer somewhat limited services and are paid for by advertisements. While the free version of Spotify has a number of restrictions, premium users gain access to shuffle play; they are not required to listen to advertising; they can skip an unlimited number of songs; they may listen offline; and they may play any track from an artist's available songs.

All of these features are available for around $10 a month. A family plan is also available. Spotify struck deals with record companies and music publishers for the ability to stream their music. The record companies and artists are paid a certain percentage of the revenue from each song streamed. Spotify insisted that 70 percent of its budget was allotted to paying labels, publishers, and distributors for music rights. However, the amount Spotify paid and the ways the company and similar streaming services affected artists' income became a source of controversy.

For its first five years, Spotify built up its user base in Europe. After its creation in 2006, the service officially launched in 2008, after the process of development and establishing rights deals with record labels. Spotify soon offered millions of songs to fit every musical taste, and thousands of new selections were added each day. More than 2 billion playlists were created, including celebrity playlists.

By 2011, the service had grown to over a million subscribers in Europe and was prepping for release in the United States. In May of that year, the company partnered with Facebook. The partnership allowed existing Facebook users to establish a Spotify account through their existing Facebook account and share music with friends.

OVERVIEW

After years of building the company up, Spotify launched in the United States in 2011. Building on the explosion of social media and its partnership with Facebook, Spotify allowed users to share musical preferences with a selected group of other users. Spotify also incorporated the popular social media concept of followers, allowing users to follow any artist, as well as other users.

As Spotify gained in popularity, it began to acquire other companies. In 2012, Spotify acquired Soundwave, a Dublin-based social network application for music lovers. Two years later, Spotify acquired the Cord Project, a New York-based operation that offers mobile messaging services. In January 2015, Spotify announced an exclusive partnership with Sony PlayStation, called PlayStation Music, which offers users streaming access for PlayStation 5, PlayStation 4, PlayStation 3, and Xperia devices. Additionally, Spotify signed partnership deals with Starbucks and Nike.

In 2017, the company struck a deal with video streaming service Hulu to offer a bundled subscription service for US college students. By early 2019, Spotify was offering a general market bundle for a limited time in which premium subscribers would receive Hulu for free. Subscribers also have access to playlists created by professional deejays for use at parties. Whether searching by musical artist or song title, the service offers users the opportunity to create their own "radio stations" of songs by selected artists.

Photo via iStock/stockcam. [Used under license.]

Initially, Spotify stood out because it is interactive, unlike other online music streaming services like Pandora. Other companies soon followed Spotify's example. Spotify was challenged by rivals including Apple Music, which had decades of trust already built up with both artists and users through its iTunes service. Other competition came from such services as Pandora, Amazon Prime Music, Google Play, Deezer, Rhapsody, and Jay Z's Tidal. Another potential threat to Spotify's success surfaced when Amazon launched its own full-catalog music streaming service, Amazon Music Unlimited, in October 2016 in the United States. Still, many observers suggested that there was room for many such services as the demand for digital music continued to increase throughout the world. According to *Billboard*, by 2018, Spotify accounted for 43 percent of all streaming music subscriptions worldwide and was valued at $8 billion.

On April 3, 2018, Spotify (SPOT) listed its shares on the New York Stock Exchange in an initial public offering (IPO), reaching a market capitalization of $26.5 billion. The IPO was considered nontraditional because Spotify offered shares directly to the public instead of indirectly via underwriters, and because the company did not impose a lockup period to prevent employees and investors from selling their shares immediately, causing some analysts to fear that the share price would become volatile with high-volume trading. Yet, despite the fact that 90 percent of Spotify's shares were immediately available for trade, trading did not become volatile and prices remained relatively stable. According to the company's website, by mid-2023 it had 220 million paying subscribers and 551 million active users in more than 180 markets.

In a further effort to compete with the likes of Apple Music, Spotify continued to diversify by making more investments in expanding its podcast offerings. By February 2019, it had been reported that the company had acquired the podcast production companies Gimlet Media and Anchor. In 2020 the service signed a major deal with popular comedian and podcaster Joe Rogan, with a reported value of over $100 million. *The Joe Rogan Experience* established itself as the most popular podcast on Spotify with over 11 million listeners to each episode.

Despite its steady growth, Spotify did attract considerable controversy. Along with ongoing complaints from some artists over royalties, the service drew scrutiny for some of its content, especially controversial podcasts such as Rogan's. This issue was highlighted during the coronavirus disease 2019 (COVID-19) pandemic that began in 2020, as a variety of podcasts on Spotify were criticized for spreading misinformation about the disease. Other examples of controversial material included misinformation about the 2020 presidential election and various conspiracy theories. While Spotify stated that it made efforts to remove harmful material that violated its terms of agreement, many critics argued that it failed to do enough to prevent the circulation of false and damaging information.

In January 2022, over two hundred health officials and professors sent an open letter asking Spotify to better regulate COVID-19 misinformation, including by making its rules and policies more transparent. Soon after, musician Neil Young publicly criticized Spotify for allowing Rogan to spread falsehoods about COVID-19 and asked the company to remove his music in protest.

Several other prominent artists joined Young's boycott, notably including Joni Mitchell. In response, Spotify reiterated its commitment to removing any dangerous content and published its rules of conduct, but it also defended Rogan and its overall opposition to censorship. Rogan himself suggested he would attempt to feature more balanced viewpoints on his show but would also continue to spotlight controversial opinions. The incident ignited widespread debate over misinformation and the limits of free speech, feeding into the broader culture wars of the time.

Spotify continued to expand in 2022 and 2023, acquiring several entertainment and advertising companies and forging partnerships with brands such as Roblox. In July of 2023, a new controversy arose upon Spotify's announcement that it would be raising prices for its premium subscribers in more than fifty markets, including the United States. The pricing increase applied to individual subscriptions as well as to the company's duo, family, and student offerings.

IMPACT

Spotify was established in the post-Napster environment in which artists and labels demanded that copyrights be acknowledged and that they receive compensation for sales of their products. Despite the fact that Napster was forced to shut down, its presence ushered in changes in the music industry. Music users had become insistent that the price of compact discs (CDs) was unreasonable, and they demanded new ways of purchasing music. Spotify offered an alternative to the purchase of CDs by negotiating deals with all major music studios, following a model that Ek called access-based, versus transaction-based. The model proved immensely successful, effectively changing the way music is consumed for many people by ushering in the era of streaming.

Yet Spotify also proved controversial, with some artists insisting that they saw very little return for their songs being offered on the platform. A few prominent artists, perhaps most notably pop musician Taylor Swift, publicly denounced the service, and pulled all of their music from Spotify. (Swift pulled her music in 2014, but she returned to Spotify in 2017.) Though the debate as to whether Spotify is helpful or harmful to artists continues, it is evident that the company had a huge effect on the music industry and making online streaming profitable.

—Elizabeth Rholetter Purdy and Jake Nicosia

Further Reading

"About Spotify." *Spotify*, newsroom.spotify.com/companyinfo. Accessed 25 July 2023

"Adjusting Our Spotify Premium Prices." *Spotify*, 24 July 2023, newsroom.spotify.com/2023-07-24/adjusting-our-spotify-premium-prices. Accessed 25 July 2023.

Andrew, Levshon. *Reformatted: Code, Networks, and the Transformation of the Music Industry*. Oxford UP, 2014.

Beer, Jeff. "What It Really Means That Spotify Has Lost Its Second Top Marketing Executive in a Week." *Fast Company*, 18 Sept. 2018, www.fastcompany.com/90238502/what-it-really-means-that-spotify-has-lost-its-second-top-marketing-executive-in-a-week. Accessed 25 July 2023.

"Company Overview of Spotify Technology S.A." *Bloomberg*, www.bloomberg.com/research/stocks/private/snapshot.asp?privcapId=225595077. Accessed 25 July 2023.

Goodman, Shalom. "Neil Young, Joe Rogan Podcast and Spotify: What to Know." *Wall Street Journal*, 2 Feb. 2022, www.wsj.com/articles/joe-rogan-podcast-neil-young-spotify-what-to-know-11643663216. Accessed 25 July 2023.

Heater, Brian. "Why Spotify Is Betting Big on Podcasting." *TechCrunch*, 6 Feb. 2019, techcrunch.com/2019/02/06/why-spotify-is-betting-big-on-podcasting. Accessed 25 July 2023.

Levine, Robert. "Billboard Cover: Spotify CEO Daniel Ek on Taylor Swift, His 'Freemium' Business Model and Why He's Saving the Music Industry." *Billboard*, 5 June 2015, www.billboard.com/articles/business/6590101/daniel-ek-spotify-ceo-streaming-feature-tidal-apple-record-labels-taylor-swift. Accessed 25 July 2023.

Linksy, Dorian. "Is Daniel Ek, Spotify Founder, Going to Save the Music Industry...or Destroy It?" *The Guardian*, 10 Nov. 2013, www.theguardian.com/technology/2013/nov/10/daniel-ek-spotify-streaming-music. Accessed 25 July 2023.

Miller, James. "Spotify Controversy: What You Need to Know." *Siskiyou*, 23 Feb. 2022, siskiyou.sou.edu/2022/02/23/spotify-controversy-what-you-need-to-know. Accessed 25 July 2023.

Morris, Jeremy Wade. *Selling Digital Music, Formatting Culture*. U of California P, 2015.

Pisani, Bob. "Spotify's IPO Disrupted Wall Street. What Lies Ahead Now for Unicorns Looking to Go Public." *CNBC Disruptor/50*, 22 May 2018, www.cnbc.com/2018/05/22/spotifys-ipo-disrupted-wall-street-what-lies-ahead-now-for-unicorns-looking-to-go-public.html. Accessed 25 July 2023.

Seabrook, John. "Revenue Streams." *New Yorker*, 24 Nov. 2014. www.newyorker.com/magazine/2014/11/24/revenue-streams. Accessed 25 July 2023.

Suhr, H. Cecilia. *Evaluation and Credentialing in Digital Music Communities: Benefits and Challenges for Learning and Assessment*. MIT Press, 2014.

Storytelling

ABSTRACT

Stories describing events in images and in oral and written language are at the heart of world cultures. Stories serve as entertainment, education, cultural tradition, and morality instruction. Scientific research reinforces the idea that the human brain is wired for storytelling, because people have been telling stories as a means of creative expression and comprehension of the world since Paleolithic humans painted the animals and warriors on the walls of the Lascaux caves in southwestern France twenty thousand years ago.

Advanced technology has transformed storytelling methods, bringing narratives from cave walls to movies, television, video games, and the internet. But despite increasingly sophisticated technology, the human need to tell individual and collective stories remains the same. Storytelling is a way for people to tap into their common humanity and contribute their individual experiences and insights to the crucible of the human condition.

BACKGROUND

Throughout the centuries people have told stories orally and through writing, shaping their methods and mechanics with the tools of their times. Ancient cultures painted the symbols of their stories on cave walls as prompts for the storyteller, who often used music and dance to help bring it to life. Storytellers have used carved tree trunks, sand, leaves, tattoos, and other methods of containing genealogical and other information to record stories in pictures.

Stories in the oral tradition are preserved by passing them down from generation to generation; they are constantly changing as they are told and retold. Early humans created oral myths that were sometimes later transmuted to written language. Common elements of these stories across many different cultures include ways to explain the changing faces of nature, descriptions of the origins of people, and the exploits of people with superhuman attributes. As stories developed over generations, they often became central to the identity of the storyteller's culture, laying the foundations for religious beliefs and mythology that would in turn influence later stories and storytelling rituals. Despite the worldwide popularity of written, televised, and digitalized stories, people continue to memorize oral stories that are important to their family and cultural history and bequeathed them to the next generations.

From the earliest practice of carving cuneiform symbols on clay tablets, people have used their written language to record their stories. Albert Bates Lord, professor of Slavic and comparative literature at Harvard University from 1950 to 1983, emphasized the intimate connections and changing natures of oral and written storytelling. When he examined the texts of such time-honored tales as the *Odyssey* and *Beowulf*, he discovered that a major part of the stories were composed of text that had been improvised during the oral recitation of the stories.

In his book *The Singer of Tales*, Lord demonstrated that European and Asian epics had established a tradition of both oral performance and oral composition, and he argued for a distinction between the oral storytellers and the scribes who later wrote the stories down. He contended that the stories consisted of individual variations, not one definitive text. Storytellers use formulas to create or repeat stories from common stocks of phrases that they have themselves heard and used before, generally changing only details. Similarly, the themes of oral stories are drawn from general examples, often grounded in universal religious or psychological experience. This view of storytelling was developed as the fields of anthropology, sociology, and psychology grew in the twentieth century, with notable contributions in mythological studies such as *The Golden Bough* by James George Frazer and *The Hero with a Thousand Faces* by Joseph Campbell.

OVERVIEW

Stories have developed and changed alongside the evolution of language, and they have been recorded on such diverse materials as bamboo, pottery, clay tablets, bark cloth, silk, and paper. The invention of the printing press and the rise of widespread printed media shaped societies' relationships with storytelling, as oral traditions were largely supplanted. The concept of authorship and related ideas such as copyright shifted people's view of stories toward individualism. By the twentieth century, stories were recorded on film and later digitally by computers, further pushing the boundaries of what is considered storytelling.

Twenty-first century storytellers have broadened traditional storytelling forms such as fairy tales, mythology, legends, and fables to include historical and educational topics, personal anecdotes and memoirs, and business and cultural teachings. Technology innovations such as video and conference calling systems, the internet, and social media have made it possible for individuals to easily collaborate on group stories across any physical distance, pushing back on the idea of stories as generated by a single author. Interactive web documentaries use narrative storytelling techniques to involve viewers, and interactive video games have expanded the platforms on which stories are created and told.

Stories are the foundation of television shows, movies, plays, radio shows, novels, comics, urban legends, songs, and many works of nonfiction. Some experts divide stories into four categories: those that are funny, those that surprise and delight, those that touch deep emotions, and those that move the spirit.

Institutions from corporations to churches have discovered that using the basic storytelling elements of conflict, character, and plot engage people and help them achieve their objectives. For example, managers in business may use narrative conversations instead of confrontation to resolve conflicts and overcome challenges. Collective narration in group discussions can influence participants and help unify the group by linking the past, present, and future. Storytelling helps people reason and persuade others. Using stories instead of abstract arguments and statistics helps provide context and make complex situations understandable.

Storytelling is used in contemporary advertising because stories provide graphic visual representations, they are easy to remember, and they allow sellers to create strong emotional bonds with customers. The film industry originated the idea of using multimedia techniques to create worlds for stories to evolve. Commercial examples include the "Happiness Factory" that Coca Cola created and the "Hidden Valley" home of the Jolly Green Giant.

Several storytelling organizations originated in the United States in the 1970s, including the International Storytelling Center (ISC) and the National Association for the Perpetuation and Preservation of Storytelling (NAPPS), which later became the National Storytelling Network (NSN). A professional organization that organizes resources for storytellers and festival planners, the NSN connects people to and through storytelling and advocates the preservation and growth of the art of storytelling. The ISC sponsors the annual National Storytelling Festival in Jonesborough, Tennessee, which originated in October 1973 and inspired a storytelling renaissance across the United States. Many storytelling festivals take place worldwide, and World Storytelling Day is celebrated every year on the day of the spring equinox in the northern hemisphere (which is the first day of the autumn equinox in the southern hemisphere).

—*Kathy Warnes*

Further Reading

Berger, Ross. *Dramatic Storytelling & Narrative Design: A Writer's Guide to Video Games and Transmedia*. CRC Press, 2019.

Brine, Kelly Gordon. *The Art of Cinematic Storytelling: A Visual Guide to Planning Shots, Cuts, and Transitions*. Oxford UP, 2020.

Denning, Stephen. *The Leader's Guide to Storytelling: Mastering the Art and Discipline of Business Narrative*. 2nd ed., Jossey-Bass, 2011.

Gitner, Seth. *Multimedia Storytelling for Digital Communicators in a Multiplatform World*. 2nd ed., Routledge, 2022.

Gottschall, Jonathan. *The Storytelling Animal: How Stories Make Us Human*. Mariner, 2013.

Miller, Carolyn Handler. *Digital Storytelling: A Creator's Guide to Interactive Entertainment*. 4th ed., CRC Press, 2019.

Simmons, Annette. *The Story Factor*. Reprint ed., Basic Books, 2019.

Storr, Will. *The Science of Storytelling: Why Stories Make Us Human and How to Tell Them Better*. Abrams, 2020.

T

Television Technologies

ABSTRACT

Television technologies are concerned with the design, development, operation, and production of televisions and television programming. The field includes further aspects of television production including content creation, signal broadcast, and set up and maintenance of electrical equipment. Television technologies have widespread influence in almost every area of modern life beyond their stated applications in entertainment and education. The television industry continues to evolve rapidly as new broadcast and television technologies emerge.

BACKGROUND

Television technologies are concerned with the design, development, operation, and production of televisions and television programming. The field includes further aspects of television production including content creation, signal broadcast, and set up and maintenance of electrical equipment. The core components of a television are broadcast audio and video signals, a receiver, a speaker, and a display device.

Television is a twentieth-century technological development. The term "television" was coined in France in 1900 from the Greek *tele* and Latin *vision*, which translates to "far sight." The beginnings of television can be traced back to three major discoveries: the photoconductivity of the element selenium in 1873, the invention of a scanning disc in 1884, and the demonstration of televised moving images in 1926. In 1923, the first television tube was patented by Russian engineer Vladimir Zworykin, although he never built a working model of the system. By 1925, the first moving picture was transmitted in Washington, DC, and soon after, the Federal Radio Commission (FRC) was created to regulate television and radio broadcasting. In 1926, the National Broadcasting Company (NBC) was founded by the Radio Corporation of America (RCA) as the first major broadcast network. By then the first United States-to-Europe broadcasts were sent and scheduled television broadcasts began. At this time, most commercially made television sets were electromechanical with a radio enhanced by a neon tube and a mechanically spinning disc to produce a small image. In 1934, the FRC was replaced by the Federal Communications Commission (FCC), the government agency that regulates radio, television, wire, satellite, and cable broadcasts.

By this time, the first commercially made electronic television sets with cathode-ray tubes became available. On these televisions, all broadcast images were live and in black and white. From 1940 to 1941, broadcasts included the first basketball game, the first political convention, and the first television commercial. By 1945, there were nine commercial television stations in operation in the United States, and televisions were available for purchase at department stores. In 1947, Kinescopes (picture tubes) were first used, and President Harry S. Truman used television to address the public. By 1950, there were 8 million television sets in the United States, and color television broadcasts were prevalent by 1953. In 1956, videotaping was first used in a television broadcast. By 1966, all of NBC's news broadcasts were in color. In 1975, cable television expanded to cover a wider area and offer more channels, and in 1976 satellite-television transmission began with At-

Image via iStock/Vadim Sazhniev. [Used under license.]

lanta's WTCG. By 1977, television was present in 71 million homes (97 percent) in the United States. Another milestone in television occurred in 2009, when all television stations were required by FCC regulation to switch from analogue to digital signals for their content broadcasts.

OVERVIEW

The core components of a television are broadcast audio and video signals, a receiver, a speaker, and a display device. The video signal is composed of the program or television content to be viewed. Television content is most commonly prerecorded. Some visual content is filmed live, such as some news programs. Whether prerecorded or live, twenty-first-century images are predominantly broadcast in color. Prior to the advent of digital signals, analogue signals created color by broadcasting red, green, and blue images in fast succession or combined, which appeared to the viewer to be a single full-color image. Color is conveyed via digital signals as pixels representing specific levels of red, blue, and green.

Also integral to television production is the audio signal, which is the sound broadcast sent with the video signals. The video and audio signals are sent via transmitter and can be broadcast over several distribution systems: cable, satellite, or the airwave transmissions. All of these signals are gathered by the receiver found in each television and transformed into images and sound for viewing. Depending on the television set, the images viewed are projected onto a cathode-ray tube, liquid crystal display (LCD), plasma (gas-charged display), or other display device.

The development of improved television technologies has allowed televisions to become thinner and lighter with sharper images. In addition to improvements in the physical televisions, there have also been improvements in broadcast technology, including a shift toward the use of digital broadcast signals. Digital signals transmit high-quality sound and high-resolution images via electronic code and use less power than the traditional analogue. In 2009, the FCC mandated that all terrestrial stations move from analogue to digital broadcast signals. This change required the addition of an improved external receiver to standard televisions or purchase of a television able to receive digital signals. In addition to improved images and television sounds, television speakers have also been updated such that improved digital audio can be funneled through home-theater systems.

APPLICATIONS AND PRODUCTS

Journalism. The use of television as a means of conveying current events and news in the 1950s revolutionized the way the general public obtained information. Through television, the average person could watch extended interviews and speeches by world leaders, creating an almost personal connection between the viewer and the person speaking. It was impossible to achieve the same effect through radio or the newspaper. The exposure to images of local, national, and international events in progress added impact to news stories and documentaries. One of the first documentaries to use television images in this fashion was *Harvest of Shame*, a searing indictment of migrant labor practices that was broadcast in 1960. Further examples of the impact of television in journalism include coverage of the first Moon landing in 1969; the wars in Vietnam, Iraq, and Afghanistan; and the devastation wrought by the Indian Ocean tsunami, Hurricane Katrina, and earthquakes in countries such as Haiti, New Zealand, and Japan.

Sporting Events. Television has made sporting events easily accessible to almost anyone. The application of high-resolution television technology has built up an industry of sports announcers and commentators, sports bars for viewing games, and an incredible amount of marketing via television commercials. Outside of the games themselves, television has changed the sports culture, as it spurred on the growth of sponsorship and increased the impact of the sports agent. Most sponsorships and endorsements are made possible by television commercials featuring athletes in addition to close-ups of endorsed gear during games. Such athletes supplement their salaries with sponsorships and endorsements ranging from sports drinks and shoes to personal-care products. The players themselves are marketed by their agents to attract sponsors and increase their income and visibility.

Marketing. One of the most effective modes of advertising and marketing is a mass-media campaign, which is often initiated on television. The object of a mass-media campaign is to reach the largest audience possible with a specific message. Beyond mass marketing, an industry of market-research companies that concentrate on television and Nielsen ratings has emerged to help deliver ads to target audiences during particular television shows or time slots.

Nielsen ratings are sets of television-viewer data based on the careful monitoring of the television-viewing habits of a chosen cross section of households. This system began with requesting chosen families to keep a television log or diary of minutes of each show watched by family members and has moved to electronic monitoring devices that can determine which family member is watching a particular show and when he or she changes the channel. Through data such as that produced by Nielsen ratings, a particular advertising rate can be set for a television commercial based on viewer composition and numbers. An extreme example of this is the incredibly high cost of advertising time during the Super Bowl. Since so many people watch the Super Bowl, most companies are willing to spend the exorbitant fee to advertise their products.

Entertainment. The development of television technologies over time has provided many entertainment options, from the early moving pictures to specialized gaming systems, in the comfort of the home. The use of advances in television can provide a more realistic or personalized experience. The expansion of television channel options through cable and satellite television allows viewers to watch shows focused on their hobby or interest to a degree not previously available. A sports enthusiast can watch a channel such as ESPN for sports news and games, while someone interested in cooking can spend time watching the Food Network or Cooking Channel. Personalization of the television-viewing experience further expanded with the development of the digi-

Image via iStock/seamartini. [Used under license.]

tal video recorder (DVR), which allows multiple programs to be recorded automatically for viewing at a later time.

The movement away from watching real-time television programs has expanded with the growth of streaming services such as Netflix, Hulu, Amazon Prime Video, Disney+, Apple TV+, and Max, which enable users to watch specific shows, films, and other content on their computers, phones, smart TVs, and other devices. The advent of streaming services likewise led to a corresponding rise in streaming-based television providers that offer a more traditional television-viewing experience, including Hulu + Live TV, fuboTV, Sling TV, and several others.

Surveillance and Security. The ability to use television technologies to track and identify the movements of individuals or intruders often relies heavily upon images. Closed-circuit television (CCTV) refers to a particular application of television using cameras to transmit images to a selected number of televisions for viewing and recording. CCTV is employed by banks to monitor its automatic teller machines or by other industries to monitor the function of an industrial process during the night shift. In many convenience stores, malls, and airports CCTV is used to watch for suspicious activity and deter crime. On a smaller scale, digital video monitors (nanny cams) can be placed in a baby's bedroom to provide information to parents on how their child is being cared for when they are not there.

Education. Television can be used at home or in the classroom to provide additional visual depictions of particular topics. As students become more media savvy, use of streaming video, games, and interactive features can increase interest in subjects such as history and science. A classic example of educational television is PBS's *Sesame Street*, which was first broadcast in 1969. Television programming also educates through the use of documentaries about world events of historic significance such as World War II. As TV viewers expanded their options from network television into cable and satellite TV, the number of channels entirely devoted to children's television programming, such as Nickelodeon, the Disney Channel, and Sprout, has increased. Programming on public television and educational cable shows are targeted at preschool-age and elementary-school-age children using cartoons, such as *Sid the Science Kid*, which focuses on science; *WordGirl*, which is meant to address vocabulary; and *Liberty's Kids*, an animated history show about the Revolutionary War. As child-focused channels dominate the educational market, other channels have worked on expanding educational television further into the teen and adult market with shows such as the Discovery Channel's *Mythbusters*, which provided a view into scientific method and engineering challenges while the format and topics keep viewers interested.

CAREERS AND COURSEWORK

Given the wide-ranging scope of television technologies, there are many industry careers that directly correlate to the creation and distribution of television content. Entry-level requirements for these positions will vary significantly by position. Examples of positions in television technologies are television program production personnel, news-related positions such as journalists and TV anchors, on-air talent, sales and marketing jobs, management, and technical roles. Depending on the location and size of market, job categories are more fluid, and employees can wear several hats. For example, at a small television station, a manager might also be in charge of marketing and production. Generally, those with a bachelor's degree have the competitive edge to obtain and advance in a position; however, many of the specialized technical positions do not require a specific post-high school degree, as hiring is based on internships, knowledge, and experience. Frequently, an entry-level position may be easier to obtain in a small market, such as a local television news station in a small town or city, although promotion often requires a move to a larger city or network. Given the wide spectrum of careers in television technologies, a sampling of jobs and coursework requirements follows.

Television-production jobs range from supervisor-level positions, such as a director, who manages the overall production, to more targeted technical positions, such as a postproduction assistant or sound designer. Given the specialized needs of scripts, lighting, sound, direction, and editing there are many jobs in television production. Supervisory occupations usually require a bachelor's degree with a major in communications, film and video, or media arts, while more technical jobs do not.

Televised news makes up a significant chunk of local and national television programming, and there are several cable channels entirely devoted to news and current events shows. Some news programs are live and require mobile teams of on-site reporters, camera operators, and associated technical positions to travel and report news. Other news shows are filmed in a studio and require the standard news team. Anchorpersons and the individuals who present the news from studios often have bachelor's degrees in journalism or mass communications. Reporters, who may also present the news, most often have bachelor's degrees in journalism, English, communications, or political science. Technical positions range from supervisory positions, which require a bachelor's degree, to lower-level positions that do not. Many of the technical positions, such as broadcast technician or sound engineering technician, are acquired through formal internships or coursework at a technical school. They may not require a formal bachelor's degree, but specialized education in broadcast technology focused on engineering or electronics is helpful in obtaining a position.

Sales and marketing positions in television are varied and may involve developing large-scale marketing plans for the media company, working with advertisers, or direct sales of television services or technologies. Most of these positions require a bachelor's degree in marketing, communications, or a related field although entry-level sales positions may not.

Television writers create the content for television including settings, situations, dialogue, plot lines, and characters. Writers create and develop the idea for a particular television show or movie and work to obtain network support for the production. In some cases, writers already work within the television company and work in-house producing ideas for shows. In other cases, writers are external to the television networks and must present their ideas for possible production. Some writers also wear the hat of producer, which entails working on multiple aspects of the television show from development of the show's concept to hiring actors and supervising the show's filming. There are a variety of jobs from the supervisory producer down to staff writers. Writers most often have a bachelor's degree or even a master's in some sort of English discipline. Some universities offer bachelor's and master's degrees in writing for screen and television or television writing and producing. Other universities offer courses on creative writing for entertainment. Internships and a job as a writer's or producer's assistant in television may also help a writer learn about the television industry and make important contacts as he or she moves into a full-time writer's role.

SOCIAL CONTEXT AND FUTURE PROSPECTS

The ability to record and view people and events will continue to change as new television technologies allow individuals and companies to explore the world further. Future technological innovations will change privacy options as the industry progresses and more information is gathered about television-viewing habits.

The increasing amount of time spent by adults and children watching television is associated with a decrease in exercise and an increase in obesity. In children, it is also associated with development delays, particularly in terms of physical and mental milestones. The American Academy of Pediatrics has suggested that televisions should not be placed in children's bedrooms and that they should limit television watching to no more than one to two hours per day. The nonprofit Campaign for a Commercial-Free Childhood sponsors events such as Screen-Free Week as a springboard for lifestyle changes that decrease reliance on screens for entertainment.

Television technologies are changing from delivery of television shows and movies on standard television sets to devices such as computers and cellular phones. In addition, television is quickly moving from the standard, linear format of television programming to a nonlinear, on-demand mechanism that allows viewers to record and watch shows, fast-forwarding through commercials, at a time different from the time when the program was originally broadcast. Television and the internet have also combined to allow individuals to watch many recent TV programs shown on channels not included in one's cable or satellite subscription. All of these changes have increased the number of individuals watching a wide variety of television programming.

—*Dawn A. Laney*

Further Reading

Basin, Ken. *The Business of Television*. Routledge, 2019.
Butler, Jeremy G. *Television: Visual Storytelling and Screen Culture*. 5th ed., Routledge, 2018.
Chalaby, Jean. *Television in the Streaming Era*. Cambridge UP, 2023.
Chamberlain, P. "From Screen to Monitor." *Engineering & Technology*, vol. 3, no. 15, 2008, pp. 18-21.
Cowan, Ruth Schwartz, and Matthew H. Hersch. *A Social History of American Technology*. 2nd ed., Oxford UP, 2017.
Edgerton, Gary R. *The Columbia History of American Television*. Columbia UP, 2009.
Engelman, Ralph. *Friendlyvision: Fred Friendly and the Rise and Fall of Television Journalism*. Columbia UP, 2011.
Morrow, Robert W. *Sesame Street and the Reform of Children's Television*. Johns Hopkins UP, 2006.
Starks, Michael. *The Digital Television Revolution: Origins to Outcomes*. Palgrave Macmillan UK, 2013.

3D Printing

ABSTRACT

Three-dimensional (3D) printing, also known as additive manufacturing (AM), comprises several automated processes for building 3D objects from layers of plastic, paper, glass, or metal. 3D printing creates strong, light 3D objects quickly and efficiently.

BACKGROUND

Three-dimensional (3D) printing, also called "additive manufacturing" (AM), builds 3D objects by adding successive layers of material onto a platform. AM differs from traditional, or subtractive, manufacturing, also called "machining." In machining, material is removed from a starting sample until the desired structure remains. Most AM processes use less raw material and are therefore less wasteful than machining.

The first AM process was developed in the 1980s, using liquid resin hardened by ultraviolet (UV) light. By the early twenty-first century, several different AM processes had been developed. Most of those processes use liquid, powder, or extrusion techniques. Combined with complex computer modeling and robotics, AM could launch a new era in manufacturing in which even complex mechanical objects could be created by AM.

OVERVIEW

3D printing begins with a computer-aided design (CAD) drawing or 3D scan of an object. Those drawings or scans are usually saved in a digital file format

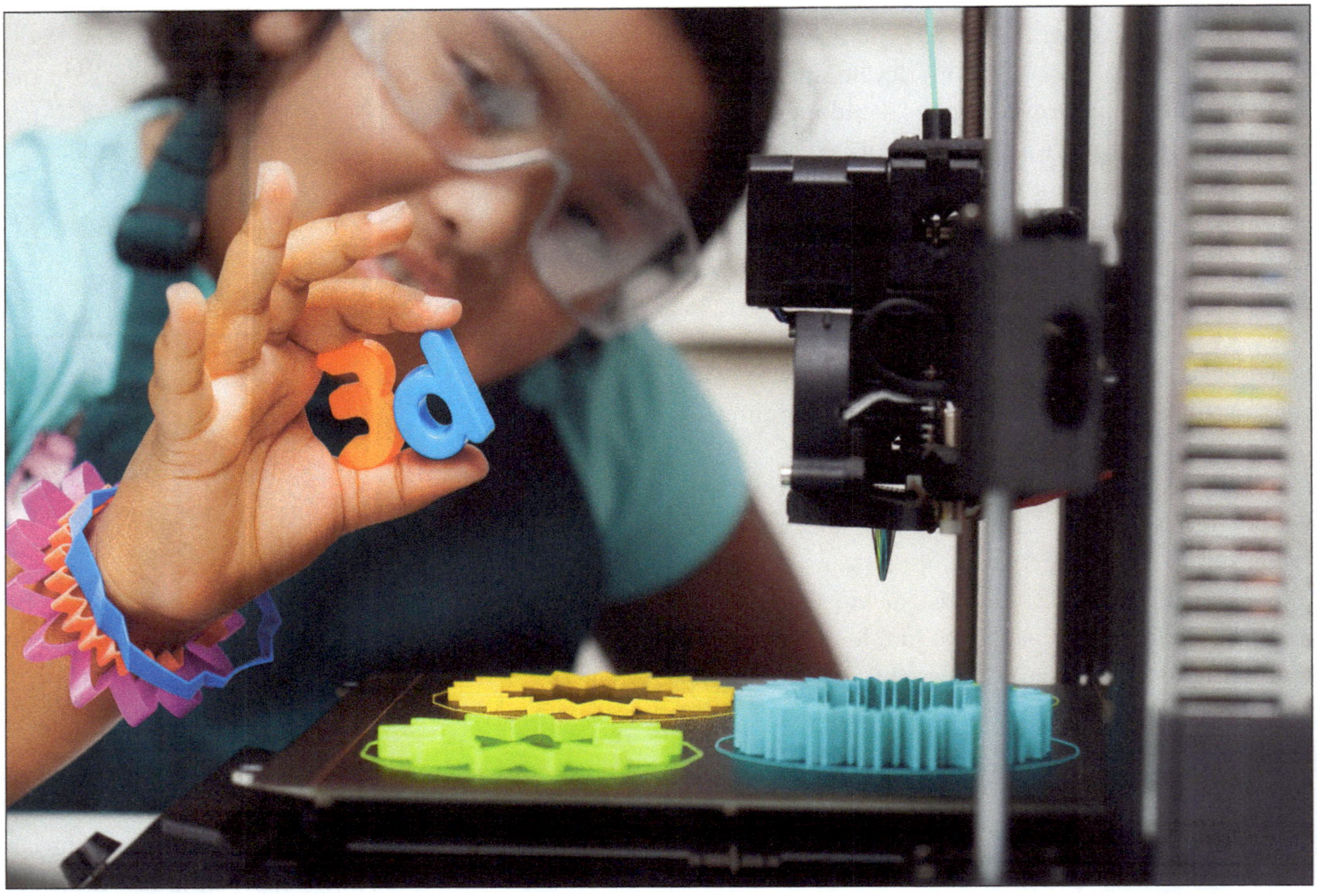

Photo via iStock/mmg1design. [Used under license.]

known as SLA, originally short for "stereolithography" but since given other meanings, such as "surface tessellation language." STL files "tessellate" the object—that is, cover its surface in a repeated pattern of shapes. Though any shape can be used, STL files use a series of nonoverlapping triangles to model the curves and angles of a 3D object. Errors in the file may need repair. "Slices" of the STL file determine the number and thickness of the layers of material needed.

LIQUID 3D PRINTING

The earliest AM technique was stereolithography (SLA), patented in 1986 by Chuck Hull. SLA uses liquid resin or polymer hardened by UV light to create a 3D object. A basic SLA printer consists of an elevator platform suspended in a tank filled with light-sensitive liquid polymer. A UV laser hardens a thin layer of resin. The platform is lowered, and the laser hardens the next layer, fusing it to the first. This process is repeated until the object is complete. The object is then cleaned and cured by UV. This technique is also called "vat photopolymerization" because it takes place within a vat of liquid resin. Various types of SLA printing processes have been given alternate names, such as "photofabrication" and "photosolidification."

POWDER-BASED 3D PRINTING

In the 1980s, engineers at the University of Texas created an alternate process that uses powdered solids instead of liquid. Selective layer sintering (SLS), or powder bed fusion, heats powdered glass, metal, ceramic, or plastic in a powder bed until the material is "sintered." To sinter something is to cause its particles to fuse through heat or pressure without liquefying it. A laser is used to sinter thin layers of the powder selectively, with the unfused powder underneath giving structural support. The platform is lowered and the powder compacted as the laser passes over the object again.

EXTRUSION PRINTING

Material extrusion printing heats plastic or polymer filament and extrudes it through nozzles to deposit a layer of material on a platform. One example of this process is called "fused deposition modeling" (FDM). As the material cools, the platform is lowered, and another layer is added atop the last layer. Creating extruded models often requires the use of a structural support to prevent the object from collapsing. Extrusion printing is the most affordable and commonly available 3D printing process.

ALTERNATIVE METHODS

Several other 3D printing methods are also possible. In material jetting, an inkjet printer head deposits liquefied plastic or other light-sensitive material onto a surface, which is then hardened with UV light. Another inkjet printing technique is binder jetting, which uses an inkjet printer head to deposit drops of glue-like liquid into a powdered medium. The liquid then soaks into and solidifies the medium. In directed energy deposition (DED), metal wire or powder is deposited in thin layers over a support before being melted with a laser or other heat source. Sheet lamination fuses together thin sheets of paper, metal, or plastic with adhesive. The resulting object is then cut with a laser or other cutting tool to refine the shape. This method is less costly but also less accurate than others.

THE FUTURE OF 3D PRINTING

While AM techniques have been in use since the 1980s, engineers believe that the technology has not yet reached its full potential. Its primary use has been in rapid prototyping, in which a 3D printer is used to quickly create a 3D model that can be used to guide production. In many cases, 3D printing can create objects that are stronger, lighter, and more customizable than objects made through machining. Printed parts are already being used for planes, race cars, medical implants, and dental crowns, among other items. Because AM wastes far less material than subtractive manufacturing, it is of interest for conservation, waste management, and cost reduction. The technology could also democratize manufacturing, as small-scale 3D printers allow individuals and small businesses to create products that traditionally required industrial manufacturing facilities. However, intellectual property disputes could also occur more often as 3D printing becomes more widespread.

3D printing has also become a relevant technology for some digital and multimedia artists. For example, an artist may create a 3D model of an object or character using digital sculpting software and then print a physical version of that model. Artists working in film, television, or web video production may also make use of 3D printing technology when creating props and costume components for their projects.

—Micah L. Issitt

Further Reading

Fuentes, Lauren, and Narée Asherian. "3D Printed Art: 3D Printing Makes Its Mark." *All3DP*, 25 Apr. 2023, all3dp.com/2/3d-printed-art-best-projects. Accessed 13 July 2023.

Hutchinson, Lee. "Home 3D Printers Take Us on a Maddening Journey into Another Dimension." *Ars Technica*, 27 Aug. 2013, arstechnica.com/gadgets/2013/08/home- 3d-printers-take-us-on-a-maddening-journey-into-another-dimension. Accessed 13 July 2023.

Matulka, Rebecca, and Matty Greene. "How 3D Printers Work." *Energy.gov*, 19 June 2014, www.energy.gov/articles/how-3d-printers-work. Accessed 13 July 2023.

"The Printed World." *Economist*, 10 Feb. 2011, www.economist.com/briefing/2011/02/10/the-printed-world. Accessed 13 July 2023.

TikTok

ABSTRACT

TikTok is a social media app that allows users to share short-form video content. Owned by Chinese technology company ByteDance, the application was first launched in China under the name Douyin before being released to international markets as TikTok in 2017. The app started out as a music-based app on which users lip synced to clips of popular songs. It later evolved to include talent-based, comedy-based, and informational videos that are typically short in form. TikTok has grown in popularity among Western nations since 2018, when it became the most downloaded app in the United States. Despite its popularity, some tech experts worry that the app is a cybersecurity risk for users due to the app's built-in biometrics and location-tracking features. News outlets have also called out TikTok for its use of censorship in certain parts of the world.

BACKGROUND

Prior to the official launch of TikTok in the United States, the app was known by a different name in China, where it was created. The app was first launched as Douyin by the Chinese internet technology company ByteDance in 2016. Content initially involved users uploading videos of themselves lip syncing to various songs. As the app developed, users also uploaded videos featuring comedic scenes and talent exhibitions. Douyin was extremely popular from the start, garnering millions of users within months. Within a year, the app had 100 million users.

After one year of operation, ByteDance decided to take their product to international markets. The app was relaunched as TikTok in September 2017 for countries outside of China, retaining its Douyin moniker within China. It was free to download and quickly became one of the most downloaded apps across the globe. ByteDance acquired a similar short-form social video networking app named Musical.ly in late 2017 and shut down the app's operation the following year while migrating all of the app's users over to TikTok.

The app was similar to Musical.ly, allowing users to upload very short videos featuring background music tracks. By the end of 2018, TikTok had been downloaded more than 80 million times in the United States alone. On a global level, the app had been downloaded more than 800 million times, and this number did not include Chinese Android phone users. These numbers surpassed download totals for popular social networking sites Facebook, Instagram, and Snapchat. A number of celebrities also began using the app, including Jimmy Fallon and Tony Hawk.

By the following year, the app had grown so popular that it had become a common household name.

Image via iStock/iNueng. [Used under license.]

Photo via iStock/5./15 WEST. [Used under license.]

The app was available in more than 150 global markets and in more than forty languages. It boasted hundreds of millions of monthly active users around the world, with more than 40 million monthly active users in the United States alone. In September 2019, the National Football League (NFL) announced it had agreed to a partnership with TikTok that included the launch of an official NFL TikTok account that users could access around the world for a variety of NFL content.

OVERVIEW

The TikTok app is a mobile app that users download onto their phones. Users create accounts with unique usernames on which they can upload short videos of themselves or other content. This content usually features background music that can be edited within the app. An example of a song edit is adjusting the play speed to either increase or decrease the pace of the song. Users have the option of choosing from a wide selection of inbuilt musical tracks, many of which are popular songs. A common form of TikTok content is a short video of a user lip-syncing to a popular song, often while in the midst of an unrelated activity. Users then add a filter to the video if desired and post the content to their account. Users can also post short comedic videos, videos recording a talent or activity, and informational videos.

User accounts can be set to public or private. Public accounts can be viewed by anyone on TikTok. Users also have the option of setting their accounts to private so only those accounts the user has authorized are able to access his or her content. Privacy settings can also be applied to individual posts, with users having the option to make their video public, private, or viewable by friends only. Users can also interact with each other on TikTok. The app has a "react" feature that allows users to film their own reactions to another user's videos. Users can also perform duets together by using the "duet" feature, which allows users to film one video next to another. TikTok has commenting and messaging features for interacting with users. TikTok also features a user feed called the "for you" page, which recommends videos to users based on videos they have previously watched. The "for you" page has age restrictions, however, and any user under the age of sixteen is not permitted to be featured on the page.

Since its launch, TikTok has become a hugely popular social media app that has shown itself capable of producing viral content. In 2019, a rapper and singer named Lil Nas X posted a video of an original song he wrote called "Old Town Road" that went on to become one of the biggest singles of the year. Media outlets credited TikTok for raising awareness of the song, which went on to receive six Grammy Award nominations in 2020 and became one of many songs to find mainstream popularity thanks in part to its early popularity on TikTok. The app also led to the viral popularity of new social media influencers, including Baby Ariel, Michael Le, Addison Rae, Charli and Dixie D'Amelio, and Spencer X, as well as a rise in influencer marketing on the platform.

Despite its success, some tech experts worried that the app could be considered a security risk given its usage of biometrics and location tracking, data that was directly conveyed to the app's Chinese parent company. The company later became a target of censorship by the Chinese government when users began posting anti-Communist content in the wake of the 2018 Hong Kong protests. Amid growing censorship concerns, the app began reviewing its content-moderation policies.

Such concerns led the app to be temporarily banned in Indonesia in July 2018; blocked from internet access in Bangladeshi in November 2018; and preliminarily banned in India in April 2019, followed by a complete ban in June 2020. In August 2020, President Donald Trump issued an executive order that would ban TikTok from the United States if ByteDance did not sell the app to a US distributor by September 15.

On September 13, ByteDance announced that it would not sell TikTok's US operations to Microsoft, which had made an offer to buy the operation the previous month. The same day, it was reported that TikTok had entered a partnership with the US software company Oracle in an attempt to address concerns over data security. However, the company reaffirmed that it would not sell or give the source code of the app to any US buyers.

Critics questioned the legality of the executive order throughout 2020, and in June of the 2021, Trump's successor, President Joe Biden, signed an executive order that revoked the ban. TikTok remained controversial in the years that followed, in part because of ongoing concerns about the privacy and security of user data. Some critics likewise expressed concerns about the proliferation of misinformation on TikTok as well the prevalence of videos promoting unhealthy beauty standards, consumerist shopping habits, and other potentially harmful content. Efforts to ban the app continued in the United States and in April 2023, the government of Montana passed legislation banning the app in that state.

—Cait Caffrey and Jake D. Nicosia

Further Reading

"8 Lessons from the Rise of Douyin (Tik Tok)." *Technode*, 15 June 2018, technode.com/2018/06/15/8-lessons-douyin. Accessed 13 July 2023.

Graziani, Thomas. "How Douyin Became China's Top Short-Video App in 500 Days." *Walk the Chat*, walkthechat.com/douyin-became-chinas-top-short-video-app-500-days. Accessed 13 July 2023.

Kelly, Makena. "Biden Revokes and Replaces Trump Orders Banning TikTok and WeChat." *Verge*, 9 June 2021, www.theverge.com/2021/6/9/22525953/biden-tiktok-wechat-trump-bans-revoked-alipay. Accessed 13 July 2023.

McCabe, David. "Montana Legislature Approves Outright Ban of TikTok." *New York Times*, 14 Apr. 2023, www.nytimes.com/2023/04/14/technology/montana-tiktok-ban-passed.html. Accessed 13 July 2023.

Perez, Sarah. "The NFL Joins TikTok in Multi-Year Partnership." *TechCrunch*, 3 Sept. 2019, techcrunch.com/2019/09/03/the-nfl-joins-tiktok-in-multi-year-partnership. Accessed 13 July 2023.

_____. "TikTok Surpassed Facebook, Instagram, Snapchat & YouTube in Downloads Last Month." *TechCrunch*, 2 Nov. 2018, techcrunch.com/2018/11/02/tiktok-surpassed-facebook-instagram-snapchat-youtube-in-downloads-last-month. Accessed 13 July 2023.

Rayome, Alison DeNisco. "10 Most-Downloaded Apps of the 2010s: Facebook Dominates the Decade." *CNET*, 16 Dec. 2019, www.cnet.com/news/10-most-downloaded-apps-of-the-decade-facebook-dominated-2010-2019. Accessed 13 July 2023.

Spangler, Todd. "Musical.ly Is Going Away: Users to Be Shifted to ByteDance's TikTok Video App." *Variety*, 2 Aug. 2018, variety.com/2018/digital/news/musically-shutdown-tiktok-bytedance-1202893205. Accessed 13 July 2023.

Touch Screens

ABSTRACT

Anyone who has used a tablet or smartphone of any kind is familiar with touch screens. Touch the screen and move the finger about to select items displayed on the screen or to make things happen. With a touch screen, touching the screen at the desired location often replaces the movement of the mouse—its navigation or button-click actions. Afterward, the touch screen becomes an output device for displaying the selected action, just as any other screen. This is where the screen becomes the viewing medium to see what the user has accessed.

BACKGROUND

A touch screen is both an input device and an output device for the computer within the particular device being used. Whether it is a laptop, a desktop, a tablet, a cell phone, an automobile, or a machine control panel, touch screens may be used. In this use, the fingertip effectively becomes the cursor used to select the action or content desired. In ordinary computer use, the user requires a mouse to control the cursor and navigate about the screen. Once a user finds the area of the screen desired, the user clicks a button on the mouse (or trackpad) to select the action the cursor indicates.

Using a touch screen for digital natives seems simple, indeed. Behind the scenes, however, there is a lot going on users are unaware of, and digital immigrants often find that the logic behind touch-screen gestures is not as intuitive as those who are familiar with the technology believe.

OVERVIEW

How a touch screen identifies where it is being touched can be determined in a variety of ways: electrical resistance, electrical capacitance, pressure, infrared (IR) light, and ultrasonic waves, for example. In all cases, the method used locates the position of the input touch signal by calculating it using an X-Y grid built into the overall construction of the screen. The screen itself is not simply a sheet of thin, tempered glass but a multilayer construct, the outermost layer of which is made of very tough glass that serves as a protective layer. In a touch screen that uses electrical resistance to determine position, the thin glass deflects sufficiently under pressure to bring two underlying conductive layers into contact with each other. This closes an electrical circuit between them and changes the electrical resistance accordingly. By alternating between the X and Y components of the grid repeatedly, the location of the touch is instantly isolated and identified.

A "capacitance screen" works in much the same way as a touch screen but uses electrical capacitance rather than electrical resistance. When two conducting materials are in close proximity but are separated by a nonconducting material such as a thin layer of air, there is a build-up of electrical charge. In a touch screen, two thin conductive sheets are separated by a nonconductive sheet, creating the condition for capacitance. When the screen is touched, the capacitance charge is shorted at that location, changing the overall capacitance within the structure. The location is calculated in the same way from the X-Y grid, and in either type of screen the user can touch with either a fingertip or a stylus. A stylus resembles an ordinary pen but has a soft rubber bulb at the end rather than an ink delivery ballpoint. The rubber tip acts as a nonconducting layer between the metal casing of the stylus and the conductive glass surface of the touch screen, creating a capacitance trigger point where it touches the screen or providing enough pressure to trigger the resistance change in a resistance-based touch screen.

While capacitance- and resistance-based touch screens are the most common, there are many other means of identifying position in the X-Y grid of a touch screen. Some use an X-Y array of IR emitters. Another type of touch screen uses piezoelectricity to identify location. Some materials produce an electrical output when they are subjected to pressure. This allows piezoelectric glass to

Photo via iStock/frantic00. [Used under license.]

be used in the construction of a touch screen. At the point of contact a piezoelectric output is produced and is subsequently detected by the change it induces in the state of the X-Y grid. Still another means of touch detection uses triangulation of the small sonic pulses. The triangulation is carried out by at least three small transducers situated about the perimeter of the screen. Other detection systems that do not use this microacoustic method identify the touch position continuously, whether the indicator is moving or not.

—Richard M. Renneboog

Further Reading

Campbell-Kelly, Martin, William F. Aspray, Jeffrey R. Yost, Honghong Tinn, and Gerardo Con Díaz. *Computer: A History of the Information Machine*. 4th ed., Routledge, 2023.

Gray, Leon. *How Does a Touchscreen Work?* Rosen, 2013.

Gray, Tony. *Projected Capacitive Touch: A Practical Guide for Engineers*. Springer International, 2019.

Lonbere, Philip. *A History of Communication Technology*. Routledge, 2021.

O'Leary, Timothy, Linda O'Leary, and Daniel O'Leary. *Computing Essentials 2023*. McGraw-Hill, 2022.

Rogers, Scott. *Swipe This!: The Guide to Great Touchscreen Game Design*. Wiley, 2012.

White, Michele. *Touch Screen Theory: Digital Devices and Feelings*. MIT Press, 2022.

TRANSMEDIA STORYTELLING

ABSTRACT

Transmedia storytelling is rooted in the distant past, when storytellers first began to use media in addition to words to craft their tales. It has expanded through the cross-platform development of franchised stories and through the participatory culture made possible by the internet. Educators have found the form particularly suited to fostering skills widely considered essential for students' success in college and the job market and have adopted it for classroom use.

BACKGROUND

The history of storytelling stretches back to a time when tales existed only in the oral tradition. For millennia, storytelling, whether oral or written, was text-based, but the late twentieth century saw tellers of tales incorporate technology into a new form known as transmedia storytelling, a weaving together of the elements of story across multiple media.

Transmedia storytelling is difficult to define in exact terms because it is an evolving concept that few agree on its tenets. The only reliable quality all transmedia storytelling has is that, simply put, it incorporates multiple media. As such, it can be as traditional as oral stories told around a fire or as contemporary as internet gaming.

The term transmedia storytelling was coined in stages. Henry Jenkins, a media scholar and a leading voice in transmedia studies, credits Marsha Kinder, another media scholar, with first using "transmedia" in 1991 to describe the cross-media systems that had developed around popular children's characters such as the Muppet Babies and the Teenage Mutant Ninja Turtles. Jenkins coined the term transmedia storytelling in 2003 when he was director of the Comparative Media Studies graduate degree program at the Massachusetts Institute of Technology (MIT). He argued that the generation who had grown up making connections in core narratives that moved across games, television programs, movies, and books, with no one medium viewed as superior to others, was likely to demand this feature in the entertainment these consumers enjoyed as adults. He later defined transmedia storytelling as a narrative that "unfolds across multiple media platforms, with each new text making a distinctive and valuable contribution to the whole."

The most immediate association with transmedia storytelling may be the entertainment products that have used multiple platforms to tell a story, but the twenty-first century has also seen educators employ transmedia storytelling in the classroom as a teaching and learning tool. They have discovered that transmedia narratives develop multiple literacies—textual, visual, and media. Building these literacies through transmedia play, a larger term than transmedia storytelling, fosters imagination, experimentation, and critical thinking.

Students engaged in transmedia play with stories and with nonnarrative puzzles and video games become not merely consumers of information but also participants in creating new information through connections, explorations, collaborations, and other forms of imaginative—and productive—play. Such play is not limited to the regular classroom; it can also occur in after-school programs, on field trips, and at home. Hence, it is often considered an important mode of articulating and experiencing one's world.

OVERVIEW

Jenkins offers a detailed analysis of *The Matrix* franchise as a definitive example of transmedia story telling. The franchise consists of four films by the American film directors, screenwriters, and producers the Wachowskis: *The Matrix* (1999), *The Matrix Reloaded* (2003), *The Matrix Revolutions* (2003), and *The Matrix Resurrections* (2021). It also includes *The Animatrix* (2003), a series of nine animated short films set in the world of the Matrix.

The Wachowskis were also heavily involved in the creation of a series of comic books set in the world of the *Matrix*, written and illustrated by different artists, including Geof Darrow, concept artist on all three *Matrix* movies. Video games include *Enter the Matrix*, produced in conjunction with the second and third films, *The Matrix: Path of Nemo*, and a massively multiplayer online role-playing game. From prerelease ads that sent consumers to the web seeking more information through internet discussion groups, each piece of the world contributes something new and connects to other elements. Only viewers who had played the video game could interpret some clues in the films, and backstory was accessible only to those who had watched animated shorts via web downloads or on digital video disc (DVD). The multiple texts work together to create what Jenkins describes as "a narrative so large it cannot be contained within a single medium."

Once Jenkins explained his concept of transmedia storytelling, other scholars pointed out that although it may have been an unintentional move by their creators, other narratives had engaged in transmedia storytelling prior to *The Matrix*. *Star Trek*'s move from television series to feature films, toys, animation, games, novels, conferences, and a themed attraction in Las Vegas spanned several decades. Another wildly successful, albeit unintentional, example of transmedia storytelling is the *Harry Potter* franchise. What began as a seven-book children's fantasy series by British author J. K. Rowling expanded into a multibillion-dollar franchise that includes feature films, music, toys, clothing, food items, video games, videos, podcasts, mobile apps, a theme park, museum exhibits, conferences, ancillary books, and websites. Foremost among the last is *Pottermore*, an interactive joint venture of Rowling and Sony that offers exclusive content, including alternate endings and original short stories about the characters by Rowling.

The entertainment company Marvel provided another example in the early twenty-first century, with various film adaptations of its comic-book titles becoming some of the most popular and highest-grossing films of all time. While the central superhero characters had existed in comics, animated works, toys, and other forms for decades, the films in the so-called Marvel Cinematic Universe (MCU), including *Iron Man* (2008) and *The Avengers* (2012), provided updated versions. Meanwhile, some characters original to the films were later featured in comics, television, and other media, showing that the flow of adaptation went both ways. In addition, both network television shows and series on the streaming services Netflix and Disney+ were developed, set in the same "universe" as the films and occasionally overlapping with actors, characters, and plot points. The MCU had enormous influence on the concept of franchises, impacting both Marvel's own comics line and the transmedia efforts of rival film studios.

Some transmedia storytelling is rooted in transformations of much older texts. The novels of Jane Austen have fueled a small industry of story transformations, with Austen and her best-known characters reimagined in guises ranging from sleuths to vampire slayers in print novels, graphic novels, digital books, feature films, television episodes, and anime. In 2012, Austen's most popular novel, *Pride and Prejudice*, entered the world of twenty-first-century technology and transmedia storytelling with Bernie Su and Hank Green's *The Lizzie Bennet Diaries*. That story cuts the five Bennet sisters to three and transforms them into modern young women: Jane, a fashion merchandiser; Lizzie, a grad student in mass communications; and Lydia, a party girl at a community college. It also changes Charlotte Lucas to Charlotte Lu, who finds a fulfilling career at Collins & Collins, makes Bingley into Bing Lee, a wealthy medical student, and transforms Pemberley from Darcy's estate to his digital company. The series consists of more than nine hours of content on over 160 videos that ran on five YouTube channels plus 35 social media profiles. The story was further extended through characters' contributions to Facebook, Twitter, and other social media that allowed for interaction with fans.

ROOTS OF TRANSMEDIA PLAY

Children, real and fictional, have routinely engaged in transmedia play. Some of the most famous children's books of the nineteenth century show young characters acting out their own versions of older stories. Louisa May Alcott's March sisters in *Little Women* (1868-69) are encouraged to see themselves as John Bunyan's Pilgrim in their lives and are said to have acted out scenes from *The Pilgrim's Progress* (1678) when they were younger. The imaginative title character of Mark Twain's *Tom Sawyer* (1876) pretends to be Robin Hood (Tom's source is probably the 1841 *Robin Hood and His Merry Foresters* by Stephen Percy and Joseph Cundall). In the early twentieth century, the title character of Lucy Maud Montgomery's *Anne of Green Gables* (1908) dramatically and riskily reenacts a scene from Sir Walter Scott's 1833 poem "The Lady of Shalott." Davy Crockett's coonskin cap and his rifle "Old Betsy," treasured by millions of children in the mid-1950s, may have been more marketing ploys by Walt Disney, who created the television series, than transmedia. However, the let's-pretend versions of Crockett's adventures acted out in backyards across the United States, often to the accompaniment of theme song "The Ballad of Davy Crockett," certainly expanded the original and thus became transmedia storytelling.

DIGITAL BOOKS

The digital book is a prime example of an educational tool that encourages students to develop critical thinking, creativity, communication, and collaboration. These skills, known as the 4Cs, are emphasized by the framework developed by the Partnership for 21st Century Skills, an advocacy group of businesses (Time Warner, Ford, Microsoft, Cisco Systems, Dell, Verizon, and others); education-related organizations (American Association of School Librarians, American Federation of Teachers, Educational Testing Service, Pearson Education, and others); foundations (Intel Foundation and Oracle Education Foundation); and media groups (Corporation for Public Broadcasting, Cable in the Classroom, and more) that fosters the inclusion of technology in education. They are also key skills in the Common Core state standards.

Inanimate Alice, a digital book, has been effective worldwide in engaging students in a transmedia experience that involves an array of literacies—verbal, digital, media, visual, information, and critical. The online episodic story, created by author Kate Pullinger, digital artist Chris Joseph, and series producer Ian Harper, features a girl named Alice growing up as she travels around the world. Created as entertainment, Inanimate Alice was quickly adopted by teachers who saw the transmedia learning potential of the book that required students to determine meaning based on text, images, sounds, and actions and provided tools to create their own content, making it possible to incorporate text, images, and sound into interactive web presentations that could then be shared through applications for iPad, iPhone, Kindle Fire, and Android. *Inanimate Alice*, with its game-playing elements has been effective also in motivating reluctant readers, those with poor comprehension skills and those who have little interest in reading.

Not all transmedia storytelling has to be high-tech. One K-6 media specialist encouraged students to create a "storyworld" after reading *Weslandia*, a Newberry Medal-winning novel by Paul Fleischmann. Their storyworld included a Facebook page where students wrote and posted updates as the protagonist Wesley. It also included original music composed in a music class to accompany the fictional character's flute playing, and kids created games based on the novel created in physical education class.

Transmedia storytelling has proved to be a particularly effective format for young adult literature. An early and innovative example is *Cathy's Book: If Found Call 266-8233* by Sean Stewart and Jordan Weisman, published by Running Press, an imprint of Perseus Books Group, in 2006. The young adult (YA) mystery, which uses alternative reality game elements, includes Cathy's doodles and notes written in the margins of the book and incorporates an evidence package of letters, phone numbers, pictures, and birth certificates. The book was almost derailed by criticism from such heavy hitters as consumer advocate Ralph Nader and novelist Jane Smiley over its use of Cover Girl products in exchange for promotional ads, but it found its audience. Readers thronged to the interactive website to discuss what happened after the book's end. The paperback version, released in 2008, removed all references to Cover Girl products. Two sequels followed: *Cathy's Key* (2008) and *Cathy's Ring* (2009). A revamped version of the original book was released in 2010 as an app for the iPod Touch and iPhone.

Chopsticks (2012), a collaboration between author Jessica Anthony and designer Rodrigo Corral, is a transmedia text that in the enhanced e-book version immerses the reader in the lives of the characters by providing access to photo albums, postcards, paintings, highlighted pages of favorite books, video clips, favorite music, and instant messages. Readers are even allowed to shuffle the pages of the book, thereby changing the order of the story and creating a personalized version of the book. Amanda Havard's paranormal YA *The Survivors* (2011), the pilot for Chafie Press's trademark book app, enhances the story by including historical and mythological background information with images of documents dating back to the seventeenth century, a music soundtrack (including three original songs by Havard) music videos, and photographs and Google satellite maps of more than fifty locations that are part of the book.

COPYRIGHT CONFUSION AND FAN FICTION

Given the many directions in which stories move as they cross platforms, it is hardly surprising that confusion and conflicts over copyrights sometimes develop. A classic example of the potential for confusion is *Splinter of the Mind's Eye* (1978) by Alan Dean Foster, the first *Star Wars* expansion novel published. The story's sexual tension between Luke Skywalker and Princess Leia was not a problem at the time, but five years later, in the film *Return of the Jedi*, Leia and Luke were revealed as being twins. To prevent a recurrence of contradictions, Howard Roffman, the executive in charge of what became known as Lucasfilm's *Star Wars* Expanded Universe, established a rule that anything new added to the world could not violate what came before. This policy served as a model for the intricate, complex storytelling that became the standard for transmedia stories.

The popularity of fan fiction has complicated copyright issues. Fan fiction has existed at least since the first decades of the twentieth century, when admirers of Jane Austen added new storylines to her fictional worlds, but it proliferated with easy access to the internet. Fan fiction sites such as fanfiction.net, the Archive of Our Own (AO3), and Wattpad have millions of users who post stories based on the characters and settings from popular novels and other media in dozens of languages. Some such sites also permit the posting of fan art, interactive projects, audio recordings of individual stories, and other forms of fan-created media that draw inspiration from traditional media and popular culture.

Lawrence Lessig, the Roy L. Furman Professor of Law at Harvard Law School and a former director of the Edmond & Lily Safra Center for Ethics at Harvard University, argued that the technology that makes possible fan fiction, fan art, and remix videos that go viral is the result of the democratization of technology. He views the twentieth century as a period in which commercial interests stifled creativity and promoted a read-only culture. Digital technology in his view has reinstated a read-write culture that fosters creativity, especially among the young. He advocates a more liberal application of copyright laws in which creators concede noncommercial use of their content to avoid rigid dichotomy between any use of protected content and a wholesale ignoring of copyright protections.

—*Wylene Rholetter*

Further Reading

Berger, Ross. *Dramatic Storytelling & Narrative Design: A Writer's Guide to Video Games and Transmedia.* CRC Press, 2019.

Dalby, James, and Matthew Freeman, editors. *Transmedia Selves: Identity and Persona Creation in the Age of Mobile and Multiplatform Media.* Routledge, 2023.

Evans, E. *Transmedia Television: Audiences, New Media, and Daily Life.* Routledge, 2011.

Freeman, Matthew, and Renira Rampazzo Gambarato. *The Routledge Companion to Transmedia Studies.* Routledge, 2020.

Hovious, A. S. *Transmedia Storytelling: The Librarian's Guide.* Libraries Unlimited, 2016.

Gutierrez, Peter. "Every Platform Tells a Story." *School Library Journal*, vol. 58, no. 6, 2012, pp. 32-34.

Hovious, A. "Inanimate Alice." *Teacher Librarian*, vol. 42, no. 2, 2014, pp. 42-46.

Jenkins, Henry. "Searching for the Origami Unicorn: The Matrix and Transmedia Storytelling." *Convergence Culture: Where Old and New Media Collide*, State U of New York P, 2006, pp. 93-130.

McDonald, R., and J. Parker. "When a Story Is More Than a Paper." *Young Adult Library Services*, vol. 11, no. 4, 2013, pp. 27-32.

Ramasubramanian, S. "Racial/Ethnic Identity, Community-Oriented Media Initiatives, and Transmedia Storytelling." *Information Society*, vol. 32, no. 5, 2016, pp. 333-42.

Rodrigues, P., and J. Bidarra. "Transmedia Storytelling and the Creation of a Converging Space of Educational Practices." *International Journal of Emerging Technologies in Learning*, vol. 9, no. 6, 2014, pp. 42-44.

Twitch

ABSTRACT

Twitch is a video livestreaming service operated by Amazon subsidiary Twitch Interactive. It is primarily known for hosting video-game livestreaming. Twitch was first created

Twitch logo. Image via Wikimedia Commons. [Public domain.]

in 2011 as a spin-off of Justin.tv, a popular general-interest streaming platform. In addition to featuring numerous channels that specialize in livestreaming video games such as League of Legends, Minecraft, and Fortnite, Twitch hosts eSports competition broadcasts, music broadcasts, and a vast array of other content.

BACKGROUND

Twitch is built on the concept of streaming. Streaming is a method of distributing specially encoded content onto computers, tablets, smartphones, and other electronic devices through broadband or wireless internet connections. Content can be streamed in either video or audio format, though video streaming allows for a more robust experience and is generally the preferred format on most streaming websites.

Streaming first emerged as a unique concept just after podcasting took hold as a popular form of consumable media. While podcasting was an effective format for audio broadcasts, it proved to have serious drawbacks when it came to video broadcasts. At the time, technological limitations meant that it was difficult to send a strong, smooth video of reasonable quality over the internet. However, as broadband technology improved and devices capable of processing constant streams of data became widely available, streaming became a more viable alternative to podcasting where video was concerned.

Streaming is an approach to broadcasting that is similar to, but also decidedly unique from, podcasting and other broadcast formats. In terms of how it works, streaming can be said to fall somewhere between podcasting and more traditional broadcast formats such as television and radio. Television and radio programming is organized and broadcast according to a strict schedule. This means that the viewer or listener must tune in to a particular station at an established time to consume the content of their choice.

Podcasting, on the other hand, affords the consumer the freedom to watch or listen to whatever they want at any time or place they so desire. Streaming falls neatly into the middle of these two approaches. While livestreaming does typically follow a set schedule that requires the viewer to tune in at an appointed time, the fact that it is broadcast over the internet means that the stream can be accessed from anywhere as long as the viewer has an internet-connected device capable of connecting to said stream.

Twitch made its debut just as ongoing technological improvements allowed streaming to become a viable broadcast format. Thanks to this timing and its early embrace of video-game content, Twitch quickly became one of the internet's premiere streaming destinations.

The genesis of Twitch was tied directly to an experimental online reality show called *Justin.tv*. Launched in 2007, *Justin.tv* was a unique show that essentially followed internet personality Justin Kan as he went about his daily life. Kan worked on *Justin.tv* alongside cofounder Emmet Shear. In building the *Justin.tv* show and website, Kan and Shear unintentionally created a new online video platform that allowed users to livestream content to an interactive audience.

When *Justin.tv* began to meet with success, Shear decided to broaden the site's focus to include video games. A long-time gamer himself, Shear believed that adapting the platform to serve as a place where other gamers could connect with high-level players could be of great interest to the broader gaming community. The subsequent introduction of gaming content caused the popularity of the Justin.tv platform to soar. It was such a hit, in fact, that Shear and Kan decided to spin the gaming portion of Justin.tv off into its own service. That decision ultimately led to the creation of Twitch in 2011.

OVERVIEW

Twitch is a video livestreaming service that primarily focuses on gaming content. In essence, it is an online platform that allows users to broadcast live feeds for others to watch. Twitch can be accessed through the official Twitch website as well as any of the Twitch apps that are available for smartphones, video-game consoles, tablets, and other devices. Users do not have to pay to watch standard broadcasts and are not required to create an account. Users who do create an account gain access to a variety of additional features, such as the ability to subscribe to their favorite channels or participate in chatrooms.

Beyond that, finding something to watch on Twitch requires little more than a few clicks. In addition to a typical search feature, users can find new Twitch channels by checking out the channel suggestions listed on the website's main page or browsing through channel categories. Since many Twitch streamers also maintain an active presence on social media platforms such as Twitter and Facebook, it is also possible to discover new Twitch channels without even visiting Twitch itself.

Using Twitch can be particularly advantageous for those who stream their own content. While most Twitch streamers simply choose to stream for fun, others do so to make money. In fact, some streamers make a full-time living just by streaming on the platform. The majority of those who stream for profit earn money through paid subscriptions, donations, sponsorships, and affiliate sales. A smaller number of streamers earn money by becoming Twitch Partners and Affiliates. Partners and Affiliates are special Twitch accounts that have monetized broadcasts. To qualify to become a Partner or Affiliate, streamers must be able to meet certain requirements related to popularity and follower count.

While video-game content continues to be Twitch's biggest draw, other types of content have also become popular on the platform. Twitch hosts a variety of talk shows, cooking shows, and much more. The popular Creative category provides a place for users to showcase their talents in everything from music and art to programming and cosplay. Another popular category is in real life (IRL), which features streamers documenting various aspects of their daily lives and chatting with users in real time.

Twitch has become enormously popular over time, eventually growing to boast more than 7 million unique broadcasters and millions upon millions of regular viewers by 2023. Using the trailblazing Twitch platform, many Twitch broadcasters have cultivated large followings and built wildly successful businesses that have earned them a great deal of money. Moreover, Twitch itself has grown to become one of the internet's most popular and most-visited websites since its inception.

—Jack Lasky

Further Reading

Asarch, Steven. "What Is Twitch? Understanding the Explosive Live-Streaming Service." *Newsweek*, 3 May 2018, www.newsweek.com/2018/05/11/twitch-909594.html. Accessed 13 July 2023.

Morris, Tee. *Twitch for Dummies*. Wiley, 2019.

"Number of Active Streamers on Twitch Worldwide from January 2018 to April 2023." *Statista*, May 2023, www.statista.com/statistics/746173/monthly-active-streamers-on-twitch. Accessed 13 July 2023.

Stephenson, Brad. "Twitch: Everything You Need to Know." *Lifewire*, 25 May 2023, www.lifewire.com/what-is-twitch-4143337. Accessed 13 July 2023.

Ward, Tom. "The Biggest Gamer in the World Breaks Down Twitch for Us." *Forbes*, 1 May 2018, www.forbes.com/sites/tomward/2018/05/01/the-biggest-gamer-in-the-world-breaks-down-twitch-for-us/#36132d0d5bb5. Accessed 13 July 2023.

Twitter

ABSTRACT

Twitter is an online social networking tool that enables registered users to send and read short messages known as tweets. Users may also share photographs, follow accounts of interest to them, and send private messages. Technology entrepreneur Elon Musk purchased Twitter for $44 billion in October of 2022, and in July of the following year, he rebranded the company and service as X.

BACKGROUND

Jack Dorsey, an undergraduate student at New York University, developed the concept of an individual using a short message service (SMS) to communicate with a small group. He originally referred to the project as twttr. Work on this idea began in March 2006. Initially, Dorsey and contractor Florian Weber developed the first Twitter prototype as an internal service for the podcasting company Odeo and its employees. The full public version was released on July 15, 2006.

Due to the limitations of SMS technology at the time of Twitter's launch, messages sent via the service, known as tweets, were limited to 140 characters in length. That limit would be raised to 280 characters in 2017. Despite such limits—or perhaps because of them—Twitter gained substantial popularity throughout the decade following its creation. By June 2015, the service had more than 300 million active monthly users, and by December 2022, that number had risen to 368 million.

OVERVIEW

Twitter has various features, with the most obvious being able to create and read tweets. The default setting is that they are visible to anyone online, but users can restrict their messages so that only followers can see them. Twitter users can tweet from the actual Twitter website or through external devices such as applications (apps) for a smartphone. Also, users can retweet messages and subscribe to view other users' feed, which is known as "following."

One major trend on Twitter is the use of hashtags. A hashtag uses the number symbol (#) and groups words together to indicate the topic of the tweet. This also allows other users to search for tweets by topic, and allows Twitter to list trending topics, which are popular topics shown on the sidebar of Twitter. Users may make a topic trend either by a concerted effort or because of recent events.

Users can also use web-based interfaces that allow them to post from multiple accounts, maintain various column views, track activities (like who is following the user), and schedule tweets to be posted in the future. Users may use Twitter to direct message (DM) one another, as long as both parties are following one another. Only the recipient, and not the general public, may see a DM. However, the message still restricts the recipient to 280 characters.

TWITTER AND SOCIETY

Over the years, individuals have begun to use Twitter in various different ways. For example, people have used Twitter to organize protests and social movements, connect with celebrities, receive emergency notifications, and read news. Also, the emergence of live tweeting has changed the way people sometimes watch television. Live tweeting usually consists of a television show's actor, producer or director tweeting about the show as fans are watching it.

Similarly, Twitter has changed the way fans interact with celebrities. Many celebrities use Twitter to promote themselves and reach out to fans, posting updates on various topics such as where they are vacationing and promoting a new television show. Fans can tweet at a celebrity, or other user, by using the at sign (@) in front of a username. This also means that celebrities can tweet back at a fan, which often allows more fan access. Obviously, this can also create security and privacy risks for public figures because the general public has access to greater information about the public figure.

Twitter has also changed how people engage in politics and elections. The 2016 election saw the rise of Twitter being used for campaigning, as well as "Twitter bots" attempting to influence the election. Twitter admitted that more than 50,000 Russian-linked accounts used Twitter to post automated materials about the 2016 US election. In January 2018, congressional Democrats called on Twitter and Facebook to investigate the use of politically motivated Twitter bots.

As is true of most online services, privacy and the use of data are a concern in regard to Twitter. For example, Twitter collects data from users when they tweet, but it also collects data when those users visit other sites. For example, many websites have embedded tweet buttons, which, even without tweeting the website, alerts Twitter to the fact that you have visited the website. Twitter has admitted that it uses this information to recommend people to follow in Twitter.

Twitter logo, 2007-2023. Image courtesy of Twitter, via Wikimedia Commons.

In 2013, Twitter acquired MoPub, a company that places ads within various mobile apps. This creates an advantage for Twitter because it allows advertisers not only to track internet usage but also to track it across all devices. Data security experts have also raised security concerns about this type of tracking because hackers could obtain a multitude of information through MoPub. Twitter's track record on privacy, however, has been largely unchallenged. It allows users to opt out of tracking functions and respects the do-not-track settings in browsers. Also, when government officials have attempted to subpoena Twitter users' data, Twitter has resisted exposing the data. The Electronic Frontier Foundation even named Twitter one of the best large technology companies for protecting data in 2013.

In October of 2022, technology entrepreneur Elon Musk completed his $44 billion acquisition of Twitter. In the months that followed, which also included mass layoffs and firings of Twitter employees, Musk made several major changes to the service, including introducing a subscription-based account-verification system that operated under the name Twitter Blue. In July of 2023, Musk announced that he planned to rebrand Twitter under the name X and replace all aspects of the service's branding, including its blue bird logo, with branding that reflected the new name. The announcement was met with criticism from many users and media outlets.

—Melissa A. Gill and Ian Gill

Further Reading

Brian, Matt. "Twitter Praised for Protecting User Data as Apple and Yahoo Falter in Privacy Report." *Verge*, 1 May 2013, www.theverge.com/2013/5/1/4289052/twitter-top-ranked-eff-privacy-report. Accessed 26 July 2023.

Curtis, Craig R., Michael C. Gizzi, and Michael J. Kittleson. "Using Technology the Founders Never Dreamed Of: Cell Phones as Tracking Devices and the Fourth Amendment." *University of Denver Criminal Law Review*, vol. 4, no. 61, 2014, pp. 61-101.

Dixon, S. "Number of Twitter Users Worldwide from 2019 to 2024." *Statista*, 14 Dec. 2022, www.statista.com/statistics/303681/twitter-users-worldwide. Accessed 26 July 2023.

Harkinson, Josh. "Here's How Twitter Can Track You on All of Your Devices." *Mother Jones*, 24 Sept. 2013, www.motherjones.com/politics/2013/09/twitter-could-threaten-your-privacy-more-facebook. Accessed 26 July 2023.

Krietzberg, Ian. "Twitter to X: The Complete Timeline of Elon Musk's Twitter Purchase." *TheStreet*, 26 July 2023, www.thestreet.com/technology/twitter-to-x-the-complete-timeline-of-elon-musks-twitter-purchase. Accessed 26 July 2023.

Swaine, Jon. "Twitter Admits Far More Russian Bots Posted on Election Than It Had Disclosed." *The Guardian*, 19 Jan. 2018, theguardian.com/technology/2018/jan/19/twitter-admits-far-more-russian-bots-posted-on-election-than-it-had-disclosed. Accessed 26 July 2023.

Vanian, Jonathan. "Musk Risks Even More Damage to Twitter's Business as the Messaging App Changes Name to X." *CNBC*, 24 July 2023, www.cnbc.com/2023/07/24/elon-musk-risks-more-damage-to-twitter-business-after-name-change-to-x.html. Accessed 26 July 2023.

TYPOGRAPHY

ABSTRACT

Typography is the art of designing proportionally spaced lettering arranged in the form of words, sentences, and blocks of text—also known as type. The central aim of typography is to create legible, readable, and aesthetically appealing textual material that clearly communicates its central message and is designed appropriately for its audience. The typographer is concerned, for instance, with issues having to do with the shape, size, and spacing of individual letterforms, as well as with the character and effect of different typefaces. The applications of typography are ubiquitous in daily life and can be found in every form of written communication, from books, newspapers, and advertising billboards to street signs and email messages.

BACKGROUND

Typography is the art and practice of designing, selecting, and placing letterforms into words, sentences, and blocks of type so that they may either be printed or electronically displayed in a digital format, such as a web page. Typography has its origins in a very specific technological innovation, and therefore, its history, unlike that of many fields of science, has a definite start date. The birth of typography was 1450, the year in or around the invention of the first printing press by German printer Johannes Gutenberg. Gutenberg's press depended on a technology called movable type—a method of printing in which each letter was cast on a separate piece of metal and could be moved around to arrange, or set, words and sentences. Gutenberg used his movable-type printing press to create the first printed book, the Gutenberg Bible.

Less than a hundred years after the invention of the printing press, printed books were reaching large numbers of readers. Specialized presses had also been developed to enable the printing of music scores. As a result, before long, the efficiency of the printing press rendered the art of calligraphy obsolete in most of Western Europe, including Germany, Italy, France, and England. Still, the roots of typography are clear: The three families of type that came to dominate Western printing (gothic, roman, and italic) were all adapted from the writing styles of calligraphers.

In the nineteenth century, mechanical type—in which blocks of text for printing could be produced by manipulating an early version of a keyboard rather than by individually setting each letter of type—made the process of printing even faster and more practical. Later, photo typesetting transformed printing yet again. This technology involved projecting characters onto photographic film using large rotating drums.

Undoubtedly, the great advancement in typography that took place during the twentieth century was the development of digital type. With this technology, characters do not have a physical form but are stored as bitmaps (maps of colored and blank pixels) or converted into mathematical formulae that describe how the lines and curves of each figure should be displayed (known as outline or vector fonts). Digital type has made desktop publishing—producing electronic versions of typographic documents such as newspapers, posters, and even entire books—the easiest and most inexpensive way to distribute written communications.

OVERVIEW

Typography is a functional art, meaning that if practiced well, it does not call attention to itself. Instead, it is a means of facilitating the expression of an idea or ideas. For example, in environmental print such as a sign indicating a traffic rule about where to turn, the task of ensuring that the rule itself is clearly conveyed is obviously more important than leading the viewer to admire the font in which it is displayed. Nevertheless, whether a written communication is attractive and easy to read are important factors in determining how well its message is understood and whether a viewer will choose to continue reading a given piece of writing through to the end.

The two fundamental principles of typography—the basic qualities that a typographer is always striving to achieve—are readability and legibility. Readability can be defined as how easy a block of text is to read, while legibility can be defined as how quickly and easily individual words or phrases can be recognized by the reader. Myriad factors influence how easy a block of text is to read and how recognizable words and letters are. Among these are x-height, weight, font size, leading, kerning, margin size, the color of type, the style of the lettering (italics, bold, and so on), and the choice of font or typeface, including whether a serif or sans serif font is used. In addition, factors having to do with the

A specimen sheet of the Trajan typeface, which is based on the letter forms of capitalis monumentalis or Roman square capitals used for the inscription at the base of Trajan's Column, from which the typeface takes its name, Image via Wikimedia Commons [Public domain.]

physiology of human vision—such as whether a reader has impaired eyesight or is colorblind—can affect readability and legibility.

HOW IT WORKS

Typeface Anatomy. The standard typographic features associated with any given typeface are known collectively as anatomy (because the names of body parts are often used to describe the parts of letters). Typeface anatomy is useful because it enables typographers to break down a character into its individual parts to analyze, manipulate, or modify each component in isolation. The terms baseline, meanline, x-height, and cap height are all used to help define the range of physical space occupied by characters. The baseline is an imaginary line denoting where characters seem to sit. A few letters, such as *g* and *q*, have parts that extend far below the baseline. The meanline is the invisible line that rests just on top of most lowercase letters, such as the tips of the arms in a lowercase *y*. It is important for the baseline and meanline to have a consistent relationship to the letters within a font. The x-height is the distance between the baseline and the meanline and corresponds to the height of the lowercase letter *x*. The cap height is the distance between the baseline and the tops of uppercase letters such as *H*.

Many other elements of typeface anatomy pertain to the characters themselves. Ascenders and descenders, for instance, are any letter or symbol parts that extend above or below the meanline or baseline. The stem of a letter is its main body; bars are any horizontal or diagonal lines that are open on at least one side, such as the three parallel lines in an uppercase *E*. If a bar connects two lines, as in the middle stroke of an uppercase *H*, it is known as a crossbar. If it seems to serve as a base for it, as in the bottom stroke of an uppercase *L*, it is known as a leg. Bowls are circular lines that form a defined space, such as in the letters *b* and *d*. The inside of a bowl is known as a counter.

Ultimately, the anatomy of a typeface has a profound influence and effect on the reader, including how readable and legible the text produced is. For example, typefaces with prominent ascenders and descenders and large, open bowls and counters, as well as serif fonts (fonts in which letter strokes are finished with a slight projection), are generally considered to be more readable than others. In addition, a typographer designing a font to be as legible as possible must attend to the frequent confusion that arises between certain characters, including *i* and *j* and the letter *o* and the number *0*.

Page Layout. Besides the qualities of the typeface itself, the way in which words and sentences are arranged on the page is an important component of typographical design. Some conventions that seem intuitive were in fact the result of careful thought on the part of typographers. For instance, the widespread use of the indent to indicate a new paragraph dates to the seventeenth century. Earlier, various other techniques had been employed to mark a new paragraph, including the placement of the first letter of the new paragraph in the margin rather than in the main body of the text.

Typographers are greatly concerned with using white space to proper effect. For example, when laying out a page of text, it is important not to use leading (the vertical spaces between lines of text) that is too wide, because this can make it harder for the reader's eye to travel from one line to the next. However, leading that is too narrow can make text less readable, as the ascenders in one line may become confused with the descenders of the previous line.

Kerning is another important aspect of spacing and layout. To kern is to selectively adjust the amount of space between particular letters or words in a block of type to achieve a desired effect. For example, a typographer may decrease the spaces between letters as the size of type on a page increases in order to keep the overall shape of the words uniform. Increasing the spaces between words can also help slow the reader down as he or she moves through the text.

Digital Type. Digital type is the creation of electronic documents and other pieces of text (as opposed to the traditional production of print in physical form). Digitized fonts can be stored in two ways: One is as bitmaps, or rasters. These are coded maps that deliver instructions to a computer program, telling it to turn on a specific set of pixels on the screen to form a particular character. The major drawback of bitmap fonts is that when they are scaled up, or displayed in a larger size, the jagged edges of the individual pixels become more and more evident, making typefaces look uneven rather than smooth. Most digital type is produced using outline, or vector, fonts. These are sets of data in mathematical form that deliver instructions to a computer program, describing particular characters as points on an outline. Because the lines between the points are not fixed, vector fonts can be scaled up or down without causing distortions.

APPLICATIONS AND PRODUCTS

Print Publishing. Print publishing—the production of materials such as newspapers, magazines, and books—would not be possible without the effective use of typography. For example, the text found in newspapers and magazines is commonly set in a series of justified vertical columns, a layout that allows publishers to squeeze in as many words of copy as possible without the page seeming cramped or cluttered. Because text in a newspaper or magazine serves so many purposes, parts of the copy are formatted in different ways so as to draw attention to their functions. Headlines are set in larger, bold type, for instance, while captions may be set in smaller type or italicized. The demands of book publishing are slightly different. A reader generally holds a book at arm's length, looking at it straight on without moving his or her head from side to side. This enables the typographer to space letters more closely together without compromising readability.

Desktop Publishing. Copy designed to be read on the upright, brightly backlit, sometimes rather small screen of a computer requires different typographical techniques than that designed to be printed on paper. For example, one difference between the two formats is the relatively low resolution of most contemporary computer screens—somewhere between 72 and 300 pixels per inch—compared with the resolution of printed materials—somewhere around 2,500 dots per inch. This constraint led typographers to the realization that although serif fonts are easier to read on the printed page, sans serif fonts are much easier to read on a screen.

Advertising. Typography plays a pivotal role in every form of advertising, from billboards to subway posters to brochures, logos, and television tag lines. The way a designer chooses to handle the type, or copy, in an advertisement can have an enormous impact on the bottom line of the business placing the advertisement. Far more than in a book or a newspaper (where the content is usually more important to consumers than the way it is presented), in an advertisement, typographical choices can either drive away or draw in a viewer. One of the biggest typographical decisions a company can make is how it presents its brand name in a logo. If done successfully and used consistently, the color, weight, spacing, and font used to display a brand name can become so recognizable that the typeface itself, even if seen in a different context, is enough of a signature to create instant brand recognition on the part of the viewer. For example, the connection consumers made between the brand identity of Swedish furniture giant Ikea and the typeface the company used (a customized version of the sans serif font Futura) was so powerful that in 2009, when Ikea changed the typeface it used on posters and in its catalog to Verdana, a font with a taller x-height and more rounded ascenders, many consumers complained vigorously.

Signage. Way-finding signage, which is any form of visual communication intended to direct people to where they need to go, is one of the most functional applications of typography. In this context, unlike in advertising, the message conveyed by the sign rather than the mood or emotion suggested by the chosen typeface is obviously of paramount importance. Still, typographical choices are just as crucial. The words on a way-finding sign need to be absolutely readable, and if possible, they should be legible to the viewer as quickly as possible, because many such signs are read by someone passing by on foot or in a vehicle. Typography was used to improve the effectiveness of signage when the US Federal Highway Administration (FHA) undertook a redesign of highway road signs. The old typeface, known as E-Modified, was found to be too difficult to comprehend, especially when upper and lowercase letters were mixed and when viewed by older drivers with poor eyesight. The FHA hired a design firm to create an entirely new typeface, one that had letters with about the same weight but with interior shapes—bowls and counters—that were larger and more open. In addition, lowercase letters in the new font, known as Clearview, were heavier, distinguishing them more clearly from uppercase letters. In 2004, several states began replacing their old signs with Clearview signs. Studies of the readability and legibility of the signs have found that drivers can determine their meanings from up to 200 feet away.

Visual Art and Design. Amazing things can happen when typography meets art. One trend involves artists using intricate typographic designs to create portraits. These artists create shapes and faces by using letters and words instead of drawing lines. Graffiti is another artistic application of typography. Many street artists become known for the distinctive fonts they use in their lettering, or tags. Typography is also sometimes used as a component of visual design in printed poetry. The twentieth-century American poet E. E. Cummings made frequent use of typographic elements such as odd spacing between words and lines and unexpected punctuation marks.

IMPACT ON INDUSTRY

Industry and Business. Besides the publishing and advertising industries (in both print and online media), whose primary business functions are intimately connected with the world of typography, virtually every business sector that makes use of written communication is affected in some way by typographical technologies and trends or seeks to make use of typography to further its goals. Every public relations department, for example, makes careful choices about the typefaces it uses to represent the person, product, or service it is trying to market. The Gotham font used by Barack Obama's presidential campaign team in 2008 on all its posters and other graphics is a good example. The typeface's clean, modern lines helped reinforce the candidate's message of fresh ideas and change. Another industry that has come to rely on sound principles of digital typography is the personal electronic device sector, companies that produce products such as smartphones and personal music players. Because the screens on such devices are so small, clever typographical techniques are required to make the text they display readable.

Major Corporations. In the age of digital type, software companies are by far the biggest producers of typographical tools and applications. One of the biggest commercial font companies (known as foundries just like the metal-casting factories that make physical type) is Monotype, which grew substantially through its acquisition of numerous smaller foundries and font retailers. Computer technology companies such as Apple and Microsoft are among the biggest licensers of the font families produced by Monotype and other foundries.

CAREERS AND COURSEWORK

In the past, typography was a specialized expertise on which a professional could base his or her entire career. It has become one of a suite of skills associated with a career in fields such as graphic design, advertising, communications, packaging design, and software design. Preparation for such a career should begin with any courses that are available in computer science at the high school level, such as computer graphics, software engineering, and computer-aided design. Traditional fine arts courses in drawing, sketching, printmaking, and the graphic arts are also appropriate. A number of schools of the arts offer typography certificate, diploma, or associate degree programs; however, students who intend to complete a bachelor's degree should consider concentrating in areas such as computer science, computer graphics, graphic design, or visual communications. Perhaps the most important skills required for any professional working with typography are those that pertain to specific pieces of software commonly used in the industry.

SOCIAL CONTEXT AND FUTURE PROSPECTS

Typography is a powerful instrument of communication. As such, it has the potential to influence the beliefs, behaviors, and ideas of its audience in both positive and negative ways. Choices about fonts, spacing, and layout on a poster urging residents of a city to use less water, for instance, could have an important impact on the effectiveness of an environmental intervention. By the same token, however, smart typographical techniques used to create propaganda materials or pamphlets containing hate speech could help spread incendiary ideas more effectively. Every typographical artifact, from an email exchange between friends to a billboard by the side of the highway—no matter the message—creates a medium of expression that conveys far more than just the meanings of the words it contains. With the advent of desktop publishing and software tools that enable anyone to design a customized font and publish his or her own writings, typography is swiftly becoming a democratized field that is virtually without barriers to entry.

—M. Lee

Further Reading

Baines, Phil, and Andrew Haslam. *Type and Typography*. 2nd ed., Laurence King, 2005.

Bringhurst, Robert. *The Elements of Typographic Style*. 3rd ed., Hartley and Marks, 2005.

Hunt, Richard. *Advanced Typography: From Knowledge to Mastery*. Bloomsbury Visual Arts, 2020.

Jury, David. *What Is Typography?* Rotovision, 2006.

Kane, John. *A Type Primer*. 2nd ed., Laurence King, 2020.

Wagner, Karin. *From ASCII Art to Comic Sans: Typography and Popular Culture in the Digital Age*. MIT Press, 2023.

V

Video Editing

ABSTRACT

Video editing is the process of altering, combining, or otherwise manipulating video frames or sequences of video frames after recording. Early video editing was done with time-consuming and visually limited linear editing techniques. Revolutionary advances in digital technology ushered in the era of nonlinear video editing.

BACKGROUND

Video editing is the process of altering, combining, or otherwise manipulating video frames or sequences of video frames after recording. Cameras create video by recording a series of still images, or "frames," which, when rapidly displayed in sequence, create the appearance of a moving picture. The speed at which video frames are recorded and displayed is the frame rate. This is measured in frames per second (FPS). Early recording technologies used frame rates as low as 16 FPS. By contrast, digital video cameras such as the GoPro HERO11 can record at speeds as high as 240 FPS. To achieve the semblance of smooth motion, frame rates of 24 to 30 FPS are used. Higher frame rates are said to provide clear, more lifelike images, while lower ones produce jerkier motion or blurrier video.

Photo via iStock/EvgeniyShkolenko. [Used under license.]

The frames recorded digitally are composed of bits. A bit is the smallest unit of data that can be stored and manipulated. The amount of data encoded, in bits, for each second of video is the bit rate. The greater the bit rate is, the higher the resolution and, thus, the quality. However, higher bit rates also result in larger video files.

OVERVIEW

In the past, video was edited using linear video editing, a technique for reordering sequences of video frames captured on analogue video tape. Linear video editing was initially achieved by physically cutting videotape into sections and then splicing the sections together in the desired sequence. By the 1960s, electronic tape editing technology allowed different sections of videotape to be combined mechanically onto a single videotape. Even with this advance, however, linear video editing remained difficult and time consuming.

The advent of digital technology revolutionized video editing. Computer hardware and video-editing software enabled the use of nonlinear editing. Nonlinear editing allows individual video frames to be accessed, altered, moved, copied, or deleted regardless of the order in which they were recorded. The frames can then be combined into a single video sequence using software.

Nonlinear editing enabled video editors to employ a range of new techniques that would have been impossible to achieve previously. One such technique is video scratching, where a sequence of video frames is manipulated to match the rhythm of a piece of music. Another is digital compositing. In this process, video frames from different sources can be combined with each other and with still images and computer graphics to create the illusion that they were all recorded at once. This is related to vision mixing, in which various video sources are combined into a single recording. An example is the combining of footage from multiple cameras at a live event, such as a concert, for a seamless broadcast viewing experience.

An important tool used in video editing is the edit decision list (EDL). An EDL is an ordered list of video frames or frame sequences ("clips") that contains information such as time data. That information allows the video editor to find individual clips quickly and efficiently during editing.

Digital video formats store varying quantities and qualities of data. Choosing the appropriate file format can make a difference when the audio speed amplitude is manipulated, or compressed and restored. With so many video formats available, it is important to choose editing software that is compatible with one's file format of choice.

VIDEO EDITING IN A DIGITAL WORLD

The development of affordable personal computers (PCs), digital video cameras, and smartphones has brought video editing from the studio to the realm of ordinary people. Video-editing software includes applications for PCs, tablets, and smartphones, and some programs, such as Apple's iMovie, are compatible with multiple categories of device. When used along with social media services such as YouTube, Facebook, Instagram, and TikTok, video-editing software allows users to easily share videos with anyone who has an internet connection. Thus, anyone with a smartphone can be a video writer, director, editor, and movie studio. This has led to a powerful new form of personal and artistic expression that will continue to grow and change as new video technologies are developed and adopted.

—Maura Valentino

Further Reading

Anderson, Gary H. *Video Editing and Post Production: A Professional Guide*. Focal, 1999.

Ascher, Steven. *The Filmmaker's Handbook: A Comprehensive Guide for the Digital Age*. Plume, 2012.

Brine, Kelly Gordon. *The Art of Cinematic Storytelling: A Visual Guide to Planning Shots, Cuts, and Transitions*. Oxford UP, 2020.

Brown, Blain. *The Basics of Filmmaking: Screenwriting, Producing, Directing, Cinematography, Audio, and Editing.* Routledge, 2020.

Dancyger, Ken. *The Technique of Film and Video Editing: History, Theory, and Practice.* Focal, 2013.

Rubin, Michael. *Nonlinear—A Field Guide to Digital Video and Film Editing.* Triad, 2000.

Schenk, Sonja, and Ben Long. *The Digital Filmmaking Handbook.* 7th ed., Foreing Films, 2021.

Video-Game Design and Programming

ABSTRACT

Video-game design and programming combines artistic and computer programming skills to create products for a variety of gaming platforms, including smartphones, personal computers (PCs), and consoles such as the Xbox and PlayStation. Whereas the products of video-game design and programming were once primarily electronic amusements for children and adolescents, a number of factors have resulted in video games being used for other purposes, such as medical or military training, fitness, and physical rehabilitation. The popularity of video games has also resulted in the growth of game design and programming curricula in education.

BACKGROUND

Video-game design and programming is the field of computer science involved in the creation of video games to be played across a variety of electronic and digital platforms. This field combines artistic skills such as computer graphics, writing, and traditional drawing with computer programming skills to produce games ranging from "casual games," a term denoting simple games designed for quick diversions, to elaborate games featuring immersive worlds, multiple levels, complex storylines and character development, and a fair amount of skill development for play.

The first video games were created by computer scientists and programmers after World War II. However, it was not until the 1960s and 1970s that the combination of graphics and audio capabilities produced what the modern user would recognize as a standard video game. For many years, the earliest games, such as *Spacewar!*, *Pong*, and *Tank*, provided a generic model for many copycat games. However, by the mid-1980s, many genres of video games had been created, including first-person shooters, adventure games, role-playing games (RPGs), and even the earliest versions of online games, most of which were adaptations of popular board or pencil-and-paper games.

The massive growth of this industry since the 1980s is in part because of the continuous improvements made to the hardware and software with which games are created and on which they are played. Personal computers and communications devices (e.g., smartphones) follow the rule of Moore's law. According to the law, which is named after Gordon E. Moore, the cofounder of Intel, the power and speed of computer chips doubles about every eighteen to twenty-four months. Thus, a computer that sold for a thousand dollars in the 1990s has far less processing power and speed than a computer selling for the same price in the 2020s. This increase in processing power allows games to be created using far more elaborate programming and design tools than were used in the past. Thus, video games have become far more realistic than before, although the cost has remained roughly the same.

Since their early history, video games have grown from being simple, entertaining diversions that could be experienced on occasion (such as arcade machines found in restaurants, bars, and arcades in the 1970s) to full-scale entertainment systems that are playable from almost any type of consumer electronics device with audiovisual capabilities. The mass adoption of video games in contemporary culture has resulted in the creation of a highly influential industry that rivals the sound-recording and motion-picture industries and may even surpass those industries in terms of profitability and user bases.

OVERVIEW

The development of video games can be divided into several phases, from initial concept to design and programming to final product. Each of these phases has a number of processes.

Concept. The initial idea for a video game is known as the concept. The design document, whether a single sheet of paper or a complicated, multipart document, will spell out all the important information that the entire design and programming team will follow for the duration of the game's development. The simplicity of the initial concept will not necessarily have any impact on how complex the final product will be, but it can guide certain decisions along the way. These include the size of the design team, the tools and resources to be used, the platform for which the game is being designed, and who the target audience will be. Early phases of game development involve decisions about the intended plot and style of the game, the language(s) of the game, the design of the game world and environments, and the characters that will populate the game world. Much of this is guided by designers and programmers as they oversee game development during each of its phases.

Design and Programming. Once a concept has been envisioned, the next phases involve the production of the game in the desired video-game format and platform, such as a video-game console or personal computer (PC). This involves many different steps with many levels of expertise: programming of game elements for the chosen platform, selection of the game engine or software used to program the game and the modeling and other tools to create characters and environments, and creation of the earliest models of the game so it may be tested. The roles of each development team member at this point are fairly fixed, and they work together to make sure that each element of the game matches the other elements to ensure proper game play (all experiences during a player's interaction with game systems). Major considerations during this phase include determining which game engine or software and audio and graphics tools to use. The programming languages commonly used in game design include C++, Java, and Python. Some companies make development kits available to budding game designers. At some point in this process, a version of the game is tested to ensure that software defects, also known as bugs, are eliminated and that the game works as intended.

Crunch Time and Postproduction. Once a video game is near its final, or completion, phase, it is adjusted by designers and programmers and undergoes further testing and quality assurance. Although the game is continuously tested and adjusted by programmers and designers throughout the development phases, it is up to professional game testers to see if the game works from a user's perspective. A marketing team is usually brought in at this level to prepare for the launch and release of the game. As the video-game industry is a multibillion-dollar industry, these later phases can be just as important as the early phases of design and development.

APPLICATIONS AND PRODUCTS

The games that existed before the boom in popularity of video games in the late 1970s—the era of video arcades and home systems—were relatively simple programs that had little compatibility beyond the specific gaming systems for which they were designed, such as the large, boxy coin-operated machines found at arcades. Therefore, these games were considered novelty pastimes that were enjoyed in specific places at specific times and by a fairly small percentage of the population. However, the evolution of video games, like that of motion pictures, is based on continuous adjustments to the technology used to create and consume them. The consequence of these improvements over time is a shift in the way that video games are perceived and an expansion of the market for these games to include consumers other than traditional gamers, who were typically male adolescents. Such stereotypes no longer apply, however. According to the "Essential Facts about the Video Game Industry from the Entertainment Software Association," 74 percent of US households owned at least one video-game player. In total, 266.6 million Americans reported playing video games. Of video-game consumers, 20 percent were under the age of eighteen years, 38 percent were aged eighteen to thirty-four years, 14 percent were aged thirty-five to forty-four, 12 percent were aged forty-five to fifty-four, 9 percent were fifty-five to sixty-four, and 7 percent were older than sixty-five. This shift in demographics has prompted a significant diversification of the video-game industry in terms of platforms, content, and design.

Video Games for All. The primary purpose of video-game design and programming is the creation of consumer goods for entertainment. Video games, once primarily entertainment targeted at male adolescents and young men, are increasingly played by people of all ages and are being adapted to train and educate people as well as to entertain them. In the 1990s and the early twenty-first century, many new applications were created that not only elevated the gaming experience but also enabled video games to be used for purposes other than pure entertainment. Video games have become valuable tools in diverse areas such as user interfaces, storytelling, training and education, the arts, marketing, medicine, physical rehabilitation, and the military.

Beginning in the late 1980s and early 1990s, with the advent of video-game production houses—and video-game franchises such as *Doom*, *Final Fantasy*, and *Super Mario Bros.*, which generated millions of dollars—game designers and programmers have constantly pushed the boundaries of computer science and graphic arts. Innovations in video-game design and programming include the use of smartphones and personal electronic devices as important new platforms for gaming and the introduction of new systems designed to appeal to nontraditional gamers, such as the Nintendo Wii and the Nintendo Switch. These innovations reflect the creativity of game makers in carving out new markets for themselves.

Innovative Platforms. Video-game programmers and designers have increasingly developed games to run on different platforms and use novel controllers, thereby drawing in consumers who historically have not been regularly associated with video gaming. In the case of handheld devices such as the Apple iPhone and Android smartphones, consumers recognize the utility of having a device that can be used not only for telecommunications but also for entertainment. Extended battery life, ease of purchasing mobile games, and user-friendly interface designs are also factors in drawing new users.

Similarly, the simple addition of a wireless, motion-sensitive game controller called the Wii Remote to the Nintendo Wii resulted in a new gaming environment for users and spawned a number of new game designs and concepts. Users could now control the game with physical gestures, prompting the development of games that encourage physical activity. The Wii Fit, introduced in 2008, contained a balance board that allows users to engage in yoga, aerobics, strength training, and balance games. The interactive nature of these fitness games has been noted by health and physical rehabilitation professionals, resulting in a burgeoning research field in which more and more innovative gaming technologies are being applied to serious health issues.

Games as Educational Tools. An important new field for video-game designers and programmers is serious games, or those with a primary purpose other than pure entertainment. The idea of a video-game environment being used to train pilots and soldiers (two of the more common occupations long associated with gaming) is not new, but the increasing realism, control, and complexity of the tools available for training make video games more and more attractive for a variety of applications. In the medical and health fields, for example, motion-sensitive controllers are useful for training physicians in surgical techniques and aiding in the rehabilitation of physical mobility in elderly patients, as well as those with limited physical movement because of congenital or age-related conditions.

Serious games have been used to teach people about such social issues as the Arab-Israeli conflict and the guerrilla war in Darfur (started in 2003). For example, in the game *Darfur Is Dying* (2006), players could take on the identities of members of a refugee family to learn more about the hazards of everyday life in that region. Other serious games deal with peak oil scenarios, climate change, and a host of other situations that could benefit from more public awareness. In particular, the relationship between game design and military recruitment highlights a possible trend in serious games and occupational fields, in that in addition to scenario simulations, games may increasingly become a part of recruitment and training.

CAREERS AND COURSEWORK

One important result of the surge in video gaming since the 1980s and 1990s is the creation of video-game design and programming curricula at universities and colleges. The overall framework for a degree in a video-game-related field is based largely on the student's focus. Most programs will include courses designed to create a general understanding of video games, such as the history of video games and an introduction to game design and game programming. After these lower-division courses, the focus narrows to more specific courses in aspects of design, such as level design, production and documentation, and methods for working with artists and programmers. Game programmers, on the other hand, continue coursework in computer science as applied to video-game development, as well as specific programming tools for game development. A number of colleges offer game design and programming degrees, with some even offering master's degrees in management and production.

The career track for a video-game designer or programmer could start with a job at a small company with only a handful of employees who create just a few games a year. In this environment, a designer should expect to work closely with other team members and to have occasional duties in other areas, such as programming. The alternative is to be hired by a much larger company in which there may be many departments simultaneously working on different game titles. In this environment, one should expect to maintain a high degree of focus on just one skill set. In either environment, the same levels of skill and workmanship are highly valued and can be quite rewarding for a talented game developer.

SOCIAL CONTEXT AND FUTURE PROSPECTS

To understand why the video-game industry has grown as rapidly as it has, one must look at the far-ranging applications that each new generation of game design and programming technology makes possible. Video games have a wider range of uses than ever before and additional practical applications are likely to be developed as gaming increasingly becomes an important part of everyday life for millions of people around the world. While some virtual-reality technology ranging in affordability for gaming was available to consumers by the 2020s, future prospects included more immersive environments that expand game playing into levels of interaction and sensory experience that previously could be found only in science fiction. Other possibilities include powerful, yet relatively inexpensive displays and controllers, as well as tools that allow almost anyone to create their own video games on a typical home computer.

Video gaming became even more prevalent during the coronavirus disease 2019 (COVID-19) pandemic as it was increasingly used among people of all ages to stay connected during stay-at-home orders and social distancing requirements. According to the Entertainment Software Association, the average age range for a video-game player was thirty-one years old in 2021. About 77 percent of video-game players said they played with other people, including friends and family members, in 2021. Moreover, 42 percent of players said they had met a close friend or even a significant other through video games. Video games are increasingly being used for nontraditional activities, such as health and fitness regimens. These trends point to a broad change in the acceptance level of video gaming across all age groups.

—Craig Belanger

Further Reading

Adams, Ernest. *Break into the Game Industry: How to Get a Job Making Video Games*. McGraw-Hill/Osborne, 2003.

_____. *Fundamentals of Game Design*. 2nd ed., New Riders, 2010.

Cox, Kate. "It's Time to Start Treating Video Game Industry Like the $21 Billion Business It Is." *Consumerist*, 9 June 2014, consumerist.com/2014/06/09/its-time-to-start-treating-video-game-industry-like-the-21-billion-business-it-is. Accessed 22 Dec. 2014.

"Essential Facts about the Video Game Industry." *Entertainment Software Association*, 2021, www.theesa.com/wp-content/uploads/2021/08/2021-Essential-Facts-About-the-Video-Game-Industry-1.pdf. Accessed 15 Mar. 2022.

Isbister, Katherine. *How Games Move Us: Emotion by Design*. MIT Press, 2016.

Kamenetz, Anya. "Why Video Games Succeed Where the Movie and Music Industries Fail." *Fast Company*, 7 Nov. 2013, www.fastcompany.com/3021008/why-video-games-succeed-where-the-movie-and-music-industries-fail. Accessed 22 Dec. 2014.

Koster, Raph. *A Theory of Fun for Game Design*. Paraglyph, 2004.

Lecky-Thompson, Guy W. *Video Game Design Revealed*. Charles River Media, 2007.

Marcovitz, Hal. *Video Games*. Lucent, 2010.

Rogers, Scott. *Level Up!: The Guide to Great Video Game Design*. 2nd ed., Wiley, 2014.

Schell, Jesse. *The Art of Game Design: A Book of Lenses*. 2nd ed., CRC Press, 2014.

Smith, Noah. "The Giants of the Video Game Industry Have Thrived in the Pandemic: Can the Success Continue?" *The Washington Post*, 12 May 2020, www.washingtonpost.com/video-games/2020/05/12/video-game-industry-coronavirus. Accessed 26 Feb. 2021.

Thompson, Jim, Barnaby Berbank-Green, and Nic Cusworth. *Game Design: Principles, Practice, and Techniques*. Quarto, 2007.

VIDEO-GAME DESIGNER

ABSTRACT

Video-game designers create the concept and gameplay of specific games. They may be employed by production studios or work as freelancers. Students aspiring to enter the profession should study art, mathematics, and computer science and should also familiarize themselves with a wide range of video games.

BACKGROUND

The video-game designer is responsible for creating the concept and the gameplay of a specific game. The designer, often one of several such professionals on a design team, comes up with a concept and creates detailed descriptions of the various aspects of the game. When production on the game begins, they work with the programmers and artists to ensure that their ideas are being properly executed. They make use of scripting languages in order to test their ideas in the game prototype.

OVERVIEW

Most game designers work in office environments, although those who are employed on a contract or freelance basis may work from home. The typical workweek for a designer is forty hours, although when a new game is being released, the designer may have to work extra hours. Salaried designers do not typically earn overtime for these additional hours, but freelance and hourly workers often do.

Video-game designers should be interested in and knowledgeable about computer science and programming. They should know basic scripting languages, such as Lua and Python. They are also generally creative people and may have backgrounds in visual art or in narrative art forms such as writing. They should also be familiar with a wide range of video games.

DUTIES AND RESPONSIBILITIES

The game-design process begins when the designer or designers create the concept for a new game. The designer or design team then typically writes a proposal which outlines the basic idea, story, and characters of the game, as well as practical considerations such as schedule and budget. Once the proposal has been approved, the designers meet regularly, to compose more detailed descriptions of the various aspects of the game, including plot, characters, and gameplay. The design team meets regularly to figure out which descriptions work best for the game and decide what to implement moving forward. These descriptions are then integrated into a guiding text known as the game-design document. In some instances, games are developed independently by a single designer who then shops his or her proposal to different development companies, but in a large number of instances, the game is conceived and designed entirely in house by a team of designers working for a game company.

During the preproduction phase and into the production phase, the designers also test various elements by implementing them into a prototype of the game. For this, they make use of various scripting languages in order to insert these elements into the game. Because scripting languages do not require much coding, it is relatively easy for designers to test elements using these languages without undue difficulty.

Once the basics of the plot, characters, and gameplay are established, the production phase of the game begins and the process is turned over to the programmers and artists whose job it is to successfully bring to life the concepts of the designers. Throughout this process, the designers work closely with these other professionals to ensure quality implementation. If any of the concepts do not work properly during the production phase, the designers most go back and edit or review the game-design document.

On a typical design team, there are a number of different designers, all of whom take on slightly different roles. The lead designer is responsible for collecting and organizing the design team's ideas into a coherent game document. They are also charged with managing the schedules and tasks of the other design team members. In addition, they serve as go-betweens between the design team and the other departments to ensure that everything is running smoothly and the design team's ideas are being successfully carried out.

Another key member of the design team is the content designer or designers. Their task is to develop the game's plot and characters. These designers are allowed a lot of creative leeway but must also work within the parameters established by the larger team. They do much of their work in preproduction but often have to make edits to the game document as the production process comes along.

Other members of the design team include the game mechanic designers who focus on more specific aspects of gameplay, and the level designers, who are charged with creating the game's specific environment, levels, and missions. They also map the location of objects and opponents in a level. Depending on the size of the production, a company may employ some or all of these different types of

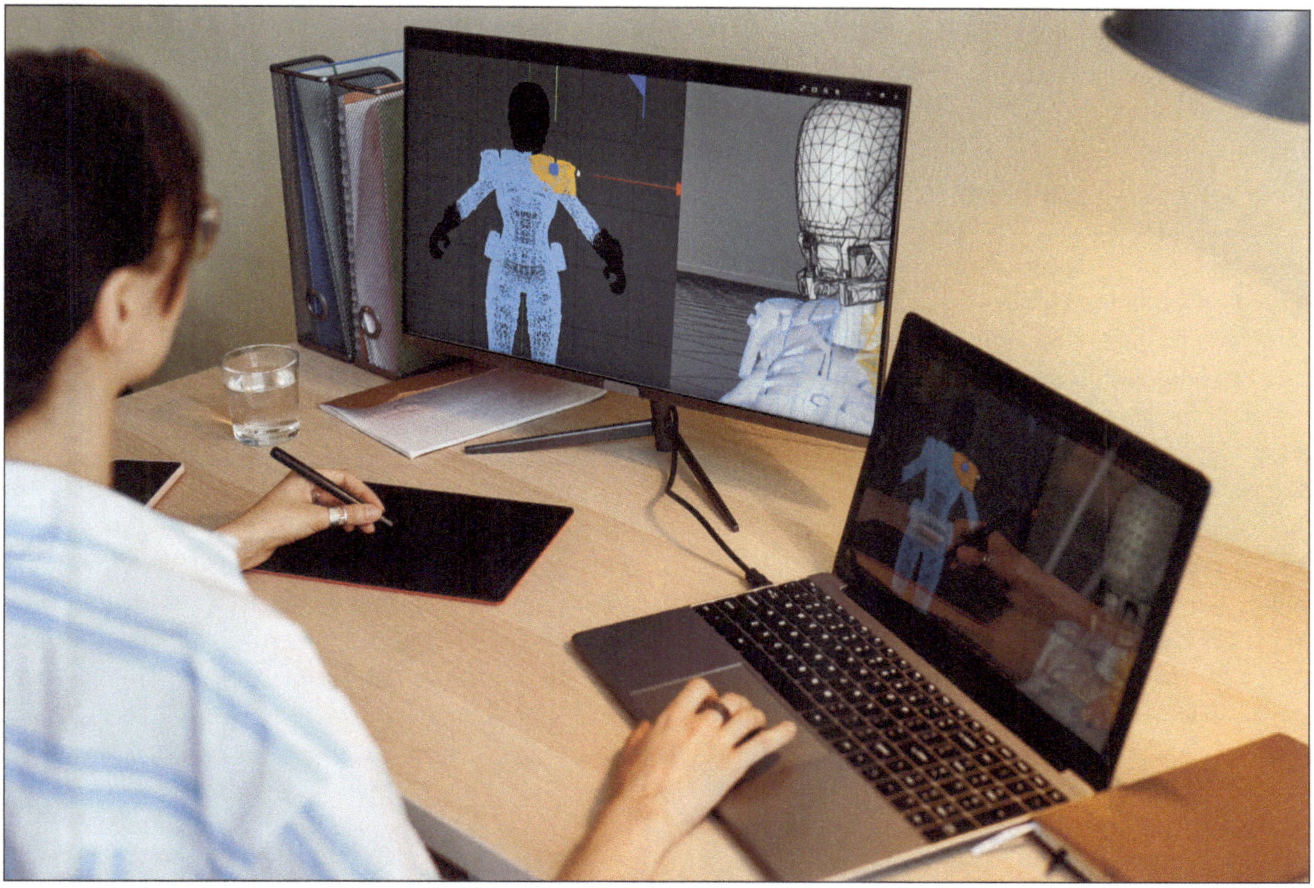

Photo via iStock/SeventyFour. [Used under license.]

designers. Smaller productions may combine several functions into single positions.

WORK ENVIRONMENT

Physical Environment. Video-game designers typically work in an office environment, although those working on a freelance or contract basis may work remotely from home. Designers generally work a forty-hour workweek, although this can be expanded as deadlines approach. The work of game designers is not physically demanding, although the work can be stressful.

Human Environment. Video-game designers often work as part of a design team, so good communication skills are necessary. Members of the design team generally perform different functions, but they need to be able to complement each other and work as a cohesive whole. They may also work with members of other departments, including the programming and art teams, so they must be flexible as well.

Technological Environment. Video-game designers typically do not require the same in-depth programming skills that developers and programmers do, but they do need to be familiar with scripting languages, such as Lua, in order to implement elements into the game prototype. They also must be familiar with a wide range of technically complex video games.

EDUCATION AND TRAINING

High School/Secondary. High school students interested in a career in video-game design should study art, including drawing, photography, and design;

math, including geometry; and computer science, including graphic design and drafting. They should also become experts in a wide range of video games. Learning good communication skills is important at this stage as well.

College/Postsecondary. Although associate's degree programs in game design are available, most employers typically require a bachelor's degree. Degrees in game design, game development, or computer science are helpful. It is necessary that the game designer, at the least, learn the basics of computer programming, particularly scripting languages. The degree of technical knowledge also varies based on the specific type of design the candidate wishes to perform. Level designers, for example, may need to be familiar with 3D modeling programs.

Adult Job Seekers. Video-game designers often get their start working in different areas of the video game world, serving as members of other teams such as quality assurance, programming or art. Learning as many different aspects of the game creation process as possible puts candidates in the best position to secure a job on a design team. Designers might also attempt to break into the industry by designing a small game of their own or modifying an existing game. The designer can then use this work as a portfolio and shop him or herself around.

Professional Certification and Licensure. Video-game designers do not typically pursue certification or licensure, although a certificate in one or more programming languages may be helpful.

Additional Requirements. Video-game designers must be strong team-players and have excellent communications skills. They must be highly creative people but capable of exercising that creativity within a specific framework. They should also possess basic computer programming skills, with a working knowledge of the most common scripting languages.

—*Andrew Schenker*

Further Reading

Berger, Ross. *Dramatic Storytelling & Narrative Design: A Writer's Guide to Video Games and Transmedia*. CRC Press, 2019.

Salmond, Michael. *Video Game Design: Principles and Practices from the Ground Up*. Bloomsbury Academic, 2020.

Wardyga, Brian J. *The Video Games Textbook: History, Business, Technology*. 2nd ed., CRC Press, 2023.

Video-Game Quality Assurance Tester

ABSTRACT

Video-game quality assurance (QA) testers are responsible for making sure that the video games developed by their company function properly. They search for flaws, glitches, or inconsistencies that must be fixed prior to the game's release. Students aspiring to enter the profession should study subjects such as computer science, game development, and programming in addition to familiarizing themselves with a wide range of video games.

BACKGROUND

Video-game QA (quality assurance) testers are responsible for making sure that video games developed by their company function properly. They typically play the games they are assigned exhaustively, inspecting every possibility of game play, to try and find any flaws, glitches, or inconsistencies. Their goal is to try and "break" the game, to find the flaws in the game that prevent its continued functioning. They then log these findings in a bug report and report their findings to the testing lead or to the development team.

OVERVIEW

QA testers work indoors at a gaming console or computer. They may work at the development studio or remotely. Although the job is not physically or mentally demanding, it can be highly repetitive and can prove taxing for this reason. Testers usually work forty hours a week and may need to work overtime when their deadlines approach.

The first requirement of being a QA tester is to be passionate about video games, although this is far from the only prerequisite. An interest in problem solving and puzzles is helpful, as is a strong knowledge of computer science and programming.

DUTIES AND RESPONSIBILITIES

Simply put, it is the responsibility of the QA tester to try and "break" the game. The tester plays the assigned game over and over, from as many angles as possible. As the tester plays, he or she attempts to find any code that doesn't work and that makes the game go on the fritz. They look for glitches and nonfunctioning effects, but they also make sure the game functions the way it is designed to function and that the gameplay unfolds according to plan.

The QA tester is usually assigned a video game that has not yet been released to the public and is then given a deadline to test it. As they play, they are responsible for assessing functionality, content, performance, playability, and hardware compatibility. When they discover any defect, they record it in detail in a bug report, a tracking database. They may also be responsible for debugging and troubleshooting, although this task may be given to other members of the QA team or to the development team. In either case, the tester has to be able to provide a detailed enough report on the bug that others can find and fix it. Once any bugs are fixed, the tester is then responsible for verifying the fixes.

For many more complicated games, one tester cannot cover the whole game, and the work is divided up between several testers, with each responsible for different aspects of the game. In these situations, lead testers often coordinate the team of testers and manage the bug report. The lead tester works closely with the developers and the programmers and is often responsible for compiling representative samples of game footage to send to regulatory bodies. Major development companies generally employ a team of testers.

WORK ENVIRONMENT

Physical Environment. QA testers typically work in an office or a remote office environment. The work can be tedious and repetitive.

Human Environment. QA testers spend much of their time studying video games, but they must also be good communicators, as they frequently interact with other members of the QA team and with other teams as well. They must be able to report their findings and collaborate with other testers and coworkers. Lead testers must also manage a team of other testers and so must be good leaders as well as communicators.

Technological Environment. QA testers need to be highly familiar with complex, sophisticated video games that demand a high level of technical knowledge. They must also have a familiarity with computer programming and development in order to be able to track and describe bugs.

EDUCATION AND TRAINING

High School/Secondary. The most important prerequisite for becoming a quality assurance tester is to become familiar with as wide a range of video games as possible. It is also advisable to learn as much about computer science, developing, and programming as possible, and any courses in these fields offered at the high school level should be taken. Classes in subjects such as English or writing can help improve communications skills and should also be taken.

College/Postsecondary. While most QA testing positions do not require a bachelor's degree, some employers prefer it, particularly for testing leads. Formal education, though, is generally considered less important than is familiarity with the video-game industry. For those who do wish to pursue a bachelor's degree as a way of improving their chances at securing work in the industry, subjects such as computer science and programming, software development, and graphic design may prove useful.

Adult Job Seekers. Most QA testers work for video-game development studios. Job postings are listed on their websites as well as on job sites such as Indeed. The more experience a person has in the video-game field, the more likely they are to succeed in securing a job as a tester. The game industry operates on a cyclical calendar, with the demand for work dependent on the company's seasonal release schedule.

Professional Certification and Licensure. QA testers typically do not have certifications or licenses, though they may be certified in individual programming languages which may enhance their job prospects.

Additional Requirements. The most important quality for a QA tester to possess is a love of and extensive knowledge of video games. In addition, learning specific technical skills, such as programming languages, or writing a bug report, can be very helpful. Candidates may also improve their chances by participating in public beta testing of games or interning at gaming companies. Good communication and problem-solving skills are a must for any candidate.

—*John Pritchard*

Further Reading

Ali, Harun H. *The Pocket Mentor for Video Game Testing.* CRC Press, 2023.

Howell, Chris. *Modern Game Testing: A Pragmatic Guide to Test Planning and Strategy.* Modern Game Testing, 2022.

Isbister, Katherine, and Celia Hodent, editors. *Game Usability: Advice from the Experts for Advancing UX Strategy and Practice in Videogames.* 2nd ed., CRC Press, 2022.

Video-Game Writer

ABSTRACT

Video-game writers compose scripts and narratives for video games. They may be self-employed or may work for a video-game production studio. Students aspiring to enter the profession should complete coursework in English, history, and computer science and familiarize themselves with a wide range of video games.

BACKGROUND

Video-game writers compose scripts and narratives for video games. Working closely with the design team, the video-game writer creates the text that the game follows. Depending on the setting, the writer has more or less involvement in shaping the overall story arc of the game. Video-game writers must be able to think nonlinearly, as they often have to compose the text for a range of possible scenarios that do not necessarily follow an orderly logic.

OVERVIEW

Most video-game writers work in an office environment. They may have an office in the studio of the company for which they are working or, since most video-game writers work on a freelance basis, they may work at a home office, or even a coffee shop.

Video-game writers enjoy working with language and ideas. They like creating a fictional world and bringing the game player into that world. Like all writers, video-game writers are highly creative people and often think about the world in unconventional ways. Video-game writers must also be deeply interested in a wide range of video games and be able to bring to life a number of different game scenarios.

DUTIES AND RESPONSIBILITIES

There are several different types of video-game writers, all of whom focus on different aspects of the game's production. Writers of all types work closely with the video-game designer or creative director. In some productions, the writer is called a narrative designer. In these situations, the narrative designer is given a lot of control in terms of responsibility. He or she is responsible for crafting the narrative arc of the story as well as writing the specific words that shape the story. In many cases, though, the designer or other professionals create the storyline and writers only enact the narrative.

In games where this is the case, the writing work is carried out primarily through script writers. These professionals create both the dialogue spoken by the characters and the text that appears both onscreen and in voice-overs. The script writer may also collaborate with the voice actors to make sure they carry out their dialogue correctly. Script writers often have to write a number of different scenarios that may occur so they have to be highly flexible and think nonlinearly as they may need to approach a single situation from a number of different areas. Among the most important elements to video-game writing is the ability to properly engage in worldbuilding. For many games, the creation of a believable and consistent universe is more important than the narrative. Video-game writers can expect to constantly edit and revise their work as necessary.

Other writing jobs in the video-game industry include pitch writers, who compose pitches for specific video games in order to sell them, and script doctors, who are brought in to tweak scripts or to perform specific smaller tasks.

Image via iStock/SpicyTruffel. [Used under license.]

WORK ENVIRONMENT

Physical Environment. Video-game writers often work on a freelance basis and sometimes work from home. They may also work in the office of the production company for which they are employed. Video-game writing provides little in the way of physical stress, though it can be mentally demanding, particularly as deadlines approach.

Human Environment. Many video-game writers work as part of a team with other writers and must be able to communicate effectively with these other writers. They also typically work under the direction of a game designer or creative director with whom they must collaborate to craft the game's basic narrative.

Technological Environment. Video-game writers, like other writers, typically use word processing software such as Microsoft Word to do their writing. They may also require some knowledge of development software in order to input their text. They should be familiar with a wide range of video games as well.

EDUCATION AND TRAINING

High School/Secondary. High school students can best prepare for a career as a video-game writer by completing coursework in English, history, and computer science. Advanced coursework in a field of particular interest can prepare students for writing knowledgably and coherently about that field. Participation in extracurricular activities such as debate clubs, school papers, or school television and radio programs can also help students develop the skills needed for a career in writing and editing. Students should also familiarize themselves with a wide range of video games.

College/Postsecondary. Video-game writing positions typically require a bachelor's degree. Recommended fields of study include creative writing, English, literature, computer science, and communications. The most important requirement for the position is strong creative writing skills, so candidates must ensure these are top notch, either through coursework or independent practice.

Adult Job Seekers. Employers typically look for people with extensive writing experience when hiring a game writer. Most game writing candidates maintain a professional portfolio with clips of their best writing. Many game writing jobs also require one or more professional writing credits. These typically can be in any field of writing. Additional experience as a copywriter or editor or in another aspect of the video-game world can help a candidate stand out in a crowded field.

Professional Certification and Licensure. Certification or licensure is not required to be employed as a video-game writer.

Additional Requirements. Video-game writers must possess a love of the language and a commitment to quality writing. They should study as many games as possible, analyzing the game's narrative and dialogue. They should be well-read in a variety of genres and should be versatile as a writer.

—*Andrew Schenker*

Further Reading

Berger, Ross. *Dramatic Storytelling & Narrative Design: A Writer's Guide to Video Games and Transmedia.* CRC Press, 2019.

Nicklin, Hannah. *Writing for Games: Theory and Practice.* CRC Press, 2022.

Wardyga, Brian J. *The Video Games Textbook: History, Business, Technology.* 2nd ed., CRC Press, 2023.

VIDEO-ON-DEMAND (VOD)

ABSTRACT

Video-on-demand (VoD) is technology that supplies video content to users upon request. They can then view the content immediately or at some later time. Furthermore, users can view the content on many different types of devices, from televisions to smartphones to gaming consoles. To access VoD content, users must have a subscription to one of the many existing VoD providers. VoD has become popular throughout the world, particularly in the United States.

BACKGROUND

Voice-on-Demand (VoD) content is offered through specific providers. Generally, individuals need to purchase a monthly subscription to such a provider to access that provider's VoD content. The content, which typically includes movies and television shows, is stored on servers in digital format. When users choose the content they would like to view, the server then distributes that content to them.

VoD content can be either streamed for viewing in real time (immediately) or downloaded for viewing in the future. Whichever option users choose, they can view the content on numerous types of devices, including smart televisions, computers, tablets, smartphones, and gaming consoles. This means that users can watch the content anytime and anywhere they choose. Furthermore, users are able to fast forward, rewind, and pause the content.

Although VoD is mostly used for entertainment purposes, it can also be used for educational purposes. For example, companies can make use of VoD training videos for their employees. VoD can also be used during a videoconference, which is a live conference that typically includes the transmission of audio, video, images, and text between two or more people in different locations. VoD can provide richer presentations during a videoconference.

OVERVIEW

Many VoD providers are available. Netflix is one of the world's most popular, with 232.5 million subscribers worldwide as of the first quarter of 2023. The provider offers a wide array of VoD movies and television shows. In addition to content created by other production companies, Netflix also offers original programming, including television series such as *Stranger Things*, *Emily in Paris*, and *The Crown*. With an internet connection, subscribers can stream any movie or show on a variety of devices.

As of 2023, the cost of subscribing to Netflix was $15.49 per month for the standard plan. For that price, subscribers could stream an unlimited amount of content. The premium plan cost $19.99 per month and included the ability to stream in high definition (HD) and on multiple devices. In addition to those plans, Netflix offered an advertising-supported plan for $6.99 per month beginning in late 2022.

In 2019, the Walt Disney Company launched its own VoD service called Disney+ (sometimes referred to as Disney Plus). The service provided a platform for all classic Disney content, as well as its newer acquisitions, such as films and television series within the Marvel and Star Wars franchises. By the first quarter of 2023, Disney+ had topped 161 million subscribers and emerged as a serious challenger to Netflix. Among other plans, Disney+ offered a bundle deal that gave users access not only to Disney+ but also to the VoD service Hulu and the sports-oriented streaming service ESPN+.

In addition to Netflix, Disney+, and Hulu, popular VoD providers of the 2020s included Amazon Prime Video, Apple TV+, Max (previously HBO Max), the NBCUniversal-owned service Peacock, and Paramount Plus. Alongside offering VoD content, some services expanded their operations, offering streaming subscribers access to traditional television channels, live sports broadcasts, and other content.

INFLUENCE

VoD has had a tremendous impact on the home entertainment industry, especially in the United States. As of 2022, about 85 percent of American homes had at least one subscription to a VoD provider. Not only do many Americans have a subscription to a VoD provider, but many have done away with—or

Photo via iStock/simpson33. [Used under license.]

simply have never had—a subscription to a conventional television programming provider such as a cable or satellite provider. In addition to appealing to viewers, some original VoD programs and films earned critical acclaim and nominations for major awards, signaling a shift in the cultural perception of prestige television and film.

—Michael Mazzei

Further Reading

Anderson, Casey. "What Is VOD Technology?" *Chron*, smallbusiness.chron.com/vod-technology-14311.html. Accessed 25 July 2023.

Brachmann, Steve. "Video on Demand Continues to Revolutionize TV, Movie Industries." *IPWatchdog*, 28 Mar. 2015, www.ipwatchdog.com/2015/03/28/video-on-demand-revolutionize-tv-movie/id=55900. Accessed 25 July 2023.

Iqbal, Mansoor. "Disney Plus Revenue and Usage Statistics (2023)." *Business of Apps*, 24 May 2023, www.businessofapps.com/data/disney-plus-statistics. Accessed 25 July 2023.

Jackson, Kourtnee. "Disney Plus vs. Netflix: How to Narrow Down to One Streaming Service." *CNET*, 19 Nov. 2022, www.cnet.com/tech/services-and-software/disney-plus-vs-netflix-how-to-narrow-done-to-one-streaming-service. Accessed 25 July 2023.

"Number of Disney Plus Subscribers Worldwide from 1st Quarter 2020 to 2nd Quarter 2023." *Statista*, May 2023, www.statista.com/statistics/1095372/disney-plus-number-of-subscribers-us. Accessed 25 July 2023.

"Number of Netflix Paid Subscribers Worldwide from 1st Quarter 2013 to 2nd Quarter 2023." *Statista*, July 2023, www.statista.com/statistics/250934/quarterly-number-of-netflix-streaming-subscribers-worldwide. Accessed 25 July 2023.

"What Is Netflix and How Does It Work?" *TechBoomers*, 15 May 2016, techboomers.com/t/what-is-netflix. Accessed 13 July 2023.

Zandt, Florian. "The Most Popular Streaming Services in the US." *Statista*, 20 July 2023, www.statista.com/chart/25382/most-used-video-streaming-platforms. Accessed 13 July 2023.

Virtual Reality

ABSTRACT

The applied science of virtual reality (VR) engages in the design and engineering of and research related to special immersive interactive computer systems. These VR systems synthesize environments, or worlds, which are simulations of reality that are usually rendered using three-dimensional (3D) computer images, sounds, and force feedbacks. Virtual reality applications are used for pilot and astronaut training, entertainment, communication, teleoperation, manufacturing, medical and surgery training, experimental psychology, psychotherapy, education, science, architecture, and the arts. This technology, which submerges humans into altered environments and processes, intensifies experience and imagination, thereby augmenting research and education. Virtual reality training systems can simplify and improve manufacturing and maintenance, while simultaneously reducing risk exposure.

BACKGROUND

A virtual reality (VR) system is an interactive technology setup (software, hardware, peripheral devices, and other items) that acts as a human-to-computer interface and immerses its user in a computer-generated three-dimensional (3D) environment. Virtual reality is the environment or world that the user experiences while using such a system. Although the term "virtual" implies that this simulated world does not actually exist, the term "reality" refers to the user's experience of the simulated environment as being real. The more that the senses are involved in a compelling fashion, the more genuine the perceived experience will be, and the more intense the imagination. Most VR systems stimulate the senses of sight, hearing, touch, and other tactile-kinesthetic sense perceptions, such as equilibrioception, torque, and even temperature. Less often they include smell, and some have experimented with taste. Virtual reality must be almost indistinguishable from reality in some applications such as pilot training, but it can often differ significantly from the real world in, for example, games.

Virtual reality in a narrow sense (a computer-generated simulation that exists virtually but not materially) is not the same as augmented reality (enhanced reality) or telepresence (in the sense of teleconferencing). Augmented reality technology improves the perception of and supplements the knowledge about existing entities or processes (highlighting data of interest while abstracting the less important information). Telepresence (as teleconferencing) refers to a remote virtual re-creation of a real situation (e.g., to enable audio and visual interactions between people at diverse places). However, a wider notion of VR includes notions of both augmented reality and telepresence.

The idea of simulated reality is often traced back to the ancient Greek philosopher Plato's allegory of the cave in *Politeia* (fourth century BCE; *Republic*, 1701). In the allegory, spectators observe images (shadows) of objects on a cave wall that they take for real objects. The term "virtual" is derived from the Latin word "virtue" (which means goodness or manliness). "Virtual" then means existing in effect or in essence but not in actuality. The notion of VR can be traced back to the French theater director, actor, playwright, and illustrator Antonin Artaud who described theater as *la réalite virtuelle* in his influential *Théâtre et son double* (1938; *The Theatre and Its Double*, 1958). Computer scientist and artist Myron W. Krueger coined the technical term "artificial reality" in his book of the same title, published in 1983.

Computer scientist and artist Jaron Lanier popularized the notion of VR as a technical term in the 1980s. Many artists, science-fiction authors, and directors incorporated concepts of computer-generated, simulated, or augmented reality in their creations. One of the most prominent examples is the holodeck, an advanced form of VR featured on the television program *Star Trek: The Next Generation* (1987-94).

Virtual reality technology dates back as far as the 1960s. Late in that decade, computer scientist Ivan Sutherland created the first head-mounted device, which was capable of tracking the user's viewing direction. In the 1970s, Sutherland and David Evans developed a computer graphic scene generator. In the 1970s and 1980s, force feedback was incorporated into tactile input devices such as gloves and interactive wands. Lanier and Thomas Zimmerman developed sensing gloves, which recognized finger and hand movements.

This kind of VR technology was intended to improve flight simulators and applications for astronaut training. In 1981, the National Aeronautics and Space Administration (NASA) combined commercially available Sony liquid crystal display (LCD) portable television displays with special optics for a prototype stereo-vision head-mounted device called the Virtual Visual Environment Display (VIVED). NASA then created the first VR system, which combined a host computer, a graphics computer, a noncontact user position-tracking system, and VIVED.

Scott Fisher and Elizabeth Wenzel developed hardware for 3D virtual sound sources in 1988. Also in the late 1980s, Lanier built a VR system for two simultaneous users that he named Reality Built for Two (RB2). Fisher incorporated sound systems, head-mounted device technology, and sensor gloves into one system called the Virtual Interactive Environment Workstation (VIEW), also used by NASA.

Photo via iStock/Damir Khabirov. [Used under license.]

The first conference on VR, "Interface for Real and Virtual Worlds," was held in Montpellier, France, in March 1992. That same year, scientists, engineers, and medical practitioners assembled in San Diego, California, for the "Medicine Meets Virtual Reality Conference." In September 1993, the Institute for Electrical and Electronics Engineers (IEEE) organized its first VR conference in Seattle.

The first commercial VR video games debuted in the 1990s, but attempts by both SEGA and Nintendo to make the technology profitable failed. In 2012, the company Oculus Rift brought virtual-reality gamers to personal computer (PC) gamers. The company proved successful and was later bought by Facebook (later known as Meta) for $2 billion. The technology exploded in popularity in the late 2010s, and by 2023, devices such as the Meta Quest 2 and the HTC Vive Pro 2 had become leaders within the growing VR device market.

OVERVIEW

To create a realistic computer-generated world, several high-end technologies must be integrated into a single VR system. This kind of high-end system is used at university, military, governmental, and private research laboratories. Adequate computing speed and power, fast image and data processors, broad bandwidth, and sophisticated software are essential. Other requirements include high-tech input-output devices or effectors (such as head-mounted devices), 3D screens, surround-sound systems, and tactile devices (such as wired gloves and suits, tracking systems, and force feedback devices, including motion chairs and multidirectional treadmills).

Input. The input devices used for VR systems—sensing gloves, trackballs, joysticks, wands, treadmills, motion sensors, position trackers, voice recognizers, and biosensors—are typically more complex than those used for personal computers. Biosensors recognize eye movement, muscle activity, body temperature, pulse, and blood pressure, all of which are vital to surgical applications. Position trackers and motion sensors identify and monitor the user's position and movements. The tracking systems used in VR systems are mechanical, optical, ultrasonic, or magnetic devices. Steering wheels, joysticks, or wands are used for pilot training and games. Sophisticated devices for research and experiments (such as those for molecular modeling) offer six-degrees-of-freedom input. Such input devices allow the computer to adjust the virtual environment according to the data received from the user. When motion sensors and position trackers detect the user's movement, the data are processed by the computer in real time, and the display, also in real time, has to accurately render the image (such as the interior of a building). If slow data processing creates a time lag, the user may experience simulator sickness or motion sickness (nausea or dizziness), especially if the user's senses register conflicting data. In a VR parachute training session, for example, equilibrioception might be in conflict with visual perception if movement is represented faster by the display than by the force feedback system.

Output. Output devices are intended to stimulate as many senses as possible for a high degree of immersion. The important visual output devices are head-mounted devices, LCDs, and projectors and screens. Developers compete to make these displays the most immersive. All displays must be able to render 3D images. Other output devices are for sound and touch. Sound systems can assist in conveying the impression of 3D space. Force feedback systems give the user the sense of physical resistance (essential in surgical VR systems), torque, tilt, and vibration (appreciated for games and essential for pilot training).

The Cave Automatic Virtual Environment (CAVE) is a surround-sound, surround-screen, projection-based room-sized VR system developed by the Electronic Visualization Laboratory at the University of Illinois at Chicago (the name is trademarked by the University of Illinois Board of Regents). Users put on lightweight stereo glasses and walk around inside the room, interacting with virtual objects. One user is an active viewer, controlling the projection, and the others are passive viewers, but all users can communicate while in the CAVE. The system was designed to help with visualization of scientific concepts.

Hardware and Software. Standard personal computers have both limited memory capacity and limited performance capability for running professional VR applications. Therefore, high-end hardware and software have been developed for specific purposes such as games, research programs, and pilot, combat infantry, and medical training. Computer languages such as C++, C#, Java, JavaScript, and Python are used for programming VR software. Virtual reality computers handle tasks such as data input and output and the interaction, integration, and recomposition of all the data required in virtual environment management. Because a VR system needs high computing, processing, and display speeds, the VR computer architecture can at times use several computers or multiple processors.

APPLICATIONS AND PRODUCTS

Aerospace and Military. One of the early applications of VR was in pilot training. Twenty-first-century flight simulators are convincingly close to real flight experiences, although the simulation of acceleration and zero gravity are still challenges. In the military, VR applications are used not only by pilots, paratroopers, and tank drivers but also by battle strategists and combat tacticians to enhance safety training and analyze battle maneuvers and positions. These military applications make targeting more precise, thereby reducing human casualties and collateral damage. Virtual reality systems are also used to evaluate new weapons systems. In space exploration, VR systems help astronauts prepare for zero-gravity activities such as the repair of solar panels on the outside of a spaceship. Virtual reality training systems give the user the opportunity to review and evaluate specific sequences in a training session or the entire session.

Entertainment and Games. The military took commercial games and adapted them to create flight simulators and other professional VR applications, which, in turn, were adapted to use as games. By the beginning of the 2020s, increased interest in VR game applications and the virtual games industry meant that some arcades dedicated to VR were popping up in countries worldwide, and they were also making significant headway into the home and in the mobile entertainment market. While several companies had become involved in the VR gaming market and had produced more advanced VR systems and headsets by that time, more work was being done on creating the right kind of games to complete the best virtual gaming experience.

Some companies lease VR game equipment (such as training applications for golf, racing, and shooting) to customers for entertainment or to corporations for internal team-bonding experiences. Some VR applications enable users to journey through fantastic and futuristic worlds. The common equipment for VR games—depending on the quality and level of sophistication of the games—has included head-mounted devices, multiple screens, tracking systems, omnidirectional treadmills for walking, force feedback or rumbling seats (or platforms) for flight and race simulations, batons for tennis, and guns, wands, or sticks for shooting. The more sensory feedback provided, the more immersive the VR game application.

Education. The more senses that are involved in the process of learning and training, the more a student is able to become engaged in the subject matter and the better the educational impact, especially when it comes to learning skills or practical content. The employment of a VR system enables data and images to become interactive, colorful, 3D, and accompanied by sound. For example, a visitor to a virtual museum is able to virtually touch the artifacts, taking them from their shelves, turning them around, and viewing them from different perspectives. In an immersive experience, a user can witness, seemingly first-hand, historical events such as famous battles as they take place. Virtual reality applications exist in almost every area of education, including sports (such as golfing), sciences (such as astronomy, physics, biology, and chemistry), the humanities (such as history), and vocational training (such as medical procedures and mechanical engineering techniques).

Art. Although classic artworks can be immersive, few of them are interactive. People are not allowed to touch exhibits in most art galleries and museums. In 1993, the Solomon R. Guggenheim Museum became the first major museum to dedicate an entire exhibition to VR art. The exhibition featured VR installations from Jenny Holzer, Thomas Dolby, and Maxus Systems International. Virtual reality technology enables artists to blur or combine genres (such as music, graphic arts, and video) and involves the viewer in the creation of art. Every viewer or user of the artwork perceives the work in a different way depending on his or her input into the VR work of art. A VR art exhibit can be programmed to produce sound or visual feedback according to, for example, a visitor's footsteps or voice input from an audience. Artworks thus become interactive, and viewers become cocreators of the art. VR technology is also utilized by multimedia artists and animators for content creation and making 3D storyboards.

Science, Engineering, and Design. Virtual reality systems can advance the creative processes involved in science, engineering, and design. Results can be tested, evaluated, shared, and discussed with other VR users. Chemists at the University of North Carolina, Chapel Hill, used VR systems for modeling molecules, and similar systems have been used to "observe" atoms. Buildings, automobiles, and mechanical parts can be designed on computers and evaluated using 3D modeling tools and visualization techniques in a VR environment. An architect can take a client on a virtual walk through a building before it is constructed, or the aerodynamics of a new automobile design can be evaluated before manufacturing a model or prototype. Virtual reality applications also can simulate crash tests. The use of VR applications reduces cost, waste, and risk.

Business. Business applications include stores on the internet featuring virtual showrooms and 360-degree, 3D views of products; this 360-degree view capability was also increasingly used in the real estate industry in the late 2010s and into the 2020s. Another important application is teleconferencing. In contrast to the traditional conference call, VR applications may permit multisensory evaluations of new products. Virtual reality systems also help network, combine, and display data from diverse sources to analyze financial markets and stock exchanges. In these situations, VR applications serve as decision support systems.

Medicine, Therapy, and Rehabilitation. Experienced surgeons as well as physicians in training use VR systems to practice surgery. The images for such training programs are taken from X-rays, computed tomography (CT) scans, and magnetic resonance imaging (MRI). Virtual operations can be recorded and repeated as many times as desired, in contrast to practice operations on animals or human corpses, which cannot be repeated. Force feedback gloves give practitioners the realistic feel and touch needed, for example, to determine how much force is needed for certain incisions.

Virtual reality environments are being developed for use in training patients as well as doctors. Virtual human limbs are being prototyped and studied for use in training patients to use and become comfortable with prosthetic limbs, as well as for patient rehabilitation. Some of the virtual prosthetics in development appear as avatars in a VR environment and are able to accept both kinematic and neural control inputs from the patient.

In psychotherapy, VR applications are sometimes used as an alternative therapy to help treat clients with disorders such as phobias, anxiety, and post-traumatic stress disorder (PTSD). Clients often undergo desensitization treatment, which is especially useful in treating phobias and PTSD. Virtual reality allows clients who fear things such as enclosed spaces (claustrophobia), dirt and germs (mysophobia), or snakes (ophidiophobia) to gradually confront their fears without being exposed to the actual condition or object. Virtual reality applications can also help clinicians understand certain psychological problems better by enabling them to experience what the client or the patient experiences. For example, a psychiatrist, psychologist, or counselor may take a virtual bus ride in the role of a schizophrenic client, during which they experience some simulated symptoms of this disorder, such as distorted images viewed through the windows of the bus or strange voices that seem to appear from nowhere.

CAREERS AND COURSEWORK

Because the development of VR systems and applications involves many areas of applied science, students pursuing careers related to VR systems should be versatile. A bachelor of science in a field such as information technology, mechanical engineering, or electrical engineering is an appropriate foundation for a career in VR, but most positions in the field require multidisciplinary skills and experience with designing and developing 3D software. Therefore, the best approach is to take additional courses in other fields.

For high-level positions or to work as a researcher in industry or academia, a master of science degree or a doctorate is recommended. Work can also be found with governmental entities (such as NASA) and the military as well as manufacturers of VR systems and products (such as head-mounted devices, screens, force feedback systems, and tracking systems). The producers of VR software and hardware need skilled professionals such as electrical and mechanical engineers, roboticists, software developers, and information technologists. As of 2023, job opportunities related to VR technology and research are still somewhat limited because of the field's highly technical and specialized nature and also because VR applications are not yet in widespread use. It is generally believed, however, that as the VR industry continues to grow, product managers and business development/sales representatives will be needed in addition to programmers and engineers.

SOCIAL CONTEXT AND FUTURE PROSPECTS

Some experts believe that VR, like the computer and the internet, soon will become a commonplace and indispensable part of everyday life, but others do not see the potential for such a wide implementation. Most agree, however, that once computing speed and power, broad bandwidth, and peripheral systems become more affordable for average consumers, VR will be more widely used. Beyond science, education, and other professional applications, the entertainment industry is thought most likely to want affordable VR innovations. Critics of VR applications point to personal and societal risks such as isolation, desocialization, and alienation, but advocates emphasize the technology's proven potential for augmenting people's lives. Both critics and advocates agree that experiences in VR can alter people's perceptions of and responses to the real world. These changes are intentional and welcome in most cases but sometimes they take place in an unintended and potentially dangerous manner. Airplane pilots have reportedly made mistakes that could be linked to the limitations of training with a flight simulator, which, for example, is incapable of realistically simulating acceleration. However, VR applications are valuable in highly technical and precise areas of medicine, industry, business, and research and are likely to gain more users, despite the potential risks and the expense.

By the 2020s, the amount of wide-scale investment that had been made by a number of companies in the development of VR technology had resulted in greater accessibility to lower-cost systems. After the coronavirus disease 2019 (COVID-19) pandemic was declared in March 2020, analysts noted that the versatility of the technology had become even more prominent as people remained at home more and reduced physical interactions in an effort to slow the spread of the virus. While businesses such as museums turned more to virtual, online tours during lockdowns, the application of VR in the area of mental health was especially highlighted, as the technology was viewed as a beneficial alternative to in-person therapy.

—Roman Meinhold

Further Reading

Alsop, Thomas. "Virtual Reality (VR)—Statistics & Facts." *Statista*, 8 June 2023, www.statista.com/topics/2532/virtual-reality-vr. Accessed 13 July 2023.

Al-Jumaily, A., and R. A. Olivares. "Bio-Driven System-Based Virtual Reality for Prosthetic and Rehabilitation Systems." *Signal, Image and Video Processing*, vol. 6, no. 1, 2012, pp. 71-84.

Bailey, David Evans. "Ten Cool Applications for Virtual Reality That Aren't Just Games." *Conversation*, 22 Mar. 2016, theconversation.com/ten-cool-applications-for-virtual-reality-that-arent-just-games-56365. Accessed 13 July 2023.

Craig, Alan B., William R. Sherman, and Jeffrey D. Will. *Developing Virtual Reality Applications. Foundations of Effective Design*. Morgan Kaufmann, 2009.

Donally, Jaime. *The Immersive Classroom: Create Customized Learning Experiences with AR/VR*. International Society for Technology in Education, 2021.

Loeffler, John. "The History and Science of Virtual Reality Headsets." *Interesting Engineering*, 28 Feb. 2019, interestingengineering.com/the-history-and-science-of-virtual-reality-headsets. Accessed 13 July 2023.

Narula, Herman. *Virtual Society: The Metaverse and the New Frontiers of Human Experience*. Currency, 2022.

Paris, Francesca. "A Strange New World: Has Virtual Reality Gaming Lived Up to Its Promise?" *WBUR*, 24 Feb. 2020, www.wbur.org/hereandnow/2020/02/24/virtual-reality-orion-13-games. Accessed 13 July 2023.

Rizzo, Albert "Skipp," and Stéphane Bouchard, editors. *Virtual Reality for Psychological and Neurocognitive Interventions*. Springer-Verlag New York, 2019.

Spiegel, Brennan. "Virtual Reality and the COVID Mental Health Crisis." *Scientific American*, 15 Nov. 2020, www.scientificamerican.com/article/virtual-reality-and-the-covid-mental-health-crisis. Accessed 13 July 2023.

"US VR and AR Users, 2019-2023." *Insider Intelligence | eMarketer*, 22 Feb. 2021, www.emarketer.com/chart/244851/us-vr-ar-users-2019-2023-millions. Accessed 13 July 2023.

Vince, John. *Introduction to Virtual Reality*. Springer-Verlag London, 2004.

Weiss, Patrice L. (Tamar), Emily A. Keshner, and Mindy F. Levin, editors. *Virtual Reality for Physical and Motor Rehabilitation*. Springer-Verlag New York, 2014.

Voice Actor

ABSTRACT

Voice actors lend their voices to a variety of media, ranging from films to video games. They are typically self-employed and work in a studio environment. Students aspiring to enter the profession should pursue studies in subjects such as speech and the dramatic arts.

BACKGROUND

Voice actors are creative professionals who lend their voices to a variety of media, including movies, television shows, audiobooks, video games, podcasts, and web videos. They typically provide the voices for animated characters as well as narration. Above all, voice actors must be versatile, as they may be called upon to take on a variety of assignments. Many voice actors have a background in live acting as well.

Photo via iStock/elebeZoom. [Used under license.

OVERVIEW

Voice actors generally work in a studio environment where they record the voiceovers in a controlled situation. Although they typically work on a freelance basis, most voice actors are based in southern California or in large cities with vibrant entertainment industries, such as New York.

Acting in general, including voice acting, requires a deep commitment and a creative personality. Actors are typically interested in literature and a wide range of art forms. They enjoy performing and are physically expressive.

DUTIES AND RESPONSIBILITIES

Although some voice actors voice long-running characters, for example on popular television cartoons, most voice work is piecemeal and inconsistent. Voice actors work in a variety of settings, and the specific nature of the work depends on the type of setting. Among the most common industries in which a voice actor may work are film, television, audiobooks, and video games. Voice actors may voice specific characters, may provide voiceover narration, or, in the case of an audiobook, simply read a text. Voice actors must be highly expressive and bring the character or text to life.

In films, television shows, and video games, actors may provide several types of voice work. Working under the supervision of the audio engineer, the voice actor may provide the project's narration, whether it is delivered by a main narrator—a character within the film, series, or game—or an absent narrator. The voice actor may also voice a character within the project and must often possess a strong understanding of the character's motivations and backstory, which enables the actor to bring the character to life. Films, television projects, and video games typically employ a range of voice actors to portray the different characters.

WORK ENVIRONMENT

Physical Environment. Most voice actors work in a controlled studio environment. Some voice actors may work from a home studio, but most often the studio is owned by the production company, publisher, or other entity responsible for producing the property.

Human Environment. Voice actors work extensively with directors and audio engineers who supervise their work. In some cases, they must also interact with other voice actors, participating together in the creation of a scene.

Technological Environment. Voice actors use a variety of microphones and other audio equipment. Although they are not intimately involved in the technical aspects of the production, they need to know how different equipment alters their vocal qualities and to use this to their advantage.

EDUCATION AND TRAINING

High School/Secondary. High school students interested in pursuing a career in voice acting should study literature, writing, and the dramatic arts. They should also participate in as many extracurricular theater groups and school performances as possible, in order to gain acting experience before finishing their education. They should take courses in speech, speechmaking, and other performance-related classes.

College/Postsecondary. Most voice actors have a background in live acting and so pursue the same course of studies as other actors, including coursework in theater arts, dramatic arts, or English literature in college or at a conservatory. They also audition for and perform in university plays, musicals, or other projects to supplement their studies and to gain additional performance experience. During the summer, many postsecondary actors enroll in regional theater or repertory companies, where they perform in various plays and often earn points toward their equity memberships.

After obtaining an undergraduate degree, actors can enroll in a master's degree or master of fine arts program at select universities, where they can continue to study their craft while meeting and interacting with other actors and faculty members in the entertainment industry.

Adult Job Seekers. Adults seeking employment in voice acting often move to Los Angeles or to other large metropolitan areas, such as New York, where the majority of work is located. Voice actors often have a background in other forms of acting. To get their foot in the door, most voice actors create a demo tape for which they will need access to professional recording equipment. Many voice actors have agents who can help get them work, but many more do not. Contacting studios or video-game development companies directly and sending them a demo tape can be a good way to make connections.

Professional voice actors usually seek membership with labor unions to receive benefits and protection. The Screen Actors Guild-American Federation of Television and Radio Artists (SAG-AFTRA) represents voice actors and advocates for members to receive guaranteed minimum rates, health insurance, and retirement.

Professional Certification and Licensure. Actors who perform professionally may become members of one of the unions affiliated with the Associated Actors and Artists of America. These include the Screen Actors Guild, the Actors' Equity Association, or the American Federation of Television and Radio Artists. Those who are union members must pay periodic dues.

Additional Requirements. The field of voice acting is highly competitive and unpredictable, with a small percentage of professional actors actually obtaining work each year. Voice actors must be patient and willing to supplement their acting work with other, less desirable temporary and odd jobs until their careers progress.

—*Andrew Schenker*

Further Reading

Alburger, James R. *The Art of Voice Acting: The Craft and Business of Performing for Voiceover*. 7th ed., Focal Press, 2023.

Clark, Elaine A. *Voice-Overs for Podcasting: How to Develop a Career and Make a Profit*. Allworth, 2020.

Lowenthal, Yuri, and Tara Platt. *Voice-Over Voice Actor: The Extended Edition*. Bug Bot Press, 2018.

W

Web Developer

ABSTRACT

Web developers generate the look and feel of a website. They are also responsible for a website's performance and capacity. Students aspiring to enter the profession should pursue studies in subjects such as web design, computer science, graphic design, and communication.

BACKGROUND

Website developers generate the look and feel of a website according to the client's needs and the principles of a chosen design theme. These developers create the public appearance, functionality, and general design, or interface, of websites. Website developers brainstorm both the creative and technical design of websites. They design a basic web page structure, or architecture, and select fonts, colors, graphics, and other visual elements, to apply creatively to that architecture. They are also responsible for a website's performance and capacity, meaning how fast it runs and how much data and user traffic it can handle.

OVERVIEW

Having a company or organizational website has become nearly universal. Web developers work in many settings, including software or graphic design firms, advertising agencies, and large and small corporations. They may also be employed in government, hospitals, schools, and colleges, and a multitude of different types of organizations. Some are self-employed. In larger firms or departments, the website developer is part of a team of creative and technical professionals who share responsibility for websites. Their "clients" are divisions or departments inside the company. Web developers spend their days working on computers. The forty-hour work week is most prevalent, although sometimes developers must work long hours to meet deadlines.

Designing websites requires creativity, color and design skills, technical knowledge about computer code languages and software, and strong organizational skills. Developers must also communicate effectively with team members and clients. They must keep up with technology and trends and be willing to update websites as necessary. They are fast, creative problem-solvers who understand the needs of business and their clients. Many also enjoy the competition inherent in their industry.

DUTIES AND RESPONSIBILITIES

Sometimes web developers begin work with the acquisition of a domain name (web address) and a suitable website host provider (server). Other projects might consist of updating an existing website. These and other details are usually worked out during meetings with clients and may also be discussed with colleagues.

One of the most important steps in designing a new site is planning its structure, or architecture. Before beginning work on the actual page, the developer compiles a list of all the necessary components, including databases, shopping carts, calendars, and directories. He or she then decides how each element will best fit into the overall structure of the website.

The website developer next creates an attractive layout for individual pages. This might involve coding a cascading style sheet (CSS), which allows colors, fonts, and other aesthetic elements to be automatically applied to all pages. The developer imports files, such as the company logo, graphics, and navigational buttons, to the website and arranges them in the layout. Sometimes the website developer creates these graphics files.

The website developer might then add coding language to make the page dynamic, mapping out illustrations, adding hyperlinks to text, and so on. Alternatively, he or she might create the site in a "what you see is what you get" (WYSIWYG) editor that requires less technical knowledge. The website developer ensures that links work properly and that the site displays correctly on various monitors, with different browsers, and on varied mobile devices. When the developer is satisfied with the work and the client has approved it, the developer publishes the website on the internet. Additional responsibilities of a website developer sometimes include website maintenance, depending on the client's expectations.

OCCUPATION SPECIALTIES

Web Architects or Programmers. Web architects or programmers are responsible for the overall technical construction of the website. They create the basic framework of the site and ensure that it works as expected. Web architects also establish procedures for allowing others to add new pages to the website and meet with management to discuss major changes to the site.

Web Designers. Web designers are responsible for how a website looks. They create the site's layout

Photo via iStock/valentinrussanov. [Used under license.]

and integrate graphics, applications (such as a retail checkout tool), and other content into the site. They also write web-design programs in a variety of computer languages, such as hypertext markup language (HTML) or JavaScript.

Webmasters. Webmasters maintain websites and keep them updated. They ensure that websites operate correctly and test for errors such as broken links. Many webmasters respond to user comments as well.

WORK ENVIRONMENT

Physical Environment. Web developers typically work in offices or studios, either alone or with other developers and programmers. They spend long hours working at a computer. Self-employed web developers frequently work out of home offices.

Human Environment. Most web developers either report to art directors or technical managers. They may supervise part-time staff or interns. In many cases, they work on teams that include illustrators, photographers, videographers, copywriters, and programmers. Web developers who work for a large company or organization may interact with marketing and advertising specialists and the many different people who are responsible for web content. Self-employed web developers necessarily are the sole point of contact for clients.

Technological Environment. Web developers should be familiar with HTML, CSS, WYSIWYG editors, and other basic website design programming languages and tools. They should be comfortable using art-creation and graphic-design programs, basic office software, and various operating systems, browsers, and displays. Developers must use scanners, printers, digital cameras, smartphones, and other electronic equipment. Those who do more substantive programming should learn more specialized skills in other programs. Some developers need to be familiar with web analytics, while others may need to know computer animation and modeling programs.

EDUCATION AND TRAINING

High School/Secondary. Aspiring web developers may benefit from taking a college preparatory program with an emphasis in English, mathematics, speech communication, and computer science, and additional electives in graphic design, fine art, photography, video, and other subjects that develop the imagination. Learning new techniques and computer programs outside of school hours is essential. Prospective web developers should consider volunteer or part-time work designing websites for local individuals and businesses.

College/Postsecondary. There is no specific postsecondary degree or certificate required by all employers; however, most employers prefer some type of certification. Building a professional portfolio is vital. There are many different learning opportunities that will meet individual employment needs and provide the education needed to prepare an attractive portfolio. Programs in web design are offered through college continuing education programs and in business, technical, and commercial art schools.

Students may opt to enter an associate's or bachelor's degree program in graphic design with an emphasis in web design. They can also pursue an undergraduate degree in computer science or information technology, with additional courses in art and design. Disciplines such as business or marketing can also be advantageous areas of study for aspiring web developers. Independent study in website design, internships, workshops offered by software developers, and distance education courses are other options.

Adult Job Seekers. Web design can be an attractive occupation for adults, especially those with some design or computer aptitude and an interest in working from home. Those who need to update their skills or learn new techniques can choose from many courses offered in the evenings, weekends, or online.

Advancement is dependent upon experience, education, and talent. Developers who acquire more specialized software skills may be given more sophisticated, prestigious jobs or additional responsibilities, such as programming or creating multimedia content. Some developers may move into supervisory positions or start their own design firms.

Professional Certification and Licensure. There are no mandatory licenses or standardized certificates; however, individual schools, vendors, and professional associations offer various certificates. Some of these meet the standards set for the Certified Web Professional (CWP) program established by the World Organization of Webmasters (WOW).

Additional Requirements. Web developers must be internet savvy and enjoy keeping up with the latest trends in web design and technology. They should be creative, detail-oriented individuals capable of learning various software languages, programs, and computer operating systems relatively quickly, often without formal training. Web developers need strong communication and business skills, and they must work well independently or in teams on deadline-oriented projects. Those who wish to establish their own design firms should also have business acumen and strong marketing skills.

—*Sally Driscoll*

Further Reading

Costello, Vic. *Multimedia Foundations: Core Concepts for Digital Design.* 3rd ed., Focal Press, 2023.

Felke-Morris, Terry. *Web Development and Design Foundations with HTML5.* 10th ed., Pearson, 2020.

Hardy, David Leicester. *Introduction to Digital Media Design: Transferable Hacks, Skills, and Tricks.* Bloomsbury Visual Arts, 2022.

WEB GRAPHIC DESIGN

ABSTRACT

Web graphic design is the use of graphic design techniques in designing websites. Web graphic designers must balance the marketing aspects of a website with aesthetic design criteria. They also attempt to increase the likelihood that the website will be found in search results and therefore be an effective advertising tool.

BACKGROUND

Web graphic design is a subfield of graphic design that focuses on designing for the web. It typically involves a blend of graphic design techniques and computer programming. Many websites are used as marketing materials for businesses and organizations. Consequently, web designers often incorporate business logos and other promotional materials. They also use search engine optimization (SEO) techniques to increase the likelihood that the web page will be found by search engines.

When creating graphics for use on the web, designers typically use any of a number of graphic-design programs, such as Adobe Photoshop and Corel's PaintShop Pro. In some cases, however, web graphic designers may choose to utilize three-dimensional (3D) modeling software to create 3D graphics or icons for use online.

Modern web design began in the 1990s, with the creation and adoption of hypertext markup language (HTML). This markup language specifies the location and appearance of objects displayed on a web page. In the mid-1990s, web designers began using HTML tables. These consist of static cells that can be arranged on a page to specify the location of text or objects. By placing tables within tables, designers could create a richer experience for websites.

The programming language JavaScript was developed in 1995 to code website behavior. This made individual web pages interactive for the first time. With JavaScript, web designers could embed image galleries, add functions such as drop-down menus, and make websites respond visually when a user clicked on text or images. Flash, developed in 1996, allowed designers to add animated elements to websites. However, not all browsers supported Flash, so it was not as universally useful as JavaScript.

The next major step in web graphic design was the introduction of tableless web design. This approach to web design does not rely on HTML tables. Instead, it uses style sheets to format pages. The Cascading Style Sheets (CSS) language, first introduced in 1996, allows designers to separate the visual elements of the design from the content. CSS can determine the appearance and spacing, while individual elements are still described in HTML. Personal Home Page (PHP), another language used in web design, was developed in 1994 but gained in popularity during the late 1990s. It provides further options for making web pages interactive. When used with HTML, PHP allows websites to create new content based on user information and to collect information from visitors.

As of the third decade of the twenty-first century, most web design is still based on a combination of CSS and HTML. Programming in PHP and the database server software MySQL provide options for greater flexibility and interactivity. Revised versions of web design languages, such as HTML5, offer more ways of incorporating multimedia, better support for multilingual websites, and a wider array of aesthetic options for designers.

OVERVIEW

Most websites, whether business, advocacy, news, or personal sites, serve as both marketing and informational tools. A website that represents a specific brand will incorporate iconic logos, logotypes, or combination marks to aid brand recognition. An iconic logo is a symbol or emblem that represents a person, business, or organization. A logotype is a company or brand name rendered in a unique or proprietary font and style. Combination marks combine icons with logotypes.

In addition to being informative, a website should be visually appealing and easy to understand. Skilled use of typography helps web designers achieve these goals. For example, while logotypes are designed to catch the eye, important information should be presented in a font that is aesthetically pleasing yet unobtrusive. A font that draws attention to itself will detract from the message of the text. Typography techniques such as this help web designers make websites easy to read and navigate.

In addition, web designers must often be familiar with SEO, which relies on elements such as keywords and links to ensure that users can find a website using a search engine. SEO techniques are most effective when incorporated into the overall design of a website. As such, many web graphic designers also help clients optimize their websites for better search results.

By the 2010s, web design had begun to focus on designing for the mobile web, creating websites and e-commerce sites that could be viewed and accessed using mobile devices such as smartphones. Many do-it-yourself (DIY) website builders began offering mobile web design templates and conversions.

USER-GENERATED WEB DESIGN

The popular website WordPress is actually a content management system (CMS) built on PHP and MySQL. It debuted in 2003 as a blogging website, offering users templates and tools to design blogs. The site soon became popular as a basic web design system, allowing users with limited knowledge of HTML or other coding to build basic and functional, if aesthetically rudimentary, websites. Soon web designers began tailoring WordPress sites for customers. This has essentially served as a shortcut for the web design industry. As of 2022, according to the firm Envisage Digital, WordPress was among the most popular web design systems in the world, accounting for more than half of all CMS-based websites on the internet.

The debut of WordPress launched a new era in DIY web graphic design. Companies such as Squarespace, Wix, and Weebly soon began offering users the ability to design websites quickly and easily, without the need to understand programming languages. Most such sites provide users templates in the form of wireframes. Wireframes are structural layouts that specify the location of images, text, and interactive elements on a web page but are not themselves interactive. Users can then insert their own images and text and, depending on the underlying program, also rearrange the layout of the wireframe. Like most elements of online commerce and marketing, web graphic design is moving toward user-generated and user-influenced design in order to open up to broader audiences.

—Micah L. Issitt

Further Reading

Allanwood, Gavin, and Peter Beare. *User Experience Design: Creating Designs Users Really Love*. Fairchild, 2014.

Hagen, Rebecca, and Kim Golombisky. *White Space Is Not Your Enemy: A Beginner's Guide to Communicating Visually through Graphic, Web & Multimedia Design*. 3rd ed., A. K. Peters/CRC Press, 2016.

Hardy, David Leicester. *Introduction to Digital Media Design: Transferable Hacks, Skills, and Tricks*. Bloomsbury Visual Arts, 2022.

Schmitt, Christopher. *Designing Web & Mobile Graphics: Fundamental Concepts for Web and Interactive Projects*. New Riders, 2013.

Tupper, Shelby Putnam. "The Best Graphic Design Software for 2023." *PCMag*, 5 Jan. 2023, www.pcmag.com/picks/the-best-graphic-design-software. Accessed 13 July 2023.

Wagner, Karin. *From ASCII Art to Comic Sans: Typography and Popular Culture in the Digital Age*. MIT Press, 2023.

Williams, Brad, David Damstra, and Hal Stern. *Professional WordPress: Design and Development*. 3rd ed., Wiley, 2015.

"WordPress Market Share in 2022." *Envisage Digital*, 2022, www.envisagedigital.co.uk/wordpress-market-share. Accessed 13 July 2023.

Windows Operating System

ABSTRACT

Microsoft Windows is the most dominant group of operating systems available for personal computers. Operating systems are software programs that support and manage the basic functioning of the computer. Operating systems provide a graphical interface through which users can easily execute tasks by clicking on icons or typing into a command prompt. Since its initial release in 1985, Windows has evolved with consistently improved functionality that has kept the brand at the forefront of the market for decades. Despite occasionally being the target of vocal criticism, Windows remained the most popular operating system for personal computers by 2023.

BACKGROUND

Prior to the mid-1980s, most early computers ran on simplistic operating systems that required users to input complex text-based commands into command prompts. Starting in 1981, the most common of these rudimentary operating systems was the Microsoft Disk Operating System (MS-DOS). Although MS-DOS was reasonably functional and reliable for its time, it was quickly overshadowed when Apple Computer's Macintosh debuted in 1984. Rather than using a text-based operating system like MS-DOS, the Macintosh featured one the earliest examples of a graphical interface-based operating system that allowed users to interact with the machine by clicking on icons.

The arrival of the graphical interface represented a major step forward in personal computing and presented a significant challenge for Microsoft. Undaunted, the company responded by developing its own graphical interface-based operating system. Referred to internally as Interface Manager, the new operating system featured a sort of virtual desktop on which programs ran in boxes called windows. These windows eventually became the inspiration for the operating system's official name when it was released in 1985. Even though the early Windows system offered mouse support; included features such as drop-down menus and scroll bars; and came loaded with a range of basic programs such as Paint, Notepad, and Calculator, Windows 1.0 was generally considered inferior to the Macintosh operating system and was a commercial failure.

OVERVIEW

Undeterred by the initial negative response to Windows 1.0, Microsoft worked to create a better product. In an effort to emulate its competitor's success, Microsoft entered a licensing agreement that allowed it to incorporate some of Apple's graphical interface features in Windows 2.0. With these added features, improved graphics, and the inclusion of additional software, Windows 2.0 clearly improved upon its predecessor and enjoyed a more positive reception upon its release in 1987.

While Windows 2.0 proved to be a step in the right direction, Microsoft's first real operating system breakthrough came with the 1990 debut of Windows 3.0 and the enhanced Windows 3.1 two years later. Touting an entirely new file management system, visual customization, alternative operating modes, and more bundled software, the new Windows was an immediate hit, selling 10 million copies in just two years. Crucially, Microsoft also released a software development kit with Windows 3.0 that made it easier for third-party software developers to write programs to run on the operating system. This effort helped to solidify the place of Windows in the personal computing market.

With the internet emerging as the driving force in the computer industry, Microsoft unveiled Windows 95 in 1995. This product featured a totally redesigned operating system with a unique look and feel. Unlike earlier releases that ran on top of MS-DOS and required a manual launch from the

Image via Wikimedia Commons. [Public domain.]

user, Windows 95 loaded automatically when the computer was started. Most notably, Windows 95 included the first taskbar and Start menus, which offered users easy access to programs, documents, and settings. Windows 95 also provided a gateway to the internet through the inclusion of the internet browser, Internet Explorer. With these new features, Windows 95 was a success, selling 7 million copies in only five weeks. Three years later, Microsoft returned with Windows 98, a release that was essentially an improved version of Windows 95.

Through the early 2000s, Microsoft released several new versions of Windows, including Windows 2000, which was an updated iteration of the professional Windows NT line, and Windows Millennium Edition. The latter was notoriously unstable, largely because it was rushed to the marketplace in anticipation of the debut of Windows XP in 2001. Windows XP was designed to address many of the issues that plagued Microsoft's previous releases and, as a result, offered significantly improved speed, stability, and file management. Thanks to these key tweaks, Windows XP became one of Microsoft's most widely used products.

Windows Vista (2007) was the next version of the product. Though this product boasted the strongest security system and most impressive graphics that had yet been seen in a Windows operating system, it was critically panned. Much of the criticism was focused on Vista's extensive compatibility issues, which convinced many XP users to delay upgrading. Microsoft released Windows 7 in 2009. With better compatibility and even more new features, Windows 7 resolved the Vista issues and satisfied critics and customers alike.

MOVE TOWARD TOUCH SCREENS

Acknowledging the growing consumer interest in tablets and smartphones, Microsoft radically changed its approach when it released Windows 8 in 2012. Specifically tailored for compatibility with touch-screen devices, Windows 8 rejected the traditional Start menu in favor of a tile-based layout. While tablet and smartphone users welcomed this design, traditional desktop and laptop users did not.

Hoping to reconcile its new aesthetic design with the user desire for traditional functionality, Microsoft introduced Windows 10 in 2015. The was the first version of their OS to be offered as a free upgrade. Windows 10 also introduced a new service model in which the OS would continually receive updates to various features and functions. This model is similar to that employed by Microsoft's competitor Apple. Windows 10 was also the first version of Windows that had a universal application architecture. This meant that apps on Windows 10 could be used on smartphones, tablets, and the Xbox One gaming system as well. This package included options for both the Start menu and the new tile scheme. Windows 10 also featured the debut of Cortana, Microsoft's first digital personal assistant. In 2021, Microsoft released Windows 11. Much like its predecessor, the new operating system was made available to users of the prior version as a free update.

DOMINANCE AND REINVENTION

Just under 75 percent of desktop personal computers (PCs) worldwide ran some version of Windows by January of 2023, according to the website *Statcounter*. The majority (about 69 percent) ran Windows 10, while about 18 percent ran Windows 11. Some PCs also continued to run older Windows OSs, including Windows 7 and Windows 8.1. Windows is sold around the world and comes formatted for a variety of languages. In addition, language interface packs are available for free download and offer support for languages not found in full versions. Each pack requires a base language that can be activated within the OS after installation.

Given the rising popularity of handheld computing devices, such as tablets and smartphones, the global computer market is no longer based solely on the PC market. While Microsoft continued to dominate the PC market into 2023, the company controlled only about 0.02 percent of the global mobile and tablet OS markets by the start of that year, falling far behind Android and Apple iOS (iPhone operating system). As a result, beginning with the release of Windows 10, Microsoft made changes to its basic strategy. The company reduced its focus on proprietary software. Instead, the company increasingly embraced the potential of open-source software. Future versions of Windows will likely show an increased focus on cloud-based computing, intuitive touch-screen technology, and alternative interface controls such as voice activation and multitouch gestures.

—Jack Lasky and Micah L. Issitt

Further Reading

Brown, Michael. "Microsoft Windows Is 30: A Short History of One of the Most Iconic Tech Products Ever." *International Business Times*, 20 Nov. 2015, www.ibtimes.com/microsoft-windows-30-short-history-one-most-iconic-tech-products-ever-2194091. Accessed 13 July 2023.

Calore, Michael. "A History of Microsoft Windows." *Wired*, 10 Dec. 2008, www.wired.com/2008/12/wiredphotos31. Accessed 13 July 2023.

Gibbs, Samuel. "From Windows 1 to Windows 10: 29 Years of Windows Evolution." *The Guardian*, 2 Oct. 2014, www.theguardian.com/technology/2014/oct/02/from-windows-1-to-windows-10-29-years-of-windows-evolution. Accessed 13 July 2023.

Holcombe, Jane, and Charles Holcombe. *Survey of Operating Systems*. 6th ed., McGraw-Hill Education, 2020.

McLellan, Charles. "The History of Windows: A Timeline." *ZDNet*, 14 Apr. 2014, www.zdnet.com/article/the-history-of-windows-a-timeline. Accessed 13 July 2023.

"Operating System Market Share Worldwide-June 2023." *Statcounter*, June 2023, gs.statcounter.com/os-market-share/desktop/worldwide/#monthly-202206-202306. Accessed 13 July 2023.

Panay, Panos. "Introducing Windows 11." *Microsoft*, 24 June 2021, blogs.windows.com/windowsexperience/2021/06/24/introducing-windows-11. Accessed 13 July 2023.

Warren, Tom. "Windows Turns 35: A Visual History." *Verge*, 20 Nov. 2020, www.theverge.com/2015/11/19/9759874/microsoft-windows-35-years-old-visual-history. Accessed 13 July 2023.

"What Is the History of Microsoft Windows?" *Indiana University Knowledge Base*, 18 Jan. 2018, kb.iu.edu/d/abwa. Accessed 13 July 2023.

Wireframes

ABSTRACT

A wireframe visually represents the basic elements of a proposed website before a web designer actually builds it. Creating a wireframe ensures that both the designer and the client can agree on the basic layout and functionality of the site. It also allows them to map out how the end user will ultimately interact with the site.

BACKGROUND

A wireframe is a visual, noninteractive representation of a proposed website. A web designer may use a wireframe to determine how best to lay out the elements of the site. If the site is being designed for a client, the designer may show the client a wireframe to get approval for the final product.

A website is essentially a sophisticated information storage bank. Ultimately, it is designed for the end users, who will need easy, efficient access to the information it contains. To that end, a web designer will block out the basic elements of the proposed website first, most often drawing them on paper. This drawing is the wireframe. Though it may be no more than a set of squares, circles, columns, and lines that represent the individual elements of the pages, it allows the designer and the client to review the site concept together before it is built. Any changes in the website format can be easily made at this stage.

A wireframe also allows the designer to sort the content of the proposed website into an information hierarchy. The designer can then determine how best to arrange it on the page. Part of web design is knowing what information users expect to find and where they are most likely to look for it. Eye-tracking studies have determined that upon accessing a website, a user will first look at the top-left corner of the page, scan across the top, then look down the left side and scan partway across the screen. This essentially forms an F-shape. Thus, crucial information should be concentrated mainly in the upper left portion of the page. Such insights into how users interact with and experience websites are an important part of the design field known as interaction design.

OVERVIEW

When designing a website, particularly for a business or other organization, certain elements are standard. These elements may include:

- The client's logo, to identify the business or organization in a recognizable way

- A navigation bar, to provide access to the different pages of the site
- A front-page image or slide show of images
- A search box
- A registration or login box, if the site allows (or requires) users to register
- A header containing an index of the major elements of the website
- A footer containing contact information, privacy guidelines, and other legal information

Many websites also include external content such as news headlines, weather reports, or social media feeds. In most cases, the creator of this external content generates a portlet or widget that designers can simply plug into the code of the site. While portlets and widgets work in similar ways, they differ mainly in their execution. A portlet is executed on the server, so that the code it generates is not visible to the end user. A widget is executed in the browser.

A wireframe should incorporate any and all elements that will be included in the final website. In most cases, the wireframe serves as a blueprint, with no distracting aesthetic features such as font, color, or images. The emphasis is on functionality—that is, how the proposed site will accommodate the end user. This type of wireframe is also called a "low-fidelity wireframe."

Once the client is on board with the concept, the designer may produce a more sophisticated, high-fidelity wireframe. This type of wireframe is usually created using dedicated software. It represents the final website in much more detail. It incorporates such design elements as images, color schemes, and typography. Sequences of images are used to represent how users will interact with the site. However, the wireframe itself is still noninteractive.

Sometimes, between the high-fidelity wireframe and the final product, a designer will produce an interactive prototype. While a prototype will usually look very similar to the final product, its appearance matters less than its functionality. A prototype should simulate the user experience with the completed website. Prototypes are ideal for early user testing. Any problems with the interface can be addressed before the designer begins coding for the final site.

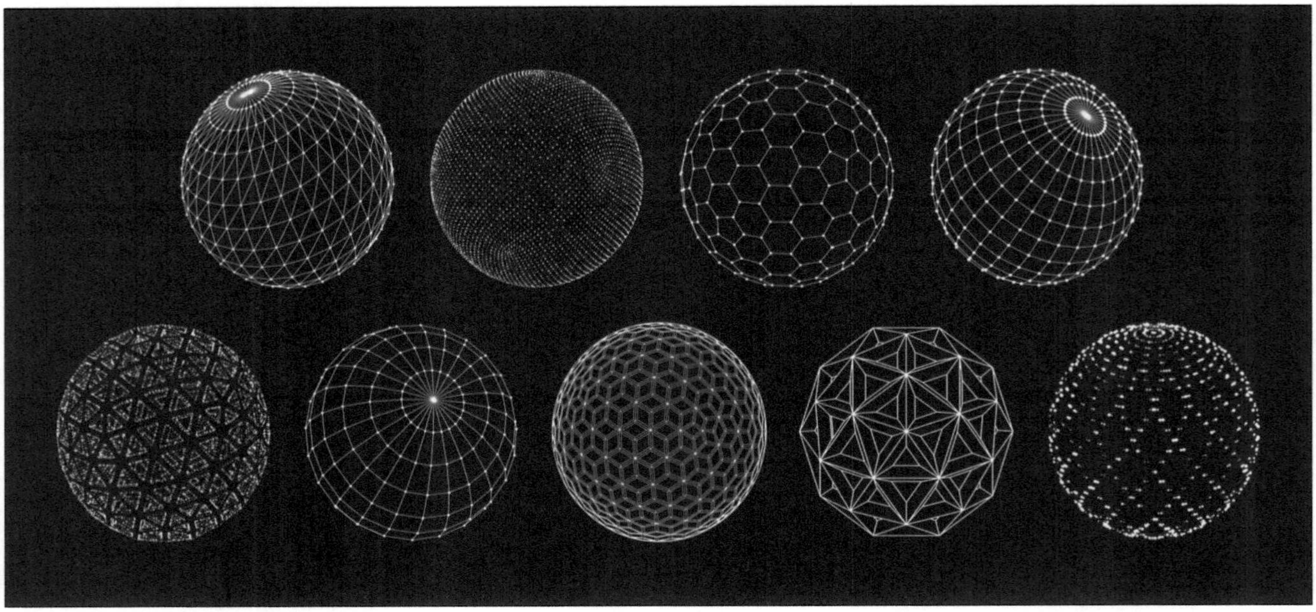

Image via iStock/seamartini. [Used under license.]

IMPACT OF WIREFRAMES

Much as an architect creates a blueprint for a skyscraper or a director storyboards a film, a web designer uses a wireframe to spot potential problems before building the final product. This saves both the designer's time and the client's money and, ultimately, shields the end user from frustration—all critical factors in the highly competitive world of the internet.

—Joseph Dewey

Further Reading

Greenberg, Saul, et al. *Sketching User Experiences: The Workbook*. Morgan, 2012.

Hamm, Matthew J. *Wireframing Essentials: An Introduction to User Experience Design*. Packt, 2014.

Hardy, David Leicester. *Introduction to Digital Media Design: Transferable Hacks, Skills, and Tricks*. Bloomsbury Visual Arts, 2022.

Klimczak, Erik. *Design for Software: A Playbook for Developers*. Wiley, 2013.

Krug, Steve. *Don't Make Me Think, Revisited: A Common Sense Approach to Web Usability*. 3rd ed., New Riders, 2014.

Marsh, Joel. *UX for Beginners: A Crash Course in 100 Short Lessons*. O'Reilly, 2016.

Miller, Brian D. *Principles of Web Design*. Allworth, 2022.

Norman, Don. *The Design of Everyday Things*. Rev. and expanded ed. Basic, 2013.

Osborn, Tracy. *Hello Web Design: Design Fundamentals and Shortcuts for Non-Designers*. No Starch, 2021.

Pernice, Kara. "F-Shaped Pattern of Reading on the Web: Misunderstood, but Still Relevant (Even on Mobile." *Nielsen Norman Group*, 12 Nov. 2017, www.nngroup.com/articles/f-shaped-pattern-reading-web-content. Accessed 13 July 2023.

Writer and Editor

ABSTRACT

Writers and editors are employed in all realms of business and industry. They may be employed by specific companies or agencies or may work on a freelance basis. Students aspiring to enter the profession should pursue studies in English, history, social studies, the arts, and science.

BACKGROUND

Writers and editors are employed in all realms of business and industry. In addition to journalism, publishing, and media (i.e., radio and television), employment for writers can be found in government, marketing, law, entertainment, and sales. Writers employed by local, state, or federal governments may craft legislation or produce speeches and press releases for elected representatives. Every industrial sector, be it the automobile industry, health care, education, retail, agriculture, or mining, utilizes writers to communicate with colleagues and clients and develop messaging regarding their productivity and business plan. Freelance writing and editing—that is, writing and editing under temporary contract—is common. Many freelancers work for online publishers, producing content for clients that adheres to specific guidelines. In general, writers create original content and editors review and revise that content, but there is significant overlap between these two roles.

OVERVIEW

Most writers and editors work in an office environment. Writers and editors in the media often work in the field, gathering data and interviewing people for news reports. Many freelance writers and editors work from a home office. Some freelance writers work at rented office spaces.

Writers and editors enjoy working with language and ideas. They enjoy the challenge of communicating complex ideas in a way that is readily digestible to a specific audience. Writers and editors have a penchant for grammar and the intricacies of publishing formats and editorial guidelines. Those who are employed by a specific industry or business sector should have a passion for that area of communication and commerce. For example, sports writ-

Photo via iStock/Antonio_Diaz. [Used under license.]

ers need to be knowledgeable about a particular sport's rules and regulations, teams, and players. Individuals interested in writing public policy or producing content for the news media should be interested in government, politics, and current events.

DUTIES AND RESPONSIBILITIES

The daily life of a writer/editor is highly dependent upon the field in which he or she is employed. For example, writers and editors employed in a marketing department—copywriters—may research a particular product line before beginning to write about it for a particular client or consumer market. Speechwriters and those working in the legal or political field will research archival material and conduct interviews with voters and policymakers. Other writing and editing work is more routinized. Writers and editors working for publishing companies traditionally follow a product development schedule, whether the product is a book, magazine, newspaper, or online publication. Technical writers produce product manuals, assembly instructions, or troubleshooting guidelines. The work of a freelance writer and editor will vary day-to-day depending on the project. In some cases, a writer/editor is involved in the planning and preparation of a product and may have a hand not only in developing the text but also in choosing photos, illustrations, and other elements. Duties sometimes include website maintenance, depending on the client's expectations.

OCCUPATION SPECIALTIES

Copywriters. Copywriters prepare advertisements to promote the sale of a good or service. They often work with a client to produce advertising themes, jingles, and slogans.

Copy Editors and Proofreaders. Copy editors review copy for errors in grammar, punctuation, and spelling and check the copy for readability, style, and agreement with editorial policy. They also may confirm sources and verify facts. Proofreaders check typeset pages for any remaining errors.

Staff Writers and Editors. Staff writers and editors prepare material for newspapers, magazine, books, or news broadcasts. They generally work under an executive or managing editor. They draft or receive copy and may suggest or choose graphics and images to accompany it. Depending on their seniority, they may be responsible for a particular subject area or type of editorial project.

Executive and Managing Editors. Executive editors plan the contents and budget of publications and supervise their preparation. In most cases, they decide what gets published and what does not. Managing editors, similarly, work with executives, department heads, and editorial staff to formulate policies, coordinate department activities, establish production schedules, solve publication problems, and make organizational changes.

Contributing Editors. Contributing editors are professional authors who contribute original articles to magazines or newspapers on a regular or semiregular basis.

Fiction and Nonfiction Writers. Fiction and nonfiction writers are professional authors who write original material for publication and seek to develop a readership for their work.

Screenwriters. Screenwriters are authors who write scripts for motion pictures or television. They may produce original stories, characters, and dialogue or adapt a book into a movie or television script. Some may produce content for radio broadcasts and other types of performance.

WORK ENVIRONMENT

Physical Environment. Freelance or contract writers and editors work primarily from home offices or in designated sections of their homes. Freelance work has no set hours or specified work schedule, and freelancers often work atypical hours and on weekends. Some long-term contracts require that writers or editors work at the company who is hiring them, which would require the writer or editor to work in an office setting during regular business hours for the length of the project they have been hired to complete.

Writers or editors who are hired as full-time employees for a company or organization work in office settings and during standard business hours and days.

Human Environment. Writers and editors interact frequently with clients and colleagues and good communication skills are essential to their work. While many writers and editors work alone, nearly all communicate regularly with colleagues and clients about project-specific guidelines and goals.

EDUCATION AND TRAINING

High School/Secondary. High school students can best prepare for a career as a writer or editor by completing coursework in English, history, social studies, the arts, and science. Advanced coursework in a field of particular interest can prepare students for writing knowledgably and coherently about that field. Participation in extracurricular activities such as debate clubs, school papers, or school television and radio programs can also help students develop the skills needed for a career in writing and editing.

College/Postsecondary. Postsecondary education is often a requirement for vacancies in the writing and editing field. Postsecondary coursework that can contribute to the numerous skills and vast frame of reference required of writers and editors includes education, literature, history, government, international business, economics, politics, and government.

Adult Job Seekers. There are selected opportunities for adult job seekers interested in writing and editing. Working knowledge or experience in a particular field, such as education, marketing, or retail, represent skills that can be transferable to writing and editing work. Editors and writers working in one area, such as newspapers, can also sometimes cross over into another, such as book publishing. Oftentimes, however, the best way to make the switch is to seek a temporary or part-time assignment first.

Professional Certification and Licensure. Certification or licensure is not required to be employed as an editor or writer. The majority of hiring companies and organizations require that applicants have at least an undergraduate degree with a concentration in either English or another field that pertains to the position needing to be filled.

Additional Requirements. Writers and editors must possess a love of the language and a commitment to quality writing. Writers and editors often work alone or from their homes, so individuals who want to explore this line of work should be comfortable in solitary settings.

—Price Grisham

Further Reading

Carroll, Brian. *Writing and Editing for Digital Media*. 4th ed., Routledge, 2019.

Kuehn, Scott A., and James Andrew Lingwall. *The Basics of Media Writing: A Strategic Approach*. CQ Press, 2017.

Lawrence, Dan. *Digital Writing: A Guide to Writing for Social Media and the Web*. Broadview Press, 2022.

Y

YouTube

ABSTRACT

YouTube is a popular video streaming website that displays uploaded video files created or disseminated by its users. The types of videos uploaded to YouTube include user-created content, such as homemade videos and videoblogs; content created by independent production studios specifically for distribution through YouTube; and short video clips from films, television, and copyrighted and noncopyrighted materials. YouTube has become very popular since its introduction in 2005 and is the video streaming website against which most others are measured. Since its launch, YouTube has become a touchstone in cultural debates about the internet. There has been controversy over the uploading of copyrighted materials. However, the popularity of YouTube has also brought favorable attention to some media outlets when users upload portions of copyrighted content. YouTube also offers television streaming services and exclusive content. These are available, like YouTube's commercial-free version, as a for-pay subscription, and they have challenged traditional television services offered by cable and satellite companies.

BACKGROUND

YouTube was founded in February 2005 by three former PayPal employees: Chad Hurley, Steve Chen, and Jawed Karim. Venture capital firm Sequoia Capital funded the launch of YouTube in December 2005. The first video featured on YouTube was uploaded by "jawed" (presumably cofounder Jawed Karim) on April 23, 2005. The video is eighteen seconds in length and features an uneventful moment at a zoo. By July 2006, YouTube was receiving 20 million visitors per month. Of the 9 billion video clips watched in the United States that same month, more than a quarter of them (2.5 billion videos at an average length of 2.7 minutes each) were watched on YouTube.

YouTube was immediately successful: the ease of use and open policy of the website, which allowed almost all forms of content (excluding copyrighted, pornographic, violent, or otherwise inappropriate content), combined with the fact that it was a free service, attracted millions of users. By August 2008, the site boasted 74 million visitors per month. By December 2014, more than 1 billion unique users were visiting the site each month; and by 2023, the number of monthly active users exceeded 2.5 billion.

Videos featured on YouTube, most of which are ten minutes or less in length, span the spectrum of content, including instructional and how-to videos, clips from popular television shows, amateur and professionally produced motion pictures and music videos, home movies, film trailers, videos created especially for YouTube, and much more.

YouTube's rapid ascent to the top tier of internet websites in terms of popularity and number of visits was achieved partially as a result of several incidents that brought it to the attention of the wider public. Soon after launch, clips from popular television shows, such as a sketch from the comedy show *Saturday Night Live* and the 2006 Winter Olympics (both of which were owned by television broadcasting company NBC), made their way onto YouTube, allowing users to watch and share specific parts of those programs that they felt warranted attention on their own. This brought about charges of copyright infringement and subsequent litigation by NBC and others, something YouTube immediately began addressing in various ways, including a policy stating that it will remove any copyrighted materials when and if the copyright holder asks, and also by asking users to police the site for inappropriate content.

However, copyrighted material abounds on YouTube. Many media critics acknowledge that it is in the best interests of some copyright holders to allow such infringement because of the potential viral marketing benefits of being hosted on YouTube, which makes this content available to many viewers who might not otherwise be exposed to these products. One television show that appeared to take advantage (unofficially) of this type of publicity was *The Daily Show with Jon Stewart*, one of several shows airing on cable television network Comedy Central that were featured regularly on YouTube.

In 2007 YouTube launched its Partners Program. The program allows content creators who join to earn part of the revenue paid to YouTube by advertisers. The more popular the video, the more the partner is paid. YouTube also promotes the videos on its platform.

Many frequent YouTube posters have become famous as a result of the popularity of their videos. Early examples included Jessica Rose, an actress whose videos under the name of lonelygirl15 were thought to be the real video diary entries of a teenage girl until it was disclosed that they were Rose's fictional creations; "Chad Vader," a *Star Wars* fan film series featuring a grocery store employee and his cohorts; and Smosh, a pair—later a large collective—of internet short film creators who produce comedy sketches and parodies of music videos.

Some YouTube celebrities who found success during the site's first decade of existence went on to be featured in mainstream media and subsequently translated their popularity into lucrative multimedia deals outside YouTube. Pop band OK Go won a 2006 Grammy Award for a video that had originally

Image via iStock/contrastaddict. [Used under license.]

achieved notoriety on YouTube rather than more traditional music video channels. In December 2014, Tad Friend reported in the *New Yorker* that in 2013 DreamWorks bought AwesomenessTV for 33 million dollars. AwesomenessTV is a talent management firm for YouTube celebrities. Felix Kjellberg, a Swedish video gamer known as PewDiePie, signed with the Disney-owned Maker Studios in December 2012; however, this partnership was terminated in 2017 after the YouTube star posted anti-Semitic comments to his channel. Lilly Singh, who launched a YouTube channel featuring comedy sketches and music videos in 2010, went on to host the late-night talk show *A Little Late with Lilly Singh*, which aired on NBC between 2019 and 2021. Many additional YouTube creators would continue to find success on the platform into the 2020s, with some becoming multimillionaires, establishing thriving production companies, and coming to be regarded as celebrities or industry experts in their own right.

Shortly after cofounding the company, Karim left YouTube to pursue a graduate degree at Stanford University. Hurley stepped down as chief executive officer (CEO) and was replaced by Salar Kamangar. Early Google employee and executive Susan Wojcicki took over as YouTube CEO in February 2014. Chen joined Google Ventures in 2014. Wojcicki remained YouTube's CEO until early 2023, when she announced that she was stepping down from the position. She was succeeded as CEO by Neal Mohan.

OVERVIEW

YouTube's interface is relatively easy to use but is designed to allow a complex array of activity. Users do not need to download software to use YouTube, but access to a high-speed internet connection does optimize the video streaming capabilities of the website. If using a mobile device or smart television, a user may choose to access YouTube content through a dedicated app rather than through a web browser. Users may respond to videos by clicking on the "like" buttons below the videos, subscribing to the creators' YouTube channels, or leaving comments. They can save videos to watch later or share them with others and can also flag videos for review if they contain content that is prohibited under YouTube's terms of service.

YouTube was purchased by leading internet search engine company Google, Inc. (later known as Alphabet), in November 2006, for $1.65 billion. At the time of the purchase, YouTube held 46 percent of the online video market. As an independently operated subsidiary of Google, YouTube was a direct competitor with Google's own video service, Google Video (discontinued in 2012) as well as other leading video sharing websites of that era, which included Yahoo Video, Myspace, VideoEgg, Blip.tv, DailyMotion, and Break. Such sites were all part of a burgeoning trend toward user-generated websites, which was a part of a larger movement known as Web 2.0. "Web 2.0" is a term that describes the movement away from fixed, unchanging content toward websites that encourage and grow in functionality as a result of user participation. Social networking websites are an example of Web 2.0.

To encourage more interaction by users, YouTube introduced TestTube, a series of experimental web tools. TestTube features included a way to add licensed music to videos, a remixer for adding titles or otherwise altering videos, active sharing to communicate to other users which videos one had watched, and chatting with other users while watching videos.

YouTube experienced an increase in professionally produced channels in 2012. There were channels designed for a number of topics, celebrities, and news outlets, including Justin Bieber, Nike Football, and ABC News. These channels were beginning to take precedence over more obscure videos, some of which—such as a popular video of a sneezing baby panda—have received millions of views. Google announced plans to establish fifty original channels in 2012. To promote these channels and videos, a $200 million marketing campaign was planned with the intent of moving YouTube down the path to replacing television. As of December 2014, the site quoted a Nielsen report stating that US adults between the ages of eighteen and thirty-four watched YouTube more than they watched any cable network. YouTube also reported that it had more than a million YouTube Partners, creators who earn income from the videos they post on YouTube, and, of these, thousands earned six figures annually.

In August 2015, Google announced plans to launch YouTube Gaming to provide a site devoted solely to video-gaming content. The service allowed users to view livestream videos for more than 25,000 different games as well as receive personalized recommendations for YouTube gaming channels to follow. Google developed the site to directly compete with the popular livestreaming Twitch, which is owned and operated by Amazon. YouTube Gaming remained operational until May of 2019, when it was discontinued.

In October 2015, Google announced another YouTube expansion in its launch of YouTube Red, which was a paid, monthly service that allowed subscribers to view any video, with the exception of paid content on YouTube, without being interrupted by advertisements. The service also included the added incentive of free monthly subscriptions to Google Play Music. Similarly, anyone who purchased a subscription to Google Play Music at that time would receive a free subscription to YouTube Red. In 2018, YouTube Red was discontinued and replaced with two separate services: YouTube Premium, which offered access to ad-free videos as well as to content known as YouTube Originals, and the music-streaming app YouTube Music.

YouTube continued to introduce new services throughout the late 2010s and early 2020s, often seeking to compete with and offer alternatives to popular services offered by other technology companies. In April 2017, YouTube entered the field of streaming television with the launch of YouTube TV, offering subscribers to that new service access to live television channels as well as on-demand content. In July 2021, following a period of beta testing in select markets, YouTube introduced YouTube Shorts, a new format of video content that was intended to compete with the short-form videos popular on apps such as TikTok.

—Craig Belanger and Marlanda English

Further Reading

Artero, Juan P. "Online Video Business Models: YouTube vs. Hulu." *Palabra Clave*, vol. 13, no. 1, 201, pp. 111-23.

Bloom, Kristen, and Kelly Marie Johnston. "Digging into YouTube Videos: Using Media Literacy and Participatory Culture to Promote Cross-Cultural Understanding." *Journal of Media Literacy Education*, vol. 2, no. 2, 2013, p. 3.

Boorstin, Julia. "YouTube Officially Launches YouTube TV in Select Markets." *CNBC*, 5 Apr. 2017, www.cnbc.com/2017/04/05/youtube-officially-launches-youtube-tv-in-select-markets.html. Accessed 13 July 2023.

Chen, Chih-Ping. "Exploring Personal Branding on YouTube." *Journal of Internet Commerce*, vol. 12, no. 4, 2013, pp. 332-47.

Etherington, Darrell. "People Now Watch 1 Billion Hours of YouTube Per Day." *Tech Crunch*, 28 Feb. 2017, techcrunch.com/2017/02/28/people-now-watch-1-billion-hours-of-youtube-per-day. Accessed 13 July 2023.

Friend, Tad. "Letter from California: Hollywood and Vine." *New Yorker*. Condé Nast, 15 Dec. 2014, www.newyorker.com/magazine/2014/12/15/hollywood-vine. Accessed 13 July 2023.

Gallardo-Camacho, Jorge. "The Low Interaction of Viewers of Video Clips on the Internet: The Case Study of YouTube Spain." *Revista Latina de Comunicación de Social*, vol. 13, no. 65, 2010, pp. 1-14.

Iqbal, Mansoor. "YouTube Revenue and Usage Statistics." *Business of Apps*, 18 July 2023, www.businessofapps.com/data/youtube-statistics. Accessed 13 July 2023.

Konijn, Elly A., et al. "YouTube as Research Tool—Three Approaches." *CyberPsychology, Behavior & Social Networking*, vol. 16, no. 5, 2013, pp. 1-7.

Lavaveshkul, Liz. "How to Achieve 15 Minutes (or More) of Fame through YouTube." *Journal of International Commercial Law & Technology*, vol. 7, no. 4, 2012, pp. 370-85.

Matney, Lucas. "YouTube Has 1.5 Billion Logged-In Monthly Users Watching a Ton of Mobile Video." *Tech Crunch*, 22 June 2017, techcrunch.com/2017/06/22/youtube-has-1-5-billion-logged-in-monthly-users-watching-a-ton-of-mobile-video. Accessed 13 July 2023.

Miller, Claire C. "YouTube to Serve Niche Tastes by Adding Channels." *New York Times*, 7 Oct. 2012, www.nytimes.com/2012/10/08/business/media/YouTube-to-serve-niche-tastes-by-adding-channels.html. Accessed 13 July 2023.

Mitroff, Sarah, and Taylor Martin. "Everything You Need to Know About YouTube Red." *CNET*, 27 Oct. 2017, www.cnet.com/how-to/youtube-red-details. Accessed 13 July 2023.

Morreale, Joanne. "From Homemade to Store Bought: Annoying Orange and the Professionalization of YouTube." *Journal of Consumer Culture*, vol. 14, no. 1, 2014, pp. 113-28.

Robertson, Adi. "The YouTube Gaming App Is Shutting Down This Week." *Verge*, 27 May 2019, www.theverge.com/2019/5/27/18641413/youtube-gaming-standalone-app-shutdown-may-30th. Accessed 13 July 2023.

Rosenbaum, Steven. "YouTube Reveals a Curated Future." *Forbes*, 10 Oct. 2012, www.forbes.com/sites/stevenrosenbaum/

2012/10/10/YouTube-reveals-a-curated-future-2. Accessed 13 July 2023.

Sarker, Samit. "YouTube Gaming Launches Aug. 26 with Website and Mobile Apps." *Polygon*, 25 Aug. 2015, www.polygon.com/2015/8/25/9208245/youtube-gaming-launch-date-web-android-ios. Accessed 13 July 2023.

Saul, Roger. "KevJumba and the Adolescence of YouTube." *Educational Studies*, vol. 46, no. 5, 2010, pp. 457-77.

Savov, Vlad. "Google Announces YouTube Music and YouTube Premium." *Verge*, 17 May 2018, www.theverge.com/2018/5/17/17364056/youtube-music-premium-google-launch. Accessed 13 July 2023.

Shreekumar, Advik. "Rise of the YouTube Star." *Harvard Political Review*, vol. 41, no. 1, 2014, p. 16.

Spangler, Todd. "YouTube Shorts, Video Giant's TikTok Copycat, Is Rolling Out in 100-Plus Countries." *Variety*, 13 July 2021, variety.com/2021/digital/news/youtube-shorts-global-launch-1235018403. Accessed 13 July 2023.

Wiggers, Kyle. "YouTube CEO Susan Wojcicki Steps Down, Will Assume Advisory Role at Google and Alphabet." *TechCrunch*, 16 Feb. 2023, techcrunch.com/2023/02/16/youtube-ceo-susan-wojcicki-steps-down-will-assume-advisory-role-at-google-and-alphabet. Accessed 13 July 2023.

Bibliography

"3D-4D Building Information Modeling." *US General Services Administration*, 6 Apr. 2022, www.gsa.gov/real-estate/design-construction/3d4d-building-information-modeling. Accessed 13 July 2023.

"8 Lessons from the Rise of Douyin (Tik Tok)." *Technode*, 15 June 2018, technode.com/2018/06/15/8-lessons-douyin. Accessed 13 July 2023.

"11 Ways VR and AR Stand to Impact Advertising, Marketing and PR." *Forbes*, 31 Aug. 2021, www.forbes.com/sites/forbesagencycouncil/2021/08/31/11-ways-vr-and-ar-stand-to-impact-advertising-marketing-and-pr/?sh=25e7879c5201. Accessed 13 July 2023.

Abbasov, Iftikhar B., editor. *Computer Modeling in the Aerospace Industry*. Wiley/Scrivener Publishing, 2020.

Abdey, James. *Business Analytics: Applied Modelling and Prediction*. Sage Publications, 2023.

"About Spotify." *Spotify*, newsroom.spotify.com/companyinfo. Accessed 25 July 2023.

"About the Licenses." *Creative Commons*, 2023, creativecommons.org/licenses. Accessed 13 July 2023.

"About Us." *Bandcamp*, 2023, bandcamp.com/about. Accessed 13 July 2023.

Abraham, Nikhil. "Building Mobile Web Apps." *Dummies*, 26 Mar. 2016, www.dummies.com/programming/building-mobile-web-apps. Accessed 13 July 2023.

Abu-Mostafa, Yaser S. "Machines That Think for Themselves." *Scientific American*, July 2012, pp. 78-81.

"Ad Age Advertising Century: Timeline." *Ad Age*, 29 Mar. 1999, adage.com/article/special-report-the-advertising-century/ad-age-advertising-century-timeline/143661. Accessed 13 July 2023.

Adams, Ernest. *Break into the Game Industry: How to Get a Job Making Video Games*. McGraw-Hill/Osborne, 2003.

_____. *Fundamentals of Game Design*. 2nd ed., New Riders, 2010.

Adams, J. Michael, and Penny Ann Dolin. *Printing Technology*. 5th ed., Delmar, 2002.

Adamson, Austin. "Being a S.O.L.I.D. Engineer." *OrdinaryEngineer*, 27 July 2020, medium.com/ordinaryengineer/solid-principles-4a3525ae0ac8. Accessed 13 July 2023.

"Adjusting Our Spotify Premium Prices." *Spotify*, 24 July 2023, newsroom.spotify.com/2023-07-24/adjusting-our-spotify-premium-prices. Accessed 25 July 2023.

"Advertising Expenditure in North America from 2017 to 2024, by Medium." *Statista*, Nov. 2022, www.statista.com/statistics/1347750/ad-spend-in-north-america-media. Accessed 13 July 2023.

Aktas, Calalettin. *The Evolution and Emergence of QR Codes*. Cambridge Scholars, 2017.

Alburger, James R. *The Art of Voice Acting: The Craft and Business of Performing for Voiceover*. 7th ed., Focal Press, 2023.

Alexander, Bryan. *The New Digital Storytelling: Creating Narratives with New Media*. ABC-CLIO, 2011.

Ali, Harun H. *The Pocket Mentor for Video Game Testing*. CRC Press, 2023.

Al-Jumaily, A., and R. A. Olivares. "Bio-Driven System-Based Virtual Reality for Prosthetic and Rehabilitation Systems." *Signal, Image and Video Processing*, vol. 6, no. 1, 2012, pp. 71-84.

Al-Khatib, Aceil, and Jaime A. Teixeira da Silva. "Stings, Hoaxes and Irony Breach the Trust Inherent in Scientific Publishing." *Publishing Research Quarterly*, vol. 32, no. 3, 2016, pp. 208-19.

Allanwood, Gavin, and Peter Beare. *User Experience Design: Creating Designs Users Really Love*. Fairchild, 2014.

Allemang, Dean, James Hendler, and Fabien Gandon. *Semantic Web for the Working Ontologist: Effective Modeling for Linked Data, RDFS, and OWL*. 3rd ed., ACM, 2020.

Allen, Stuart. *Photojournalism and Citizen Journalism: Co-operation, Collaboration and Connectivity*. Routledge, 2019.

Almasy, Steve. "Thailand Baby Killing: Facebook Removes Video." *CNN*, 26 Apr. 2017, www.cnn.com/2017/04/25/asia/thailand-baby-killed-facebook-live-trnd. Accessed 13 July 2023.

Alpaydin, Ethem. *Machine Learning*. Rev. ed., MIT Press, 2021.

Alsop, Thomas. "Virtual Reality (VR)-Statistics & Facts." *Statista*, 8 June 2023, www.statista.com/topics/2532/virtual-reality-vr. Accessed 13 July 2023.

Altavilla, Dave. "Apple Further Legitimizes Augmented Reality Tech with Acquisition of Metaio." *Forbes*, 30 May 2015, www.forbes.com/sites/davealtavilla/2015/05/30/apple-further-legitimizes-augmented-reality-tech-with-acquistion-of-metaio. Accessed 13 July 2023.

Alvermann, Donna E., and Margaret C. Hagood. "Critical Media Literacy: Research, Theory, and Practice in 'New Times.'" *Journal of Educational Research*, vol. 93, no. 3, 2000, pp. 193-205.

Amadeo, Ron. "Google Maps Gets Augmented Reality Search Results." *Ars Technica*, 28 Sept. 2022, arstechnica.com/gadgets/2022/09/google-maps-gets-augmented-reality-search-results. Accessed 13 July 2023.

_____. "Report: Google Will Graciously Let Android OEMs Build Amazon Fire Devices." *Ars Technica*, 28 Oct. 2022, arstechnica.com/gadgets/2022/10/report-google-will-graciously-let-android-oems-build-amazon-fire-devices. Accessed 13 July 2023.

_____. "The (Updated) History of Android." *Ars Technica*, 31 Oct. 2016, arstechnica.com/gadgets/2016/10/building-android-a-40000-word-history-of-googles-mobile-os. Accessed 13 July 2023.

Ambrose, Gavin, Paul Harris, and Sally Stone. *The Visual Dictionary of Architecture*. AVA, 2008.

Ammori, Marvin. "The Case for Net Neutrality." *Foreign Affairs*, vol. 93, no. 4, 2014, pp. 62-73.

Anderson, Casey. "What Is VOD Technology?" *Chron*, smallbusiness.chron.com/vod-technology-14311.html. Accessed 25 July 2023.

Anderson, Deborah. "Global Linguistic Diversity for the Internet." *Communications of the ACM*, vol. 48, no. 1, 2005, pp. 27-28.

Anderson, Eric. *Social Media Marketing: Game Theory and the Emergence of Collaboration*. Springer, 2010.

Anderson, Gary H. *Video Editing and Post Production: A Professional Guide*. Focal, 1999.

Anderson, J. "Fighting to Bridge the Digital Divide." *Social Policy*, vol. 44, 2014, pp. 56-57.

Anderson, John Nathan. *Radio's Digital Dilemma: Broadcasting in the Twenty-First Century*. 2014. Routledge, 2016.

Anderson, Monica, and Madhumitha Kumar. "Digital Divide Persists Even as Lower-Income Americans Make Gains in Tech Adoption." *Pew Research Center*, 22 June 2021, www.pewresearch.org/fact-tank/2019/05/07/digital-divide-persists-even-as-lower-income-americans-make-gains-in-tech-adoption. Accessed 13 July 2023.

Anderson, Steve. "Net Neutrality: The View from Canada." *Media Development*, vol. 56, no. 1, 2009, pp. 8-11.

Andrew, Levshon. *Reformatted: Code, Networks, and the Transformation of the Music Industry*. Oxford UP, 2014.

"App Store." *Apple*, 2023, www.apple.com/app-store. Accessed 13 July 2023.

Archer, L. Bruce. "Whatever Became of Design Methodology?" *Design Studies*, vol. 1, no. 1, 1979, pp. 17-20.

Arditi, David. *Streaming Culture: Subscription Services and the Unending Consumption of Culture*. Emerald Group, 2021.

Arnold, Gordon. *Animation and the American Imagination: A Brief History*. Praeger, 2017.

Arp, Charlie. *Archival Basics: A Practical Manual for Working with Historical Collections*. Rowman & Littlefield, 2019.

ARRL Ham Radio License Manual. 5th ed., ARRL, 2022.

Artasanchez, Alberto, and Prateek Joshi. *Artificial Intelligence with Python*. 2nd ed., Packt, 2020.

Artero, Juan P. "Online Video Business Models: YouTube vs. Hulu." *Palabra Clave*, vol. 13, no. 1, 201, pp. 111-23.

Asarch, Steven. "What Is Twitch? Understanding the Explosive Live-Streaming Service." *Newsweek*, 3 May 2018, www.newsweek.com/2018/05/11/twitch-909594.html. Accessed 13 July 2023.

Ascher, Steven. *The Filmmaker's Handbook: A Comprehensive Guide for the Digital Age*. Plume, 2012.

Ashcroft, Fran. *The Analogue Approach to Digital Recording and Mixing*. Crowood Press, 2023.

Aslam, Salman. "Instagram by the Numbers: Stats, Demographics & Fun Facts." *Omnicore*, 21 June 2017, www.omnicoreagency.com/instagram-statistics. Accessed 13 July 2023.

Audry, Sofian. *Art in the Age of Machine Learning*. MIT Press, 2021.

Augstein, Mirjam, Eelco Herder, and Wolfgang Wörndl, editors. *Personalized Human-Computer Interaction*. 2nd ed., De Gruyter Oldenbourg, 2023.

Auxier, Brooke, and Monica Anderson. "Social Media Use in 2021." *Pew Research Center*, 7 Apr. 2021, www.pewresearch.org/internet/2021/04/07/social-media-use-in-2021. Accessed 26 July 2023.

Baca, Murtha, editor. *Introduction to Metadata*. 3rd ed., Getty Research Institute, 2016.

Bacus, John. *Digital Sketching: Computer-Aided Conceptual Design*. Wiley, 2020.

Bailey, David Evans. "Ten Cool Applications for Virtual Reality That Aren't Just Games." *Conversation*, 22 Mar. 2016, theconversation.com/ten-cool-applications-for-virtual-reality-that-arent-just-games-56365. Accessed 13 July 2023.

Bailey, Jonathan. "What Is Creative Commons Anyway?" *Plagiarism.org*, 24 July 2018, www.plagiarism.org/blog/2018/07/24/what-is-creative-commons-anyway. Accessed 13 July 2023

Bainbridge, William Sims, editor. *Berkshire Encyclopedia of Human-Computer Interaction: When Science Fiction Becomes Fact*. Berkshire, 2004. 2 vols.

Baines, Phil, and Andrew Haslam. *Type and Typography*. 2nd ed., Laurence King, 2005.

Ballora, Mark. *Essentials of Music Technology*. Oxford UP, 2015.

Ballou, Glen, editor. *Handbook for Sound Engineers*. 5th ed., Focal Press, 2015.

Banga, Cameron, and Josh Weinhold. *Essential Mobile Interaction Design: Perfecting Interface Design in Mobile Apps*. Addison-Wesley, 2014.

Barlow, Aaron. *The DVD Revolution: Movies, Culture, and Technology*. Praeger, 2005.

Barnett, Chance. "Top 10 Crowdfunding Sites for Fundraising." *Forbes*, 8 May 2013, www.forbes.com/sites/chancebarnett/2013/05/08/top-10-crowdfunding-sites-for-fundraising/?sh=4dd914103850. Accessed 13 July 2023.

Bas, Patrick, et al. *Watermarking Security*. Springer Singapore, 2016.

Basin, Ken. *The Business of Television*. Routledge, 2019.

Basl, John. "The Ethics of Creating Artificial Consciousness." *American Philosophical Association Newsletters: Philosophy and Computers*, vol. 13., no. 1, 2013, pp. 25-30.

Bass, Henry E., and William J. Cavanaugh, editors. *ASA at Seventy-Five*. Acoustical Society of America, 2004.

Beck, U. *Power in the Global Age: A New Global Political Economy*. Polity Press, 2006.

Beer, Jeff. "What It Really Means That Spotify Has Lost Its Second Top Marketing Executive in a Week." *Fast Company*, 18 Sept. 2018, www.fastcompany.com/90238502/what-it-really-means-that-spotify-has-lost-its-second-top-marketing-executive-in-a-week. Accessed 25 July 2023.

Bell, Mary Ann, Mary Ann Berry, and James L. Van Roekel. *Internet and Personal Computing Fads*. Haworth Press, 2004.

Bell, Tom. *Programming: A Primer; Coding for Beginners*. Imperial College Press, 2016.

Ben Manmoud-Jouini, Sihem, Christophe Midler, and Philippe Silberzahn. "Contributions of Design Thinking to Project Management in an Innovation Context." *Project Management Journal*, vol. 47, no. 2, 2016, pp. 144-56.

Bennett, James, and Tom Brown, editors. *Film and Television after DVD*. Routledge, 2008.

Berge, Zane L., and Lin Muilenburg. *Handbook of Mobile Learning*. Routledge, 2013.

Berger, Martin A. *Freedom Now!: Forgotten Photographs of the Civil Rights Struggle*. U of California P, 2013.

Berger, Ross. *Dramatic Storytelling & Narrative Design: A Writer's Guide to Video Games and Transmedia*. CRC Press, 2019.

Bergin, Michael S. "A Brief History of BIM." *ArchDaily*, 7 Dec. 2012, www.archdaily.com/302490/a-brief-history-of-bim. Accessed 13 July 2023.

Berlatsky, Noah. *Artificial Intelligence*. Greenhaven Press, 2011.

Berners-Lee, Tim. *Weaving the Web: The Original Design and Ultimate Destiny of the World Wide Web by Its Inventor*. Harper, 1999.

Berners-Lee, Tim, James Hendler, and Ora Lassila. "The Semantic Web." *Scientific American*, 1 May 2021, www.scientificamerican.com/article/the-semantic-web. Accessed 13 July 2023.

Berry, M. A. J., and G. Linoff. *Data Mining Techniques for Marketing, Sales and Customer Support*. Wiley, 1997.

"Best General-Purpose CAD Software for Enterprise Businesses." *G2*, 2023, www.g2.com/categories/general-purpose-cad. Accessed 13 July 2023.

Bevan, Kate. "Instagram Is Debasing Real Photography." *Guardian*, 19 July 2012, www.theguardian.com/technology/2012/jul/19/instagram-debasing-real-photography. Accessed 13 July 2023.

Beyer, Robert T. *Sounds of Our Times: Two Hundred Years of Acoustics*. Springer-Verlag, 1999.

Bi, Zhuming, and Xiaoqin Wang. *Computer Aided Design and Manufacturing*. Wiley-ASME Press, 2020.

Biden, Joe. "Promoting Competition in the American Economy." *Federal Register*, 14 July 2021, www.federalregister.gov/documents/2021/07/14/2021-15069/promoting-competition-in-the-american-economy. Accessed 13 July 2023.

Biffl, Stefan, and Marta Sabou, editors. *Semantic Web Technologies for Intelligent Engineering Applications*. Springer International Publishing, 2016.

Binmore, K. G. *Game Theory: A Very Short Introduction*. Oxford UP, 2007.

Bloom, Kristen, and Kelly Marie Johnston. "Digging into YouTube Videos: Using Media Literacy and Participatory Culture to Promote Cross-Cultural Understanding." *Journal of Media Literacy Education*, vol. 2, no. 2, 2013, p. 3.

Bly, Robert W. *The Copywriter's Handbook: A Step-by-Step Guide to Writing Copy That Sells*. 4th ed., St. Martin's Griffin, 2020.

Bogdon, Razuan, et al. "A Practical Experience on the Amazon Alexa Integration in Smart Offices." *MDPI*, 22 Jan. 2021, www.mdpi.com/1424-8220/21/3/734. Accessed 13 July 2023.

Booch, Grady, et al. *Object-Oriented Analysis and Design with Applications*. 3rd ed., Addison-Wesley, 2007.

Boorstin, Julia. "YouTube Officially Launches YouTube TV in Select Markets." *CNBC*, 5 Apr. 2017, www.cnbc.com/2017/04/05/youtube-officially-launches-youtube-tv-in-select-markets.html. Accessed 13 July 2023.

Borja, K., and S. Dieringer. "Streaming or Stealing? The Complementary Features between Music Streaming and Music Piracy." *Journal of Retailing & Consumer Services*, vol. 32, 2016, pp. 86-95.

Borrmann, André, Markus König, Christian Koch, and Jakob Beetz, editors. *Building Information Modeling: Technology Foundations and Industry Practice*. Springer International Publishing, 2018.

Bostrom, Nick. "Ethical Issues in Advanced Artificial Intelligence." *NickBostrom.com*, 2003, nickbostrom.com/ethics/ai.html. Accessed 13 July 2023.

Bostrom, Nick, and Eliezer Yudkowsky. "The Ethics of Artificial Intelligence." *The Cambridge Handbook of Artificial Intelligence*, edited by Keith Frankish and William M. Ramsay, Cambridge UP, 2014, pp. 316-34.

Botto, Richard. *Crowdsourcing for Filmmakers: Indie Film and the Power of the Crowd*. Routledge, 2017.

Bourdieu, P. "The Forms of Capital." *Handbook of Theory and Research in the Sociology of Education*, edited by J. C. Richardson, Greenwood Press, 1986, pp. 241-58.

Brabham, Daren C. *Crowdsourcing*. MIT Press, 2013.

Brachmann, Steve. "Video on Demand Continues to Revolutionize TV, Movie Industries." *IPWatchdog*, 28 Mar. 2015, www.ipwatchdog.com/2015/03/28/video-on-demand-revolutionize-tv-movie/id=55900. Accessed 25 July 2023.

Bradford, Alina. "Everything You Need to Master Instagram Stories." *CNet*, 24 Apr. 2018, www.cnet.com/how-to/how-to-use-instagram-stories. Accessed 13 July 2023.

"Brand Value of Instagram Worldwide from 2020 to 2022." *Statista*, May 2022, www.statista.com/statistics/1324427/instagram-brand-value. Accessed 13 July 2023.

Brant, Tom. "SSD vs. HDD: What's the Difference?" *PCMag*, 26 Aug. 2022, www.pcmag.com/news/ssd-vs-hdd-whats-the-difference. Accessed 13 July 2023.

Brian, Matt. "Twitter Praised for Protecting User Data as Apple and Yahoo Falter in Privacy Report." *Verge*, 1 May 2013, www.theverge.com/2013/5/1/4289052/twitter-top-ranked-eff-privacy-report. Accessed 26 July 2023.

"A Brief SGML Tutorial." *W3C*, www.w3.org/TR/WD-html40-970708/intro/sgmltut.html. Accessed 13 July 2023.

Brin, Sergey, and Lawrence Page. "The Anatomy of a Large-Scale Hypertextual Web Search Engine." *Stanford University InfoLab*, 1998, snap.stanford.edu/class/cs224w-readings/Brin98Anatomy.pdf. Accessed 13 July 2023.

Brine, Kelly Gordon. *The Art of Cinematic Storytelling: A Visual Guide to Planning Shots, Cuts, and Transitions*. Oxford UP, 2020.

Bringhurst, Robert. *The Elements of Typographic Style*. 3rd ed., Hartley and Marks, 2005.

Brodley, Carla E. "Challenges and Opportunities in Applied Machine Learning." *AI Magazine*, vol. 33, no. 1, 2012, pp. 11-24.

Brown, Adrian. "Graphics File Formats." *National Archives*, Aug. 2008, www.nationalarchives.gov.uk/documents/information-management/graphic-file-formats.pdf. Accessed 13 July 2023.

Brown, Andrew R. *Music Technology and Education: Amplifying Musicality*. 2nd ed., Routledge, 2015.

Brown, Blain. *The Basics of Filmmaking: Screenwriting, Producing, Directing, Cinematography, Audio, and Editing*. Routledge, 2020.

_____. *Cinematography: Theory and Practice*. 4th ed., Routledge, 2022.

Brown, Eric. "True Physics." *Game Developer*, vol. 17, no. 5, 2010, pp. 13-18.

Brown, Michael. "Microsoft Windows Is 30: A Short History of One of the Most Iconic Tech Products Ever." *International Business Times*, 20 Nov. 2015, www.ibtimes.com/microsoft-windows-30-short-history-one-most-iconic-tech-products-ever-2194091. Accessed 13 July 2023.

Bryfonski, Dedria. *The Global Impact of Social Media*. Greenhaven, 2012.

Buckingham, David. *The Media Literacy of Children and Young People: A Review of the Research Literature on Behalf of Ofcom*. Ofcom, 2005.

Bueno de Mesquita, Bruce. *The Predictioneer's Game: Using the Logic of Brazen Self-Interest to See and Shape the Future*. Random House, 2009.

Burger, Arnold S. *Debugging Embedded and Real-Time Systems*. Newnes, 2020.

Burgess, Matt. "What Is the Internet of Things? WIRED Explains." *Wired*, 16 Feb. 2018, www.wired.co.uk/article/internet-of-things-what-is-explained-iot. Accessed 13 July 2023.

Burke, Brian. *Gamify: How Gamification Motivates People to Do Extraordinary Things*. Bibliomotion, 2014.

Burns, Monica. *Deeper Learning with QR Codes and Augmented Reality: A Scannable Solution for Your Classroom*. Corwin Press, 2016.

Butcher, Kenneth S., Linda D. Crown, and Elizabeth J. Gentry, editors. *The International System of Units (SI):*

Conversion Factors for General Use. National Institute of Standards and Technology, 2006.

Butler, Jeremy G. *Television: Visual Storytelling and Screen Culture*. 5th ed., Routledge, 2018.

Bygrave, Lee A. *Data Privacy Law, an International Perspective*. Oxford UP, 2014.

_____. "The Place of Privacy in Data Protection Laws." *University of New South Wales Law Journal*, vol. 24, no. 1, 2001, pp. 277-83.

Cailliau, Robert, and Helen Ashman. "Hypertext in the Web: A History." *ACM Computing Surveys*, vol. 31, Dec. 1999, pp. 1-6.

Calhoun, Karen. *Exploring Digital Libraries: Foundations, Practice, Prospects*. Facet, 2018.

Calore, Michael. "A History of Microsoft Windows." *Wired*, 10 Dec. 2008, www.wired.com/2008/12/wiredphotos31. Accessed 13 July 2023.

Calude, Cristian S., editor. *The Human Face of Computing*. Imperial College Press, 2016.

Campbell-Kelly, Martin, William F. Aspray, Jeffrey R. Yost, Honghong Tinn, and Gerardo Con Díaz. *Computer: A History of the Information Machine*. 4th ed., Routledge, 2023.

Cardarelli, Francois. *Encyclopedia of Scientific Units, Weights and Measures: Their SI Equivalences and Origins*. Springer, 2003.

Carroll, Brian. *Writing and Editing for Digital Media*. 4th ed., Routledge, 2019.

Cartwright, William, Michael P. Peterson, and Georg F. Garner, editors. *Multimedia Cartography*. 2nd ed., Springer, 2007.

Casabona, Joe. *HTML and CSS: Visual QuickStart Guide*. 9th ed., Peachpit Press, 2020.

Castells, M. *The Internet Galaxy: Reflections on the Internet, Business and Society*. Oxford UP, 2001.

Castro, Elizabeth, and Bruce Hyslop. *HTML and CSS*. 8th ed., Peachpit Press, 2014.

Cate, Fred. "The Changing Face of Privacy Protections in the European Union and the United States." *Indiana Law Review*, vol. 33, 1999, pp. 173-232.

Cellan-Jones, Rory. "Deepfake Videos 'Double in Nine Months.'" *BBC News*, 7 Oct. 2019, www.bbc.com/news/technology-49961089. Accessed 13 July 2023.

Chaffey, Dave, Tanya Hemphill, and David Edmundson-Bird. *Digital Business and E-Commerce Management*. 7th ed., Pearson Education, 2019.

Chalaby, Jean. *Television in the Streaming Era*. Cambridge UP, 2023.

Chamberlain, P. "From Screen to Monitor." *Engineering & Technology*, vol. 3, no. 15, 2008, pp. 18-21.

Chan, Melanie. *Virtual Reality: Representations in Contemporary Media*. Bloomsbury, 2014.

Chao, Loretta. "Tech Partnership Looks beyond the Bar Code with Digital Watermarks." *Wall Street Journal*, 12 Jan. 2016, www.wsj.com/articles/tech-partnership-looks-beyond-the-bar-code-with-digital-watermarks-1452623450. Accessed 13 July 2023.

Chen, Angela. "Three Threats Posed by Deepfakes That Technology Won't Solve." *MIT Technology Review*, 2 Oct. 2019, www.technologyreview.com/s/614446/deepfake-technology-detection-disinformation-harassment-revenge-porn-law. Accessed 13 July 2023.

Chen, Chih-Ping. "Exploring Personal Branding on YouTube." *Journal of Internet Commerce*, vol. 12, no. 4, 2013, pp. 332-47.

Chen, W. "The Implications of Social Capital for the Digital Divides in America." *Information Society*, vol. 29, 2013, pp. 13-25.

Chevalier, Stephanie. "Global Retail E-Commerce Sales 2014-2026." *Statista*, 21 Sept. 2022, www.statista.com/statistics/379046/worldwide-retail-e-commerce-sales. Accessed 13 July 2023.

Ching, Francis D. K. *Architectural Graphics*. 6th ed., Wiley, 2015.

Choi, Paul J., Rod J. Oskouian, and R. Shane Tubbs. "Telesurgery: Past, Present, and Future." *Cureus*, vol. 10, no. 5, 2018, www.cureus.com/articles/12751-telesurgery-past-present-and-future. Accessed 13 July 2023.

Chokkattu, Julian. "The Top New Features in Apple's iOS 17 and iPadOS 17." *Wired*, 8 June 2023, www.wired.com/story/apple-iphone-ios-17-ipados-17-new-features. Accessed 13 July 2023.

Chokshi, Niraj. "Facebook Helped Drive a Voter Registration Surge, Election Officials Say." *New York Times*, 12 Oct. 2016, www.nytimes.com/2016/10/13/us/politics/facebook-helped-drive-a-voter-registration-surge-election-officials-say.html. Accessed 13 July 2023.

Chowdhry, Amit. "Facebook Is Going to Suppress 'Click-Bait' Articles." *Forbes*, 26 Aug. 2014, www.forbes.com/sites/amitchowdhry/2014/08/26/facebook-is-going-to-suppress-click-bait-articles/#51f7bf8b47b0. Accessed 13 July 2023.

Christian, Sue Ellen. *Everyday Media Literacy: An Analog Guide for Your Digital Life*. 2nd ed., Routledge, 2023.

Cipolla-Ficarra, Francisco. *Optimizing Human-Computer Interaction with Emerging Technologies*. IGI Global, 2018.

Cisternino, Doriana, et al. "Augmented Reality Applications to Support the Promotion of Cultural Heritage: The Case of the Basilica of Saint Catherine of

Alexandria in Galatina." *Journal on Computing and Cultural Heritage*, vol. 14, no. 4, Dec. 2021, pp. 1-30.

Clark, Elaine A. *Voice-Overs for Podcasting: How to Develop a Career and Make a Profit*. Allworth, 2020.

Clark, Jocalyn, and Richard Smith. "Firm Action Needed on Predatory Journals." *BMJ*, no. 350, 2015, p. h210.

Clark, Ruth Colvin, and Richard E. Mayer. *E-Learning and the Science of Instruction: Proven Guidelines for Consumers and Designers of Multimedia Learning*. 5th ed., Wiley, 2023.

Clifford, Catherine. "Crowdfunding Generates More Than $60,000 an Hour." *Entrepreneur*, 19 May 2014, www.entrepreneur.com/starting-a-business/crowdfunding-generates-more-than-60000-an-hour/234051. Accessed 13 July 2023.

Closs, Wyatt. "Calling Them 'Girls' Was Their First Mistake: What Happened Next Is a Great Moment in Fierceness." *Upworthy*, 11 July 2013, www.upworthy.com/calling-them-girls-was-their-first-mistake-what-happens-next-is-a-great-moment-in-fierceness. Accessed 13 July 2023.

Collins, Ben. "Facebook to Restrict Livestream Feature after Christchurch Attack." *NBC News*, 14 May 2019, www.nbcnews.com/tech/tech-news/facebook-restrict-livestream-feature-after-christchurch-attack-n1005741. Accessed 13 July 2023.

Collins, Mike. *Pro Tools for Music Production: Recording, Editing, and Mixing*. 2nd ed., Routledge, 2017.

"Comer, Douglas. *Computer Networks and Internets*. Pearson, 2015.

_____. *The Internet Book: Everything You Need to Know about Computer Networking and How the Internet Works*. 5th ed., Chapman and Hall/CRC, 2018.

"Commercial Advertisement." *University of Washington*, depts.washington.edu/chinaciv/graph/tcommain.htm. Accessed 13 July 2023.

"Company Overview of Spotify Technology S.A." *Bloomberg*, www.bloomberg.com/research/stocks/private/snapshot.asp?privcapId=225595077. Accessed 25 July 2023.

"Computational Modeling." *National Institute of Biomedical Imaging and Bioengineering*, May 2020, www.nibib.nih.gov/science-education/science-topics/computational-modeling. Accessed 13 July 2023.

"Computer Animator." *Art Career Project*, 15 July 2021, theartcareerproject.com/careers/computer-animation. Accessed 13 July 2023.

"A Computer in Your Pocket: The Rise of Smartphones." *Science Museum*, 13 Nov. 2018, www.sciencemuseum.org.uk/objects-and-stories/computer-your-pocket-rise-smartphones. Accessed 13 July 2023.

"Conceptual Art." *Tate*, www.tate.org.uk/learn/online-resources/glossary/c/conceptual-art#introduction. Accessed 13 July 2023.

Congdon-Fuller, Ashleigh, Antonio Ramirez, and Douglas Smith. *Technical Drawing 101 with AutoCAD 2022: A Multidisciplinary Guide to Drafting Theory and Practice*. SDC Publications, 2021.

Constine, Josh. "Instagram Plans June 20th Launch Event for Long-Form Video Hub." *TechCrunch*, 11 June 2018, techcrunch.com/2018/06/11/instagram-long-form-video. Accessed 13 July 2023.

Constine, Josh, and Kim-Mai Cutler. "Facebook Buys Instagram for $1 Billion, Turns Budding Rival into Its Standalone Photo App." *TechCrunch*, 9 Apr. 2012, techcrunch.com/2012/04/09/facebook-to-acquire-instagram-for-1-billion. Accessed 13 July 2023.

Conti, Robyn, and John Schmidt. "What You Need to Know about Non-Fungible Tokens (NFTs)." *Forbes Advisor*, 14 May 2021, www.forbes.com/advisor/investing/nft-non-fungible-token. Accessed 13 July 2023.

Conway, Luke. "Blockchain Explained." *Investopedia*, 1 June 2021, www.investopedia.com/terms/b/blockchain.asp. Accessed 13 July 2023.

Cook, Vickie S. "Net Neutrality: What Is It and Why Should Educators Care?" *Delta Kappa Gamma Bulletin*, vol. 80, no. 4, 2014, pp. 46-49.

Copeland, Jack. "What Is Artificial Intelligence?" *AlanTuring.net*, May 2000, www.alanturing.net/turing_archive/pages/reference%20articles/what_is_AI/What%20is%20AI05.html. Accessed 13 July 2023.

"Core Principles of Media Literacy Education in the United States." *NAMLE*, namle.net/resources/core-principles. Accessed 13 July 2023.

Cormen, Thomas H. *Algorithms Unlocked*. MIT Press, 2013.

Corrado, Edward M., and Heather Moulaison Sandy. *Digital Preservation for Libraries, Archives, and Museums*. 2nd ed., Rowman & Littlefield, 2017.

Costello, Sam. "The History of iOS, from Version 1.0 to 17.0." *Lifewire*, 5 June 2023, www.lifewire.com/ios-versions-4147730. Accessed 13 July 2023.

Costello, Vic. *Multimedia Foundations: Core Concepts for Digital Design*. 3rd ed., Focal Press, 2023.

Coulthard, Charissa. "Self-Portraits and Social Media: The Rise of the 'Selfie.'" *BBC News*, 7 June 2013, www.bbc.com/news/magazine-22511650. Accessed 13 July 2023.

Council of Europe. *Journalism at Risk: Threats, Challenges and Perspectives*. Council of Europe, 2015.

Cowan, Ruth Schwartz, and Matthew H. Hersch. *A Social History of American Technology*. 2nd ed., Oxford UP, 2017.

Cox, Debbie. "Developing and Raising Awareness of the Zine Collections at the British Library." *Art Libraries Journal*, vol. 43, no. 2, 2018, pp. 77-81.

Cox, Kate. "It's Time to Start Treating Video Game Industry Like the $21 Billion Business It Is." *Consumerist*, 9 June 2014, consumerist.com/2014/06/09/its-time-to-start-treating-video-game-industry-like-the-21-billion-business-it-is. Accessed 22 Dec. 2014.

Cox, Mike, Ellen Mulder, and Linda Tadic. *Descriptive Metadata for Television*. Focal, 2006.

Craig, Alan B., William R. Sherman, and Jeffrey D. Will. *Developing Virtual Reality Applications. Foundations of Effective Design*. Morgan Kaufmann, 2009.

Crain, Matthew. *Profit over Privacy: How Surveillance Advertising Conquered the Internet*. U of Minnesota P, 2021.

Cramer-Flood, Ethan. "Global Ecommerce Update 2021: Worldwide Ecommerce Will Approach $5 Trillion This Year." *eMarketer, Business Insider Intelligence*, Jan. 2021, www.emarketer.com/content/global-ecommerce-update-2021. Accessed 13 July 2023.

Cross, Mark. *Audio Post Production for Film and Television*. Berklee Press, 2013.

Cukier, Kenneth. *Big Data: A Revolution that Will Transform How We Live, Work, and Think*. Dolan, 2013.

Cunningham, Andrew. "Apple Previews a Trio of Apps That Will Finally Replace iTunes for Windows." *Ars Technica*, 12 Jan. 2023, arstechnica.com/gadgets/2023/01/apple-previews-a-trio-of-apps-that-will-finally-replace-itunes-for-windows. Accessed 13 July 2023.

Cunningham, William Michael. *The JOBS Act: Crowdfunding for Small Businesses and Startups*. Springer, 2012.

Curry, David. "Android Statistics (2023)." *Business of Apps*, 27 Feb. 2023, www.businessofapps.com/data/android-statistics. Accessed 13 July 2023.

Curtis, Craig R., Michael C. Gizzi, and Michael J. Kittleson. "Using Technology the Founders Never Dreamed Of: Cell Phones as Tracking Devices and the Fourth Amendment." *University of Denver Criminal Law Review*, vol. 4, no. 61, 2014, pp. 61-101.

D'Onfro, Jillian. "Here's a Reminder of Just How Huge Google Search Truly Is." *Tech Insider*, 27 Mar. 2016, www.businessinsider.com/google-search-engine-facts-2016-3. Accessed 13 July 2023.

Dalby, James, and Matthew Freeman, editors. *Transmedia Selves: Identity and Persona Creation in the Age of Mobile and Multiplatform Media*. Routledge, 2023.

Dale, Nell, and John Lewis. *Computer Science Illuminated*. 7th ed., Jones & Bartlett Learning, 2019.

Danae Human Intelligence. "What Is Digital Art? Definition and Scope of the New Media." *Digital Art Weekly*, 31 Aug. 2018, medium.com/digital-art-weekly-by-danae-hi/what-is-digital-art-definition-and-scope-of-the-new-media-f645058cfd78. Accessed 13 July 2023.

Dancyger, Ken. *The Technique of Film and Video Editing: History, Theory, and Practice*. Focal, 2013.

Dang, Sheila, Katie Paul, and Dawn Chmielewski. "Focus: Do Spam Bots Really Comprise Under 5% of Twitter Users? Elon Musk Wants to Know." *Reuters*, 13 May 2022, www.reuters.com/technology/do-spam-bots-really-comprise-under-5-twitter-users-elon-musk-wants-know-2022-05-13. Accessed 13 July 2023.

Das, Sumitabha. *Your UNIX/Linux: The Ultimate Guide*. 3rd ed., McGraw-Hill, 2012.

Dash, Anil. "NFTs Weren't Supposed to End Like This." *The Atlantic*, 2 Apr. 2021, www.theatlantic.com/ideas/archive/2021/04/nfts-werent-supposed-end-like/618488. Accessed 13 July 2023.

Dastin, Jeffrey, Krystal Hu, and Paresh Dave. "Exclusive: ChatGPT Owner OpenAI Projects $1 Billion Revenue by 2024." *Reuters*, 15 Dec. 2022, www.reuters.com/business/chatgpt-owner-openai-projects-1-billion-revenue-by-2024-sources-2022-12-15. Accessed 18 July 2023.

"Data Measurement Chart." *University of Florida*, www.wu.ece.ufl.edu/links/dataRate/DataMeasurementChart.html. Accessed 13 July 2023.

Davis, Riccardo A. "NFTs: What Are They, And How Do They Work?" *Kiplinger*, 6 May 2021, www.kiplinger.com/investing/602743/nfts-what-are-they-and-how-do-they-work. Accessed 13 July 2023.

Davis, Yvette. *Google Secrets: Do What You Never Thought Possible with Google*. Wiley, 2011.

De Courtenay, Nadine, Olivier Darrigol, and Oliver Schlaudt, editors. *The Reform of the International System of Units (SI): Philosophical, Historical and Sociological Issues*. Routledge, 2019.

De Swaan Arons, Marc. "How Brands Were Born: A Brief History of Modern Marketing." *The Atlantic*, 3 Oct. 2011, www.theatlantic.com/business/archive/2011/10/how-brands-were-born-a-brief-history-of-modern-marketing/246012. Accessed 13 July 2023.

Dean, John. *Web Programming with HTML5, CSS, and JavaScript*. Jones & Bartlett Learning, 2018.

Dekker, Annet, editor. *Curating Digital Art: From Presenting and Collecting Digital Art to Networked Co-curation.* Valiz, 2021.

Delfs, Hans, and Helmut Knebl. *Introduction to Cryptography: Principles and Applications.* 3rd ed., Springer, 2015.

Denham, Hannah. "These Are the Platforms That Have Banned Trump and His Allies." *Washington Post,* 14 Jan. 2021, www.washingtonpost.com/technology/2021/01/11/trump-banned-social-media. Accessed 26 July 2023.

Denning, Stephen. *The Leader's Guide to Storytelling: Mastering the Art and Discipline of Business Narrative.* 2nd ed., Jossey-Bass, 2011.

Dennis, Alan, Barbara Haley Wixom, and David Tegarden. *Systems Analysis and Design: An Object-Oriented Approach with UML.* 5th ed., Wiley, 2015.

"Deprecated HTML5 Tags and Attributes." *DESE,* 14 Mar. 2018, www.doe.mass.edu/nmg/html5-deprecated.html. Accessed 13 July 2023.

The Design Book. Rev. ed., Phaidon, 2020.

Deutsch, Randy. *BIM and Integrated Design: Strategies for Architectural Practice.* Wiley, 2011.

"Development History." *W3C,* 6 Jan. 2003, www.w3.org/XML/hist2002. Accessed 13 July 2023.

DeWitte, Debra J., M. Kathryn Shields, and Ralph M. Larmann. *Gateways to Art: Understanding the Visual Arts.* 4th ed., Thames & Hudson, 2023.

Dhillon, Gupreet. "Dimensions of Power and IS Implementation." *Information and Management,* vol. 41, no. 5, 2004, pp. 635-44.

Di Pietro, Francesca. *Crowdfunding for Entrepreneurs: Developing Strategic Advantage through Entrepreneurial Finance.* Routledge, 2020.

Diamond, Ethan. "Bandcamp Is Joining Epic Games." *Bandcamp Updates,* 2 Mar. 2022, blog.bandcamp.com/2022/03/02/bandcamp-is-joining-epic. Accessed 13 July 2023.

DiChristina, Mariette. "How Human Creativity Arose." *Scientific American,* 1 Mar. 2013, www.scientificamerican.com/article/how-human-creativity-arose. Accessed 13 July 2023.

Diduck, Ryan. *Mad Skills: MIDI and Music Technology in the Twentieth Century.* Repeater, 2018.

Dieny, Bernard, Ronald B. Goldfarb, and Kyung-Jin Lee. *Introduction to Magnetic Random-Access Memory.* IEEE Press/John Wiley & Sons, 2017.

"Different Types of Software." *Introduction to IT English,* web2.uvcs.uvic.ca/elc/sample/ite/u01/u1_1_03.html. Accessed 13 July 2023.

"Digital Art." *Art Story,* 2019, www.theartstory.org/movement-digital-art-history-and-concepts.htm#beginnings_header. Accessed 13 July 2023.

Dijksterhuis, Eva, and Gilbert Silvius. "The Design Thinking Approach to Projects." *Project Management Development-Practices and Perspectives,* vol. 5, no. 6, 2016, pp. 67-81.

"Distance Learning." *IES NCES,* 2022, nces.ed.gov/fastfacts/display.asp?id=80. Accessed 13 July 2023.

Dittmar, Tim. *Audio Engineering 101: A Beginner's Guide to Music Production.* 2nd ed., Routledge, 2018.

Dixon, S. "Number of Twitter Users Worldwide from 2019 to 2024." *Statista,* 14 Dec. 2022, www.statista.com/statistics/303681/twitter-users-worldwide. Accessed 26 July 2023.

Dodge, Martin, Rob Kitchin, and Chris Perkins, editors. *Rethinking Maps: New Frontiers in Cartographic Theory.* Routledge, 2009.

Domingos, Pedro. "A Few Useful Things to Know about Machine Learning." *Communications of the ACM,* vol. 55, no. 10, 2012, pp. 78-87.

Donally, Jaime. *The Immersive Classroom: Create Customized Learning Experiences with AR/VR.* International Society for Technology in Education, 2021.

Dordal, Peter L. "An Introduction to Computer Networks." *Loyola University Chicago,* 2015, intronetworks.cs.luc.edu/current/html. Accessed 13 July 2023.

Douglas, Nick. "Get Rid of YouTube's Terrible Clickbait Thumbnail Images." *Lifehacker,* 5 Aug. 2019, lifehacker.com/get-rid-of-youtubes-terrible-clickbait-thumbnail-images-1836969236. Accessed 13 July 2023.

Dowling, David. *The Gamification of Digital Journalism: Innovation in Journalistic Storytelling.* Routledge, 2020.

Downes, Larry. "Unscrambling the FCC's Net Neutrality Order: Preserving the Open Internet-But Which One?" *CommLaw Conspectus,* vol. 20, no. 1, 2011, pp. 83-128.

Dredge, Stuart. "Kickstarter's Biggest Hits: Why Crowdfunding Now Sets the Trends." *Guardian,* 17 Apr. 2014, www.theguardian.com/technology/2014/apr/17/kickstarter-crowdfunding-technology-film-games. Accessed 13 July 2023.

Duckett, Jon. *Web Design with HTML, CSS, JavaScript and jQuery.* 2 vols. John Wiley & Sons, 2014.

Dushnitsky, Gary. "What If Crowdfunding Becomes the Leading Source of Finance for Entrepreneurs or Growing Companies?" *Forbes,* 20 May 2015, www.forbes.com/sites/lbsbusinessstrategyreview/2015/05/20/what-if-crowdfunding-becomes-the-leading-source-of

-finance-for-entrepreneurs-or-growing-companies/?sh=248b28aa4faa. Accessed 13 July 2023.

Dutson, Phil. *Creating QR and Tag Codes*. Sams Publishing, 2013.

Dwoskin, Elizabeth, and Craig Timberg. "Facebook Wanted 'Visceral' Live Video: It's Getting Live-Streaming Killers and Suicides." *Washington Post*, 17 Apr. 2017, www.washingtonpost.com/business/technology/facebook-wanted-visceral-live-video-its-getting-suicides-and-live-streaming-killers/2017/04/17/a6705662-239c-11e7-a1b3-faff0034e2de_story.html. Accessed 13 July 2023.

Edgerton, Gary R. *The Columbia History of American Television*. Columbia UP, 2009.

Edwards, Benj. "The Little-Known Apple Lisa: Five Quirks and Oddities." *Macworld*, 30 Jan. 2013, www.macworld.com/article/2026544/the-little-known-apple-lisa-five-quirks-and-oddities.html. Accessed 13 July 2023.

Edwards, Jim. "Proof That Android Really Is for the Poor." *Business Insider*, 27 June 2014, www.businessinsider.in/Proof-That-Android-Really-Is-For-The-Poor/articleshow/37328668.cms. Accessed 13 July 2023.

Edwards, Paul N. *A Vast Machine: Computer Models, Climate Data, and the Politics of Global Warming*. MIT Press, 2010.

"Electronica Music Guide: The History and Sound of Electronica." *MasterClass*, 14 Sept. 2021, www.masterclass.com/articles/electronica-music-guide. Accessed 13 July 2023.

Engelman, Ralph. *Friendlyvision: Fred Friendly and the Rise and Fall of Television Journalism*. Columbia UP, 2011.

"Ensuring BotNets Are Not 'Too Big to Investigate.'" *US Department of Justice*, 22 Nov. 2016, www.justice.gov/archives/opa/blog/ensuring-botnets-are-not-too-big-investigate. Accessed 13 July 2023.

Epstein, D., E. C. Nisbet, and T. Gillespie. "Who's Responsible for the Digital Divide? Public Perceptions and Policy Implications." *Information Society*, vol. 27, 2011, pp. 92-104.

Eriksson, Maria, et al. *Spotify Teardown: Inside the Black Box of Streaming Music*. MIT Press, 2019.

Erl, Thomas, editor. *Big Data Fundamentals: Concepts, Drivers & Techniques*. Pearson, 2016.

Eskilson, Stephen. *Digital Design: A History*. Princeton UP, 2023.

"Essential Facts about the Video Game Industry." *Entertainment Software Association*, 2021, www.theesa.com/wp-content/uploads/2021/08/2021-Essential-Facts-About-the-Video-Game-Industry-1.pdf. Accessed 15 Mar. 2022.

Estelles-Miguel, Sofia, Ignacio Gil-Pechuán, and Fernando J. Garrigos-Simon. *Advances in Crowdsourcing*. Springer, 2015.

Etherington, Darrell. "People Now Watch 1 Billion Hours of YouTube Per Day." *Tech Crunch*, 28 Feb. 2017, techcrunch.com/2017/02/28/people-now-watch-1-billion-hours-of-youtube-per-day. Accessed 13 July 2023.

Evans, E. *Transmedia Television: Audiences, New Media, and Daily Life*. Routledge, 2011.

Everest, F. Alton, and Ken C. Pohlmann. *Master Handbook of Acoustics*. 7th ed, McGraw-Hill, 2021.

"Extensible Markup Language (XML)." *W3C*, 11 Oct. 2016, www.w3.org/XML. Accessed 13 July 2023.

Fairchild, Charles. *Sounds, Screens, Speakers: An Introduction to Music and Media*. Bloomsbury Academic, 2019.

Fang, Block. "PAK's NFT Artwork 'The Merge' Sells for $91.8 Million." *Penta*, 7 Dec. 2021, www.barrons.com/articles/paks-nft-artwork-the-merge-sells-for-91-8-million-01638918205. Accessed 13 July 2023.

Farber, Dan. "The Next Big Thing in Tech: Augmented Reality." *CNET*, 7 June 2013, www.cnet.com/news/the-next-big-thing-in-tech-augmented-reality. Accessed 13 July 2023.

Feder, Samuel L., and Luke C. Platzer. "FCC Open Internet Order: Is Net Neutrality Itself Problematic for Free Speech?" *Communications Lawyer*, vol. 28, no. 1, 2011, pp. 1-26.

Feiner, Lauren, and Salvador Rodriguez. "FTC Slaps Facebook with Record $5 Billion Fine, Orders Privacy Oversight." *CNBC*, 24 July 2019, www.cnbc.com/2019/07/24/facebook-to-pay-5-billion-for-privacy-lapses-ftc-announces.html. Accessed 13 July 2023.

Felke-Morris, Terry. *Web Development and Design Foundations with HTML5*. 10th ed., Pearson, 2020.

Ferguson, R. Stuart. *Practical Algorithms for 3D Computer Graphics*. 2nd ed., AK Peters/CRC Press, 2013.

Field, Kenneth. *Cartography: A Compendium of Design Thinking for Mapmakers*. ESRU, Press, 2018.

Fiell, Charlotte, and Peter Fiell. *Industrial Design A-Z*. Taschen, 2016.

Filimowicz, Michael, editor. *Foundations in Sound Design for Embedded Media: A Multidisciplinary Approach*. Routledge, 2020.

Fisher, Len. *Rock, Paper, Scissors: Game Theory in Everyday Life*. Basic Books, 2008.

Fisk, Nathan W. *Understanding Online Piracy: The Truth about Illegal File Sharing*. ABC-CLIO, 2009.

Fitter, Hetal N., Akash B. Pandey, Divyang D. Patel, and Jitendra M. Mistry. "A Review on Approaches for Handling Bezier Curves in CAD for Manufacturing." *Procedia Engineering*, vol. 97, 2014, pp. 1155-66.

FitzGerald, Jerry, Alan Dennis, and Alexandra Durcikova. *Business Data Communications and Networking*. 14th ed., Wiley, 2020.

Foote, Steven. *Learning to Program*. Pearson, 2015.

Forbes, Ted. *Visually Speaking: Mastering Photography as a Visual Language*. Cider Mill Press, 2023.

Forouzan, Behrouz. *Data Communications and Networking*. 5th ed. McGraw-Hill, 2013.

Fox, Kara. "Instagram World Social Media App for Young People's Mental Health." *CNN*, 19 May 2017, www.cnn.com/2017/05/19/health/instagram-worst-social-network-app-young-people-mental-health/index.html. Accessed 13 July 2023.

Fox, Zoe. "The Evolution of Advertising: From Stone Carving to the Old Spice Guy." *Mashable*, 26 Dec. 2011, mashable.com/2011/12/26/history-advertising/#6sdCJNghF5qK. Accessed 13 July 2023.

Frain, Ben. *Design with HTML5 and CSS*. 4th ed., Packt, 2022.

_____. *Responsive Web Design with HTML5 and CSS*. 4th ed., Packt, 2022.

Frampton, Ben. "Clickbait: The Changing Face of Online Journalism." *BBC News*, 14 Sept. 2015, www.bbc.com/news/uk-wales-34213693. Accessed 13 July 2023.

Frazel, Midge. *Digital Storytelling Guide for Educators*. ISTE, 2010.

Freeman, Eric, and David Gelernter. "Lifestreams: A Storage Model for Personal Data." *SIGMOD Record*, vol. 25, no. 1, 1996, pp. 80-86.

Freeman, Matthew, and Renira Rampazzo Gambarato. *The Routledge Companion to Transmedia Studies*. Routledge, 2020.

Freeman, Michael. *Digital Image Editing & Special Effects: Quickly Master the Key Techniques of Photoshop & Lightroom*. Focal, 2013.

Friend, Tad. "Letter from California: Hollywood and Vine." *New Yorker*. Condé Nast, 15 Dec. 2014, www.newyorker.com/magazine/2014/12/15/hollywood-vine. Accessed 13 July 2023.

Frith, Jordan. *Barcode*. Bloomsbury Academic, 2023.

Fuentes, Lauren, and Narée Asherian. "3D Printed Art: 3D Printing Makes Its Mark." *All3DP*, 25 Apr. 2023, all3dp.com/2/3d-printed-art-best-projects. Accessed 13 July 2023.

Fulwood, Nathan. "Opinion: Design Thinking in Publishing and Media." *What's New in Publishing*, 22 June 2021, whatsnewinpublishing.com/opinion-design-thinking-in-publishing-and-media. Accessed 13 July 2023.

Fung, B. "Comcast Is Expanding Its $10-A-Month Internet Program for the Poor." *Washington Post*, 4 Aug. 2014, www.washingtonpost.com/blogs/the-switch/wp/2014/08/04/comcast-is-expanding-its-10-a-month-internet-program-for-the-poor. Accessed 13 July 2023.

Furniss, Maureen. *A New History of Animation*. Thames & Hudson, 2016.

Gallardo-Camacho, Jorge. "The Low Interaction of Viewers of Video Clips on the Internet: The Case Study of YouTube Spain." *Revista Latina de Comunicación de Social*, vol. 13, no. 65, 2010, pp. 1-14.

Gambetta, Gabriel. *Computer Graphics from Scratch*. No Starch Press, 2021.

Garciìa-Barriocanal, Elena, et al., editors. *Metadata and Semantic Research*. Springer, 2011.

Garrison, D. Randy. *E-Learning in the 21st Century: A Community of Inquiry Framework for Research and Practice*. 3rd ed., Routledge, 2016.

Gartner, Richard. *Metadata in the Digital Library: Building an Integrated Strategy with XML*. Facet, 2021.

Garza, Melita M., Michael Fuhlhage, and Tracy Lucht, editors. *The Routledge Companion to American Journalism History*. Routledge, 2023.

Gaston-Bird, Leslie. *Women in Audio*. Focal Press, 2019.

Gee, James Paul. *Unified Discourse Analysis: Language, Reality, Virtual Worlds, and Video Games*. Routledge, 2015.

Gershgorn, Dave. "A California Law Now Means Chatbots Have to Disclose They're Not Human." *BotLaw, Quartz*, 3 Oct. 2018, qz.com/1409350/a-new-law-means-californias-bots-have-to-disclose-theyre-not-human. Accessed 13 July 2023.

Gibbs, Samuel. "From Windows 1 to Windows 10: 29 Years of Windows Evolution." *The Guardian*, 2 Oct. 2014, www.theguardian.com/technology/2014/oct/02/from-windows-1-to-windows-10-29-years-of-windows-evolution. Accessed 13 July 2023.

Gilbert, Jeremy, and Ewan Pearson. *Discographies: Dance Music, Culture, and the Politics of Sound*. Routledge, 1999.

Gillett, Rachel. "How the Most Successful Brands Dominate Instagram, and You Can Too." *Fast Company*, 22 Apr. 2014, www.fastcompany.com/3029395/how-the-most-successful-brands-dominate-instagram-and-you-can-too. Accessed 13 July 2023.

Gintis, Herbert. *Game Theory Evolving: A Problem-Centered Introduction to Modeling Strategic Behavior*. 2nd ed., Princeton UP, 2009.

Giri, Kaiser J., Shabir Ahmad Parah, Rumaan Bashir, and Khan Muhammad, editors. *Multimedia Security: Algorithm Development, Analysis and Applications*. Springer Nature Singapore, 2021.

Gitner, Seth. *Multimedia Storytelling for Digital Communicators in a Multiplatform World*. 2nd ed., Routledge, 2022.

Glaser, J. D. *Secure Development for Mobile Apps: How to Design and Code Secure Mobile Applications with PHP and JavaScript*. CRC Press, 2015.

Glawion, Alex. "Building the Best PC for 3D Animation (2023 Guide)." *CGDirector*, 3 Jan. 2023, www.cgdirector.com/best-computer-for-animation. Accessed 13 July 2023.

"Global Market Share Held by Leading Internet Browsers from January 2012 to May 2023." *Statista*, May 2023, www.statista.com/statistics/268254/market-share-of-internet-browsers-worldwide-since-2009. Accessed 13 July 2023.

"Global Social Media Statistics." *Datareportal*, Apr. 2023, datareportal.com/social-media-users. Accessed 13 July 2023.

Goggin, Gerard. *Apps: From Mobile Phones to Digital Lives*. Wiley, 2021.

Gold, Julian. *Object-Oriented Game Development*. Addison-Wesley, 2004.

Goldstein, Phil. "How Cities Are Forging Partnerships to Close the Digital Divide." *StateTech*, 10 Dec. 2020, statetechmagazine.com/article/2020/12/how-cities-are-forging-partnerships-close-digital-divide. Accessed 13 July 2023.

_____. "What Is Building Information Modeling Technology in Government?" *FedTech*, 16 June 2020, fedtechmagazine.com/article/2020/06/what-building-information-modeling-technology-government-perfcon. Accessed 13 July 2023.

Gómez Cruz, Edgar, and Asko Lehmuskallio, editors. *Digital Photography and Everyday Life: Empirical Studies on Material Visual Practices*. Routledge, 2016.

Gong, Yihong. *Intelligent Image Databases: Towards Advanced Image Retrieval*. Kluwer, 1998.

Gonzalez, Rafael C., and Richard E. Woods. *Digital Image Processing*. 4th ed., Pearson, 2019.

Goodman, Shalom. "Neil Young, Joe Rogan Podcast and Spotify: What to Know." *Wall Street Journal*, 2 Feb. 2022, www.wsj.com/articles/joe-rogan-podcast-neil-young-spotify-what-to-know-11643663216. Accessed 25 July 2023.

Google Dominates Search: But the Real Problem Is Its Monopoly on Data." *The Guardian*, 19 Apr. 2015, www.theguardian.com/technology/2015/apr/19/google-dominates-search-real-problem-monopoly-data. Accessed 13 July 2023.

Gordon, Sherri Mabry. *Downloading Copyrighted Stuff from the Internet: Stealing or Fair Use?* Enslow, 2005.

Gordon, Steve. *The Future of the Music Business: How to Succeed with the New Digital Technologies*. Hal Leonard, 2011.

Gordon, Whitson. "How to Build a Computer, Lesson 1: Hardware Basics." *Lifehacker*, 1 Aug. 2011, lifehacker.com/5826509/how-to-build-a-computer-from-scratch-lesson-1-hardware-basics. Accessed 13 July 2023.

Goriunova, Olga, editor. *Fun and Software: Exploring Pleasure, Paradox, and Pain in Computing*. Bloomsbury, 2014.

Gottschall, Jonathan. *The Storytelling Animal: How Stories Make Us Human*. Mariner, 2013.

Gralla, Preston. *How the Internet Works*. 4th ed., Que Publishing, 1998.

"Graphical User Interface." *Techopedia*, 28 May 2021, www.techopedia.com/definition/5435/graphical-user-interface-gui. Accessed 13 July 2023.

Gray, Leon. *How Does a Touchscreen Work?* Rosen, 2013.

Gray, Tony. *Projected Capacitive Touch: A Practical Guide for Engineers*. Springer International Publishing, 2019.

Graziani, Thomas. "How Douyin Became China's Top Short-Video App in 500 Days." *Walk the Chat*, walkthechat.com/douyin-became-chinas-top-short-video-app-500-days. Accessed 13 July 2023.

Greenberg, Saul, et al. *Sketching User Experiences: The Workbook*. Morgan, 2012.

Greengard, Samuel. *The Internet of Things*. Rev. ed., MIT Press, 2021.

Greenwald, Will. "Augmented Reality (AR) vs. Virtual Reality (VR): What's the Difference?" *PCMag*, 31 Mar. 2021, www.pcmag.com/news/augmented-reality-ar-vs-virtual-reality-vr-whats-the-difference. Accessed 13 July 2023.

Greenwood, Elizabeth. "Wait, Why Did That Woman Sit in the MoMA for 750 Hours?" *The Atlantic*, 2 July 2012, www.theatlantic.com/entertainment/archive/2012/07/wait-why-did-that-woman-sit-in-the-moma-for-750-hours/259069. Accessed 13 July 2022.

Gregory, Sam, and Eric French. "How Do We Work Together to Detect AI-Manipulated Media?" *Witness Media Lab*, 2019, lab.witness.org/projects/osint-digital-forensics. Accessed 13 July 2023.

Griffin, Andrew. "Facebook Live: Site Adds Huge New Update to Feature That Lets People Stream in Their

Timeline." *Independent*, 6 Apr. 2016, www.independent.co.uk/life-style/gadgets-and-tech/news/facebook-live-site-adds-huge-new-update-to-feature-that-lets-people-stream-in-their-timeline-a6971566.html. Accessed 13 July 2023.

Griffith, Eric. "What Is Cloud Computing?" *PCMag*, 15 Feb. 2022, www.pcmag.com/news/what-is-cloud-computing. Accessed 13 July 2023.

Griffiths, Devin C. *Virtual Ascendance: Video Games and the Remaking of Reality*. Rowman, 2013.

Guerrero Garcia, Virginia and Bella Palomo. "The Crisis of Photojournalism: Rethinking the Profession in a Participatory Media Ecosystem." *Communication and Society*, vol. 28, no. 4, 2015, pp. 33-48.

Gupta, Siddarth, and Vagesh Porwal. "Recent Digital Watermarking Approaches, Protecting Multimedia Data Ownership." *Advances in Computer Science*, vol. 4, no. 2, 2015, pp. 21-30.

Gupton, Nancy, and Patrick J. Kiger. "What's the Difference between AR, VR, and MR?" *Franklin Institute*, 21 Sept. 2017, www.fi.edu/difference-between-ar-vr-and-mr. Accessed 13 July 2023.

Gutierrez, Peter. "Every Platform Tells a Story." *School Library Journal*, vol. 58, no. 6, 2012, pp. 32-34.

Habermas, J. *The Structural Transformation of the Public Sphere*. MIT Press, 1991.

Hagen, A. N. "The Metaphors We Stream By: Making Sense of Music Streaming." *First Monday*, vol. 21, no. 3, 2016, p. 8.

Hagen, Rebecca, and Kim Golombisky. *White Space Is Not Your Enemy: A Beginner's Guide to Communicating Visually through Graphic, Web & Multimedia Design*. 3rd ed., A. K. Peters/CRC Press, 2016.

Hallas, Joel. *Basic Radio: Understanding the Key Building Blocks*. American Radio Relay League, 2005.

Hallgren, Scott W., editor. *Composing for the Screen*. Routledge, 2022.

Hamburger, Ellis. "Instagram's New Editing Features Could Make It Your Only Photo App." *Verge*, 3 June 2014, www.theverge.com/2014/6/3/5771860/instagram-new-editing-features-could-make-it-your-only-photo-app-vsco-cam-afterlight. Accessed 13 July 2023.

Hamedy, Saba. "Report: Online Piracy Remains Multi-Hundred-Million-Dollar Business." *Los Angeles Times*, 19 May 2015, www.latimes.com/entertainment/envelope/cotown/la-et-ct-report-online-piracy-digital-citizens-alliance-medialink-20150518-story.html. Accessed 13 July 2023.

Hamm, Matthew J. *Wireframing Essentials: An Introduction to User Experience Design*. Packt, 2014.

Hampton, Dave. *The Business of Audio Engineering*. 2nd ed., Hal Leonard, 2013.

_____. *So, You're an Audio Engineer: Well, Here's the Other Stuff You Need to Know*. Outskirts, 2005.

Hampton-Sosa, W. "The Impact of Creativity and Community Facilitation on Music Streaming Adoption and Digital Piracy." *Computers in Human Behavior*, vol. 69, 2017, pp. 444-53.

Hancock, James G., Hennie Haworth, Stuart Hill, and Sarah King. *The Art of Map Illustration*. Quarto Publishing, 2018.

"Hard Drive." *Computer Hope*, 18 Oct. 2022, www.computerhope.com/jargon/h/harddriv.htm. Accessed 13 July 2023.

Hardin, Brad, and Dave McCool. *BIM and Construction Management: Proven Tools, Methods, and Workflows*. 2nd ed., Wiley, 2015.

Hardy, David Leicester. *Introduction to Digital Media Design: Transferable Hacks, Skills, and Tricks*. Bloomsbury Visual Arts, 2022.

Hargrave, Marshall. "Crowdsourcing: Definition, How It Works, Types, and Examples." *Investopedia*, 20 Nov. 2022, www.investopedia.com/terms/c/crowdsourcing.asp. Accessed 13 July 2023.

Hariman, Robert. *No Caption Needed: Iconic Photographs, Public Culture, and Liberal Democracy*. U of Chicago P, 2007.

Harkinson, Josh. "Here's How Twitter Can Track You on All of Your Devices." *Mother Jones*, 24 Sept. 2013, www.motherjones.com/politics/2013/09/twitter-could-threaten-your-privacy-more-facebook. Accessed 26 July 2023.

Harris, John M. "Truthful as the Record of Heaven." *Southern Cultures*, vol. 19, no. 3, 2013, pp. 79-94.

Hartley, John, and Kelly McWilliam, editors. *Story Circle: Digital Storytelling around the World*. Blackwell, 2009.

Hasegawa, Shoichi, et al., editors. *Haptic Interaction: Science, Engineering and Design*. Springer Nature Singapore, 2018.

Haslam, Karen. "macOS Versions: Every Version Including the Latest." *Macworld*, 13 June 2023, www.macworld.com/article/672681/list-of-all-macos-versions-including-the-latest-macos.html. Accessed 13 July 2023.

Hassenfeld, Noam. "Even the Scientists Who Build AI Can't Tell You How It Works." *Vox*, 15 July 2023, www.vox.com/unexplainable/2023/7/15/23793840/chat-gpt-ai-science-mystery-unexplainable-podcast. Accessed 18 July 2023.

Hatmaker, Taylor. "Instagram Adds Shopping Tags Directly into Stories." *TechCrunch*, 12 June 2018, techcrunch.com/2018/06/12/instagram-adds-shopping-tags-directly-into-stories. Accessed 13 July 2023.

Hausman, Carl, Philip G. Benoit, and Fritz Messere. *Announcing for Broadcasting and the Internet: The Modern Guide to Performing in the Electronic Media*. Routledge, 2019.

Healy, Kieran. *Data Visualization: A Practical Introduction*. Princeton UP, 2018.

Heater, Brian. "Why Spotify Is Betting Big on Podcasting." *TechCrunch*, 6 Feb. 2019, techcrunch.com/2019/02/06/why-spotify-is-betting-big-on-podcasting. Accessed 25 July 2023.

Heath, Alex. "This Is Meta's AR/VR Hardware Roadmap for the Next Four Years." *Verge*, 28 Feb. 2023, www.theverge.com/2023/2/28/23619730/meta-vr-oculus-ar-glasses-smartwatch-plans. Accessed 13 July 2023.

Heaven, Douglas. "Not Like Us: Artificial Minds We Can't Understand." *New Scientist*, 7 Aug. 2013, www.newscientist.com/article/mg21929290-700-not-like-us-artificial-minds-we-cant-understand. Accessed 13 July 2023.

Heckman, Rocky. *Designing Platform Independent Mobile Apps and Services*. Wiley-IEEE Computer Society Press, 2016.

Heidmets, Rao. *The Animation Textbook*. CRC Press, 2022.

Heikkilä, Melissa. "The Algorithm: AI-Generated Art Raises Tricky Questions about Ethics, Copyright, and Security." *MIT Technology Review*, 20 Sept. 2022, www.technologyreview.com/2022/09/20/1059792/the-algorithm-ai-generated-art-raises-tricky-questions-about-ethics-copyright-and-security. Accessed 13 July 2023.

Heisler, Yoni. "The History and Evolution of iOS, from the Original iPhone to iOS 9." *BGR*, 19 Dec. 2018, bgr.com/2016/02/12/ios-history-iphone-features-evolution. Accessed 13 July 2023.

Helander, Martin. *A Guide to Human Factors and Ergonomics*. 2nd ed., CRC Press, 2006.

Hendricks, John Allen, and Bruce Mims. *The Radio Station: Broadcasting, Podcasting, and Streaming*. 10th ed., Routledge, 2018.

Henrie, Curtis R., Lisa R. Halverson, and Charles R. Graham. "Measuring Student Engagement in Technology-Mediated Learning: A Review." *Computers & Education*, vol. 90, 2015, pp. 36-53.

Hern, Alex. "Mark Zuckerberg's Letter Annotated: What He Said and What He Didn't." *Guardian*, 17 Feb. 2017, www.theguardian.com/technology/ng-interactive/2017/feb/17/mark-zuckerberg-facebook-letter-annotated-what-he-said-what-he-didnt. Accessed 13 July 2023.

Herrman, John. "What Happens When Facebook Goes the Way of Myspace?" *New York Times*, 12 Dec. 2018, www.nytimes.com/2018/12/12/magazine/what-happens-when-facebook-goes-the-way-of-myspace.html. Accessed 13 July 2023.

Hey, Tony, and Gyuri Pápay. *The Computing Universe: A Journey through a Revolution*. Cambridge UP, 2015.

Hider, Philip. *Information Resource Description: Creating and Managing Metadata*. American Library Association, 2012.

Highfield, Roger. "Fast Forward to Cartoon Reality." *Telegraph*, 13 June 2006, www.telegraph.co.uk/technology/3346141/Fast-forward-to-cartoon-reality.html. Accessed 13 July 2023.

Highfield, Tim. *Social Media and Everyday Politics*. Polity, 2016.

Hill, Janette R., Llyan Song, and Richard E. West. "Social Learning Theory and Web-Based Learning Environments: A Review of Research and Discussion Implications." *American Journal of Distance Education*, vol. 23, no. 2, 2009, pp. 88-103.

Hill, Kashmir. "Facebook's Mark Zuckerberg: 'We've Made a Bunch of Mistakes.'" *Forbes*, 29 Nov. 2011, www.forbes.com/sites/kashmirhill/2011/11/29/facebooks-mark-zuckerberg-weve-made-a-bunch-of-mistakes. Accessed 13 July 2023.

Hinduja, Sameer. *Music Piracy and Crime Theory*. LFB, 2005.

History of Cryptography: An Easy to Understand History of Cryptography. Thawte, 2013.

"History of Photography." *PBS*, www.pbs.org/wgbh/americanexperience/features/eastman-history-photography. Accessed 13 July 2023.

"History of Universal Design." *Centre for Excellence in Universal Design*, 2020, universaldesign.ie/what-is-universal-design/history-of-ud. Accessed 13 July 2023.

"The History of Web Browsers." *Mozilla*, www.mozilla.org/en-US/firefox/browsers/browser-history. Accessed 13 July 2023.

Hitzler, Pascal, Markus Krotzsch, and Sebastian Rudolph. *Foundations of Semantic Web Technologies*. Taylor & Francis, 2010.

Hoffstein, Jeffrey, Jill Pipher, and Joseph H. Silverman. *An Introduction to Mathematical Cryptography*. 2nd ed., Springer, 2014.

Hofmann, Chris, Marcia Knous, and John V. Hedke. *Firefox and Thunderbird Garage*. Prentice Hall Professional Technical Reference, 2005.

Holcombe, Jane, and Charles Holcombe. *Survey of Operating Systems*. 6th ed., McGraw-Hill Education, 2020.

Holden, Joshua. *The Mathematics of Secrets: Cryptography from Caesar Ciphers to Digital Encryption*. Princeton UP, 2018.

Holland, Oscar. "How NFTs Are Fueling a Digital Art Boom." *CNN*, 10 Mar. 2021, www.cnn.com/style/article/nft-digital-art-boom/index.html. Accessed 13 July 2023.

Hollander, P. "Popular Culture, the New York Times and the New Republic." *Society*, vol. 51, 2014, pp. 288-96.

Holleley, Douglas. *Photo-Editing and Presentation: A Guide to Image Editing and Presentation for Photographers and Visual Artists*. Clarellen, 2009.

Hollows, Joanne. *Media Studies: A Complete Introduction*. Hodder & Stoughton, 2016.

Holmes, Thomas B. *Electronic and Experimental Music: Pioneers in Technology and Composition*. Routledge, 2002.

Horner, Veronica. "How Big Data Analytics Is Changing the Media Industry." *SEI*, 5 Oct. 2020, www.sei.com/insights/article/how-big-data-analytics-is-changing-the-media-industry. Accessed 13 July 2023.

Horodyski, John. *Metadata Matters*. CRC Press, 2022.

Horspool, Nigel, and Nikolai Tillmann. *Touchdevelop: Programming on the Go*. Apress, 2013.

Horstmann, Cay. *Big Java: Early Objects*. 7th ed., John Wiley & Sons, 2018.

Hoser, Tania. *Introduction to Cinematography: Learning through Practice*. Routledge, 2018.

Hosken, Dan. *An Introduction to Music Technology*. 2nd ed., Routledge, 2014.

Hovious, A. "Inanimate Alice." *Teacher Librarian*, vol. 42, no. 2, 2014, pp. 42-46.

_____. *Transmedia Storytelling: The Librarian's Guide*. Libraries Unlimited, 2016.

"How Computers Work: The CPU and Memory." *University of Rhode Island*, homepage.cs.uri.edu/faculty/wolfe/book/Readings/Reading04.htm. Accessed 13 July 2023.

"How Cryptography and Web3 Can Help Restore Trust in Digital Media." *Stanford Engineering*, 17 June 2022, engineering.stanford.edu/magazine/how-cryptography-and-web3-can-help-restore-trust-digital-media. Accessed 13 July 2023.

"How to: Use PGP for Windows." *Surveillance Self-Defense*, 17 June 2018, ssd.eff.org/module/how-use-pgp-windows. Accessed 13 July 2023.

Howard, Nicole. *The Book: The Life Story of a Technology*. Greenwood Press, 2005.

Howe, Jeff. *Crowdsourcing: Why the Power of the Crowd Is Driving the Future of Business*. Crown, 2009.

Howell, Chris. *Modern Game Testing: A Pragmatic Guide to Test Planning and Strategy*. Modern Game Testing, 2022.

Hoy, Meredith. *From Point to Pixel: A Genealogy of Digital Aesthetics*. Dartmouth College P, 2017.

"HTML Tutorial." *W3Schools*, www.w3schools.com/html. Accessed 13 July 2023.

Huber, David Miles, and Emiliano Caballero. *Modern Recording Techniques: A Practical Guide to Modern Music Production*. 10th ed., Focal Press, 2023.

Huddleston, Rob. *HTML, XHTML, and CSS: Your Visual Blueprint for Designing Effective Web Pages*. John Wiley & Sons, 2009.

Hughes, Alex. "ChatGPT: Everything You Need to Know about OpenAI's GPT-4 Tool." *BBC Science Focus*, 16 Mar. 2023, www.sciencefocus.com/future-technology/gpt-3. Accessed 18 July 2023.

Hughes, Brian. "Technology Becomes Us: The Age of Human-Computer Interaction." *Huffington Post*, 20 Apr. 2016, www.huffingtonpost.com/brian-hughes/technology-becomes-us-the_b_9732166.html. Accessed 13 July 2023.

Hughes, John F. *Computer Graphics: Principles and Practice*. Addison-Wesley, 2014.

Humphreys, Ashlee. *Social Media: Enduring Principles*. Oxford UP, 2016.

Hunt, Richard. *Advanced Typography: From Knowledge to Mastery*. Bloomsbury Visual Arts, 2020.

Husak, Douglas. *Overcriminalization: The Limits of the Criminal Law*. Oxford UP, 2008.

Hutchinson, Andrew. "Instagram Announces Closure of Separate IGTV App, Removal of In-Stream Video Ads." *SocialMediaToday*, 28 Feb. 2022, www.socialmediatoday.com/news/instagram-announces-closure-of-separate-igtv-app-removal-of-in-stream-vide/619532. Accessed 13 July 2023.

Hutchinson, Lee. "Home 3D Printers Take Us on a Maddening Journey into Another Dimension." *Ars Technica*, 27 Aug. 2013, arstechnica.com/gadgets/2013/08/home-3d-printers-take-us-on-a-maddening-journey-into-another-dimension. Accessed 13 July 2023.

Hutchison, Tom, Paul Allen, and Amy Macy. *Record Label Marketing*. Taylor & Francis, 2012.

Iglesias, Beatriz, and Josep Valor. "The Good and Bad Sides of Clickbait." *IESE*, 1 Mar. 2023, blog.iese.edu/the-media-industry/2023/03/01/the-good-and-bad-sides-of-clickbait. Accessed 13 July 2023.

"An Illustrated History of Mac OS X." *Tower*, 12 Jan. 2016, www.git-tower.com/blog/history-of-osx. Accessed 13 July 2023.

"Influencer Advertising-Worldwide." *Statista*, www.statista.com/outlook/amo/advertising/influencer-advertising/worldwide. Accessed 13 July 2023.

Ingram, Mathew. "Facebook Killing Another Example of Live Video Feature's Dark Side." *Fortune*, 17 Apr. 2017, fortune.com/2017/04/17/facebook-killing. Accessed 13 July 2023.

"Instagram Accounts with the Most Followers Worldwide as of July 2023." *Statista*, July 2023, www.statista.com/statistics/421169/most-followers-instagram. Accessed 13 July 2023.

"Intel's First Microprocessor: Its Invention, Introduction, and Lasting Influence." *Intel*, www.intel.com/content/www/us/en/history/museum-story-of-intel-4004.html. Accessed 13 July 2023.

"Introduction to Image Files Tutorial." *Boston University Information Services and Technology*, www.bu.edu/tech/support/research/training-consulting/online-tutorials/imagefiles/image101-1. Accessed 13 July 2023.

"iOS: A Visual History." *Verge*, 16 Sept. 2013, www.theverge.com/2011/12/13/2612736/ios-history-iphone-ipad. Accessed 13 July 2023.

Iqbal, Mansoor. "Disney Plus Revenue and Usage Statistics (2023)." *Business of Apps*, 24 May 2023, www.businessofapps.com/data/disney-plus-statistics. Accessed 25 July 2023.

_____. "YouTube Revenue and Usage Statistics." *Business of Apps*, 18 July 2023, www.businessofapps.com/data/youtube-statistics. Accessed 13 July 2023.

Iredale, Gwyneth. "List of 10 Most Expensive NFTs Ever Sold." *101 Blockchains*, 27 Apr. 2021, 101blockchains.com/most-expensive-nfts. Accessed 13 July 2023.

Isaac, Mike. "Facebook Renames Itself Meta." *New York Times*, 28 Oct. 2021, www.nytimes.com/2021/10/28/technology/facebook-meta-name-change.html. Accessed 13 July 2023.

_____. "Instagram May Change Your Feed, Personalizing It with an Algorithm." *New York Times*, 15 Mar. 2016, www.nytimes.com/2016/03/16/technology/instagram-feed.html. Accessed 13 July 2023.

Isbister, Katherine. *How Games Move Us: Emotion by Design*. MIT Press, 2016.

Isbister, Katherine, and Celia Hodent, editors. *Game Usability: Advice from the Experts for Advancing UX Strategy and Practice in Videogames*. 2nd ed., CRC Press, 2022.

Jacko, Julie A., editor. *The Human-Computer Interaction Handbook: Fundamentals, Evolving Technologies, and Emerging Applications*. 3rd ed., CRC Press, 2012.

Jackson, Charles L. "Wireless Efficiency Versus Net Neutrality." *Federal Communications Law Journal*, vol. 63, no. 2, 2011, pp. 445-80.

Jackson, Daniel. *The Essence of Software Design: Why Concepts Matter for Great Design*. Princeton UP, 2023.

Jackson, Kourtnee. "Disney Plus vs. Netflix: How to Narrow Down to One Streaming Service." *CNET*, 19 Nov. 2022, www.cnet.com/tech/services-and-software/disney-plus-vs-netflix-how-to-narrow-done-to-one-streaming-service. Accessed 25 July 2023.

James, Mike. "The Triumph of Deep Learning." *I Programmer*, 14 Dec. 2012, www.i-programmer.info/programming/article-intelligence/5206-the-triumph-of-deep-learning.html. Accessed 13 July 2023.

Jansen, Mark, and Paula Beaton. "5G vs. 4G: How Does the Newest Network Improve on the Last?" *DigitalTrends*, 22 Apr. 2022, www.digitaltrends.com/mobile/5g-vs-4g. Accessed 13 July 2023.

Jarmul, Katharine. *Practical Data Privacy: Enhancing Privacy and Security in Data*. O'Reilly Media, 2023.

Jarnow, Jesse. "Apple's iTunes Is Alienating Its Most Music-Obsessed Users." *Wired*, 17 Nov. 2015, www.wired.com/2015/11/itunes-alternatives. Accessed 13 July 2023.

Jenkins, Henry. "Searching for the Origami Unicorn: The Matrix and Transmedia Storytelling." *Convergence Culture: Where Old and New Media Collide*, State U of New York P, 2006, pp. 93-130.

Jenkinson, Mark. *Photography Careers: Finding Your True Path*. Focal Press, 2016.

Jensen, Robin E., Erin F. Doss, and Rebecca Ivic. "Metaphorical Invention in Early Photojournalism: New York Times Coverage of the 1876 Brooklyn Theater Fire and the 1911 Shirtwaist Factory Fire." *Critical Studies in Media Communication*, vol. 28, no. 4, 2011, pp. 334-52.

Jimenez, Jorge, et al. "Destroy All Jaggies." *Game Developer*, vol. 18, no. 6, 2011, pp. 13-20.

Johns, Adrian. *Piracy: The Intellectual Property Wars from Gutenberg to Gates*. U of Chicago P, 2009.

Johnson, Jeff. *Designing with the Mind in Mind*. 2nd ed., Morgan Kaufmann, 2014.

Jokinen, Jussi P. P. "Emotional User Experience: Traits, Events, and States." *International Journal of Human-Computer Studies*, vol. 76, 2015, pp. 67-77.

"Joseph Beuys." *Walker*, walkerart.org/collections/artists/joseph-beuys. Accessed 13 July 2023.

Juang, B. H., and Lawrence R. Rabiner. "Automatic Speech Recognition-A Brief History of the Technology Development." *University of California, Santa Barbara*, 2004, web.ece.ucsb.edu/Faculty/Rabiner/ece259/Reprints/354_LALI-ASRHistory-final-10-8.pdf. Accessed 13 July 2023.

Jury, David. *What Is Typography?* Rotovision, 2006.

Kamber, Michael. *Photojournalists on War: The Untold Stories from Iraq*. U of Texas P, 2013.

Kamenetz, Anya. "Why Video Games Succeed Where the Movie and Music Industries Fail." *Fast Company*, 7 Nov. 2013, www.fastcompany.com/3021008/why-video-games-succeed-where-the-movie-and-music-industries-fail. Accessed 22 Dec. 2014.

Kane, John. *A Type Primer*. 2nd ed., Laurence King, 2020.

Kastrenakes, Jacob. "'Six Strikes' Anti-Piracy Initiative Ends after Failing to Scare Off 'Hardcore' Pirates." *Verge*, 30 Jan. 2017, www.theverge.com/2017/1/30/14445596/six-strikes-piracy-system-failed-ending. Accessed 13 July 2023.

Kato, Hiroki, Keng T. Tan, and Douglas Chai. *Barcodes for Mobile Devices*. Cambridge UP, 2010.

Katz, Jonathan, and Yehuda Lindell. *Introduction to Modern Cryptography*. 3rd ed., Chapman and Hall, 2020.

Katz, Mark. *Capturing Sound: How Technology Has Changed Music*. Rev. ed., U of California P, 2010.

Kefauver, Alan P., and David Patschke. *Fundamentals of Digital Audio*. A-R Editions, 2007.

Kelly, Makena. "Biden Revokes and Replaces Trump Orders Banning TikTok and WeChat." *Verge*, 9 June 2021, www.theverge.com/2021/6/9/22525953/biden-tiktok-wechat-trump-bans-revoked-alipay. Accessed 13 July 2023.

_____. "House Approves Save the Internet Act That Would Reinstate Net Neutrality." *Verge*, 10 Apr. 2019, www.theverge.com/2019/4/10/18304522/net-neutrality-save-the-internet-act-house-of-representatives-approval. Accessed 13 July 2023.

Kensek, Karen M., and Douglas E. Noble. *Building Information Modeling: BIM in Current and Future Practice*. Wiley, 2014.

Kent, Alexander, and Peter Vujakovic, editors. *The Routledge Handbook of Mapping and Cartography*. Routledge, 2017.

Kenton, Will. "Copyright." *Investopedia*, 10 June 2023, www.investopedia.com/terms/c/copyright.asp. Accessed 13 July 2023.

Kerlow, Isaac. *The Art of 3D Computer Animation and Effects*. 4th ed., Wiley, 2009.

Kim, J., C. Nam, and M. H. Ryu. "What Do Consumers Prefer for Music Streaming Services?: A Comparative Study between Korea and US." *Telecommunications Policy*, vol. 41, no. 4, 2017, pp. 263-72.

King, Andrew, Evangelos Himonides, and S. Alex Ruthman, editors. *The Routledge Companion to Music, Technology, and Education*. Routledge, 2019.

Kirkpatrick, David. *The Facebook Effect: The Inside Story of the Company That Is Connecting the World*. Simon, 2011.

Kizza, Joseph Migga. *Ethical and Social Issues in the Information* Age. 5th ed., Springer, 2013.

Klein, Christopher. "Why Coca-Cola's 'New Coke' Flopped." *History.com*, 23 May 2023, www.history.com/news/hungry-history/the-new-coke-flop-30-years-ago. Accessed 13 July 2023.

Klein, Ezra. "Mark Zuckerberg on Facebook's Hardest Year, and What Comes Next." *Vox*, 2 Apr. 2018, www.vox.com/2018/4/2/17185052/mark-zuckerberg-facebook-interview-fake-news-bots-cambridge. Accessed 13 July 2023.

Klimczak, Erik. *Design for Software: A Playbook for Developers*. Wiley, 2013.

Knott, Daniel. *Hands-On Mobile App Testing*. Pearson Education, 2015.

Koblin, Aaron. "Q&A: Aaron Koblin: The Data Visualizer." Interview by Jascha Hoffman. *Nature*, vol. 486, no. 7401, 2012, p. 33.

Kojiæ, Miloš, et al. *Computer Modeling in Bioengineering: Theoretical Background, Examples and Software*. Wiley, 2008.

Kolb, Jason, and Jeremy Kolb. *The Big Data Revolution*. Applied Data Labs, 2013.

Kolko, Jon. "Design Thinking Comes of Age." *Harvard Business Review*, Sept. 2015, hbr.org/2015/09/design-thinking-comes-of-age. Accessed 13 July 2023.

Kong, S. C. "A Curriculum Framework for Implementing Information Technology in School Education to Foster Information Literacy." *Computers & Education*, vol. 51, 2008, pp. 129-41.

Konijn, Elly A., et al. "YouTube as Research Tool-Three Approaches." *CyberPsychology, Behavior & Social Networking*, vol. 16, no. 5, 2013, pp. 1-7.

Koops, Bert-Jaap. "The Trouble with European Data Protection Law." *International Data Privacy Law*, vol. 4, no. 4, 2014, pp. 250-61.

Koster, Raph. *A Theory of Fun for Game Design*. Paraglyph, 2004.

Kraak, Menno-Jan, and Ferjan Ormeling. *Cartography: Visualization of Spatial Data*. 3rd ed., Prentice-Hall, 2010.

Krietzberg, Ian. "Twitter to X: The Complete Timeline of Elon Musk's Twitter Purchase." *TheStreet*, 26 July 2023, www.thestreet.com/technology/twitter-to-x-the-complete-timeline-of-elon-musks-twitter-purchase. Accessed 26 July 2023.

Krug, Steve. *Don't Make Me Think, Revisited: A Common Sense Approach to Web Usability*. 3rd ed., New Riders, 2014.

Kuehn, Scott A., and James Andrew Lingwall. *The Basics of Media Writing: A Strategic Approach*. CQ Press, 2017.

Kuhn, Karl F. *In Quest of the Universe*. 6th ed., Jones and Bartlett, 2010.

Kumar, Kaushik, Divya Zindani, and J. Paulo Davim. *Design Thinking to Digital Thinking*. Springer Nature Switzerland, 2020.

Kumar, Raghvendra, Rohit Sharma, and Prasant Kumar Pattnaik, editors. *Multimedia Technologies in the Internet of Things Environment*. Springer, 2021.

La Counte, Scott. *The Ridiculously Simple Guide to Surfing the Internet with Google Chrome*. SL Editions, 2020.

Laalaoui, Yacine, and Nizar Bouguila, editors. *Artificial Intelligence Applications in Information and Communication Technologies*. Springer International Publishing, 2015.

Lacy, Sarah. *Once You're Lucky, Twice You're Good: The Rebirth of Silicon Valley and the Rise of Web 2.0*. Gotham Books, 2008.

Lakatos, Anna. *The Music Producer's Creative Guide to Ableton Live 11*. Packt, 2023.

Lambert, Joe, and Brooke Hessler. *Digital Storytelling: Capturing Lives, Creating Community*. 5th ed., Routledge, 2018.

Lanagan, James, and Alan Smeaton. "Video Digital Libraries: Contributive and Decentralized." *International Journal on Digital Libraries*, vol. 12, no. 4, 2012, pp. 159-78.

Langford, Michael, et al. *Langford's Basic Photography*. 10th ed., Focal Press, 2015.

Langton, Loup. *Photojournalism and Today's News: Creating Visual Reality*. Wiley, 2009.

Lash, S., and J. Urry. *Economies of Signs and Space*. SAGE, 1994.

Laudon, Kenneth C., and Carol Guercio Traver. *E-Commerce 2019*. 15th ed., Pearson Education, 2020.

Lavaveshkul, Liz. "How to Achieve 15 Minutes (or More) of Fame through YouTube." *Journal of International Commercial Law & Technology*, vol. 7, no. 4, 2012, pp. 370-85.

Law, Averill M. *Simulation Modeling and Analysis*. 6th ed., McGraw-Hill, 2023.

Law, Marcus. "From ELIZA to ChatGPT: The Evolution of Chatbots Technology." *Technology*, 7 Dec. 2022, technologymagazine.com/articles/from-eliza-to-chatgpt-the-evolution-of-chatbots-technology. Accessed 18 July 2023.

Lawrence, Dan. *Digital Writing: A Guide to Writing for Social Media and the Web*. Broadview Press, 2022.

Lawton, Kevin, and Dan Marom. *The Crowdfunding Revolution: How to Raise Venture Capital Using Social Media*. McGraw. 2012.

Layton, Julia, and Jonathan Strickland. "How iTunes Works." *HowStuffWorks*, electronics.howstuffworks.com/itunes.htm. Accessed 13 July 2023.

Leach, Neil. *Architecture in the Age of Artificial Intelligence: An Introduction to AI for Architects*. Bloomsbury Visual Arts, 2021.

Lebert, Marie. "eBooks: 1998-The First eBook Readers." *Project Gutenberg News*, 16 July 2011, www.gutenbergnews.org/20110716/ebooks-1998-the-first-ebook-readers. Accessed 13 July 2023.

Lebow, Sara. "Worldwide Digital Ad Spend Will Top $600 Billion This Year." *Insider Intelligence*, 31 Jan. 2023, www.insiderintelligence.com/content/worldwide-digital-ad-spend-will-top-600-billion-this-year. Accessed 13 July 2023.

Lecky-Thompson, Guy W. *Video Game Design Revealed*. Charles River Media, 2007.

Lee, Alexander. "With the Metaverse on the Horizon, Its Creators Reflect on the Importance of Diversity in the Virtual World to Come." *Digiday*, 17 Aug. 2021, digiday.com/marketing/with-the-metaverse-on-the-horizon-its-creators-reflect-on-the-importance-of-diversity-in-the-virtual-world-to-come. Accessed 13 July 2023.

Lee, Kent D. *Foundations of Programming Languages*. 2nd ed., Springer, 2017.

Lee, Timothy B. "Why It's Time for Uber to Get Out of the Self-Driving Car Business." *Ars Technica*, 27 Mar. 2018, arstechnica.com/cars/2018/03/ubers-self-driving-car-project-is-struggling-the-company-should-sell-it. Accessed 13 July 2023.

Lee, W., C. Chen, J. Huang, and J. Liang. "A Smartphone-Based Activity-Aware System for Music Streaming Recommendation." *Knowledge-Based Systems*, vol. 131, 2017, pp. 70-82.

Leifer, Larry J., Christoph Meinel, and Hasso Plattner. *Design Thinking: Understand-Improve-Apply*. Springer, 2011.

LeJeune, Urban A., and Jeff Duntemann. *Netscape and HTML Explorer*. Coriolis Group Books, 1995.

Lessig, Lawrence. *Remix: Making Art and Commerce Thrive in the Hybrid Economy*. Penguin, 2008.

Lettnin, Djones, and Markus Winterholer, editors. *Embedded Software Verification and Debugging*. Springer-Verlag New York, 2017.

Leurs, Bas, and Isobel Roberts. "What Do We Mean by Design?" *Design Council*, 9 Nov. 2018, www.designcouncil.org.uk/news-opinion/what-do-we-mean-design. Accessed 13 July 2023.

Levin, Avner, and Mary Jo Nicholson. "Privacy Law in the United States, the EU, and Canada: The Allure of the Middle Ground." *University of Ottawa Law & Technology Journal*, vol. 2, no. 2, 2005, pp. 357-95.

Levine, Robert. "Billboard Cover: Spotify CEO Daniel Ek on Taylor Swift, His 'Freemium' Business Model and Why He's Saving the Music Industry." *Billboard*, 5 June 2015, www.billboard.com/articles/business/6590101/daniel-ek-spotify-ceo-streaming-feature-tidal-apple-record-labels-taylor-swift. Accessed 25 July 2023.

Levy, Stephen. "Almost 50 Years into the Crypto Wars, Encryption's Opponents Are Still Wrong." *Wired*, 21 July 2023, www.wired.com/story/plaintext-50-years-into-the-crypto-wars-encryptions-opponents-are-still-wrong. Accessed 22 July 2023.

_____. *In the Plex: How Google Thinks, Works, and Shapes Our Lives*. Simon, 2011.

Lewrick, Michael, Patrick Link, and Larry Leifer. *The Design Thinking Toolbox*. Wiley, 2020.

Libby, Kristina. "This Bill Hader Deepfake Video Is Amazing: It's Also Terrifying for Our Future." *Popular Mechanics*, 13 Aug. 2019, www.popularmechanics.com/technology/security/a28691128/deepfake-technology. Accessed 13 July 2023.

Liedtka, Jeanne. "Why Design Thinking Works." *Harvard Business Review*, Sept. 2018, hbr.org/2018/09/why-design-thinking-works. Accessed 13 July 2023.

Liggett, Susan, Rae Earnshaw, and Jill Townsley. *Creativity in Art, Design and Technology*. Springer Cham, 2023.

Lima, Cristiano. "Want Our Metadata? Get a Warrant, Rep. Ted Lieu Says." *Washington Post*, 20 Apr. 2022, www.washingtonpost.com/politics/2022/04/20/want-our-metadata-get-warrant-rep-ted-lieu-says. Accessed 13 July 2023.

Lima, Manuel. *Visual Complexity: Mapping Patterns of Information*. Princeton Architectural, 2013.

Lincoln, Sian, and Brady Robards. "10 Years of Facebook." *New Media & Society*, vol. 16, no. 7, 2014, pp. 1047-50.

Lingwood, Gyles, and Mark Shaw. *Copywriting: Successful Writing for Design, Advertising, and Marketing*. 3rd ed., Laurence King, 2022.

Linksy, Dorian. "Is Daniel Ek, Spotify Founder, Going to Save the Music Industry...or Destroy It?" *The Guardian*, 10 Nov. 2013, www.theguardian.com/technology/2013/nov/10/daniel-ek-spotify-streaming-music. Accessed 25 July 2023.

Loeffler, John. "The History and Science of Virtual Reality Headsets." *Interesting Engineering*, 28 Feb. 2019, interestingengineering.com/the-history-and-science-of-virtual-reality-headsets. Accessed 13 July 2023.

Logue, D., and M. Edwards. "Across the Digital Divide." *Stanford Social Innovation Review*, vol. 11, 2013, pp. 66-71.

Lohr, Steve. "At Google, Cube Culture Has New Rules." *New York Times*, 5 Dec. 2005, www.nytimes.com/2005/12/05/technology/at-google-cube-culture-has-new-rules.html. Accessed 13 July 2023.

Lonbere, Philip. *A History of Communication Technology*. Routledge, 2021.

Long, Ben. *Complete Digital Photography*. 9th ed., CDP Press, 2018.

Long, Simon. *An Introduction to C & GUI Programming*. Raspberry Pi Press, 2019.

Love, Chris. *ASP.NET 3.5 Website Programming: Problem-Design-Solution*. John Wiley & Sons, 2010.

Lovink, Geert. *Networks without a Cause: A Critique of Social Media*. Polity, 2011.

Lowenthal, Yuri, and Tara Platt. *Voice-Over Voice Actor: The Extended Edition*. Bug Bot Press, 2018.

Lu, Weisheng, and Chimay J. Anumba, editors. *Research Companion to Building Information Modeling*. Edward Elgar, 2022.

Lucaites, John Louis, and Robert Hariman. "Visual Rhetoric, Photojournalism, and Democratic Public Culture." *Rhetoric Review*, vol. 20, no. 1/2, 2001, pp. 37-43.

Luke, Carmen. "As Seen on TV or Was That My Phone? New Media Literacy." *Policy Futures in Education*, vol. 5, no. 1, 2007, pp. 50-58.

Luther, Jason, Frank Farmer, and Steve Parks. "Special Issue Editors' Introduction: The Past, Present, and Future of Self-Publishing: Voices, Genres, Publics." *Community Literacy Journal*, vol. 12, 2017, pp. 1-4.

MacCormick, John. *Nine Algorithms That Changed the Future: The Ingenious Ideas That Drive Today's Computers*. Princeton UP, 2012.

"The Magic of the Internet." *Imgur*, imgurinc.com. Accessed 13 July 2023.

Manning, Peter. *Electronic and Computer Music*. Oxford UP, 2004.

Mantlero, Alessandro. "The EU Proposal for a General Data Protection Regulation and the Roots of the 'Right to Be Forgotten.'" *Computer Law & Security Review*, vol. 29, no. 3, June 2013, pp. 229-35.

Manzo, V. J. *Foundations of Music Technology*. Oxford UP, 2015.

Marcovitz, Hal. *Video Games*. Lucent, 2010.

Margolin, Victor. *World History of Design: Two-Volume Set*. Bloomsbury, 2015.

Markelo, Steve. *Microsoft Edge: A Beginner's Guide to the Windows 10 Browser*. Conceptual Kings, 2015.

Markowitz, Eric. "How Instagram Grew from Foursquare Knock-Off to $1 Billion Photo Empire." *Inc.*, 10 Apr. 2012, www.inc.com/eric-markowitz/life-and-times-of-instagram-the-complete-original-story.html. Accessed 13 July 2023.

Marr, Bernard. "The Best Examples of What You Can Do with ChatGPT." *Forbes*, 1 Mar. 2023, www.forbes.com/sites/bernardmarr/2023/03/01/the-best-examples-of-what-you-can-do-with-chatgpt. Accessed 18 July 2023.

Marschner, Steve, et al. *Fundamentals of Computer Graphics*. 5th ed., CRC Press, 2021.

Marsh, Joel. *UX for Beginners: A Crash Course in 100 Short Lessons*. O'Reilly, 2016.

Marsland, Stephen. *Machine Learning: An Algorithmic Perspective*. Taylor, 2009.

Masic, I., and E. Suljevic. "An Overview of E-Health Implementation in Countries, Members of the European Union." *Acta Informatica Medica*, vol. 15, 2007, pp. 242-45.

Masters, Marc. *High Bias: The Distorted History of the Cassette Tape*. U of North Carolina P, 2023.

Matineau, Paris. "What Is a Bot?" *Wired*, 16 Nov. 2018, www.wired.com/story/the-know-it-alls-what-is-a-bot. Accessed 13 July 2023.

Matney, Lucas. "YouTube Has 1.5 Billion Logged-In Monthly Users Watching a Ton of Mobile Video." *Tech Crunch*, 22 June 2017, techcrunch.com/2017/06/22/youtube-has-1-5-billion-logged-in-monthly-users-watching-a-ton-of-mobile-video. Accessed 13 July 2023.

"Matsui, Markey, and Wyden Introduce Legislation to Reinstate Net Neutrality, Reverse Trump-Era Deregulation." *Doris Matsui*, 28 July 2022, matsui.house.gov/media/press-releases/matsui-markey-and-wyden-introduce-legislation-reinstate-net-neutrality-reverse. Accessed 13 July 2023.

Matulka, Rebecca, and Matty Greene. "How 3D Printers Work." *Energy.gov*, 19 June 2014, www.energy.gov/articles/how-3d-printers-work. Accessed 13 July 2023.

Matusitz, J., and G. M. Breen. "E-health: A New Kind of Telemedicine." *Social Work in Public Health*, vol. 23, 2007, pp. 95-113.

Mayer, Grace. "OpenAI Cofounder Responds to Elon Musk's Criticism that ChatGPT Is Too 'Woke:' 'We Made a Mistake.'" *Business Insider*, 9 Mar. 2023, www.businessinsider.com/openai-cofounder-greg-brockman-responds-elon-musk-woke-criticism-chatgpt-2023-3?IR=T. Accessed 18 July 2023.

Mayer, Richard E., and Logan Fiorella, editors. *The Cambridge Handbook of Multimedia Learning*. 3rd ed., Cambridge UP, 2022.

McCabe, David. "Montana Legislature Approves Outright Ban of TikTok." *New York Times*, 14 Apr. 2023, www.nytimes.com/2023/04/14/technology/montana-tiktok-ban-passed.html. Accessed 13 July 2023.

McCauley, Renée, et al. "Debugging: A Review of the Literature from an Educational Perspective." *Computer Science Education*, vol. 18, no. 2, 2008, pp. 67-92.

McConnell, Robert, James Haynes, and Richard Warren. "Understanding ASCII Codes." *NADCOMM*, 14 May 2011, www.nadcomm.com/index3fbb.html?p=89. Accessed 13 July 2023.

McCormick, Ty. "Gamification: A Short History." *Foreign Policy*, 24 June 2013, foreignpolicy.com/2013/06/24/gamification-a-short-history. Accessed 13 July 2023.

McDaniel, Adam. *HTML 5: Your Visual Blueprint for Designing Rich Web Pages and Applications*. John Wiley & Sons, 2011.

McDonald, R., and J. Parker. "When a Story Is More Than a Paper." *Young Adult Library Services*, vol. 11, no. 4, 2013, pp. 27-32.

McDowell, Maghan. "These Platforms Want to Be the Farfetch of Digital Fashion." *Vogue Business*, 25 Jan. 2021, www.voguebusiness.com/technology/these-platforms-want-to-be-the-farfetch-of-digital-fashion. Accessed 13 July 2023.

McElhearn, Kirk. "iTunes: A Look at Apple's Media App and Its Influence on an Industry." *Macworld*, 8 Jan. 2016, www.macworld.com/article/3019878/software/15-years-of-itunes-a-look-at-apples-media-app-and-its-influence-on-an-industry.html. Accessed 13 July 2023.

McGarry, Caitlin. "How Apple Plans to Make iCloud, Siri, and iTunes Better." *Macworld*, 6 Oct. 2016, www.macworld.com/article/3128754/software/how-apple-plans-to-make-icloud-siri-and-itunes-better.html. Accessed 13 July 2023.

McGonigal, Jane. *Reality Is Broken: Why Games Make Us Better and How They Can Change the World*. Penguin, 2011.

McLellan, Charles. "The History of Windows: A Timeline." *ZDNet*, 14 Apr. 2014, www.zdnet.com/article/the-history-of-windows-a-timeline. Accessed 13 July 2023.

McLoughlin, Ian. *Computer Systems: An Embedded Approach*. McGraw-Hill, 2018.

McMorrough, Julia. *The Architecture Reference & Specification Book*. Rev. ed., Rockport, 2018.

Media Arts, Design and Technology Computer Requirements." *NC State College of Design*, May 2023, academics.design.ncsu.edu/it/docs/artdesign-computer-requirements. Accessed 13 July 2023.

"Media Literacy: A Definition and More." *Center for Media Literacy*, www.medialit.org/media-literacy-definition-and-more. Accessed 13 July 2023.

"Mehta, Prateek. *Creating Google Chrome Extensions*. Springer Science + Business Media, 2016.

Menezes, Alfred J., Paul C. van Oorschot, and Scott A. Vanstone. *Handbook of Applied Cryptography*. CRC Press, 1996.

Meola, Andrew. "Smart Buses, Trains, Cars & Planes: How IoT Will Create Private and Public Smart Transportation." *Business Insider*, 30 Jan. 2020, www.businessinsider.com/smart-transportation-iot. Accessed 13 July 2023.

Merchant, Brian. *The One Device: The Secret History of the iPhone*. Little, 2017.

Mertes, Alyssa. "The History of TV Commercials: From Super Bowl Ads to Funny Ads We Still Love Watching." *Quality Logo*, 13 Apr. 2023, www.qualitylogoproducts.com/promo-university/history-of-tv-ads.htm. Accessed 13 July 2023.

Metz, Cade. "OpenAI to Offer New Version of ChatGPT for a $20 Monthly Fee." *New York Times*, 1 Feb. 2023, www.nytimes.com/2023/02/01/technology/openai-chatgpt-plus-subscription.html. Accessed 13 July 2023.

Mezrich, Ben. *The Accidental Billionaires: The Founding of Facebook, a Tale of Sex, Money, Genius and Betrayal*. Doubleday, 2009.

Middleton, Paul, and Steven Gurevitz. *Music Technology Workbook: Key Concepts and Practical Projects*. Focal Press, 2008.

Miller, Arthur I. *The Artist in the Machine: The World of AI-Powered Creativity*. MIT Press, 2019.

Miller, Brian D. *Principles of Web Design*. Allworth, 2022.

Miller, Carolyn Handler. *Digital Storytelling: A Creator's Guide to Interactive Entertainment*. 4th ed., CRC Press, 2019.

Miller, Charles, and Aaron Doering. *The New Landscape of Mobile Learning: Redesigning Education in an App-Based World*. Routledge, 2014.

Miller, Claire C. "YouTube to Serve Niche Tastes by Adding Channels." *New York Times*, 7 Oct. 2012, www.nytimes.com/2012/10/08/business/media/YouTube-to-serve-niche-tastes-by-adding-channels.html. Accessed 13 July 2023.

Miller, James. *Game Theory at Work: How to Use Game Theory to Outthink and Outmaneuver Your Competition*. McGraw-Hill, 2003.

_____. "Spotify Controversy: What You Need to Know." *Siskiyou*, 23 Feb. 2022, siskiyou.sou.edu/2022/02/23/spotify-controversy-what-you-need-to-know. Accessed 25 July 2023.

Miller, Lisa C. *Make Me a Story: Teaching Writing through Digital Storytelling*. Stenhouse, 2010.

Miller, Steven Jack. *Metadata for Digital Collections*. 2nd ed., ALA Neal-Schuman, 2022.

Milmo, Dan. "ChatGPT Reaches 100 Million Users Two Months after Launch." *Guardian*, 2 Feb. 2023, www.theguardian.com/technology/2023/feb/02/chatgpt-100-million-users-open-ai-fastest-growing-app. Accessed 13 July 2023.

Milton, John, and Silvia Cobelo. *Translation, Adaptation and Digital Media*. Routledge, 2023.

Minsky, Marvin, and Seymour A. Papert. Perceptrons: An Introduction to Computational Geometry. Reissue ed., MIT Press, 2017.

Miranda, Carolina A. "Newsletter: QR Codes in Museums Can Be a Blessing. They Are Also a Curse." *Los Angeles Times*, 22 Jan. 2022, www.latimes.com/entertainment-arts/newsletter/2022-01-22/essential-arts-qr-codes-in-museums-blessing-and-curse-essential-arts. Accessed 13 July 2023.

Mitchell, Tom M. *Machine Learning*. McGraw, 1997.

Mitroff, Sarah, and Taylor Martin. "Everything You Need to Know About YouTube Red." *CNET*, 27 Oct. 2017, www.cnet.com/how-to/youtube-red-details. Accessed 13 July 2023.

Mityushev, Vladimir, Wojciech Nawalaniec, and Natalia Rylko. *Introduction to Mathematical Modeling and Computer Simulations*. CRC Press, 2018.

Mizruchi, Susan L., editor. *Libraries and Archives in the Digital Age*. Palgrave Macmillan, 2020.

"Mobile Operating System Market Share Worldwide-June 2023." *Statcounter*, June 2023, gs.statcounter.com/os-market-share/mobile/worldwide/#monthly-202206-202306. Accessed 13 July 2023.

Moe, Wendy, and David A. Schweidel. *Social Media Intelligence*. Cambridge UP, 2014.

Mon, Sierra. "Becoming a Concept Artist for a Hollywood Film." *ArtStation*, 17 July 2017, magazine.artstation.com/2017/07/concept-artist-hollywood-film. Accessed 13 July 2023.

Monckton, Paul. "Instagram Makes Biggest Ever Photo Changes." *Forbes*, 27 Aug. 2015, www.forbes.com/sites/paulmonckton/2015/08/27/instagram-makes-biggest-ever-photo-changes. Accessed 13 July 2023.

Moretti, Marcus. "Before Mac OS X, There Was OS 1 Through 9: A History of Apple's Operating System." *Business Insider*, 10 July 2012, www.businessinsider.com/mac-os-i-through-x-2012-7?op=1. Accessed 13 July 2023.

Morgan, Jacob. "A Simple Explanation of 'The Internet of Things.'" *Forbes*, 13 May 2014, www.forbes.com/sites/jacobmorgan/2014/05/13/simple-explanation-internet-things-that-anyone-can-understand. Accessed 13 July 2023.

Moritz, William. "Digital Harmony: The Life of John Whitney, Computer Animation Pioneer." *Animation World Magazine*, Aug. 1997, www.awn.com/mag/issue2.5/2.5pages/2.5moritzwhitney.html. Accessed 13 July 2023.

Morozov, Evgeny. *The Net Delusion: The Dark Side of Internet Freedom*. Public Affairs, 2011.

Morreale, Joanne. "From Homemade to Store Bought: Annoying Orange and the Professionalization of YouTube." *Journal of Consumer Culture*, vol. 14, no. 1, 2014, pp. 113-28.

Morris, Jeremy Wade. *Selling Digital Music, Formatting Culture*. U of California P, 2015.

Morris, Tee. *Twitch for Dummies*. Wiley, 2019.

Morrison, Foster. *The Art of Modeling Dynamic Systems: Forecasting for Chaos, Randomness, and Determinism*. 1991. Dover, 2008.

Morrison, Jim. "How Doctors Are Using Artificial Intelligence to Battle Covid-19." *Smithsonian Magazine*, 5 Mar. 2021, www.smithsonianmag.com/science-nature/how-doctors-are-using-artificial-intelligence-battle-covid-19-180977124. Accessed 13 July 2023.

Morrow, Robert W. *Sesame Street and the Reform of Children's Television*. Johns Hopkins UP, 2006.

Moscaritolo, Angela. "5G Will Save You Almost 24 Hours of Download Time Per Month." *PCMag*, 16 Oct. 2018, www.pcmag.com/news/5g-will-save-you-almost-24-hours-of-download-time-per-month. Accessed 13 July 2023.

Moscaritolo, Angela, and Chloe Albanesius. "Zuckerberg's Vision for Facebook: A Global Community Backed by AI." *PC Magazine*, 16 Feb. 2017, www.pcmag.com/news/351809/zuckerbergs-vision-for-facebook-a-global-community-backed. Accessed 13 July 2023.

Moss, Frank. *The Sorcerers and Their Apprentices: How the Digital Magicians of the MIT Media Lab Are Creating the Innovative Technologies That Will Transform Our Lives*. Crown Business, 2011.

"Most Popular Social Networks Worldwide as of January 2023, Ranked by Number of Monthly Active Users." *Statista*, Jan. 2023, www.statista.com/statistics/272014/global-social-networks-ranked-by-number-of-users. Accessed 13 July 2023.

Mowery, David C., and Timothy Simcoe. "Is the Internet a US Invention? An Economic and Technological History of Computer Networking." *Research Policy*, vol. 31, Dec. 2002, pp. 1369-87.

Müller, Katja. *Digital Archives and Collections: Creating Online Access to Cultural Heritage*. Berghahn, 2021.

Mund, Ben. "CAD/CAM Options for Increased Shop Productivity." *SME*, 27 Jan. 2021, www.sme.org/technologies/articles/2021/february/cadcam-options-for-increased-shop-productivity. Accessed 13 Jul. 2023.

Munday, Jeremy, Sara Ramos Pinto, and Jacob Blakesley. *Introducing Translation Studies*. 5th ed., Routledge, 2022.

Murray, Brian H. *Defending the Brand: Aggressive Strategies for Protecting Your Brand in the Online Arena*. AMACOM, 2004.

Musciano, Chuck, and Bill Kennedy. *HTML and XHTML: The Definitive Guide*. 6th ed., O'Reilly, 2006.

Myers, Glenford J., Tom Badgett, and Corey Sandler. *The Art of Software Testing*. 3rd ed., Wiley, 2012.

Narula, Herman. *Virtual Society: The Metaverse and the New Frontiers of Human Experience*. Currency, 2022.

National Information Standards Organization. *Understanding Metadata*. NISO Press, 2004.

Neapolitan, Richard E. *Foundations of Algorithms*. 5th ed., Jones, 2015.

Needham, Paul, and Michael Joseph. *Adventure and Art: The First One Hundred Years of Printing*. Rutgers UP, 1999.

Nematollahi, Mohammad Ali, Chalee Vorakulpipat, and Hamurabi Gamboa Rosales. *Digital Watermarking: Techniques and Trends*. Springer Singapore, 2017.

Newitz, Annalee. "Facebook Fires Human Editors, Algorithm Immediately Posts Fake News." *Ars Technica*, 29 Aug. 2016, arstechnica.com/business/2016/08/facebook-fires-human-editors-algorithm-immediately-posts-fake-news. Accessed 13 July 2023.

Newman, Jared. "With Android Lollipop, Mobile Multitasking Takes a Great Leap Forward." *Fast Company*, 6 Nov. 2014, www.fastcompany.com/3038213/with-android-lollipop-mobile-multitasking-takes-a-great-leap-forward. Accessed 13 July 2023.

"NFT Sales Drop 83% in a Year ($2.37bn Drop in Revenue)." *Yahoo!Finance*, 16 Jan. 2023, finance.yahoo.com/news/nft-sales-drop-83-2-114048086.html. Accessed 13 July 2023.

Nhan, Johnny. *Policing Cyberspace: A Structural and Cultural Analysis*. LFB, 2010.

Nicklin, Hannah. *Writing for Games: Theory and Practice*. CRC Press, 2022.

Nicosia, Laura. *Educators Online: Preparing Today's Teachers for Tomorrow's Digital Literacies*. Peter Lang, 2013.

Niederreiter, Harald, and Chaoping Xing. *Algebraic Geometry in Coding Theory and Cryptography*. Princeton UP, 2009.

Nielsen, Jakob. *Multimedia and Hypertext: The Internet and Beyond*. SunSoft Press, 1995.

Nolan, James. *Interpretation: Techniques and Exercises*. 2nd ed., Multilingual Matters, 2012.

"Non-Fungible Tokens (NFT)." *Ethereum*, 2 June 2021, ethereum.org/en/nft. Accessed 13 July 2023.

Norman, Don. *The Design of Everyday Things*. Rev. and expanded ed. Basic, 2013.

Norman, Jeremy M. "Ivan Sutherland Creates the First Graphical User Interface." *Historyofinformation.com*, historyofinformation.com/detail.php?id=805. Accessed 13 July 2023.

Nover, Scott. "The NFT Market Is Down by Almost Every Metric." *Quartz*, 14 Oct. 2022, qz.com/the-nft-market-is-down-by-almost-every-metric-1849655660. Accessed 13 July 2023.

"Number of Active Streamers on Twitch Worldwide from January 2018 to April 2023." *Statista*, May 2023, www.statista.com/statistics/746173/monthly-active-streamers-on-twitch. Accessed 13 July 2023.

"Number of Available Applications in the Google Play Store from December 2009 to June 2023." *Statista*, June 2023, www.statista.com/statistics/266210/number-of-available-applications-in-the-google-play-store. Accessed 13 July 2023.

"Number of Disney Plus Subscribers Worldwide from 1st Quarter 2020 to 2nd Quarter 2023." *Statista*, May 2023, www.statista.com/statistics/1095372/disney-plus-number-of-subscribers-us. Accessed 25 July 2023.

"Number of Internet of Things (IoT) Connected Devices Worldwide from 2019 to 2021, with Forecasts from 2022 to 2030." *Statista*, July 2022, www.statista.com/statistics/1183457/iot-connected-devices-worldwide. Accessed 13 July 2023.

"Number of Netflix Paid Subscribers Worldwide from 1st Quarter 2013 to 2nd Quarter 2023." *Statista*, July 2023, www.statista.com/statistics/250934/quarterly-number-of-netflix-streaming-subscribers-worldwide. Accessed 25 July 2023.

Nyholm, Sven, and Jilles Smids. "The Ethics of Accident-Algorithms for Self-Driving Cars: An Applied Trolley Problem?" *Ethical Theory & Moral Practice*, vol. 19, no. 5, 2016, pp. 1275-89.

O'Connell, Brian. "History of Google: How It Began and What's Happening Beyond 2019." *Street*, 31 Dec. 2018, www.thestreet.com/technology/history-of-google-14820930. Accessed 13 July 2023.

O'Leary, Mike. *Cyber Operations: Building, Defending, and Attacking Modern Computer Networks*. 2nd ed., Apress, 2019.

O'Leary, Timothy, Linda O'Leary, and Daniel O'Leary. *Computing Essentials 2023*. McGraw-Hill, 2022.

O'Marah, Kevin, and Pierfrancesco Manenti. "The Internet of Things Will Make Manufacturing Smarter." *Industry Week*, 14 Aug. 2015, www.industryweek.com/manufacturing-smarter. Accessed 13 July 2023.

Ohler, Jason B. *Digital Storytelling in the Classroom: New Media Pathways to Literacy, Learning, and Creativity*. 2nd ed., Sage, 2013.

Okada, Alexandra, Simon J. Buckingham Shum, and Tony Sherborne, editors. *Knowledge Cartography: Software Tools and Mapping Techniques*. Springer, 2008.

Olson, Parmy. "Facebook Wants Users to Help It Weed Out Fake News." *Forbes*, 6 Dec. 2016, www.forbes.com/sites/parmyolson/2016/12/06/facebook-users-fake-news. Accessed 13 July 2023.

"OpenAI Funding." *Crunchbase*, 2023, www.crunchbase.com/organization/openai/company_financials. Accessed 18 July 2023.

"Operating System Market Share Worldwide-June 2023." *Statcounter*, June 2023, gs.statcounter.com/os-market-share/desktop/worldwide/#monthly-202206-202306. Accessed 13 July 2023.

Oppliger, Rolf. *Secure Messaging on the Internet*. Artech House, 2014.

Oremus, Will. "I Didn't Type This Article." *Slate*, 23 Apr. 2014, slate.com/technology/2014/04/the-end-of-typing-speech-recognition-technology-is-getting-better-and-better.html.

Ortiz, Sabrina. "What Is GPT-4? Here's Everything You Need to Know." *ZDNet*, 15 Mar. 2023, www.zdnet.com/article/what-is-gpt-4-heres-everything-you-need-to-know. Accessed 18 July 2023.

Osborn, Tracy. *Hello Web Design: Design Fundamentals and Shortcuts for Non-Designers*. No Starch, 2021.

Osnos, Evan. "Can Mark Zuckerberg Fix Facebook Before It Breaks Democracy?" *New Yorker*, 17 Sept. 2018, www.newyorker.com/magazine/2018/09/17/can-mark-zuckerberg-fix-facebook-before-it-breaks-democracy. Accessed 13 July 2023.

Oswald, Vanessa. *Indie Rock: Finding an Independent Voice*. Greenhaven Publishing LLC, 2018.

"Our Story." *Pixar*, 2023, www.pixar.com/our-story-pixar. Accessed 13 July 2023.

Oxborrow, Ian. "What Is ChatGPT and Why Is It Controversial?" *National*, 24 Jan. 2023, www.thenationalnews.com/business/technology/2023/01/25/chatgpt-what-why-controversial. Accessed 18 July 2023.

Paar, Christof, Jan Pelzl, and Tim Güneysu. *Understanding Cryptography: A Textbook for Students and Practitioners*. Springer Cham, 2023.

Palmer, Daniel. *Photography and Collaboration: From Conceptual Art to Crowdsourcing*. Routledge, 2017.

Panay, Panos. "Introducing Windows 11." *Microsoft*, 24 June 2021, blogs.windows.com/windowsexperience/2021/06/24/introducing-windows-11. Accessed 13 July 2023.

Parent, Rick. *Computer Animation: Algorithms and Techniques*. 3rd ed., Elsevier, 2012.

Paris, Francesca. "A Strange New World: Has Virtual Reality Gaming Lived Up to Its Promise?" *WBUR*, 24 Feb. 2020, www.wbur.org/hereandnow/2020/02/24/virtual-reality-orion-13-games. Accessed 13 July 2023.

Parisi, Tony. *Learning Virtual Reality: Developing Immersive Experiences and Applications for Desktop, Web, and Mobile*. O'Reilly, 2015.

Park, Jung-ran. *Metadata Best Practices and Guidelines*. Routledge, 2011.

Parker, Matt. *Things to Make and Do in the Fourth Dimension: A Mathematician's Journey through Narcissistic Numbers, Optimal Dating Algorithms, at Least Two Kinds of Infinity, and More*. Farrar, 2014.

Passman, Donald S. *All You Need to Know about the Music Business*. 11th ed., Simon & Schuster, 2023.

Patel, Ruchika, and Parth Bhatt. "A Review Paper on Digital Watermarking and Its Techniques." *International Journal of Computer Applications*, vol. 110, no. 1, 2015, pp. 10-13.

Patt, Yale, and Sanjay Patel. *Introduction to Computing Systems: From Bits & Gates to C/C++ & Beyond*. 3rd ed., McGraw-Hill, 2019.

Patterson, David A., and John L. Hennessy. *Computer Organization and Design: The Hardware/Software Interface*. 6th ed., Morgan Kaufmann, 2020.

Paul, Christiane. *Digital Art*. 3rd ed., Thames & Hudson, 2015.

Paul, Katie, and Akash Sriram. "Meta's Twitter Rival Threads Surges to 100 Million Users Faster Than ChatGPT." *Reuters*, 10 July 2023, www.reuters.com/technology/metas-twitter-rival-threads-hits-100-mln-users-record-five-days-2023-07-10. Accessed 13 July 2023.

Payling, Dave. *Electronic Visual Music: The Elements of Audiovisual Creativity*. Focal Press, 2023.

Pearson, Luke Caspar, and Sandra Youkhana. *Videogame Atlas: Mapping Interactive Worlds*. Thames & Hudson, 2022.

Peláez, Carlos Alberto, and Andrés Solano. "A Practice for the Design of Interactive Multimedia Experiences Based on Gamification: A Case Study in Elementary Education." *Sustainability*, vol. 15, no. 3, 2023, doi.org/10.3390/su15032385. Accessed 13 July 2023.

Pentland, Alex. "Saving Big Data from Itself." *Scientific American*, vol. 311, no. 2, 2014, pp. 65-67.

Pepitone, Julianne, and David Goldman. "The Evolution of iTunes." *CNN Money*, 26 Apr. 2013, money.cnn.com/gallery/technology/2013/04/25/itunes-history.

Perez, Sarah. "The NFL Joins TikTok in Multi-Year Partnership." *TechCrunch*, 3 Sept. 2019, techcrunch.com/2019/09/03/the-nfl-joins-tiktok-in-multi-year-partnership. Accessed 13 July 2023.

_____. "Redesigned Version of iTunes Launches with iCloud Music Glitches." *Tech Crunch*, 13 Sept. 2016, techcrunch.com/2016/09/13/redesigned-version-of-itunes-launches-with-icloud-music-glitches. Accessed 13 July 2023.

_____. "TikTok Surpassed Facebook, Instagram, Snapchat & YouTube in Downloads Last Month." *TechCrunch*, 2 Nov. 2018, techcrunch.com/2018/11/02/tiktok-surpassed-facebook-instagram-snapchat-youtube-in-downloads-last-month. Accessed 13 July 2023.

Pernice, Kara. "F-Shaped Pattern of Reading on the Web: Misunderstood, but Still Relevant (Even on Mobile." *Nielsen Norman Group*, 12 Nov. 2017, www.nngroup.com/articles/f-shaped-pattern-reading-web-content. Accessed 13 July 2023.

Peters, Jay, and Alex Cranz. "Google Is Shutting Down Stadia." *Verge*, 29 Sept. 2022, www.theverge.com/2022/9/29/23378713/google-stadia-shutting-down-game-streaming-january-2023. Accessed 13 July 2023.

Peterson, Bryan. *Understanding Exposure: How to Shoot Great Photographs with Any Camera*. 4th ed., Amphoto Books, 2016.

Petkauskas, Vilius. "Report: Number of Expert-Crafted Video Deepfakes Double Every Six Months." *Cybernews*, 28 Sept. 2021, cybernews.com/privacy/report-number-of-expert-crafted-video-deepfakes-double-every-six-months. Accessed 13 July 2023.

Petty, Luke. "What Is Computer Aided Manufacturing? And How Is It Leading the Digital Site Revolution?" *Chapman Taylor*, 19 May 2016, www.chapmantaylor.com/en/insights/article/what-is-computer-aided-manufacturing-and-how-is-it-leading-the-digital-site/en. Accessed 13 July 2023.

Petzold, Charles. *Code: The Hidden Language of Computer Hardware and Software*. 2nd ed., Microsoft Press, 2022.

Philips, Winfred. "Introduction to Logic for ProtoThinker." *MindProject*, Apr. 1999, www.mind.ilstu.edu/curriculum/protothinker/logic_intro.php. Accessed 13 July 2023.

Pick, J. B., and R. Azari. "Global Digital Divide: Influence of Socioeconomic, Governmental, and Accessibility Factors on Information Technology." *Information Technology for Development*, vol. 14, 2008, pp. 91-115.

Pierce, Allan D. *Acoustics: An Introduction to Its Physical Principles and Applications*. 3rd ed., Springer-International Publishing, 2019.

Pinch, T. J., and Karin Bijsterveld. *The Oxford Handbook of Sound Studies*. Oxford UP, 2013.

Pinola, Melanie. "Speech Recognition through the Decades: How We Ended Up with Siri." *PCWorld*, 2 Nov. 2011, www.pcworld.com/article/477914/speech_recognition_through_the_decades_how_we_ended_up_with_siri.html. Accessed 13 July 2023.

Piore, Adam. "The Quest to Build a Silicon Brain." *Discover*, 24 May 2013, www.discovermagazine.com/mind/the-quest-to-build-a-silicon-brain. Accessed 13 July 2023.

Pisani, Bob. "Spotify's IPO Disrupted Wall Street. What Lies Ahead Now for Unicorns Looking to Go Public." *CNBC Disruptor/50*, 22 May 2018, www.cnbc.com/2018/05/22/spotifys-ipo-disrupted-wall-street-what-lies-ahead-now-for-unicorns-looking-to-go-public.html. Accessed 25 July 2023.

Platz, Cheryl. *Design Beyond Devices: Creating Multimodal, Cross-Device Experiences*. Rosenfeld Media, 2020.

Pless, Vera. *Introduction to the Theory of Error-Correcting Codes*. John Wiley & Sons, 2011.

Plotkin, David. *Data Stewardship: An Actionable Guide to Effective Data Management and Data Governance*. 2nd ed., Academic Press, 2020.

Pocock, Kevin. "What Is ChatGPT? What Is It Used For?" *PC Guide*, 15 Mar. 2023, www.pcguide.com/apps/what-is-chat-gpt. Accessed 18 July 2023.

Pollard, Barry. *HTTP/2 in Action*. Simon & Schuster, 2019.

Pongnumkul, S., and K. Motohashi. "A Bipartite Fitness Model for Online Music Streaming Services." *Physica A*, vol. 490, 2018, pp. 1125-37.

Porter, Jon. "Apple Announces macOS Sonoma with Game Mode and Support for Desktop Widgets." *Verge*, 5 June 2023, www.theverge.com/2023/6/5/23745460/apple-macos-14-sonoma-features-updates-wwdc-2023. Accessed 13 July 2023.

Porter, Jon, and Nick Statt. "Google Completes Purchase of Fitbit." *Verge*, 14 Jan. 2021, www.theverge.com/2021/1/14/22188428/google-fitbit-acquisition-completed-approved. Accessed 13 July 2023.

Porup, J. M. "How and Why Deepfake Videos Work-And What Is at Risk." *CSO*, 10 Apr. 2019, www.csoonline.com/article/3293002/deepfake-videos-how-and-why-they-work.html. Accessed 13 July 2023.

Posch, Maya. "On Getting a Computer's Attention and Striking Up a Conversation." *Hackaday*, 14 Nov. 2022, hackaday.com/2022/11/14/on-getting-a-computers-attention-and-striking-up-a-conversation. Accessed 13 July 2023.

"Post-Conceptualism and Digital Art: The Future of Art?" *US Advisory*, www.ugoscalia.com/post/post-conceptualism-and-digital-art-the-future-of-art. Accessed 13 July 2023.

Potter, W. James. *Media Literacy*. 10th ed., SAGE Publications, 2021.

_____. "Review of Literature on Media Literacy." *Sociology Compass*, vol. 7, no. 6, 2013, pp. 417-35.

Powell, John. *How Music Works: The Science and Psychology of Beautiful Sounds, from Beethoven to the Beatles and Beyond*. Little, Brown, 2010.

Prayaga, Lakshmi, and Hamsa Suri. *Programming the Web with ColdFusion MX 6.1 Using XHTML*. McGraw-Hill, 2004.

Pressman, Aaron. "Alexa Turns 5: What Amazon's Kindergarten-Aged Assistant Can Teach the Rest of Tech." *Fortune*, 6 Nov. 2019, fortune.com/2019/11/06/amazon-alexa-echo-5-anniversary-dave-limp-interview. Accessed 13 July 2023.

"Principles for Useable Design." *Usability Body of Knowledge*, www.usabilitybok.org/principles-for-usable-design. Accessed 13 July 2023.

Pringle, Heather. "The Origins of Creativity." *Scientific American*, Oct. 2016, www.scientificamerican.com/article/the-origins-of-creativity. Accessed 13 July 2023.

"The Printed World." *Economist*, 10 Feb. 2011, www.economist.com/briefing/2011/02/10/the-printed-world. Accessed 13 July 2023.

Pritchard, Michael. *A History of Photography in 50 Cameras*. Bloomsbury Visual Arts, 2014.

"Press Room." *IMDb*, June 2023, www.imdb.com/pressroom/stats. Accessed 13 July 2023.

Prust, Z. A., and Peggy B. Deal. *Graphic Communications: Digital Design and Print Essentials*. 7th ed., Goodheart-Willcox, 2022.

Pulkki, Ville, and Matti Karjalainen. *Communication Acoustics: An Introduction to Speech, Audio, and Psychoacoustics*. John Wiley & Sons, 2015.

Purchase, Helen C. *Experimental Human-Computer Interaction: A Practical Guide with Visual Examples*. Cambridge UP, 2012.

Purdy, Kevin. "Hosting Site Imgur Will Remove Explicit and Anonymous Content Next Month." *Ars Technica*, 20 Apr. 2023, arstechnica.com/gadgets/2023/04/hosting-site-imgur-will-remove-explicit-and-anonymous-content-next-month. Accessed 13 July 2023.

Qualman, Erik. *Socialnomics: How Social Media Transforms the Way We Live and Do Business*. Wiley, 2010.

"Quick Facts about the Digital Watermarking Alliance." *Digital Watermarking Alliance*, digitalwatermarkingalliance.org/about/quick-facts. Accessed 13 July 2023.

Quinn, Brendan. *Data Protection Implementation Guide: A Legal, Risk and Technology Framework for the GDPR*. Wolters Kluwer, 2021.

Quirk, Vanessa. "A Brief History of BIM." *ArchDaily*, 7 Dec. 2012, www.archdaily.com/302490/a-brief-history-of-bim. Accessed 13 July 2023.

Rabiger, Michael, and Mick Hurbis-Cherrier. *Directing: Film Techniques and Aesthetics*. Routledge, 2020.

Ramasubramanian, S. "Racial/Ethnic Identity, Community-Oriented Media Initiatives, and Transmedia Storytelling." *Information Society*, vol. 32,no. 5, 2016, pp. 333-42.

Raphael, J. R. "Android Versions: A Living History from 1.0 to 14." *Computerworld*, 26 June 2020, www.computerworld.com/article/3235946/android-versions-a-living-history-from-1-0-to-today.html. Accessed 13 July 2023.

Rathnam, K. *A First Course in Engineering Drawing*. Springer, 2018.

Ravens, Chal. "Good Vibrations: How Bandcamp Became the Heroes of Streaming." *The Guardian*, 25 June 2020, www.theguardian.com/music/2020/jun/25/bandcamp-music-streaming-ethan-diamond-online-royalties. Accessed 13 July 2023.

Rayome, Alison DeNisco. "10 Most-Downloaded Apps of the 2010s: Facebook Dominates the Decade." *CNET*, 16 Dec. 2019, www.cnet.com/news/10-most-downloaded-apps-of-the-decade-facebook-dominated-2010-2019. Accessed 13 July 2023.

"The Redefinition of the SI Units." *NPL*, www.npl.co.uk/si-units/the-redefinition-of-the-si-units. Accessed 13 July 2023.

"Research and Analysis: Online Copyright Infringement Tracker Survey (10th Wave) Executive Summary." *Intellectual Property Office*, gov.uk/government/publications/online-copyright-infringement-tracker-survey-10th-wave/online-copyright-infringement-tracker-survey-10th-wave-executive-summary. Accessed 13 July 2023.

Reimer, Jeremy. "A History of the GUI." *Ars Technica*, 5 May 2005, arstechnica.com/features/2005/05/gui. Accessed 13 July 2023.

Relf, Emily. "Maps Providing a Sense of Direction." *Art21*, 29 June 2021, art21.org/read/maps-providing-a-sense-of-direction. Accessed 13 July 2023.

Renda, Andrea. "Competition-Regulation Interface in Telecommunications: What's Left of the Essential Facility Doctrine." *Telecommunications Policy*, vol. 34, no. 1, 2010, pp. 23-35.

Retna, Kala S. "Thinking about 'Design Thinking': A Study of Teacher Experiences." *Asia Pacific Journal of Education*, vol. 36, suppl. 1, 2016, pp. 5-19.

Riley, Gail Blasser. *Internet Piracy. Cavendish*, 2010.

Riselvato, John. *Bandcamp for Artists and Music Labels: Marketing Tips to Sell More Music and Merch*. Amazon Digital Services LLC, 2020.

Rizzo, Albert "Skipp," and Stéphane Bouchard, editors. *Virtual Reality for Psychological and Neurocognitive Interventions*. Springer-Verlag New York, 2019.

Roberts, Dan, and Spencer Ackerman. "NSA Mass Phone Surveillance Revealed by Edward Snowden Ruled Illegal." *The Guardian*, 7 May 2015, www.theguardian.com/us-news/2015/may/07/nsa-phone-records-program-illegal-court. Accessed 13 July 2023.

Robertson, Adi. "The YouTube Gaming App Is Shutting Down This Week." *Verge*, 27 May 2019, www.theverge.com/2019/5/27/18641413/youtube-gaming

-standalone-app-shutdown-may-30th. Accessed 13 July 2023.

Robinson, Arthur H., et al. *Elements of Cartography*. 6th ed., Wiley, 1995.

Robinson, Joanna. "The Sci-Fi Guru Who Predicted Google Earth Explains Silicon Valley's Latest Obsession." *Vanity Fair*, 23 June 2017, www.vanityfair.com/news/2017/06/neal-stephenson-metaverse-snow-crash-silicon-valley-virtual-reality. Accessed 13 July 2023.

Rodrigues, P., and J. Bidarra. "Transmedia Storytelling and the Creation of a Converging Space of Educational Practices." *International Journal of Emerging Technologies in Learning*, vol. 9, no. 6, 2014, pp. 42-44.

Roettgers, Janko. "Apple Is Officially Killing iTunes for Mac, Replacing It with Three Dedicated Media Apps." *Variety*, 3 June 2019, variety.com/2019/digital/news/itunes-shutting-down-1203231598. Accessed 13 July 2023.

Rogers, Scott. *Level Up!: The Guide to Great Video Game Design*. 2nd ed., Wiley, 2014.

_____. *Swipe This!: The Guide to Great Touchscreen Game Design*. John Wiley & Sons, 2012.

Rogers, Yvonne, Helen Sharp, and Jennifer Preece. *Interaction Design: Beyond Human-Computer Interaction*. 6th ed., Wiley, 2023.

Rognerud, Jon. *Ultimate Guide to Optimizing Your Website*. 3rd ed., Entrepreneur Media, 2014.

Rolston, Clyde Philip, Amy Macy, Tom Hutchison, and Paul Allen. *Record Label Marketing: How Music Companies Brand and Market Artists in the Digital Era*. CRC Press, 2015.

Romm, Tony. "Millions of Low-Income Americans Will Receive Internet Access Rebates under New $7 Billion Broadband Stimulus Plan." *Washington Post*, 22 Dec. 2020, www.washingtonpost.com/technology/2020/12/22/internet-rebate-coronavirus-stimulus. Accessed 13 July 2023.

Rosenbaum, Steven. "YouTube Reveals a Curated Future." *Forbes*, 10 Oct. 2012, www.forbes.com/sites/stevenrosenbaum/2012/10/10/YouTube-reveals-a-curated-future-2. Accessed 13 July 2023.

Rosenberg, Matthew, and Sheera Frenkel. "Facebook's Role in Data Misuse Sets of Storms on Two Continents." *New York Times*, 18 Mar. 2018, www.nytimes.com/2018/03/18/us/cambridge-analytica-facebook-privacy-data.html. Accessed 13 July 2023.

Rosenblatt, Kalhan. "ChatGPT Banned from New York City Public Schools' Devices and Networks." *NBC News*, 5 Jan. 2023, www.nbcnews.com/tech/tech-news/new-york-city-public-schools-ban-chatgpt-devices-networks-rcna64446. Accessed 13 July 2023.

Rosencrance, Linda. "Software." *Tech Target*, Mar. 2021, www.techtarget.com/searchapparchitecture/definition/software. Accessed 13 July 2023.

Rossing, Thomas, and Neville Fletcher. *Principles of Vibration and Sound*. 2nd ed., Springer-Verlag, 2004.

Rouse, Margaret. "Mobile Web." *Techopedia*, 4 Apr. 2017, www.techopedia.com/definition/23588/mobile-web. Accessed 13 July 2023.

Rowe, Peter G. *Design Thinking*. MIT Press, 1987.

Rubin, Michael. *Nonlinear-A Field Guide to Digital Video and Film Editing*. Triad, 2000.

Rubin, Victoria L. "Deception Detection and Rumor Debunking for Social Media." *The SAGE Handbook of Social Media Research Methods*, edited by Luke Sloan and Anabel Quan-Haase, Sage Publications, 2017, pp. 342-64.

Rudel, Anthony. *Hello, Everybody: The Dawn of American Radio*. Houghton Mifflin Harcourt, 2008.

Ruiz, Rebecca R. "F.C.C. Sets Net Neutrality Rules." *New York Times*, 12 Mar. 2015, www.nytimes.com/2015/03/13/technology/fcc-releases-net-neutrality-rules.html. Accessed 13 July 2023.

Rumelhart, David E., James L. McClelland, and the PDP Research Group. *Parallel Distributed Processing: Explorations in the Microstructure of Cognition*. 1986. MIT Press, 1989. 2 vols.

Rumsey, Francis, and Tim McCormick. *Sound and Recording: An Introduction*. 8th ed., Routledge, 2021.

Russell, Stuart, and Peter Norvig. *Artificial Intelligence: A Modern Approach*. 4th ed. Pearson, 2020.

Rychagov, Michael N., Ekaterina V. Tolstaya, and Mikhail Y. Sirotenko, editors. *Smart Algorithms for Multimedia and Imaging*. Springer Cham, 2021.

Sacks, Rafael, Chuck Eastman, Ghang Lee, and Paul Teicholz. *BIM Handbook: A Guide to Building Information Modeling for Owners, Designers, Engineers, Contractors, and Facility Managers*. 3rd ed., Wiley, 2018.

Saffer, Dan. *Designing Gestural Interfaces*. O'Reilly Media, 2008.

Salmond, Michael. *Video Game Design: Principles and Practices from the Ground Up*. Bloomsbury Academic, 2020.

Saltzman, Steven. *Music Editing for Film and Television: The Art and the Process*. Focal, 2015.

Samuels, Miranda. "Tragedy of the Non-Visible: Aesthetic and Ontological Differences in Early Computer versus

Conceptual Art." *Leonardo*, vol. 56, no. 2, 2023, pp. 159-63.

Sanz, E., and G. Turlea. "Downloading Inclusion: A Statistical Analysis of Young People's Digital Communication Inequalities." *Innovation: The European Journal of Social Sciences*, vol. 25, 2012, pp. 337-53.

Sarkar, Jayanta. *Computer Aided Design: A Conceptual Approach*. CRC Press, 2017.

Sarker, Samit. "YouTube Gaming Launches Aug. 26 with Website and Mobile Apps." *Polygon*, 25 Aug. 2015, www.polygon.com/2015/8/25/9208245/youtube-gaming-launch-date-web-android-ios. Accessed 13 July 2023.

Saul, Roger. "KevJumba and the Adolescence of YouTube." *Educational Studies*, vol. 46, no. 5, 2010, pp. 457-77.

Savov, Vlad. "Google Announces YouTube Music and YouTube Premium." *Verge*, 17 May 2018, www.theverge.com/2018/5/17/17364056/youtube-music-premium-google-launch. Accessed 13 July 2023.

Schapire, Robert E., and Yoav Freund. *Boosting: Foundations and Algorithms*. MIT Press, 2012.

Schell, Jesse. *The Art of Game Design: A Book of Lenses*. 2nd ed., CRC Press, 2015.

Schellekens, Elizabeth. "Conceptual Art." *Stanford Encyclopedia of Philosophy*, 21 Aug. 2022, plato.stanford.edu/archives/fall2022/entries/conceptual-art. Accessed 13 July 2023.

Schenk, Sonja, and Ben Long. *The Digital Filmmaking Handbook*. 7th ed., Foreing Films, 2021.

Schmidt, Eric, and Jonathan Rosenberg. *How Google Works*. Grand Central, 2017.

Schmidt, Eric, and Jared Cohen. *The New Digital Age: Reshaping the Future of People, Nations, and Business*. Knopf, 2013.

Schmitt, Christopher. *Designing Web & Mobile Graphics: Fundamental Concepts for Web and Interactive Projects*. New Riders, 2013.

Schneiderhan, Erik, and Martin Lukk. *GoFailMe: The Unfulfilled Promise of Digital Crowdfunding*. Stanford UP, 2023.

Schodek, Daniel, et al. *Digital Design and Manufacturing: CAD/CAM Applications in Architecture and Design*. Wiley, 2005.

Scott, Michael L. *Programming Language Pragmatics*. 4th ed., Morgan Kaufmann, 2016.

Seabrook, John. "Revenue Streams." *New Yorker*, 24 Nov. 2014. www.newyorker.com/magazine/2014/11/24/revenue-streams. Accessed 25 July 2023.

Seidl, Martina, et al. *UML@Classroom: An Introduction to Object-Oriented Modeling*. Springer, 2015.

Sellers, Audrey. "Top Programming Languages of 2023." *Coding Dojo*, 3 Feb. 2023, www.codingdojo.com/blog/top-programming-languages. Accessed 13 July 2023.

Sexton, Timothy J. *A Concise Introduction to Engineering Graphics Including Worksheet Series B*. 6th ed., SDC Publications, 2019.

Shah, H., K. Warwick, J. Vallverdú, and D. Wu. "Can Machines Talk? Comparison of Eliza with Modern Dialogue Systems." *Computers in Human Behavior*, vol. 58, 2016, pp. 278-95.

Shao, Grace. "What 'Deepfakes' Are and How They May Be Dangerous." *CNBC*, 13 Oct. 2019, www.cnbc.com/2019/10/14/what-is-deepfake-and-how-it-might-be-dangerous.html. Accessed 13 July 2023.

Shaoqiang, Wang, editor. *Get Impressed!: The Revival of Letterpress and Handmade Type*. Hoaki, 2021.

Shapiro, Peter, editor. *Modulations: A History of Electronic Music Throbbing Words on Sound*. Caipirinha Productions, 2000.

Share, Jeff. *Media Literacy Is Elementary: Teaching Youth to Critically Read and Create Media*. Lang, 2009.

Shearer, Allan W. "Abduction to Argument: A Framework of Design Thinking." *Landscape Journal*, vol. 34, no. 2, 2015, pp. 127-38.

Shen, Jialie, John Shepherd, Bin Cui, and Ling Liu. *Intelligent Music Information Systems: Tools and Methodologies*. IGI Global, 2008.

Shepherd, Maddie. "Crowdfunding Statistics: Market Size and Growth." *Fundera*, 23 Jan. 2023, www.fundera.com/resources/crowdfunding-statistics. Accessed 13 July 2023.

Sherif, Mostafa Hashem. *Protocols for Secure Electronic Commerce*. 3rd ed., CRC Press, 2018.

Shin, Dong-Hee. "A Comparative Analysis of Net Neutrality: Insights Gained by Juxtaposing the US and Korea." *Telecommunications Policy*, vol. 38, no. 11, 2014, pp. 1117-133.

Shirky, Clay. *Here Comes Everybody: The Power of Organizing without Organizations*. Penguin, 2009.

Shoenberger, Allen. "Privacy Wars." *Indiana International & Comparative Law Review*, vol. 17, 2007, pp. 355-93.

Shreekumar, Advik. "Rise of the YouTube Star." *Harvard Political Review*, vol. 41, no. 1, 2014, p. 16.

"SI Redefinition." *NIST*, www.nist.gov/si-redefinition. Accessed 13 July 2023.

Sickels, Robert C. *The Business of Entertainment*. 3 vols. ABC-CLIO, 2009.

Siddiqi, Muzaffer A. *Dynamic RAM: Technology Advancements*. CRC Press, 2013.

Sidlow, Faith, and Kim Stephens. *Broadcast News in the Digital Age: A Guide to Reporting, Producing and Anchoring Online and on TV.* Routledge, 2022.

Silver, H. Ward. "Digital Code Basics." *Qst*, vol. 98, no. 8, 2014, pp. 58-59.

Simmons, Annette. *The Story Factor.* Reprint ed., Basic Books, 2019.

Simonite, Tom. "Prepare for the Deepfake Era of Web Video." *Wired*, 6 Oct. 2019, www.wired.com/story/prepare-deepfake-era-web-video. Accessed 13 July 2023.

Sinclair, G., and J. Tinson. "Psychological Ownership and Music Streaming Consumption." *Journal of Business Research*, vol. 71, 2017, pp. 1-9.

Singer, Graham. "The History of the Microprocessor and the Personal Computer." *TechSpot*, 1 Oct. 2020, www.techspot.com/article/874-history-of-the-personal-computer. Accessed 13 July 2023.

Singh, Lakhwinder Pal, and Harwinder Singh. *Engineering Drawing: Principles and Applications.* Cambridge UP, 2021.

Sito, Tom. *Moving Innovation: A History of Computer Animation.* MIT Press, 2013.

"SK Hynix Launches World's First DDR5 DRAM." *HPC Wire*, 7 Oct. 2020, www.hpcwire.com/off-the-wire/sk-hynix-launches-worlds-first-ddr5-dram. Accessed 13 July 2023.

Sloane, Paul, editor. *A Guide to Open Innovation and Crowdsourcing: Advice from Leading Experts.* Kogan, 2012.

Smiraglia, Richard P., editor. *Metadata: A Cataloger's Primer.* Routledge, 2012.

Smith, Craig. "Imgur Statistics and User Count (2023)." *DMR*, 8 Mar. 2023, expandedramblings.com/index.php/imgur-statistics. Accessed 13 July 2023.

Smith, Ernie. "The Story of AllMusic, the Internet's Largest, Most Influential Music Database." *Vice*, 24 Sept. 2016, www.vice.com/en/article/53djj8/the-story-of-allmusic-the-internets-largest-most-influential-music-database. Accessed 13 July 2023.

Smith, Michael D., and Rahul Telang. *Streaming, Sharing, Stealing: Big Data and the Future of Entertainment.* Reprint ed., MIT Press, 2017.

Smith, Noah. "The Giants of the Video Game Industry Have Thrived in the Pandemic: Can the Success Continue?" *The Washington Post*, 12 May 2020, www.washingtonpost.com/video-games/2020/05/12/video-game-industry-coronavirus. Accessed 26 Feb. 2021.

Soares, Marcelo M., and Francisco Rebelo. *Advances in Usability Evaluation.* CRC Press, 2013.

Soegaard, Mads, and Rikke Friis Dam, editors. *The Encyclopedia of Human-Computer Interaction.* 2nd ed., Interaction Design Foundation, 2014.

Sommer, Jeff. "What the Net Neutrality Rules Say." *New York Times*, 12 Mar. 2015, www.nytimes.com/interactive/2015/03/12/technology/net-neutrality-rules-explained.html. Accessed 13 July 2023.

Spaanjaars, Imar. *Beginning ASP.NET 3.5 in C# and VB.* Wiley Publishing, 2008.

Spangler, Todd. "Musical.ly Is Going Away: Users to Be Shifted to ByteDance's TikTok Video App." *Variety*, 2 Aug. 2018, variety.com/2018/digital/news/musically-shutdown-tiktok-bytedance-1202893205. Accessed 13 July 2023.

_____. "YouTube Shorts, Video Giant's TikTok Copycat, Is Rolling Out in 100-Plus Countries." *Variety*, 13 July 2021, variety.com/2021/digital/news/youtube-shorts-global-launch-1235018403. Accessed 13 July 2023.

Spiegel, Brennan. "Virtual Reality and the COVID Mental Health Crisis." *Scientific American*, 15 Nov. 2020, www.scientificamerican.com/article/virtual-reality-and-the-covid-mental-health-crisis. Accessed 13 July 2023.

Stallings, William. *Data and Computer Communications.* 10th ed., Prentice Hall, 2014.

Starks, Michael. *The Digital Television Revolution: Origins to Outcomes.* Palgrave Macmillan UK, 2013.

Stein, Scott. "Microsoft, Facebook and Everyone Else are Building the Metaverse. Will You Want to Live There?" *CNet*, 4 Aug. 2021, www.cnet.com/tech/computing/microsoft-facebook-and-everyone-else-are-building-the-metaverse-but-will-you-want-to-live-there. Accessed 13 July 2023.

Steinberg, S. H. *Five Hundred Years of Printing.* 1955. Dover, 2017.

Steiner, Christopher. *Automate This: How Algorithms Came to Rule Our World.* Penguin, 2012.

Steinhauer, Jennifer, and Jonathan Weisman. "U.S. Surveillance in Place since 9/11 Is Sharply Limited." *New York Times*, 2 June 2015, www.nytimes.com/2015/06/03/us/politics/senate-surveillance-bill-passes-hurdle-but-showdown-looms.html. Accessed 13 July 2023.

Stephenson, Brad. "Twitch: Everything You Need to Know." *Lifewire*, 25 May 2023, www.lifewire.com/what-is-twitch-4143337. Accessed 13 July 2023.

Stewart, Peter, and Ray Alexander. *Broadcast Journalism: Techniques of Radio and Television News.* 8th ed., Routledge, 2021.

Stinson, Liz. "Facebook Reactions, the Totally Redesigned Like Button, Is Here." *Wired*, 24 Feb. 2016, www.wired.com/2016/02/facebook-reactions-totally-redesigned-like-button. Accessed 13 July 2023.

Stockson, Eric. *Brave Browser: Blockchain Internet Browsing Made Easy*. First Rank, 2019.

Stokel-Walker, Chris. "Why Do DVDs Still Exist?" *Wired*, 27 Nov. 2021, www.wired.com/story/why-do-dvds-exist. Accessed 13 July 2023.

Stone, Stephen C. *Music Theory and Composition: A Practical Approach*. Rowman & Littlefield, 2018.

"Storage vs. Memory." *PCMag*, www.pcmag.com/encyclopedia/term/storage-vs-memory. Accessed 13 July 2023.

Storr, Will. *The Science of Storytelling: Why Stories Make Us Human and How to Tell Them Better*. Abrams, 2020.

Stricklin, Kasey. *Social Media Bots: Laws, Regulations, and Platform Policies*. CNA, Sept. 2020, www.cna.org/reports/2020/10/DIM-2020-U-028193-Final.pdf. Accessed 13 July 2023.

Strong, William J., and George R. Plitnik. *Music, Speech, Audio*. 4th ed., BYU Academic Publishing, 2013.

Su, Qingtang. *Color Image Watermarking*. Tsinghua UP/Walter de Gruyter, 2017.

Suhr, H. Cecilia. *Evaluation and Credentialing in Digital Music Communities: Benefits and Challenges for Learning and Assessment*. MIT Press, 2014.

Suie, Daniel, Sara Elwood, and Michael Goodchild, editors. *Crowdsourcing Geographic Knowledge: Volunteered Geographic Information (VGI) in Theory and Practice*. Springer, 2013.

Sulleyman, Aatif. "Facebook Live Killings: Why the Criticism Has Been Harsh." *Independent*, 27 Apr. 2017, www.independent.co.uk/life-style/gadgets-and-tech/features/facebook-live-killings-ai-artificial-intelligence-not-blame-fatalities-murders-us-steve-stephens-a7706056.html. Accessed 13 July 2023.

Sullivan, Emily Tate. "The Digital Divide Has Narrowed, but 12 Million Students Are Still Disconnected." *EdSurge*, 27 Jan. 2021, www.edsurge.com/news/2021-01-27-the-digital-divide-has-narrowed-but-12-million-students-are-still-disconnected. Accessed 13 July 2023.

"Summary of Digital Art." *Art Story*, 2019, www.theartstory.org/movement-digital-art.htm. Accessed 13 July 2023.

Summers, Danny. *Imgur: 42 Success Secrets-42 Most Asked Questions-Imgur: What You Need to Know*. Emereo Pty Ltd., 2014.

Surowiecki, James. *The Wisdom of Crowds: Why the Many Are Smarter than the Few and How Collective Wisdom Shapes Business, Economies, Societies, and Nations*. New York: Doubleday, 2004.

Swaine, Jon. "Twitter Admits Far More Russian Bots Posted on Election Than It Had Disclosed." *The Guardian*, 19 Jan. 2018, www.theguardian.com/technology/2018/jan/19/twitter-admits-far-more-russian-bots-posted-on-election-than-it-had-disclosed. Accessed 13 July 2023.

Swarts, Jason. *Wicked, Incomplete and Uncertain: User Support in the Wild and the Role of Technical Communication*. UP of Colorado, 2018.

Sylvester, Ruth, and Wendy-lou Greenidge. "Digital Storytelling: Extending the Potential for Struggling Writers." *Reading Teacher*, vol. 63, no. 4, 2009, pp. 284-95.

Tavernor, Robert. *Smoot's Ear: The Measure of Humanity*. Yale UP, 2007.

Taylor, David. *Digital Photography Complete Course*. DK, 2015.

Taylor, Jim. *Everything You Ever Wanted to Know About DVD: The Official DVD FAQ*. McGraw-Hill, 2004.

TED-Ed. "Animation Basics: The Optical Illusion of Motion." *TedEd*, ed.ted.com/lessons/animation-basics-the-optical-illusion-of-motion-ted-ed#watch. Accessed 13 July 2023.

Templeton, Brad. "Self-Driving Cars 2021: Year in Review." *Forbes*, 3 Jan. 2022, www.forbes.com/sites/bradtempleton/2022/01/03/self-driving-cars-2021-year-in-review/?sh=4aa85563773b. Accessed 13 July 2023.

Thaddeus-Johns, Josie. "What Are NFTs, Anyway? One Just Sold for $69 Million." *New York Times*, 12 May 2021, www.nytimes.com/2021/03/11/arts/design/what-is-an-nft.html. Accessed 13 July 2023.

Thatcher, Jim, et al. *Web Accessibility: Web Standards and Regulatory Compliance*. Springer, 2006.

Thompson, Jim, Barnaby Berbank-Green, and Nic Cusworth. *Game Design: Principles, Practice, and Techniques*. Quarto, 2007.

Tidwell, Jenifer, Charles Brewer, and Aynne Valencia. *Designing Interfaces: Patterns for Effective Interaction Design*. 3rd ed., O'Reilly Media, 2020.

Tidy, Joe. "YouTube Accused of Not Tackling Musk Bitcoin Scam Streams." *BBC*, 10 June 2022, www.bbc.com/news/technology-61749120. Accessed 13 July 2023.

"Timeline of Computer History." *Computer History Museum*, www.computerhistory.org/timeline/computers. Accessed 13 July 2023.

Titcomb, James. "Instagram Is Changing Its Iconic Logo-Here's Why." *Telegraph*, 11 May 2016, www.telegraph.co.uk/technology/2016/05/11/instagram-is-changing-its-iconic-logo—heres-why. Accessed 13 July 2023.

Torr, James D. *Internet Piracy*. Greenhaven, 2004.

"Towards the Semantic Web." *Cambridge Semantics*, cambridgesemantics.com/blog/semantic-university/intro-semantic-web/towards-the-semantic-web. Accessed 13 July 2023.

Trigonis, John T. *Crowdfunding for Filmmakers: The Way to a Successful Film Campaign*. Michael Wiese Productions, 2013.

Tsai, Wei-Tek, et al., editors. *Crowdsourcing: Cloud-Based Software Development*. Springer, 2015.

Tufte, Edward R. *The Visual Display of Quantitative Information*. 2nd ed., Graphics, 2007.

Tupper, Shelby Putnam. "The Best Graphic Design Software for 2023." *PCMag*, 5 Jan. 2023, www.pcmag.com/picks/the-best-graphic-design-software. Accessed 13 July 2023.

Turban, Efraim, Carol Pollard, and Gregory Wood. *Information Technology for Management: On-Demand Strategies for Performance, Growth and Sustainability*. 11th ed., John Wiley & Sons, 2018.

Tuten, Tracy L., and Michael R, Solomon. *Social Media Marketing*. Pearson, 2013.

Tyler, Dustin. "How to Use Game Theory in Video Game Development." *Game Designing*, 10 June 2023, www.gamedesigning.org/learn/game-theory. Accessed 13 July 2023.

Ulrich, Karl, Steven Eppinger, and Maria C. Yang. *Product Design and Development*. 7th ed., McGraw-Hill, 2020.

Umar, Amjad. "IT Infrastructure to Enable Next Generation Enterprises." *Information Systems Frontiers*, vol. 7, no. 3, 2005, pp. 217-56.

Underdahl, Brian. *Macromedia Flash MX: The Complete Reference*. McGraw-Hill/Osborne Media, 2003.

United States Federal Communications Commission. *Preserving the Open Internet: Report and Order*. GPO, 2010.

"US VR and AR Users, 2019-2023." *Insider Intelligence | eMarketer*, 22 Feb. 2021, www.emarketer.com/chart/244851/us-vr-ar-users-2019-2023-millions. Accessed 13 July 2023.

"User Interface Design Basics." *Usability.gov*, www.usability.gov/what-and-why/user-interface-design.html. Accessed 13 July 2023.

Valiant, Leslie. *Probably Approximately Correct: Nature's Algorithms for Learning and Prospering in a Complex World*. Basic, 2013.

Vanhemert, Kyle. "Leap Motion's Augmented-Reality Computing Looks Stupid Cool." *Wired*, 7 July 2015, www.wired.com/2015/07/leap-motion-glimpse-at-the-augmented-reality-desktop-of-the-future. Accessed 13 July 2023.

Vanian, Jonathan. "Musk Risks Even More Damage to Twitter's Business as the Messaging App Changes Name to X." *CNBC*, 24 July 2023, www.cnbc.com/2023/07/24/elon-musk-risks-more-damage-to-twitter-business-after-name-change-to-x.html. Accessed 26 July 2023.

Vasquez, Emily, and Adam Hardy. "What Is WiFi Calling? Pros, Cons & More." *Forbes Advisor*, 8 June 2023, www.forbes.com/advisor/business/software/what-is-wi-fi-calling. Accessed 13 July 2023.

Vaux, Robert. "The Definition of an Advertising Strategy." *Chron*, smallbusiness.chron.com/definition-advertising-strategy-19332.html. Accessed 13 July 2023.

Vince, John. *Introduction to Virtual Reality*. Springer-Verlag London, 2004.

"Visionary Futurist Neal Stephenson and Crypto Pioneer Peter Vessenes Announce Lamina1, the Layer-1 Blockchain for the Open Metaverse." *BusinessWire*, 8 June 2022, www.businesswire.com/news/home/20220608005423/en. Accessed 13 July 2023.

Vogels, Emily A., et al. "53% of Americans Say the Internet Has Been Essential during the COVID-19 Outbreak." *Pew Research Center*, 30 Apr. 2020, www.pewresearch.org/internet/2020/04/30/53-of-americans-say-the-internet-has-been-essential-during-the-covid-19-outbreak. Accessed 13 July 2023.

"W3C Data Activity: Building the Web of Data." *W3C*, www.w3.org/2013/data. Accessed 13 July 2023.

"W3C Semantic Web Frequently Asked Questions." *W3C*, 12 Nov. 2009, www.w3.org/RDF/FAQ. Accessed 13 July 2023.

Wacholtz, Larry. *Monetizing Entertainment: An Insider's Handbook for Careers in the Entertainment and Music Industry*. CRC Press, 2016.

Wagner, Karin. *From ASCII Art to Comic Sans: Typography and Popular Culture in the Digital Age*. MIT Press, 2023.

Wang, Chih-Hsuan, David M. Shannon, and Margaret E. Ross. "Students' Characteristics, Self-Regulated Learning, Technology Self-Efficacy, and Course Outcomes in Online Learning." *Distance Education*, vol. 34, no. 3, 2013, pp. 302-23.

Wang, Tian. "Temporal Design: An Introduction to the 4 Most Popular Animation Software for Architects." *Architizer*, architizer.com/blog/practice/tools/animation-software-for-architects. Accessed 13 July 2023.

Wang, Yongtian, Xueming Li, and Yuxin Peng, editors. *Image and Graphics Technologies and Applications*. Springer Singapore, 2020.

Ward, Tom. "The Biggest Gamer in the World Breaks Down Twitch for Us." *Forbes*, 1 May 2018, www.forbes.com/sites/tomward/2018/05/01/the-biggest-gamer-in-the-world-breaks-down-twitch-for-us/#36132d0d5bb5. Accessed 13 July 2023.

Wardyga, Brian J. *The Video Games Textbook: History, Business, Technology*. 2nd ed., CRC Press, 2023.

Ware, Colin. *Visual Thinking for Information Design*. 2nd ed., Morgan Kaufmann, 2021.

Warren, Christina. "The Evolution of Mac OS, From 1984 to Mountain Lion." *Mashable*, 17 Feb. 2012, mashable.com/2012/02/17/mac-os-timeline/#KKsyA8f26qqR. Accessed 13 July 2023.

Warren, Tom. "Windows Turns 35: A Visual History." *Verge*, 20 Nov. 2020, www.theverge.com/2015/11/19/9759874/microsoft-windows-35-years-old-visual-history. Accessed 13 July 2023.

"Web Standards." *W3C*, 2023, www.w3.org/standards. Accessed 13 July 2023.

Wei, June. *Mobile Electronic Commerce: Foundations, Development, and Applications*. CRC Press, 2015.

Wei, L., and D. Hindman. "Does the Digital Divide Matter More? Comparing the Effects of New Media and Old Media Use on the Education-Based Knowledge Gap." *Mass Communication & Society*, vol. 14, 2011, pp. 216-35.

Weisfeld, Matt. *The Object-Oriented Thought Process*. 5th ed., Addison-Wesley, 2019.

Weiss, Patrice L. (Tamar), Emily A. Keshner, and Mindy F. Levin, editors. *Virtual Reality for Physical and Motor Rehabilitation*. Springer-Verlag New York, 2014.

Weizenbaum, J. *Computer Power and Human Reason: From Judgement to Calculation*. W. H. Freeman, 1976.

Werbach, Kevin, and Dan Hunter. *For the Win: How Game Thinking Can Revolutionize Your Business*. Wharton, 2012.

"What Are the Benefits of Speech Recognition Technology?" *IEEE Signal Processing Society*, 2 Nov. 2018, signalprocessingsociety.org/publications-resources/blog/what-are-benefits-speech-recognition-technology. Accessed 13 July 2023.

"What Happened to iTunes?" *Apple*, support.apple.com/en-us/HT210200. Accessed 13 July 2023.

"What Is ASCII?" *Indiana University Knowledge Base*, 18 Jan. 2018, kb.iu.edu/d/afht. Accessed 13 July 2023.

"What Is Big Data?" *Oracle*, www.oracle.com/big-data/what-is-big-data. Accessed 13 July 2023.

"What Is Copyright?" *Copyright.gov*, www.copyright.gov/what-is-copyright. Accessed 13 July 2023.

"What Is a Creative Commons License?" *Copyright Alliance*, 2023, copyrightalliance.org/faqs/what-is-creative-commons-license. Accessed 13 July 2023.

"What Is Debugging?" *AWS*, 2023, aws.amazon.com/what-is/debugging. Accessed 13 July 2023.

"What Is a Deepfake?" *Economist*, 7 Aug. 2019, www.economist.com/the-economist-explains/2019/08/07/what-is-a-deepfake. Accessed 13 July 2023.

"What Is Design?" *University of Illinois at Chicago College of Architecture, Design, and the Arts*, design.uic.edu/what-is-design. Accessed 13 July 2023.

"What Is the History of Microsoft Windows?" *Indiana University Knowledge Base*, 18 Jan. 2018, kb.iu.edu/d/abwa. Accessed 13 July 2023.

"What Is Netflix and How Does It Work?" *TechBoomers*, 15 May 2016, techboomers.com/t/what-is-netflix. Accessed 13 July 2023.

"Whatever Happened to the Semantic Web?" *Two-Bit History*, 27 May 2018, twobithistory.org/2018/05/27/semantic-web.html. Accessed 13 July 2023.

Whitaker, Rob. *Developing Inclusive Mobile Apps*. Apress, 2020.

White, Michele. *Touch Screen Theory: Digital Devices and Feelings*. MIT Press, 2022.

Whitington, John. *How Computers Make Books: From Graphics Rendering, Search Algorithms, and Functional Programming to Indexing and Typesetting*. Manning, 2023.

Whitman, James Q. "Two Western Cultures of Privacy: Dignity versus Liberty." *Yale Law Journal*, vol. 113, no. 6, Apr. 2004, pp. 1151-221.

Wiggers, Kyle. "YouTube CEO Susan Wojcicki Steps Down, Will Assume Advisory Role at Google and Alphabet." *TechCrunch*, 16 Feb. 2023, techcrunch.com/2023/02/16/youtube-ceo-susan-wojcicki-steps-down-will-assume-advisory-role-at-google-and-alphabet. Accessed 13 July 2023.

Wiggins, Chris, and Matthew L. Jones. *How Data Happened: A History from the Age of Reason to the Age of Algorithms*. Norton, 2023.

Wilford, John N. *The Mapmakers: The Story of the Great Pioneers in Cartography from Antiquity to the Space Age*. Pimlico, 2002.

Wilke, Claus. *Fundamentals of Data Visualization: A Primer on Making Informative and Compelling Figures*. O'Reilly, 2019.

Williams, Brad, David Damstra, and Hal Stern. *Professional WordPress: Design and Development*. 3rd ed., Wiley, 2015.

Williams, Joe. *Entertainment on the Net*. Que, 1995.

Williams, Rhiannon. "Apple iOS: A Brief History." *Telegraph*, 17 Sept. 2015, www.telegraph.co.uk/technology/apple/11068420/Apple-iOS-a-brief-history.html. Accessed 13 July 2023.

Wilson, Mark. "Exclusive: A First Look Inside Google's Top-Secret Design Lab." *Fast Company*, 6 Aug. 2019, www.fastcompany.com/90378288/exclusive-a-first-look-inside-googles-top-secret-design-lab. Accessed 13 July 2023.

Wilton, Paul, and Jeremy McPeak. *Beginning JavaScript*. 5th ed., Wrox, 2015.

Winder, Catherine, and Zahra Dowlatabadi. *Producing Animation*. Focal, 2011.

Winer, Ethan. *The Audio Expert: Everything You Need to Know about Audio*. Focal Press, 2013.

Winter, Mick. *Scan Me-Everybody's Guide to the Magical World of QR Codes*. Westsong, 2011.

Witten, Ian H., Eibe Frank, Mark A. Hall, and Christopher J. Pal. *Data Mining: Practical Machine Learning Tools and Techniques*. 4th ed., Morgan Kaufmann, 2016.

Wood, David. *Interface Design: An Introduction to Visual Communication in UI Design*. Fairchild Books, 2014.

Wooley, Scott. *The Network: The Battle for the Airwaves and the Birth of the Communications Age*. HarperCollins Publishers, 2016.

Woolridge, Michael. *A Brief History of Artificial Intelligence: What It Is, Where We Are, and Where We Are Going*. Flatiron Books, 2021.

"WordPress Market Share in 2022." *Envisage Digital*, 2022, www.envisagedigital.co.uk/wordpress-market-share. Accessed 13 July 2023.

Woyke, Elizabeth. *The Smartphone: Anatomy of an Industry*. New Press, 2014.

Wroot, Jonathan, and Andy Willis, editors. *DVD, Blu-ray and Beyond: Navigating Formats and Platforms within Media Consumption*. Palgrave Macmillan, 2017.

Wuerthele, Mike. "Apple Unveils iPadOS, Adding Features Specifically to iPad." *AppleInsider*, 2 June 2019, appleinsider.com/articles/19/06/03/apple-supplements-ios-13-with-new-tablet-specific-ipad-os-branch. Accessed 13 July 2023.

Wulff, Alex. *Beginning Radio Communications: Radio Projects and Theory*. Apress, 2019.

Xie, Iris, and Krystyna Matusiak. *Discover Digital Libraries: Theory and Practice*. Elsevier, 2016.

"XML." *PC Mag*, www.pcmag.com/encyclopedia/term/55048/xml. Accessed 13 July 2023.

"XML Tutorial." *W3Schools*, www.w3schools.com/xml. Accessed 13 July 2023.

Xue, Su. *Data-Driven Image Editing for Perceptual Effectiveness*. Yale UP, 2013.

Young, Sherman. *The Book Is Dead. Long Live the Book*. U of New South Wales P, 2007.

Zager, Michael. *Music Production: A Manual for Producers, Composers, Arrangers, and Students*. Rowman & Littlefield, 2021.

Zandt, Florian. "The Most Popular Streaming Services in the US." *Statista*, 20 July 2023, www.statista.com/chart/25382/most-used-video-streaming-platforms. Accessed 13 July 2023.

Zeller, Andreas. *Why Programs Fail: A Guide to Systematic Debugging*. Kaufmann, 2009.

Zewe, Adam. "Toward Speech Recognition for Uncommon Spoken Languages." *MIT News*, 4 Nov. 2021, news.mit.edu/2021/speech-recognition-uncommon-languages-1104. Accessed 13 July 2023.

Zhao, Pengfei, et al., editors. *Advanced Graphic Communications and Media Technologies*. Springer Singapore, 2017.

Zichermann, Gabe, and Christopher Cunningham. *Gamification by Design: Implementing Game Mechanics in Web and Mobile Apps*. O'Reilly, 2011.

Zichermann, Gabe, and Joselin Linder. *The Gamification Revolution: How Leaders Leverage Game Mechanics to Crush the Competition*. McGraw, 2013.

Ziobro, Paul, and Jeff Horwitz. "Facebook Suspends Donald Trump for at Least Two Years." *Wall Street Journal*, 4 June 2021, www.wsj.com/articles/facebook-suspends-donald-trump-for-two-years-11622825480. Accessed 13 July 2023.

Glossary

3D rendering: the process of creating a two-dimensional (2D) animation using three-dimensional (3D) models

algorithm: a set of precise, computable instructions that, when executed in the correct order, will provide a solution to a certain problem

Alphabet: the parent company of Google

amplifier: a device that strengthens the power, voltage, or current of a signal

analogue signal: a continuous signal whose values or quantities vary over time

Android: a mobile operating system introduced by Google in 2008

animation variables (avars): defined variables that control the movement of an animated character or object

application program interface (API): the code that defines how two pieces of software interact, particularly a software application and the operating system on which it runs

application suite: a set of programs designed to work closely together, such as an office suite that includes a word processor, spreadsheet, presentation creator, and database application

applications: software installed on computers (or web-accessed) that allow users to perform a wide range of activities; the full term is usually used for desktop computing devices, while the shorter one (app) is used for tablets and smart devices such as phones and watches

artificial intelligence (AI): the intelligence exhibited by machines or computers, in contrast to human, organic, or animal intelligence

ASCII: the original common character code for computers using 8 bits

assistive technologies (AT): any item, piece of equipment, or product system, whether acquired commercially off the shelf, modified, or customized, that is used to increase, maintain, or improve the functional capabilities of a person with a disability

attributes: the specific features that define an object's properties or characteristics

audio codec: a program that acts as a "coder-decoder" to allow an audio stream to be encoded for storage or transmission and later decoded for playback

augmented reality: technology that inserts digital interfaces or graphics into the real world through the use of headsets, glasses, smartphones, or other devices

behavioral marketing: advertising to users based on their habits and previous purchases

bit: a single binary digit that can have a value of either 0 or 1

bit rate: the amount of data encoded for each second of video; often measured in kilobits per second (kbps) or kilobytes per second (Kbps)

blog: an online writing tool that allows an individual to post multiple written comments that are then published in reverse chronological order on the internet

born-digital materials: materials originally created in a digital format, including digital art and multimedia content, digital manuscripts, and digital records

botnet: a collection of computers that have been infected with software by computer hackers to force those computers to commit crimes, such as sending out computer viruses or unsolicited email (spam)

broadcast: an audio or video transmission sent via a communications medium to anyone with the appropriate receiver

building information modeling (BIM): the creation of a model of a building or facility that accounts for its function, physical attributes, cost, and other characteristics

byte: a group of 8 bits

cel animation: a traditional form of animation in which the movement of characters, objects, and backgrounds are created frame by frame

central processing unit (CPU): electronic circuitry that provides instructions for how a computer handles, processes, and manages data from applications and programs

character: a unit of information that represents a single letter, number, punctuation mark, blank space, or other symbol used in written language

chatbot: a computer program that mimics human conversation responses in order to interact with people through text; also called "talkbot," "chatterbot," or simply "bot"

ChatGPT: a chatbot created by technology company OpenAI and powered by artificial intelligence technology

cloud computing: a form of technology service where the user's data and applications are stored in and run on a server to which the user connects over the internet; this contrasts with the traditional model in which both data and application are housed on the local machine that the user has physical access to

cloud services: computing services that are hosted on external servers and made available to users via the internet

compressed data: data that has been encoded such that storing or transferring the data requires fewer bits of information

computer: any device that can be programmed to perform mechanical or electrical computation, processing numbers; since the middle of the twentieth century, the term has commonly been used for equipment that can be programmed using a binary number system to perform a variety of work; for a computer to process letters, words, graphic images, or maps, human programmers have to encode nonnumerical data in a numeric form

computer-aided design (CAD): the use of computer technology to create designs or virtual models of objects or structures

content management system: a software system that allows users to manage the content of a collection of data including computer files, audio files, graphics and images, electronic documents, and web content

cookies: small data files that allow websites to track users

Creative Commons: an international nonprofit organization that provides easy access to creative works that others can legally build upon and share through its range of copyright licenses known as Creative Commons licenses

cryptology: the science of encrypting or decrypting information, including creating codes for privacy or security and finding means to break a code by working from a message in an unknown code to learn the pattern

data mining: the process using algorithms to search for meaningful patterns in large collections of data, often without stating any kind of a hypothesis in advance

data source: the origin of the information used in a computer model or simulation, such as a database or spreadsheet

database: a collection of data items used for multiple purposes that is stored on a computer

declaration: a piece of source code that adds a name to the program's environment and binds it to a class of meanings, and may also define the name

deep learning: an emerging field of artificial intelligence research that uses neural network algorithms to improve machine learning

desktop publishing: the process of formatting and designing a page of text (images and words) using computer software

device: equipment designed to perform a specific function when attached to a computer, such as a scanner, printer, or projector

digital commerce: the purchase and sale of goods and services via online vendors or information technology systems

digital divide: a concept that describes the unequal access to technology that exists because of the stratification of society into different economic classes

digital immigrant: an individual born before the digital age who had to learn the intricacies of digital technology as it evolved

digital literacy: familiarity with the skills, behaviors, and language specific to using digital devices to access, create, and share content through the internet

digital native: an individual born during the digital age or raised using digital technology and communication

e-book: an electronic book consisting of text, images, or both, to be read on computers, e-readers, and other digital devices, sometimes enhanced with sound, video, animation, and so on

edit decision list (EDL): a list that catalogs the reel or time code data of video frames so that the frames can be accessed during video editing

electromagnetic spectrum: the complete range of electromagnetic radiation, from the longest wavelength and lowest frequency (radio waves) to the shortest wavelength and highest frequency (gamma rays)

Electronic Communications Privacy Act: a 1986 law that extended restrictions on wiretapping to cover the retrieval or interception of information transmitted electronically between computers or through computer networks

emulators: programs that mimic the functionality of a mobile device and are used to test apps under development

engagement: a marketing term for the active involvement of a consumer with a brand or online content

e-reader: a handheld computer designed for use in storing and displaying e-books in one or more file formats; also, e-book reader

e-waste: short for "electronic waste"; computers and other digital devices that have been discarded by their owners

Federal Communications Commission (FCC): the government agency that regulates radio, television, wire, satellite, and cable broadcasts

font: a complete set of characters (including upper and lowercase letter forms, numbers, symbols, and punctuation) sharing the same basic size and style; also known as typeface

frames per second (FPS): a measurement of the rate at which individual video frames are displayed

game loop: the main part of a game program that allows the game's physics, artificial intelligence, and graphics to continue to run with or without user input

game theory: a branch of mathematics dedicated to analyzing strategic behavior in which several people or entities must make choices even when the outcomes of their decisions rely on the choices made by others

gamification: the use of video-game logic and psychology in real-world environments

geometry: literally "earth measurement," the branch of mathematics concerned with the properties and relationships of points, lines, and surfaces and the space contained by those entities

gestures: combinations of finger movements used to interact with multitouch displays in order to accomplish various tasks; examples include tapping the finger on the screen, double-tapping, and swiping the finger along the screen

glyph: a specific representation of a grapheme, such as the letter A rendered in a particular typeface

grapheme: the smallest unit used by a writing system, such as alphabetic letters, punctuation marks, or Chinese characters

graphical user interface (GUI): a computer interface in which the user controls the computer by interacting with menus, icons, and other graphical elements

hardware: the physical parts that make up a computer; these include the motherboard and processor, as well as input and output devices such as monitors, keyboards, and mice

hexadecimal: a base-16 number system that uses the digits 0 through 9 and the letters A, B, C, D, E, and F as symbols to represent numbers

HTML: Hypertext Markup Language; used to define pages on the World Wide Web

HTML editor: a computer program for editing web pages encoded in hypertext markup language (HTML)

information technology: the use of computers and related equipment for the purpose of processing and storing data

ink-jet printing: a printing process in which the image is produced by spraying individual drops of ink through an aperture onto the intended surface

input: data supplied to some program, subprogram, operating system, machine, system, or abstraction

intellectual property: intangible assets, including creative ideas and specialized knowledge that has commercial value; some can be protected by patents, trademarks, and copyrights

interface: the function performed by the device driver, which mediates between the hardware of the peripheral and the hardware of the computer

International System of Units: also called SI units, the units of measurement used in most of the world for measurements; the seven basic SI units are the kilogram (for mass), the second (for time), the Kelvin (for temperature), the ampere (for electric current), the mole (for the amount of a substance), the candela (for luminous intensity), and the meter (for distance)

Internet of Things (IoT): a wireless network connecting devices, buildings, vehicles, and other items with network connectivity

iOS: Apple's proprietary mobile operating system, installed on Apple devices such as the iPhone and iPod Touch

iPads: small mobile computers manufactured by Apple that broke ground in the tablet computer market and pioneered mobile devices designed to enhance classroom learning

isometric drawing: a form of two-dimensional (2D) drawing in which the object being depicted is represented in such a way that it appears to be three-dimensional (3D) and positioned at a specific angle

kerning: adjustments made to the space between characters within words

keyframing: a part of the computer animation process that shows, usually in the form of a drawing, the position and appearance of an object at the beginning of a sequence and at the end

language interface packs: programs that translate interface elements such as menus and dialog boxes into different languages

leading: space between lines of text

letterpress: the process of printing from an inked elevated surface

lithography: the process of transferring an image from any surface

lossless compression: data compression that allows the original data to be compressed and reconstructed without any loss of accuracy

lossy compression: a method of decreasing image file size by discarding some data, resulting in some image quality being irreversibly sacrificed

machine learning algorithm: an algorithm that is capable of being trained and of learning based on the datasets to which it is exposed and the tasks it completes

Meta: the parent company of Facebook, Instagram, and Threads

mastering: the creation of a master recording that can be used to make other copies for distribution

metadata: data that contains information about other data, such as author information, organizational information, or how and when the data was created

metaverse: a virtual world in which users can interact with one another

microwaves: electromagnetic radiation with a frequency higher than that of radio wave but lower than that of visible light

mixing: the process of combining different sounds into a single audio recording

mobile website: a website that has been optimized for use on mobile devices, typically featuring a simplified interface designed for touch screens

modeling: the process of creating a two-dimensional (2D) or three-dimensional (3D) representation of an object, structure, or animated character

multitasking: in computing, the process of executing multiple tasks concurrently on an operating system (OS); in the mobile phone environment, allowing different apps to run concurrently, much like the ability to work in multiple open windows on a PC

multitouch gestures: combinations of finger movements used to interact with touch-screen or other touch-sensitive displays in order to accomplish various tasks; examples include double-tapping and swiping the finger along the screen

National Association of Schools of Art and Design (NASAD): a national organization that provides accreditation to colleges and other institutions offering programs in art and design

nondestructive editing: a mode of image editing in which the original content of the image is not destroyed because the edits are made only in the editing software

nonlinear editing: a method of editing video in which each frame of video can be accessed, altered, moved, copied, or deleted regardless of the order in which the frames were originally recorded

offset printing: a process in which an image is transferred from a plate to a rubber blanket that, in turn, presses the image onto a sheet

open-source software: software for which users have permission to modify the source code

operating system (OS): a specialized program that manages a computer's functions

optical storage: storage of data by creating marks in a medium that can be read with the aid of a laser or other light source

output: a means whereby data or objects are passed from a part to a wider context; for example, a program sending data to the operating system so that one can see it on the screen

peripherals: devices connected to a computer but not part of the computer itself, such as scanners, external storage devices, and so forth

piracy: in the digital context, unauthorized reproduction or use of copyrighted media in digital form

platform: the underlying computer system on which an application is designed to run

postproduction: in digital modeling, the period after a model has been designed and an image has been rendered, when an individual may manipulate the created image by adding effects or making other aesthetic changes

Pretty Good Privacy: a data encryption program created in 1991 that provides both encryption and authentication

programming languages: sets of terms and rules of syntax used by computer programmers to create instructions for computers to follow; this code is then compiled into binary instructions for a computer to execute

proprietary software: software owned by an individual or company that places certain restrictions on its use, study, modification, or distribution and typically withholds its source code

prototype: an early version of product that is still under development, used to demonstrate what the finished product will look like and what features it will include

quality assurance: the practice of improving processes and empowering employees to make decisions that enable avoidance of errors and defects

radio waves: low-frequency electromagnetic radiation, commonly used for communication and navigation

random-access memory (RAM): memory that the computer can access very quickly, without regard to where in the storage media the relevant information is located

raster: a means of storing, displaying, and editing image data based on the use of individual pixels

render farm: a cluster of powerful computers that combine their efforts to render graphics for animation applications

rendering: the process of transforming one or more models into a single image

RGB: a color model that uses red, green, and blue to form other colors through various combinations

screen-reading program: a computer program that converts text and graphics into a format accessible to visually impaired, blind, learning disabled, or illiterate users

script: a group of written signs, such as Latin or Chinese characters, used to represent textual information in a writing system

scrubbing: navigating through a recording repeatedly in order to locate a specific cue or word

search engine optimization (SEO): techniques used to increase the likelihood that a website will appear among the top results in certain search engines

serif: a slight projection at the end of the vertical stroke of a letterform; styles without such projections are called sans serif

simulation: a computer model executed by a computer system

Sketchpad: an early computer-aided design software created by American computer scientist Ivan Sutherland as part of his 1963 doctoral thesis

smartphone: a mobile telephone that has the capability of running software applications and accessing the internet

social networking services: online platforms in which one of the primary functions is to create and maintain social networks that users opt into (by adding one another as friends or following accounts)

software as a service (SaaS): a cloud service in which users subscribe to software and access that software via the internet

software patches: updates to software that correct bugs or make other improvements

software: the sets of instructions that a computer follows in order to carry out tasks; software may be stored on physical media, but the media is not the software

solid modeling: the process of creating a three-dimensional (3D) representation of a solid object

solid-state storage: computer memory that stores information in the form of electronic circuits, without the use of disks or other read/write equipment

surface modeling: the process of creating a three-dimensional (3D) representation of an object in which all surfaces are modeled but the model itself is not solid

system software: the operating system that allows programs on a mobile device to function

system: a set of interacting or interdependent component parts that form a complex whole

tableless web design: the use of style sheets rather than hypertext markup language (HTML) tables to control the layout of a web page

tablet: a mobile computer that is small enough to be held in the hand and that provides access to a wide range of software applications and the internet

3D rendering: the process of creating a two-dimensional (2D) animation using three-dimensional (3D) models

transmedia: literally, across media; more recently, a description of media projects that use multiple platforms (print, film, web, etc.) to communicate an idea, a narrative, or information with greater complexity and dynamic potential than is possible in a single medium

tweet: a message posted to the social media service Twitter

typography: the art and technique of arranging type to make language readable and appealing

user-centered design: design based on a perceived understanding of user preferences, needs, tendencies, and capabilities

user-generated content: content contributed by users of an online community or social networking service, such as uploaded images and videos, text posts/blogs/status updates, or comments

utility programs: apps that perform basic functions on a computer or mobile device such as displaying the time or checking for available network connections

variable: a symbol representing a quantity with no fixed value

vector: a means of storing, displaying, and editing image data based on the use of defined points and lines

video scratching: a technique in which a video sequence is manipulated to match the rhythm of a piece of music

video-on-demand (VoD): technology that supplies video content to users upon request

virtual reality: a form of technology that enables the user to view and interact with a simulated environment

vision mixing: the process of selecting and combining multiple video sources into a single video

web application: an application that is downloaded either wholly or in part from the internet each time it is used

widgets: small, self-contained applications that run continuously without being activated like a typical application

wireframe modeling: the process of creating a three-dimensional (3D) representation of an object in which all edges are outlined, surfaces are not modeled, and the model itself is not solid

XML: extensible markup language, a programming language used to categorize and describe data

Organizations

ACM Special Interest Group on Computer-Human Interaction (SIGCHI)
sigchi.org

American Institute of Graphic Arts (AIGA)
228 Park Avenue South
Suite 58603
New York, NY 10003
212-807-1990
aiga.org

College Art Association of America, Inc.
50 Broadway
21st Floor
New York, NY 10004
212-691-1051
collegeart.org

Design History Society
70 Cowcross Street
London EC1M 6EJ
designhistorysociety.org

Design Incubation
St. John's University
Manhattan Campus
101 Astor Place
New York, NY 10003
designincubation.com

Entertainment Software Association
601 Massachusetts Avenue NW
Suite 300
Washington, DC 20001
theesa.com

Graphic Artists Guild
2248 Broadway, #1341
New York, NY 10025
212-791-3400
graphicartistsguild.org

International Association for Computing Machinery's Special Interest Group on Computer Graphics and Interactive Techniques (ACM SIGGRAPH)
1601 Broadway
10th floor
New York, NY 10019
siggraph.org

International Association of Word and Image Studies/Association Internationale pour l'Etude des Rapports entre Texte et Image
iawis.org

International Colour Association (AIC)
aic-color.org

International Council of Graphic Design Associations (ICOGRADA)
CP 519 Succ. Place d'Arems
Montréal, Québec H2Y 3H3
Canada
1-514-875-7545
theicod.org

International Game Developers Association (IGDA)
1 Eglinton Avenue East
Suite 705
Toronto, Ontario M4P 3A1
Canada
igda.org

Inter-Society Color Council (ISCC)
712 H Street NE
Suite 1640
Washington, DC 20002
iscc.org

International Society of Typographic Designers (ISTD)
Elizabeth House
13-19 London Road
Newbury RG14 1JL
United Kingdom
istd.org.uk

Leonardo Education and Art Forum
The International Society for Arts + Sciences + Technology
Leonardo/ISAST
1201 Martin Luther King Jr Way
Suite 100
Oakland, CA 94612
leonardo.info

National Art Education Association (NAEA)
901 Prince Street
Alexandria, VA 22314
800-299-8321
arteducators.org

New Media Caucus (NMC)
Affiliate Society of the College Art Association
nmccommcomm@gmail.com
newmediacaucus.org

The One Club for Creativity
450 West 31st Street
6th Floor
New York, NY 10001
212-979-1900
oneclub.org

Society for Experiential Graphic Design (SEGD)
1900 L Street NW
Suite 710
Washington, DC 20036
202-638-5555
segd.org

Society of Illustrators
The Museum of Illustration
128 East 63rd Street
New York, NY 10065
212-838-2560
societyillustrators.org

The Society of Illustrators Artists & Designers (SIAD)
207 Regent Street
3rd Floor
Mayfair, London W1B 3HH
44-0-845-647-0150
siad.org

The Society of Typographic Aficionados (SoTA)
2500 NE Sandy Boulevard
Suite E
Portland, OR 97232
typesociety.org

Type Directors Club (TDC)
450 West 31 Street
6th Floor
New York, NY 10001
917-509-2291
tdc.org

Visual Culture Caucus
American Studies Association
202-596-1270
asastaff@thease.net
theasa.net/communities/caucuses/visual-culture-caucus

Subject Index

The Principles of... Series

Principles of Science

Principles of Anatomy
Principles of Astronomy
Principles of Behavioral Science
Principles of Biology
Principles of Biotechnology
Principles of Botany
Principles of Chemistry
Principles of Climatology
Principles of Computer Science
Principles of Computer-Aided Design
Principles of Ecology
Principles of Energy
Principles of Fire Science
Principles of Forestry & Conservation
Principles of Geology
Principles of Information Technology
Principles of Marine Science
Principles of Mathematics
Principles of Microbiology
Principles of Modern Agriculture
Principles of Pharmacology
Principles of Physical Science
Principles of Physics
Principles of Programming & Coding
Principles of Robotics & Artificial Intelligence
Principles of Scientific Research
Principles of Sports Medicine & Kinesiology
Principles of Sustainability
Principles of Zoology

The Principles of... Series

Principles of Health

Principles of Health: Allergies & Immune Disorders
Principles of Health: Anxiety & Stress
Principles of Health: Depression
Principles of Health: Diabetes
Principles of Health: Nursing
Principles of Health: Obesity
Principles of Health: Occupational Therapy
Principles of Health: Pain Management
Principles of Health: Prescription Drug Abuse

Principles of Business

Principles of Business: Accounting
Principles of Business: Economics
Principles of Business: Entrepreneurship
Principles of Business: Finance
Principles of Business: Globalization
Principles of Business: Leadership
Principles of Business: Management
Principles of Business: Marketing

Principles of Sociology

Principles of Sociology: Group Relationships & Behavior
Principles of Sociology: Personal Relationships & Behavior
Principles of Sociology: Societal Issues & Behavior

SALEM PRESS https://salempress.com (800) 221-1592